Trends in Mathematics is a series devoted to the publication of volumes arising from conferences and lecture series focusing on a particular topic from any area of mathematics. Its aim is to make current developments available to the community as rapidly as possible without compromise to quality and to archive these for reference.

Proposals for volumes can be sent to the Mathematics Editor at either

Birkhäuser Verlag
P.O. Box 133
CH-4010 Basel
Switzerland

or

Birkhäuser Boston Inc.
675 Massachusetts Avenue
Cambridge, MA 02139
USA

Material submitted for publication must be screened and prepared as follows:

All contributions should undergo a reviewing process similar to that carried out by journals and be checked for correct use of language which, as a rule, is English. Articles without proofs, or which do not contain any significantly new results, should be rejected. High quality survey papers, however, are welcome.

We expect the organizers to deliver manuscripts in a form that is essentially ready for direct reproduction. Any version of TeX is acceptable, but the entire collection of files must be in one particular dialect of TeX and unified according to simple instructions available from Birkhäuser.

Furthermore, in order to guarantee the timely appearance of the proceedings it is essential that the final version of the entire material be submitted no later than one year after the conference. The total number of pages should not exceed 350. The first mentioned author of each article will receive 25 free offprints. To the participants of the congress the book will be offered at a special rate.

Trends in Mathematics is a series devoted to the publication of volumes arising from conferences and lecture series focusing on a particular topic from any area of mathematics. Its aim is to make current developments available to the community as rapidly as possible without compromise to quality and to archive these for reference.

Proposals for volumes can be sent to the Mathematics Editor at either

Birkhäuser Verlag
P.O. Box 133
CH-4010 Basel
Switzerland

or

Birkhäuser Boston Inc.
675 Massachusetts Avenue
Cambridge, MA 02139
USA

Material submitted for publication must be screened and prepared as follows:

All contributions should undergo a reviewing process similar to that carried out by journals and be checked for correct use of language which, as a rule, is English. Articles without proofs, or which do not contain any significantly new results, should be rejected. High quality survey papers, however, are welcome.

We expect the organizers to deliver manuscripts in a form that is essentially ready for direct reproduction. Any version of TeX is acceptable, but the entire collection of files must be in one particular dialect of TeX and unified according to simple instructions available from Birkhäuser.

Furthermore, in order to guarantee the timely appearance of the proceedings it is essential that the final version of the entire material be submitted no later than one year after the conference. The total number of pages should not exceed 350. The first-mentioned author of each article will receive 25 free offprints. To the participants of the congress the book will be offered at a special rate.

Abelian Groups and Modules

International Conference in Dublin, August 10–14, 1998

Paul C. Eklof
Rüdiger Göbel
Editors

Springer Basel AG

Editors' addresses:

Paul C. Eklof
Mathematics Department
University of California at Irvine
Irvine, CA 92697-3875
USA

Rüdiger Göbel
Fachbereich 6, Mathematik und Informatik
Universität GH Essen
45117 Essen
Germany

1991 Mathematical Subject Classification 20-06, 13-06

A CIP catalogue record for this book is available from
the Library of Congress, Washington D.C., USA

Abelian groups and modules : international conference in Dublin,
August 10 – 14, 1998 / Paul C. Eklof ; Rüdiger Göbel ed. - Basel ;
Boston ; Berlin : Birkhäuser, 1999
(Trends in mathematics)

ISBN 978-3-0348-7593-6 ISBN 978-3-0348-7591-2 (eBook)
DOI 10.1007/978-3-0348-7591-2

Originally published by Birkhäuser Verlag, Basel, Switzerland in 1999

Printed on acid-free paper produced from chlorine-free pulp. TCF ∞

Softcover reprint of the hardcover 1st edition 1999

9 8 7 6 5 4 3 2 1

CONTENTS

PREFACE

This volume contains the refereed proceedings of the INTERNATIONAL CONFERENCE ON ABELIAN GROUPS AND MODULES held at the Dublin Institute of Technology in Dublin, Ireland from August 10th until August 14th, 1998. The meeting brought together more than 50 researchers and graduate students from 14 countries around the world. In a series of eight invited survey talks, experts in the field presented some active areas of research including:

- Almost completely decomposable abelian groups, Butler groups and almost free groups — the classification problem, and invariants of special classes of torsion–free abelian groups.
- Totally projective groups, their automorphism groups and their group rings — questions about unique passage between these categories.
- Radicals commuting with products.
- The Ziegler spectra of Neumann regular rings and the class (semi–) groups of Prüfer domains.
- The Krull–Schmidt property for valuation domains.

These main talks were accompanied by many other presentations of current research on abelian groups and modules. Methods from model theory, category theory, infinite combinatorics, representation theory, classical algebra and geometry were applied to the study of abelian groups and modules; on the other hand, results and methods from abelian group theory were applied to general module theory and non–commutative groups.

All this is reflected in the thirty articles in this volume, which introduce the reader to an active and attractive part of algebra that over the years has gained much from its position at crossroads of mathematics. Lively discussions at the conference influenced the final work on the presented papers; the editors hope that the papers will convey some sense of the intellectual ferment they generated and stimulate the reader to consider and actively investigate the topics and problems contained therein.

The papers included in this volume, each of which passed the critical scrutiny of two referees, deal with the topics of the survey lecture and also with the following:

- If the type set of a pure subgroup of a direct sum of finitely many subgroups of the rationals is small and not complicated, then we have a chance to classify these (Butler) groups: where is the borderline for 'not complicated'?
- A similar problem, also related to representation theory, arises for valuated p-groups of exponent p^5: is 5 the critical number?

- The countable rank automorphism groups of torsion–free groups are determined and those of arbitrary rank almost determined.
- Indecomposable abelian groups are constructed that remain indecomposable when the universe of sets is extended.
- In several papers, the 'wild behavior' of torsion–free abelian groups or modules is described, even for restricted subclasses like those with large free subgroups, those of finite rank, or modules over rings close to fields.
- A proof that classifying countable torsion–free groups is as difficult as classifying arbitrary countable structures; and that any classification scheme for finite rank torsion–free groups is inherently complex.
- An application of Corner's ideas for realizing countable rings as endomorphism rings to non–commutative groups, for investigations of automorphisms of metabelian groups.

Other papers investigate, among other topics: generalized notions of purity; generalized E–rings; vector groups; mixed self–small groups; and methods for studying general module categories.

These proceedings are headed by a memorial to Ross Beaumont, who was Reinhold Baer's first Ph.D. student. We would like to thank Jim Reid and Bill Wickless for their summary of Ross' important contributions to algebra, which include the deep theory of torsion–free groups of rank two, developed jointly with R. S. Pierce.

The Dublin conference is one link in a chain of meetings dealing with abelian groups and modules which have taken place in recent years in the United States, Italy, Germany, Australia and in Curaçao; the next link will be added in Perth, Australia in the year 2000. However, it was the first conference of this kind in Ireland, which gave Ray Mines the opportunity, in a special talk, to draw connections between James Joyce's great *Ulysses* and mathematics. This and other special events made this meeting a particularly pleasant and memorable one. We would like to thank the two organizers Brendan Goldsmith, President of DIT, and Simone Pabst, post–doctoral fellow at DIT, for running a conference which we all enjoyed very much. We also would like to thank DIT for financial support and for an exciting trip on an antique bus which explored the countryside and history of Ireland and gave many of us the chance to observe closely how sheep eat.

Paul Eklof and Rüdiger Göbel

Authors and Conference Participants

U. Albrecht, Department of Mathematics, Auburn University, Auburn, Alabama 36849, USA; albreuf@mail.auburn.edu

D. Arnold, Department of Mathematics, Baylor University, Waco, TX 76798, USA; arnoldd@baylor.edu

M. A. Avino, Department of Mathematics, Fac de Ciencias, Univ Nat Autonoma de Mexico, Circuito Exterior, DF CP 04360, CU Mexico; maad@hp.fciencias.unam.mx

L. Bican, Katedra Algebry MFF UK, Sokolovska 83, 186 75 Praha 8 Karlin, Czech Republic; bican@karlin.mff.cuni.cz

E. Blagoveshchenskaya, Department of Mathematics, St. Petersburg State Technical University, Polytechnicheskaya 29, St Petersburg 195251, Russia; kat@osipenko.stu.neva.ru

A. Blass, Department of Mathematics, University of Michigan, Ann Arbor, MI 48109-1109, USA; ablass@umich.edu

J. Clark, Department of Mathematics and Statistics, University of Otago, PO Box 56, Dunedin, New Zealand; jclark@maths.otago.ac.nz

A. L. S. Corner, Worcester College, Oxford OX1 2HB, England, UK; tony.corner@worcester.oxford.ac.uk

C. De Vivo, Dipartimento di Matematica e Applicazioni, Università Federico II di Napoli, 80100 Napoli, Italy; devivo@matna2.dma.unina.it

M. Dugas, Department of Mathematics, Baylor University, Waco, TX 76798, USA; dugasm@baylor.edu

A. Ecker, Mathematisches Institut, Universität Tübingen, Auf der Morgenstelle 10, 72076 Tübingen, Germany; andreas.ecker@uni-tuebingen.de

P. Eklof, Department of Mathematics, University of California at Irvine, CA 92697-3875, USA; peklof@math.uci.edu

S. Feigelstock, Department of Mathematics, Bar-Ilan University, 52900 Ramat-Gan, Israel; feigel@macs.biu.ac.il

A. A. Fomin, Department of Mathematics, Moscow Pedagogical State University, Krasnopudnaja ul., d. 14. Moscow 107140, Russia; afomin@ipc.ru

L. Fuchs, Department of Mathematics, Tulane University, New Orleans, LA 70118, USA; fuchs@mailhost.tcs.tulane.edu

A. Giovannitti, Department of Mathematics, State University of West Georgia, Carrollton, Georgia 30118, USA; agiovann@westga.edu

H. P. Goeters, Department of Mathematics, Auburn University, Auburn, Alabama 36849, USA; goetehp@mail.auburn.edu

R. Göbel, Fachbereich Mathematik und Informatik, Universität GH Essen, 45117 Essen, Germany; r.goebel@uni-essen.de

B. Goldsmith, Dublin Institute of Technology, 30 Upper Pembroke Street, Dublin 2, Ireland; bgoldsmith@dit.ie

J. Hausen, Department of Mathematics, University of Houston, Houston, TX 77204-3476, USA; hausen@uh.edu
G. Hennecke, Fachbereich Mathematik und Informatik, Universität GH Essen, 45117 Essen, Germany; georg.hennecke@uni-essen.de
I. Herzog, Department of Mathematics, The Ohio State University at Lima, Lima, OH 45804, USA; herzog.23@osu.edu
P. Hill, Department of Mathematics, Auburn University, Auburn, AL 36849, USA; hillpad@mail.auburn.edu
G. Hjorth Department of Mathematics, University of California at Los Angeles, CA 90095-1555, USA; greg@math.ucla.edu
J. Irwin, 9001 Starmount, Las Vegas, Nevada 89134, USA; drjmirwin@lvcm.com
P. Keef, Whitman College, Walla Walla, WA 99362, USA; keef@whitman.edu
T. Kelly, 40 Wilfied Road, Sandymount, Dublin 4, Ireland; tka@tinet.ie
O. Kolman, c/o London School of Jewish Studies, Schaller House, Albert Road, London NW4 2SJ, UK; okolman@e-math.ams.org
T. Koyama, 5-24-21-703 Koishikawa, Bunkyo-ku, Tokyo 112-0002, Japan; koyama@is.ocha.ac.ip
K. P. Krog, Department of Mathematics, Marist College, 290 North Road, Poughkeepsie, NY 12601, USA; jz9g@musicb.marist.edu
C. Leopold, Mathematisches Institut der FAU Erlangen, Bismarckstr. $1\frac{1}{2}$, 91054 Erlangen, Germany; leopold@mi.uni-erlangen.de
W. Liebert, Technische Universität München, Zentrum Mathematik, 80290 München, Germany; liebert@mathematik.tu-muenchen.de
P. Loth, Department of Mathematics, Sacred Heart University, Fairfield, Connecticut 06432, USA; lothp@sacredheart.edu
A. Mader, Department of Mathematics, University of Hawaii, 2565 The Mall, Honolulu, HI 96822, USA; adolf@math.hawaii.edu
W. May, Department of Mathematics, University of Arizona, Tucson, Arizona 85721, USA; may@math.arizona.edu
C. Meehan, 18 Dodder Court, Firhouse, Dublin 24, Ireland
C. Metelli, Dipartimento di Matematica e Applicazioni, Università Federico II di Napoli, 80100 Napoli, Italy; cmetelli@math.unipd.it
R. Mines, Department of Mathematical Sciences, New Mexico State University, Las Cruces, New Mexico 88003, USA; ray@nmsu.edu
O. Mutzbauer, Mathematisches Institut, Universität Würzburg, Am Hubland, 97074 Würzburg, Germany; mutzbauer@mathematik.uni-wuerzburg.de
L. G. Nongxa, Department of Mathematics, University of the Western Cape, Private Bag x17, Belville 7535, South Africa; gnongxa@math.uwc.ac.za
J. O'Hogain, 3 Meath Square, Dublin 8, Ireland

T. Okuyama, Toba National College of Maritime, Technology 1-1, Ikegami-Cho, Toba-Shi, Mie-Ken 5178501, Japan; okuyamat@toba-cmt.ac.jp
A. Opdenhövel, Fachbereich Mathematik und Informatik, Universität GH Essen, 45117 Essen, Germany; ansgar.opdenhoevel@uni-essen.de
N. E. O'Sullivan, Department of Mathematics, NUI Galway, Galway, Ireland, niamh.eleanor.osullivan@ucg.ie
M. Ouldbeddi, Faculté des Sciences et Techniques, Université de Nouakchott, Nouakchott B.P. 5026, Mauritania; beddima@univ-nkc.mr
S. L. Pabst, Fachbereich Mathematik und Informatik, Universität GH Essen, 45117 Essen, Germany; simone.pabst@uni-essen.de
A. T. Paras, Department of Mathematics, University of the Philippines at Diliman, 1101 Quezon City, Philippines; agnes@math01.cs.upd.edu.ph
P. Plaumann, Mathematisches Institut der FAU Erlangen, Bismarckstr. $1\frac{1}{2}$, 91054 Erlangen, Germany; peter.plaumann@mi.uni-erlangen.de
K. M. Rangaswamy, Department of Mathematics, University of Colorado, PO Box 7150, Colorado Springs, Colorado 80933-7150, USA; ranga@math.uccs.edu
R. Raphael, Department of Mathematics, Concordia University, Montreal PQ, H3G 1M8, Canada; raphael@alcor.concordia.ca
J. D. Reid, Department of Mathematics, Wesleyan University, Middletown, CT 06459, USA; jreid@wesleyan.edu
F. Richman, Florida Atlantic University, Boca Raton, FL 33431, USA; richman@fau.edu
L. Salce, Dipartimento di Matematica Pura e Applicata, Universita, Via Belzoni 7, 35131 Padova, Italy; salce@math.unipd.it
P. Schultz, Department of Mathematics, University of Western Australia, Nedlands 6907, Australia; schultz@maths.uwa.edu.au
A. Scott, "Les Beaux", Knockmark, Drumree, Co. Meath, Ireland; lscott@iol.ie
N. A. Serdiukova, Ul. Marshala Tukhachevskogo, 17 Kovp. 1, kv. 30, 123423 Moscow, Russia; abic@academ-mf.rosmail.com
S. Shelah, Institute of Mathematics, Hebrew University, Jerusalem, Israel; shelah@math.huji.ac.il
M. Siddoway, Department of Mathematics, Colorado College, Colorado Springs, CO 80903, USA; msiddoway@cc.colorado.edu
L. Strüngmann, Fachbereich Mathematik und Informatik, Universität GH Essen, 45117 Essen, Germany; lutz.struengmann@uni-essen.de
E. Toubassi, Department of Mathematics, 617 N Santa Rita, PO Box 210089, Tucson, AZ 85721, USA; elias@math.arizona.edu
J. Trlifaj, Katedra Algebry MFF UK, Sokolovska 83, 18675 Praha 8, Czech Republic; trlifaj@karlin.mff.cuni.cz
C. Vinsonhaler, Department of Mathematics, University of Connecticut, Storrs, CT 06269, USA; vinson@math.uconn.edu

E. A. Walker, Department of Mathematics, New Mexico State University, Las Cruces, NM 88003, USA; elbert@nmsu.edu
W. J. Wickless, Department of Mathematics, University of Connecticut, Storrs, CT 06269, USA; wickless@math.uconn.edu
A. Yakovlev, Math-mech Faculty, Bibliotechnaya Pl. 2, Stary Peterhof, St. Petersburg 198904, Russia; yakovlev@yak.pdmi.ras.ru
P. Zanardo, Dipartimento di Matematica Pura e Applicata, Universita, Via Belzoni 7, 35131 Padova, Italy; pzanardo@math.unipd.it

Ross Allen Beaumont
July 23, 1914 – September 28, 1996

James D. Reid and William J. Wickless

Ross Beaumont was born in Dallas, Texas, on July 23, 1914. He grew up in Detroit, Michigan, where his family had moved while he was still young and he received a Bachelor's degree and a Master of Science degree from the University of Michigan. The year 1940 was a big one in Ross's life. He received a Ph.D. in Mathematics from the University of Illinois as the first American student of Reinhold Baer, he married and he and his wife Lois moved to Seattle where Ross had been offered a position. According to Lois, neither he (as an adult) nor she "had ever before been west of Chicago". Evidently the move was a good one personally. Ross served on the faculty at the University of Washington from 1940 until his retirement in 1985. Lois is still a resident of Seattle. Ross and Lois had two children, Linda, now a well known artist in Seattle, and Thomas, now an Assistant Professor in the Department of Mental Health at the University of Minnesota. Ross died on September 28, 1996.

Reflecting the esteem in which he was held by his colleagues at the University of Washington are the facts that Ross was Chairman of the Mathematics Department for five years, President of the Faculty Senate of the University and President of the Faculty Club. He was a visitor at the University of Arizona and at the Institute for Advanced Study in Princeton.

Ross had thirteen Ph.D. students. In alphabetical order they are: James Armstrong, Richard Byrne, James Clay, Frank Cornelius, John Koehler, Charles Murley, Phillip Schultz, Stephen Tellman, Burnette Toskey, William Wickless, George Williams, Robert Wisner and Paul Yearout.

We thank Lois Beaumont for providing details of Ross's life; and we thank Donna Gardner of the Mathematics Department of the University of Washington for providing the list of Ph.D. students.

Ross's first two papers ([1],[2]) build on work of Reinhold Baer and concern projections of one group to another. Here a projection from G to H is a map f from the set of subgroups of G onto the set of subgroups of H such that for subgroups S, T of G, $S \subseteq T$ if and only if $f(S) \subseteq f(T)$. The questions are to find necessary and sufficient conditions for the existence of a projection and to relate the structures of G and H when such does exist.

In 1945 ([3]) now working in a sense parallel to Baer, he begins what will become a major theme in his research, studying abelian groups which contain proper subgroups isomorphic to themselves. Baer had previously considered the case of non-abelian groups obtaining there a general result. Beaumont sharpens the results in the abelian case.

In 1955 ([7]) Beaumont revisits non-abelian groups when he and Peterson study set-transitive permutation groups; subgroups G of the symmetric group S_n which induce transitive permutations on the subsets $X, Y \subseteq \{1, 2, \dots, n\}$ of the same size (for each relevant cardinality). They show that if $n \neq 5, 6$, or 9 and G is equal to neither S_n nor A_n then G is not set transitive. They go on to determine explicitly the set-transitive groups for the exceptional values of n. The idea of set-transitive groups had come up in the work of von Neumann and Morgenstern on Game Theory. Chevalley, in unpublished work, obtained results similar to those of Beaumont and Peterson.

Probably motivated by his work in 1945, Beaumont, beginning his long and fruitful collaboration with R. S. Pierce, returns to the "isomorphic proper subgroup" problem ([12]) in 1959, but now in the context of modules over principal ideal domains. Here the concept of partly transitive modules is introduced. For example an abelian p-group G is partly transitive if whenever elements $x, y \in G$ satisfy an obvious necessary condition on their Ulm sequences, there is a monomorphism of G into itself taking x to y. They prove that countably generated p-primary R-modules are partly transitive and conjecture that every p-primary abelian group is partly transitive.

This joint study is continued and extended in the paper [13] in which the partly invariant submodules of a torsion module over a principal ideal domain R are determined. The submodule $N \subseteq M$ is partly invariant if $\theta N \subseteq N$ for all monomorphisms θ of M into itself. They give an explicit description of a certain class of partly invariant submodules of a given module M and show that any other partly invariant submodule of M must be fully invariant. The fully invariant submodules had already been determined by Kaplansky, so this completes the picture. Their work is based on their earlier concept of partial transitivity.

A second main theme in Beaumont's work is the study of the various ring structures that a given additive abelian group supports. The papers [4],[5],[9],[10],[11], and [25] as well as the joint work [16] and [17] with Pierce reflect in varying degrees this interest. According to Hanna Neumann in her review of the original book of Fuchs on abelian groups, the "two central problems" in this area are to find all non-isomorphic rings with a given additive group; and to characterize rings with given special properties by means of the structure of their additive group[1]. The papers [4] of Beaumont,

[1]We refer to the 1958 publication. All of Chapter XII was devoted to the problem of the additive group of a ring and its relation to the ring structure. For the review see Acta Sci. Math. Szeged, **20**, 1959, pp. 100-102.

and [9] and [11] written with his students Byrne and Wisner respectively, are among the earliest contributions to the first of these fundamental problems while [25] and the joint papers [16] and [17] are important contributions to the study of the second problem. The paper [25] was influenced by results from [16] and [17] as well as work of one of the present authors (WJW) and, we might add, motivated in turn some work on "rings on groups" by the other of us (JDR). For more on [16] and [17], see below.

One of the major pieces of joint work of Beaumont and Pierce was their Memoir on rank two torsion free groups [14]. The memoir gives a thorough investigation of rank two groups. A pair of quasi-isomorphism invariants for a rank two group G is developed, consisting of the Richman type of G and an equivalence class $[(\chi, \eta)]$, where χ, η are products of p-adic integers. These invariants are tractable enough to determine whether G is quotient divisible or is strongly indecomposable and to investigate the typeset of G. The ring of quasi-endomorphisms of G can be described in terms of the invariants. Using the invariants, a characterization of when G is quasi-isomorphic to any rank two H implies that G is isomorphic to H is given.

Our final comments in this review concern the genesis of the idea of quasi-isomorphism and the role that Beaumont and Pierce played in its development. In the 1950's abelian group theory, stimulated by the books of Fuchs and Kaplansky, was undergoing a vigorous period of development. Much of the work however, at least in a positive sense, concerned p-groups. We have Kaplansky's statement about torsion free groups: "...in this strange part of the subject anything that can conceivably happen actually does happen". The original examples of Jónsson of the failure of uniqueness of decomposition into indecomposable subgroups for torsion free groups of finite rank were followed by further, more spectacular, examples, due to Corner, Fuchs and others, supporting the validity of Kaplansky's insight. The mold of isomorphism, at least as a classification vehicle for torsion free groups, had been broken.

However, Jónsson had introduced the notions of quasi-isomorphism and strong indecomposability and shown that a version of the Krull-Schmidt theorem could be salvaged. All this is well known. But it was Beaumont and Pierce who put Humpty Dumpty back together again with their introduction of, and emphasis on, the **algebra of quasi endomorphisms.** This has permeated almost all discussions since. For example the idea that a group is sort of a "quasi module" over its quasi endomorphism algebra led one of us not only to another proof of Jónsson's Theorem, but to the idea of "irreducible" groups, which has found many applications. This mix of ideas then led Walker to introduce a categorical setting that in the case of torsion free groups of finite rank has objects torsion free groups A, B (of finite rank) with morphisms $Q \otimes Hom(A, B)$, (the "quasi category") that many use today as their theater of operations. We might mention, though, a remark made to us by Pierce shortly before his death to the effect that "...you can't help but feel that something is lost in going to the quasi category...". Perhaps the last word is yet to be uttered here. (In a slightly different vein we should mention the fruitful notion of near isomorphism introduced by Lady.)

In any event we believe that the work of Beaumont and Pierce has had a profound effect on the development of torsion free abelian group theory and a definitive effect on the theory of torsion free rings. The paper [17] contains a deep analysis of the interplay between the structure of a ring and that of its additive group resulting for example in the discovery of the quotient divisible groups and the introduction of the *q.d.*-invariants which classify these groups up to quasi isomorphism. (Any ring A such that $Q \otimes A$ is a finite dimensional semi-simple Q-algebra has a *q.d.* group as additive group.) Another major result in this paper is the Beaumont-Pierce analog of the Wedderburn Principal Theorem. And the paper [16] is just beautiful. For details of all this work we refer the reader to the papers [19] and [26] of Beaumont and the joint papers [16] and [17]. For more on questions of quasi isomorphism see [20] and [21]. And we recommend as well to the reader's attention the fine recent tribute[2] of C. Vinsonhaler to R. S. Pierce.

Ross loved teaching at all levels. He was the author or co-author of three books: Introduction to Modern Algebra and Matrix Theory (with Richard Ball), 1954; Algebraic Foundations of Mathematics (with R. S. Pierce), 1963; and the book written alone Linear Algebra, 1965. On a personal note, the only course in group theory of any kind that one of us (JDR) ever had was a course on Abelian Groups that Ross taught out of the first book of that title by Fuchs.

Moreover, Ross was a generous and gifted thesis advisor. One of us (Wickless) recuses himself from the following remarks since he was a student of Ross's, but the other (who was a student of Pierce) shamelessly reports that Ross's students have distinguished themselves in many directions. Three of Ross's students (Armstrong, Cornelius and Murley) had somewhat brief mathematical careers for various reasons (Armstrong due to an unfortunate early death), yet they all did significant things: Armstrong published several papers; Cornelius is cited as having established a "quite astonishing result" by Göbel[3]; and Murley has the "Murley groups" named after him by various authors, notably Arnold. Bob Wisner was prominent in mathematical education and administration as well as research and continues to publish well into his 70's; Koehler's work in his thesis was one of the early treatments of the type-set of a torsion free group; Clay became a world leader in research on near rings and was, at the age of 34, the youngest

[2]Richard Scott Pierce, by C. Vinsonhaler, Abelian Groups and Related Topics, Contemporary Mathematics, Vol. 171, 1994, pp. 1-11.

[3]See Göbel's tribute to Fuchs in honor of Fuchs' 70th birthday in Periodica Math. Hungarica **32** (1996) pp. 13 –39

person to receive the Distinguished Senior U.S. Scientist Award from the (then West) German government[4]; and finally we are all familiar with the numerous and important contributions of Schultz and Wickless to abelian group theory. These are just some of the examples known to the authors. We apologize for inevitable oversights.

For those of us–the present writers as well as many others–who were students at the University of Washington during Ross's tenure it would be difficult to overstate the measure of our gratitude to him for his inspiration and encouragement. He was simply a wonderful person.

As one reviews the work of Beaumont over the years it is interesting to observe how the thread of his mathematical interests is woven through his own work, of course, but also how it binds together his work with Pierce. Their joint work is well known and deeply respected. It seems clear to us that Beaumont, because of his interests and his attractive personal qualities, was instrumental in bringing Pierce into the study of abelian groups. And for that alone we could all be grateful.

References

[1] *Projections of non-abelian groups upon abelian groups containing elements of infinite order*, Amer. J. Math. **64** (1942), 115–136.

[2] *Projections of the prime-power abelian group of order p^n and type $(m-1,1)$*, Bull. Amer. Math. Soc. **48** (1942), 866–870.

[3] *Groups with isomorphic proper subgroups* Bull. Amer. Math. Soc. **51** (1945), 381–387.

[4] *Rings with additive group which is the direct sum of cyclic groups*, Duke Math. J. **15** (1948), 367–369.

[5] (with H. S. Zuckerman) *A characterization of the subgroups of the additive rationals*, Pacific J. Math. **1** (1951), 169–177.

[6] (with Richard W. Ball) *Introduction to Modern Algebra and Matrix Theory*, Rinehart & Co., Inc., New York (1954) $xii + 331$ pp.

[7] (with R. P. Peterson) *Set-transitive permutation groups*, Canad. J. Math. **7** (1955), 35–42.

[8] *Matric criteria for the uniqueness of basis number and the equivalence of algebras over a ring*, Publ. Math. Debrecen **4** (1956), 35–42.

[9] (with J. Richard Byrne) *On the construction of R-modules and rings with polynomial multiplication*, Pacific J. Math. **7** (1957), 1305–1317.

[10] *Generalized rings*, Proc. Amer. Math. Soc. **9** (1958), 876–881.

[11] (with R. J. Wisner) *Rings with additive group which is a torsion-free group of rank two*, Acta Sci. Math. Szeged **20** (1959), 105–116.

[12] (with R. S. Pierce) *Partly transitive modules and modules with proper isomorphic submodules*, Trans. Amer. Math. Soc. **91** (1959), 209–219.

[4]See the University of Arizona Mathematics Department Web Site.

[13] (with R. S. Pierce) *Partly invariant submodules of a torsion module*, Trans. Amer. Math. Soc. **91** (1959), 220–230.

[14] (with R. S. Pierce) *Torsion free groups of rank two*, Mem. Amer. Math. Soc. No. **38** (1961), 41 pp.

[15] *Quasigroup algebras*, Math. Z. **79** (1962), 1–9.

[16] (with R. S. Pierce) *Subrings of algebraic number fields*, Acta Sci. Math. (Szeged) **22** (1961), 202–216.

[17] (with R. S. Pierce) *Torsion-free rings*, Illinois J. Math. **5** (1961), 61–98.

[18] (with R. S. Pierec) *Isomorphic direct summands of abelian groups*, Math. Ann. **153** (1964), 21–37.

[19] *A survey of torsion-free rings*, Topics in Abelian Groups (Proc. Sympos., New Mexico State Univ., 1962), Scott Foresman and Co., Chicago, Ill. (1963), 41–50.

[20] (with R. S. Pierce) *Quasi-isomorphism of p-groups*, Proc. Colloq. Abelian Groups (Tihany, 1963), Akadémiai Kiadő, Budapest (1964), 13–27.

[21] (with R. S. Pierce) *Quasi-isomorphism of direct sums of cyclic groups*, Acta Math. Acad. Sci. Hungar. **16** (1965), 33–36.

[22] (with R. S. Pierce) *Some invariants of p-groups* Michigan J. Math. **11** (1964), 137–149.

[23] *Abelian groups G which satisfy $G \cong G \oplus K$ for every direct summand K of G*, Studies on Abelian Groups (Symposium, Montpelier, 1967), Springer, Berlin (1968), 69–74.

[24] *A note on products of homogenous torsion free abelian groups*, Proc. Amer. Math. Soc. **22** (1969), 434–436.

[25] (with D. A. Lawver) *Strongly semisimple abelian groups*, Pacific J. Math. **53** (1974), 327–336.

[26] *Generalizations of quasi-isomorphism*, Symposium Mathematica, Vol **XII** (Convegno Gruppi Abeliani, INDAM, Rome, 1972), Academic Press, London, (1974), 299–310.

Trends in Mathematics, © 1999 Birkhäuser Verlag Basel/Switzerland

Modular group algebras and simply presented groups

Paul Hill

Abstract

We consider the connection and interplay between some well-known problems in modular group algebras and the structure of simply presented p-groups. New results in both areas are obtained.

1 Introduction

The group of units of group rings and group algebras has been the object of much study. In this paper, we consider the group of units of modular group algebras over a perfect field, where the group on which the group algebra is based is usually a primary abelian group.

The first part of our paper (consisting of sections 1 and 2) is intended as a brief survey, whereas the remaining part of the paper is intended as a research article; actually, the second part may be closer to a research announcement since proofs of all the propositions and corollaries are omitted and those of theorems sometimes abbreviated.

The underlying theme of the paper is the connection and interplay of certain problems in modular group algebras and the structure theory of simply presented p-groups. It is surprising to find that the relationship between the two has become in recent times somewhat symbiotic. It is of course no surprise at all that the theory of simply presented groups would be beneficial to an investigation of modular group algebras over such groups, but how could modular group algebras possibly enhance the already rich structure theory of simply presented p-groups? As will soon become apparent, it has been the *search* for answers to the group algebra problems, rather than the solutions themselves, that has led to new results for simply presented groups that perhaps otherwise would not have been discovered, although some of the new results have been barely hidden beneath the surface of well-known results for a long time.

Let F be a perfect field of characteristic p and let G be an abelian p-group for the same prime p; in some instances, G is allowed to be mixed, but assume that G is primary unless we state otherwise. However, for clarity, we often repeat the hypothesis that G is primary. Recall that in the context of group algebras the underlying group G is usually written multiplicatively even when it is abelian. Therefore, when we are considering the group algebra $F(G)$ or some part thereof such as its group of units, we will use the multiplicative notation for G. On the other hand, when we are dealing strictly with abelian groups, we follow custom and use the additive setting.

The group of normalized units of $F(G)$ is denoted by $V(G)$; the normalized units are those that have augmentation 1, that is, the units whose coefficients sum to 1 . The full group of units $U(G)$ of $F(G)$ can be written as the direct product

$$U(G) = F^* \times V(G),$$

where F^* is the multiplicative group of the field F. Therefore, by describing $V(G)$ we obtain a description of the full group of units of the group algebra $F(G)$. With this reduction, the coefficients are reduced to a rather passive role in what we do here, so one can assume that F is the prime field of characteristic p with no great loss of the main ideas of the paper. With the field F chosen (once for all), we denote the cokernel of G in $V(G)$ merely by G^*, so

$$G^* = V(G)/G.$$

One version of a well-known problem in group algebras known as the *direct factor problem* is the following.

Problem A. For which groups G is it true that G is a direct factor of $V(G)$?

It is easy to establish the fact that G is at least a balanced subgroup of $V(G)$ if not a direct factor. Therefore, the direct factor problem leads naturally to the following problem that we call the *simply presented problem.*

Problem B. For which groups G is it true that G^* is simply presented?

Although we do not pursue it here, another problem that is related to the preceding two and that has been considered by a number of authors (see, for example, [11, 13]) is the *isomorphism problem.*

Problem C. Must G and H be isomorphic if their group algebras $F(G)$ and $F(H)$ are isomorphic?

As our title suggests, in this paper we focus primarily on the *simply presented problem* — mindful, of course, that if G^* is simply presented then G is a direct factor of $V(G)$ because, as we said before, G is always a balanced subgroup of $V(G)$. Hence, G is a direct factor of $V(G)$ whenever G^* is simply presented.

We will discuss some of the progress made to date on the aforementioned group algebra problems later, but since we are particularly interested in the question of when G^* is simply presented and since this has proved to be such a difficult question to give a definitive answer, it might be worthwhile to pause for a moment and reflect on the following fundamental question. How can one determine whether a group — in particular, G^* — is simply presented or not?

In connecting the simply presented problem stated above to the direct factor problem, we have already been reminded of the fact that the p-group G is simply presented if and only if it is balanced projective. However, the best known characterizations of simply presented groups are included in what has been referred to in the literature as the *trinity*, which holds that the following are equivalent:

1. G is simply presented.

2. G is totally projective.

3. G satisfies Axiom 3.

In turn, Axiom 3 itself has three versions usually attributed respectively to Fuchs, Griffith, and the author (although the one attributed to Griffith is actually the *original* Axiom 3, first used by the author to prove the uniqueness theorem; it was necessary to adopt later the stronger version in order to prove the summand theorem). For the purpose of comparing different versions of Axiom 3 and for other purposes developed later, it is convenient to introduce here the following condition on a collection $\mathcal{C}$ of nice subgroups of G, which we will refer to as the *countable extension* property:

(CE) If N is a member of $\mathcal{C}$ and A is any countable subgroup of G, there is a countable extension N' of N that contains A and belongs again to $\mathcal{C}$.

There is a weak version of (CE), denoted by (WCE), obtained from (CE) by only requiring that (CE) is satisfied for $N = 0$. The three well-known versions of Axiom 3 mentioned above are:

(3.F) G has a composition sreies of nice subgroups.

(3.G) G has a collection $\mathcal{C}$ of nice subgroups that satisfies the countable extension property (CE), contains 0, and is closed with respect to unions of ascending chains.

(3.H) The same as (3.G) except that $\mathcal{C}$ is required to be closed with respect to arbitrary group unions.

Clearly, it does not matter logically whether (CE) or (WCE) is used in (3.H); historically, (CE) was used because it was retained from the original version (3.G), where it is necessary. Indeed, if we substitute (WCE) in (3.G), we obtain a genuinely weaker condition called *weak Axiom 3.* Groups satisfying this weak version of Axiom 3 are also referred to as *almost totally projective* or *almost simply presented* (see [9]). The concept of being almost simply presented is especially relevant to group algebras.

Entertaining again now the question of how we might demonstate that G^* is simply presented, if it is, we should first point out that one should not try to show that G^* is simply presented by showing that it is simply presented! In other words, as in most cases, seeking an actual manifestation of a simple presentation is not the best approach. Moreover, we would naturally be inclined to use an Axiom 3 characterization of simply presented groups, especially when we consider the fact that $V(G)$ has an abundance of nice subgroups. With this strategy in place, which version of Axiom 3 is preferred? Since the implications

$$(3.H) \Longrightarrow (3.G) \Longrightarrow (3.F)$$

are rather immediate, Fuchs' version of Axiom 3 — the existence of a composition series of nice subgroups — is ostensibly the weakest of the three axioms. Hence, it makes sense that in order to show that a p-group is simply presented we should attempt to show that it has a composition series of nice subgroups. Incidentally, as is well known, it suffices to establish a "crude" composition series with countable factors since a refinement to finite factors is gratuitous for a p-group. Therefore, to show that G^* is simply presented, we set as our goal the task of showing that it has a composition series *unless* we find some other characterization of simply presented groups that seems to offer more promise when applied to G^*. In this connection, we should recall that there are various other criteria of being simply presented when special conditions are imposed on G such as restrictions on its length or size. For example, Kulikov's criterion for coproducts of cyclic groups might still be the most useful tool in proving that G^* is simply presented when G has no elements of infinite height.

2 Prior results

We will attempt now to give a brief summary of what is known about the group algebra problems mentioned in the introduction. Recall that F is a perfect field of characteristic p and that G is an abelian p-group. We begin with the fundamental results of Warren May from his seminal paper [10].

May's Theorem (May [10]).
(a) $V(G)$ is simply presented if and only if G is.
(b) G^* is simply presented if G is.
(c) If the group algebras $F(G)$ and $F(H)$ are isomorphic, and G is simply presented, then G and H must be isomorphic.

Remarks. May actually stated and proved part (c) more generally for a p-local Warfield group G; part (c) had been proved earlier in [1] whenever *both* G and H are p-local Warfield groups. Parts (a) and (b) subsumed a number of previous results dealing with special cases including the case where G is countable and the case where G is an arbitrary coproduct of cyclic groups.

The following corollary is a noteworthy special case of May's theorem, which plays some role in our discussion. Note that even the corollary encompasses the special cases mentioned in the preceding remarks.

Corollary (May). $V(G)$ is simply presented whenever G is simply presented and has countable length.

An outline of the proof of the parts (a) and (b) of May's theorem should be helpful in understanding our subsequent discussion. First, however, we need a few basic facts.

Let H be a subgroup of G. The natural map $G \to H$ induces a map

$$V(G) \to V(G/H)$$

whose kernel is designated as $K(H)$. This mapping is epic when G is a p-group.

Proposition (May [10], Hill and Ullery [6]). If H is an arbitrary subgroup of the p-group G, the following are true.

(i) $V(H)$ and $K(H)$ are nice in $V(G)$ if and only if H is nice in G.

(ii) $G\,V(H)$ is *always* nice in $V(G)$, and consequently $G\,V(H)/G$ is nice in $G^* = V(G)/G$.

(iii) G is compatible with both $V(H)$ and $K(H)$.

(iv) $G\,V(H)$ is compatible with both $G\,V(A)$ and $V(A)$ whenever A is any subgroup of G.

Recall that A and B are compatible (written $A \| B$) if $x \in AB$ can be written as $x = ab$ where the factors $a \in A$ and $b \in B$ have as much height as the product $x = ab$.

Proof of May's Theorem (Sketch). To see why $V(G)$ is simply presented when G is simply presented, choose a composition series of nice subgroups of G. Suppose that

$$1 = N_0 \subseteq N_1 \subseteq N_2 \subseteq \cdots N_\alpha \subseteq \cdots$$

is a composition series of nice subgroups of the simply presented group G. According to the preceding proposition,

$$(1) \qquad 1 = V(N_0) \subseteq V(N_1) \subseteq \cdots \subseteq V(N_\alpha) \subseteq \cdots$$

and

$$(2) \qquad 1 = K(N_0) \subseteq K(N_1) \subseteq \cdots \subseteq K(N_\alpha) \subseteq \cdots$$

are both chains of nice subgroups of $V(G)$, but they are *not* composition series; the factors of adjacent terms tend to be large, certainly not in general finite.

By employing some basic ring theory involing nil ideals and using an ingenious scheme of identifying certain nice subgroups of $V(G)$, May was able to refine the series (2) so that it becomes a composition series. The key step in doing this is the demonstration that there is always a composition series of nice subgroups of $V(G)$ between 1 and $K(A)$ whenever A is any *finite* subgroup of G. For details, we refer to [10]. Hereafter, we refer to such a series as *May's composition series for finite kernels.*

Once we have May's composition series for finite kernels, the series (2) consisting of the $K(N_\alpha)$'s can be refined to a composition series for $V(G)$ by inserting between $K(N_\alpha)$ and $K(N_{\alpha+1})$ the subgroups of $V(G)$ corresponding to May's composition series in $V(G/N_\alpha)$ for the finite subgroup $N_{\alpha+1}/N_\alpha$ of G/N_α.

We have just indicated how a composition series for $V(G)$ can be constructed from a composition series of G. May also proved this the other way around: a composition series of $V(G)$ leads to a composition series of G. Hence, $V(G)$ is simply presented if and only if G is simply presented, which is part (a) of May's theorem.

It should be observed that in the preceding proof it is necessary to use the $K(N_\alpha)$'s as the framework of the composition series for $V(G)$, not the $V(N_\alpha)$'s.

Open Question. When G is a simply presented p-group, can the series (1) consisting of the nice subgroups $V(N_\alpha)$ also be refined to a composition series for $V(G)$?

We now turn to part (b) of May's theorem. For this we need to argue that G^* has a composition series if G does. Although it is not immediate and is not the method used by May, it turns out that the composition series constructed above for $V(G)$, which passes through the $K(N_\alpha)$'s, can be lifted to a composition series of nice subgroups of $V(G)$ between G and $V(G)$ simply by tacking G onto each term; this may be somewhat surprising since generally the join of two nice subgroups is not again a nice subgroup. The composition series of nice subgroups of $V(G)$ between G and $V(G)$ of course yields immediately a composition series of nice subgroups of $G^* = V(G)/G$. Therefore, G^* is simply presented if G is. □

Subsequent to May's results and motivated by them, W. Ullery and the author (both jointly and severally) have considered the group algebra problems. Generally, we have concentrated on the simply presented problem. Two of our results are the following.

Theorem(Hill and Ullery [6]) If G has cardinality not exceeding $\aleph_1$, then G^* is simply presented.

Theorem(Hill and Ullery [7]). If G is an isotype subgroup of a simply presented group of length not exceeding ω_1 (d.s.c), then G^* is simply presented.

Either of the two theorems just cited quickly defeats the notion that G must be simply presented in order for G^* to be simply presented (as was the case for $V(G)$). To the contrary, in a certain sense, these results suggest that G^* might be simply presented for all p-groups G. Indeed, it can be argued that the groups covered by these theorems can be about as complicated stucturally as p-groups ever get (see [5]). Moreover, the *ad hoc* proofs of the two theorems do not resolve the mystery as to why G^* is simply presented peculiarly for these groups, and this naturally leads to speculation about other groups. We mention, however, that this mystery is at least partially solved in the final section of paper, where the two theorems are placed under the same umbrella.

The first of the two preceding theorems has added significance, for it has led to a new characterization of simply presented groups of cardinality $\aleph_1$.

Theorem (Hill [3]). A p-group G of cardinality not exceeding $\aleph_1$ is simply presented if and only if G has a v-basis.

We suppress a discussion here of v-bases except to say that G^* has a v-basis for any p-group G (see [4]) and that any group with a v-basis is almost simply presented [9]. Therefore, for any p-group G, it is a fact that G^* is almost simply presented, if not simply presented.

The last theorem quoted above is, of course, an example of the interplay between problems in modular group algebras and simply presented groups, which as we said in the beginning is our focus here. We hasten to include another example. The next example provides a new characterization of simply presented groups of countable length that was discovered as a direct result of investigating group algebras — this time for mixed groups.

Theorem (Hill and Ullery [8]). If G is a mixed group whose torsion is a p-primary simply presented group of countable length, then G^* is simply presented.

This theorem was motivated by previously proven special cases covering the case where G_t, the torsion subgroup of G, is a coproduct of cyclic groups and the case where G_t is countable. However, in order to prove the theorem, the following new characterization of simply presented p-groups of countable length was required.

Theorem (Hill and Ullery [8]). A p-group G of countable length is simply presented if and only if it is the union of a sequence of subgroups each of which has only a finite number of different heights computed in G.

Remarks. Although the above characterization of simply presented p-groups of countable length could have easily been established at any point subsequent to the Axiom 3 characterization, as we have said, it was the group algebra problem that led to this result. The reader should note that the above characterization is a simple yet elegant extension of Kulikov's criterion from groups of length not exceeding ω_0 to groups of length less than ω_1. Finally, we remark that the new characterization has other applications, some of which will be demonstrated in the next sections.

3 A new characterization of simply presented

Definition 3.1 Call a subgroup an fs-subgroup if its height spectrum is a finite set when heights are computed in the containing group. ("fs" is intended for "finite spectrum".)

It is noteworthy that if H is an fs-subgroup of G, then $V(H)$ is an fs-subgroup of $V(G)$. Therefore, we have the following.

Proposition 3.2 *If G is the union of an ascending sequence of fs-subgroups G_n, then $V(G)$ is the union of the ascending sequence of fs-subgroups $V(G_n)$.*

The preceding provides a short proof of the result designated earlier as May's corollary: $V(G)$ is simply presented whenever G is simply presented and has countable length.

For further applications of the new characterization of simply presented groups of countable length, we need the following concept.

Definition 3.3 A subgroup H of G is said to be an *fs-invariant* subgroup if every subgroup of H which is an fs-subgroup of G is also an fs-subgroup of H.

Observe that both isotype subgroups of G and the subgroups of the form $p^\alpha G$, which are the opposite of isotype, are both fs-invariant subgroups of G, but of course there are others including all the bounded subgroups of G.

The following result generalizes a well–known result due to the author [2] that predates Axiom 3. The original theorem was restricted to *isotype* subgroups.

Proposition 3.4 *An fs-invariant subgroup of a simply presented p-group of countable length is itself simply presented.*

Since we can characterize simply presented p-groups of countable length in such a pleasant way using fs-subgroups, the question naturally arises as to whether or not this can also be accomplished for arbitrary simply presented p-groups. A moment of reflection should convince one that this is not accomplished by an Axiom 3 system of fs-subgroups. In fact, a group that has an Axiom 3 system of fs-subgroups must have countable length. Before investigating the question at hand, we observe some of the properties of fs-subgroups.

Proposition 3.5 *For any p-group G, fs-subgroups are always nice subgroups, and the following inclusions hold provided that G is reduced:*
$\{$ *finite subgroups* $\} \subseteq \{$ *fs-subgroups* $\} \subseteq \{$ *bounded subgroups* $\}$.

It is of interest to consider the classes of p-groups that are unions of ascending sequences of subgroups that belong, respectively, to the various classes of subgroups listed above; obviously, in the first instance, we obtain the countable groups, and in the third instance, all p-groups. In the second instance, we know now that we obtain precisely the simply presented groups of countable length.

We should mention that one potential defect of fs-subgroups is the fact that two fs-subgroups need not generate again an fs-subgroup. However, this flaw is largely mitigated by the following proposition.

Proposition 3.6 *The subgroup generated by an fs-subgroup and a finite subgroup is again an fs-subgroup, and the subgroup generated by two compatible fs-subgroups is again an fs-subgroup.*

In order to obtain a characterization of arbitrary simply presented groups (not just those of countable length) in terms of fs-subgroups, it is necessary to extend the concept of a compatible pair of subgroups to a *compatible family* of subgroups.

Definition 3.7 A family of subgroups $\{A_i\}_{i\in I}$ of G is said to be a compatible family of subgroups of G if, for each pair of subsets J and K of the index set I, the subgroups $A(J) = \langle A_j \rangle_{j\in J}$ and $A(K) = \langle A_k \rangle_{k\in K}$ are compatible.

Remarks. In the preceding definition, it suffices to verify the condition of a compatible family in case one, both not both, of the subsets J and K is a singleton set. Moreover, the other one can be assumed to be finite. On the other hand, it is easy to construct examples of three subgroups that do not form a compatible family, yet any two of the three are compatible.

We now come to yet another characterization of simply presented p-groups.

Theorem 3.8 *A p-group G is simply presented if and only if*

$$G = \bigcup_{i \in I} G_i,$$

where $\{G_i\}_{i\in I}$ is a compatible family of fs-subgroups.

Proof. First, we will explain why the condition is *necessary* in order for the p-group G to be simply presented. Suppose that G is simply presented of length μ. The argument that G has the required compatible family of fs-subgroups is by induction on μ. Clearly, if $\mu \leq \omega_0$ then G being a direct sum of cyclic groups has the desired compatible family, for we can simply take finite direct sums of the exhibited cyclic summands. Thus, we proceed by induction on μ. It is almost trivial that the induction survives isolated ordinal. Indeed, if $\mu = \lambda + 1$ and if $\{G_i/p^\lambda G\}$ is the desired family of fs-subgroups of $G/p^\lambda G$, then $\{G_i\}$ is a compatible family of fs-subgroups of G, itself. We can assume, therefore, that μ is a limit. In this case, however, we can write $G = \bigoplus_{i\in I} G_i$, where G_i has smaller length than μ. By the induction hypothesis, each G_i has a compatible family of fs-subgroups $G_{i,j}$ whose union is G_i. It is easily verified that the subgroups of the form $\oplus_{i\in N} G_{i,j}$, where N is a finite subset of I, is a compatible family of fs-subgroups of G whose union is G.

Conversely, assume that the group G has a compatible family of fs-subgroups G_i that satisfy $G = \cup_{i\in I} G_i$, and let J be an arbitrary subset of the index set I. Set $G(J) = \langle G_j \rangle_{j\in J}$. The claim is that $G(J)$ is nice in G. In order to prove this, let $x \in G \setminus G(J)$ and choose $i \in I$ so that $x \in G_i$. If $y \in G(J)$ and $\mid x + y \mid = \lambda$, then the compatibility of $G(J)$ and G_i implies that $\mid x + c \mid \geq \lambda$ for some $c \in G(J) \cap G_i$. But G_i has finite height spectrum, so $\mid x + c \mid$ must be one of those finite heights. It quickly follows that the coset $x + G(J)$ must have a proper element. Moreover, it follows from the valuated isomorphism

$$G_i + G(J)/G(J) \cong G_i/G_i \cap G(J)$$

that $G_i + G(J)/G(J)$ has finite height spectrum where heights are computed in $G/G(J)$. Therefore, any subgroup between $G(J)$ and $G_i + G(J)$ is nice in G. If we choose the index set I to be an initial segment of ordinals, we have shown that the chain of nice subgroups

$$\cdots \subseteq G(\{\alpha : \alpha < \lambda\}) \subseteq G(\{\alpha : \alpha \leq \lambda\}) \subseteq \cdots$$

can be refined to a composition series of nice subgroups of G, and this completes the proof of the theorem. □

4 Simply presented G^*

In this section, we consider the class of p-groups for which G^* is simply presented. As explained earlier, it is useful to know that G^* is simply presented since this solves affirmatively the direct factor problem, and furthermore it yields the structure of the group of units of the group algebra $F(G)$. Therefore, we introduce the following terminology.

Definition 4.1 The p-group G is said to be exemplary if G^* is simply presented.

Part (b) of May's theorem quoted in our introduction can now be restated by saying that simply presented p-groups are exemplary. The two results of Hill and Ullery that were quoted earlier can be restated by saying, in the first instance, (1) all p-groups of cardinality not exceeding $\aleph_1$ are exemplary and, in the second instance, (2) isotype subgroups of d.s.c.'s are exemplary.

The facts just quoted represent the major known results concerning the simply presented problem. Our aim now is to generalize these results by proving some closure properties for the class of exemplary groups. In this endeavor, however, we first take a strike, as it were. We do not know the answer to the following basic question.

Question 4.2 *Is the coproduct of two exemplary groups itself exemplary?*

The preceding question notwithstanding, we can establish a weak closure for infinite coproducts.

Proposition 4.3 *Suppose that G is the coproduct of a collection of subgroups $\{G_i\}_{i\in I}$. If the coproduct of every finite subcollection of the G_i's is an exemplary group, then G must be exemplary.*

The preceding proposition is actually a corollary to the fact that G is exemplary if it is the union of a smooth ascending chain of direct factors G_α each of which is exemplary. To prove the latter, we use the fact if

$$G_{\alpha+1} = G_\alpha \times C_\alpha,$$

then

$$V(G_{\alpha+1}) = V(G_\alpha) \times K_\alpha(C_\alpha),$$

where

$$K_\alpha(C_\alpha) = K(C_\alpha) \cap V(G_{\alpha+1}).$$

The above yields

$$G\,V(G_{\alpha+1})/G = G\,V(G_\alpha)/G \times G\,K_\alpha(C_\alpha)/G,$$

and G^* is the coproduct of the simply presented groups $G\,K_\alpha(C_\alpha)/G$, so G is exemplary.

In preparation for our next result we recall the definition of an elementary elongation. When translated to the multiplicative notation, it reads as follows.

Definition 4.4 A p-group G is an elementary elongation of H if $G^{\mu+1} = 1$ and $G/G^\mu \cong H$.

Theorem 4.5 *An elementary elongation of an exemplary group remains exemplary.*

Indeed, we have that the following more general result is true.

Theorem 4.6 *Suppose that N is a nice subgroup of the p-group G and that N has a composition series consisting of nice subgroups of G. Suppose, further, that G/N is exemplary. Then G is exemplary.*

Proof. Let $\{N_\alpha\}$ be a composition series of N where N_α is nice in G for each α. First, we want to show that there is a composition series consisting of nice subgroups of $V(G)$ between 1 and $K(N)$. It is not difficult to show that

$$K(N_{\alpha+1})/K(N_\alpha) \cong K(N_{\alpha+1}/N_\alpha)$$

is a valuated isomorphism, where heights are computed in the appropriate containing groups: $V(G)/K(N_\alpha)$ and $V(G/N_\alpha)$, respectively. Therefore, it is enough to show that

$$K(N_{\alpha+1}/N_\alpha)$$

has a composition series of nice subgroups of $V(G/N_\alpha)$. But May's composition series for finite kernels is applicable here, so there exists a composition series between 1 and $K(N)$ consisting of nice subgroups of $V(G)$. As in Section 2, the composition series just established between 1 and $K(N)$ lifts to a composition series from G to $G\,K(N)$ that still consists of nice subgroups of $V(G)$. Consequently, there is a composition series of $G\,K(N)/G$ consisting of nice subgroups of $G^* = V(G)/G$. Finally, since G/N is exemplary and since $G\,K(N)$ is nice in $V(G)$, the exact sequence

$$1 \to G\,K(N)/G \to V(G)/G \to V(G/N)/(G/N) \to 1$$

reveals that the middle group, G^*, has a composition series and is therefore simply presented. Thus G is exemplary. □

Corollary 4.7 *If A is simply presented and B is exemplary, then the coproduct of A and B is exemplary.*

We introduce now some interesting classes of groups. The initial classes turn out to be extremely relevant to the simply presented problem. It is understood in the following definition that *all heights are computed in the containing group.* In other words, subgroups are valuated subgroups.

Definition 4.8 The p-group G is said to be of order zero if the height spectrum of G is finite, while G is said to be of order $n+1$ if G does not have order $\leq n$ but is the union of a chain of subgroups of order $\leq n$ in G.

Notice that G has order ≤ 2 if G is the union of a chain of subgroups each of which is itself the union of a chain of subgroups having finite height spectrum where heights are computed in G.

Proposition 4.9 *An fs-invariant subgroup of a group that has order $\leq n$ must itself have order $\leq n$.*

One of our main results concerning exemplary groups is the following.

Theorem 4.10 *Any p-group G of order ≤ 2 is exemplary.*

Proof. Let G be the union of a chain of subgroups H_α each of which has order ≤ 1, where $H_\alpha \subseteq H_\beta$ if $\alpha < \beta$. In turn, let H_α be the union of a sequence of fs-subgroups $B_{\alpha,n}$ of G. Since $V(B_{\alpha+1,n})$ has finite height spectrum in $V(G)$ and since $V(B_{\alpha+1,n})$ is compatible with $G\,V(H_\alpha)$, we conclude that the subgroup

$$G\,V(H_\alpha)\,V(B_{\alpha+1,n})\,/G\,V(H_\alpha)$$

has finite height spectrum in $V(G)\,/G\,V(H_\alpha)$, so any subgroup of $V(G)$ between $G\,V(H_\alpha)$ and $G\,V(H_\alpha)\,V(B_{\alpha+1,n})$ is nice in $V(G)$. It quickly follows that there is a composition series of nice subgroups of $V(G)$ between $G\,V(H_\alpha)$ and $G\,V(H_{\alpha+1})$; indeed, there is such a composition series between $G\,V(\cup_{\alpha<\beta} H_\alpha)$ and $G\,V(H_\beta)$. Thus, there is a composition series of nice subgroups of $V(G)$ between G and $V(G)$, and consequently G is exemplary. □

Clearly, any p-group of cardinality not exceeding $\aleph_1$ has order ≤ 2 because such a group is the union of a chain of countable groups (each of which is the union of a chain of finite groups). Therefore, we have as an immediate corollary the result of [6].

Corollary 4.11 *Every p-group of cardinality not exceeding $\aleph_1$ is exemplary.*

The following is the key in establishing another large class of groups of order 2 (and consequently many more exemplary groups).

Theorem 4.12 *Any simply presented p-group of length less than ω_2 has order ≤ 2.*

Proof. As a precursor, consider the proof that if G is simply presented of length $< \omega_1$ then G has order ≤ 1. The proof that such a group G has order ≤ 1 is easily established by induction on the length μ of G. If $\mu = \lambda + 1$ is isolated, we know by the induction hypothesis that there exists an ascending sequence of fs-subgroups $\{G_n/p^\lambda G\}$ of $G/p^\lambda G$ whose union is all of $G/p^\lambda G$. Notice that G_n must be an fs-subgroup of G, so we conclude in this case that G must have order ≤ 1. If μ is a limit, we can write $G = \oplus_{m<\omega} G_m$ where G_m has length less than μ for each m. Again, by the induction hypothesis, G_m has an ascending sequence of fs-subgroups $G_{m.n}$ that lead up to G_m. Clearly,the subgroups $H_n = \oplus_{m \leq n} G_{m,n}$ is an ascending sequence of fs-subgroups of G whose union is G. Thus G has order ≤ 1. Now suppose that G is a simply presented p-group of length $\mu < \omega_2$. We want to show that G has order ≤ 2. We know that if G has length $< \omega_1$, then in fact G has order ≤ 1. Moreover, if G has length equal to ω_1, then G is the union of an ascending chain of direct summands each of which is simply presented of countable length and therefore a subgroup of order ≤ 1 in G. Hence, G has order ≤ 2. Thus we can assume that the length μ of G exceeds ω_1 but is less than ω_2. If $\mu = \lambda + 1$ is isolated, the proof that G is the union of an ascending chain of subgroups G_α , $\alpha < \omega_1$, is analogous to the corresponding case outlined above for the special case of order ≤ 1. It is, of course, based on the induction hypothesis that $G/p^\lambda G$ has such a chain. Finally, consider the case that μ is a limit. In this case, under the assumptions on μ that were made above, we can write $G = \oplus_{\alpha<\omega_1} G_\alpha$, where G_α has length greater than ω_1 but less than μ for each α. Once again by the induction hypothesis, the simply presented group G_α is the union of an ascending chain of subgroups $G_{\alpha,\beta}$, $\beta < \omega_1$, that are of order ≤ 1 in G. The chain of subgroups

$$H_\beta = \bigoplus_{\alpha \leq \beta} G_{\alpha,\beta} \qquad (\beta < \omega_1)$$

is an ascending chain of order ≤ 1 subgroups of G whose union is G, so G must have order ≤ 2. □

The preceding theorem, together with Proposition 4.9 and Theorem 4.10, yields the following.

Corollary 4.13 *An fs-invariant subgroup (in particular, an isotype subgroup) of a simply presented p-group of length less than ω_2 is exemplary.*

A special case of the preceding corollary is the other result of Hill and Ullery [7] about exemplary groups mentioned above.

Corollary 4.14 *If the p-group G is an isotype subgroup of a d.s.c., then G is exemplary.*

In conclusion, we consider the case where G has cardinality $\aleph_2$. Our results here complement those in [12].

Theorem 4.15 *Suppose that the p-group G has countable length and cardinality $\aleph_2$. If G has a filtration of simply presented groups, then G is exemplary.*

Proof. Since G has countable length, we may assume that the groups in the filtration of G are all isotype because this is the case for some subfiltration. Since the subgroups in the filtration are now isotype and simply presented of countable length, they are subgroups of order ≤ 1 in G. Therefore, G has order ≤ 2, and consequently G is exemplary. □

Corollary 4.16 *Suppose that G has cardinality $\aleph_2$ and has no elements of infinite height. If every subgroup of G having cardinality $\aleph_1$ is a coproduct of cyclic groups, then G is exemplary.*

We conclude with the following open question, which is intended in part for amusement. But more than that, as things stand now, it is a genuine test case.

Question 4.17 *Let B denote the coproduct of the cyclic groups of order p^n, one for each n, and let G be the torsion completion of B. In other words, G is the standard $\overline{B}$. If we assume the continuum hypothesis, we know that G is exemplary since it has cardinalty $\aleph_1$. Must G still be exemplary if, for example, we assume that $c = \aleph_2$?*

References

[1] D. Beers, F. Richman, and E. Walker, *Group algebras of abelian groups*, Rend. Sem. Mat. Univ. Padova **69** (1983), 41–50.

[2] P. Hill, *Isotype subgroups of direct sums of countable groups*, Illinois Jour. Math. **13** (1969), 281–290.

[3] P. Hill, *A new characterization of totally projective groups of cardinality $\aleph_1$*, Proc. Amer. Math. Soc. **123** (1995), 657–662.

[4] P. Hill, *Units of commutative modular group algebras*, J. Pure and Applied Algebra **94** (1994), 175–181.

[5] P. Hill and C. Megibben, *On the theory and classification of abelian groups*, Math. Zeit. **190** (1985), 17–38.

[6] P. Hill and W. Ullery, *A note on a theorem of May concerning commutative group algebras*, Proc. Amer. Math. Soc. **110** (1990), 59–63.

[7] P. Hill and W. Ullery, *Units of modular group algebras of isotype subgroups of d.s.c.'s*, in Abelian Groups and Modules (D. Arnold and K. M. Rangaswamy, eds) Marcel Dekker Publ. (1995), 285–290.

[8] P. Hill and W. Ullery, *On commutative group algebras of mixed groups*, Comm. Algebra **25** (1997), 4029–4038.

[9] P. Hill and W. Ullery, *Almost totally projective groups*, Czechoslovak Math. J. **46** (1996), 249–258.

[10] W. May, *Modular group algebras of simply presented abelian groups*, Proc. Amer. Math. Soc. **104** (1988), 403–409.

[11] W. May, *Isomorphism of group algebras*, Jour. Algebra **40** (1976), 10–18.

[12] W. Ullery, *Units of modular group algebras of abelian groups of cardinality* $\aleph_2$, Comm. Algebra **25** (1997), 3751–3760.

[13] W. Ullery, *Isomorphism of group algebras*, Comm. Algebra **14** (1986), 767–785.

Abelian automorphism groups of countable rank

Warren May

Abstract

The abelian automorphism groups of countable rank are largely determined. Those arising from automorphisms of torsion-free groups are completely determined.

1 Introduction

Let G denote an abelian group. We want to classify certain of the automorphism groups, Aut G. In 1964, Hallett and Hirsch [3] showed that the possible finite automorphism groups for torsion-free G were quite restricted. That this should be true is apparent once one shows that an element of Aut G having order not dividing 12 leads to torsion-free elements in Aut G. We shall investigate this interplay between torsion and torsion-free elements while restricting ourselves to the case that Aut G is abelian. Furthermore, we shall only consider those Aut G having countable torsion-free rank. These restrictions will enable us to make liberal use of properties of finitely generated commutative rings.

To give the flavor of the results, let us describe Aut G in the important case where G is torsion-free. If the torsion-free rank of Aut G is finite, then the description involves certain nonnegative integers $r(1, B)$, which are defined for every finite abelian group B of even order. We refer the reader to the definition of these numbers following Lemma 3.6.

Theorem 1.1 *Consider the automorphism groups of torsion-free abelian groups.*

(1) U is isomorphic to an abelian automorphism group of finite rank if and only if $U = \prod_{i<n} U_i$ for a finite ordinal n, such that for each i:

(a) $U_i = B_i \times A_i \times F_i$, where B_i is finite of even order;

(b) if $h_i \geq 1$ is minimal such that B_i has a cyclic direct factor of order 2^{h_i}, then $A_i = 1$ if $h_i \geq 3$, $A_i^4 = 1$ if $h_i = 2$, and $A_i^{12} = 1$ if $h_i = 1$;

(c) *F_i is torsion-free of finite rank, and contains a pure free subgroup of rank $r(1, B_i)$.*

(2) *U is isomorphic to an abelian automorphism group of rank $\aleph_0$ if and only if $U = U' \times A$, where U has an element of order $2, |U'| = \aleph_0$, and if $1 \leq h \leq \infty$ is minimal such that U has a direct factor isomorphic to $\mathbb{Z}_{2^h}$, then $A = 1$ if $h \geq 3$, $A^4 = 1$ if $h = 2$, and $A^{12} = 1$ if $h = 1$.*

The theorem follows from Theorems 3.8, 3.10 and 4.11. In part (2), the only if statement is, in fact, proved for arbitrary infinite rank, leading to the question of whether the converse also holds in greater generality. This would be a matter of modifying Theorem 4.1. There is some hope for this since the construction in that theorem can be generalized in an obvious manner. The problem is whether it actually works. If so, then one would have the abelian automorphism groups of torsion-free groups completely characterized.

All groups in this paper will be abelian, all rings R will have identity, and $U(R)$ will denote the group of units in R. If Z is a commutative ring and B is a subset of a Z-algebra, then $Z[B]$ will denote the subalgebra generated by B. On the other hand, if B is a group, then ZB will denote the group algebra of B over Z. We shall use multiplicative notation extensively, C_m will denote a cyclic group of order m written multiplicatively, and ζ_m will denote a primitive m-th root of unity. By rank, we shall always mean torsion-free rank and denote it by rk. Finally, p and p_i always denote prime numbers.

2 Torsion–free elements in automorphism groups

Let Z be an integral domain whose quotient field Q is a finite extension of $\mathbb{Q}$, and let E be a (commutative) torsion-free Z-algebra. We may regard E as embedded in $\mathbb{Q} \otimes_{\mathbb{Z}} E$, which is a Q-algebra. We will let R stand for a Z-subalgebra of E of a certain type, which we now describe. Let B be a finite subgroup of $U(E)$, and let $x^t = v$, where $x \in U(E), v \in U(Z[B])$, and $t \geq 1$. Put $R = Z[B, x]$. Note that $R = Z[B]$ is a special case by taking $x = 1$. Let $\mathbb{Q}R$ denote the $\mathbb{Q}$-subalgebra of $\mathbb{Q} \otimes_{\mathbb{Z}} E$ generated by R. Then $\mathbb{Q}R$ is in fact a Q-subalgebra, and $\mathbb{Q}R = Q[B, x]$.

Lemma 2.1 *Let R and $\mathbb{Q}R$ be as above.*

(1) *Then $\mathbb{Q}R = \prod_{i<\alpha} L_i$, where each L_i is a field of form $L_i = Q[\zeta_{m_i}, x_i]$, x_i the image of x in L_i.*

(2) *If $R_i = Z[\zeta_{m_i}, x_i]$ is the image of R in L_i, then $R \subseteq \prod_{i<\alpha} R_i$.*

(3) If $R \subseteq R'$, where R' is another Z-subalgebra of E of the same type, then $\prod_{i<\alpha} R_i \subseteq \prod_{j<\beta} R'_j$, and $\prod_{i<\alpha} L_i \subseteq \prod_{j<\beta} L'_j$, where each R_i (respectively, L_i) is embedded diagonally in a subproduct of $\prod_{j<\beta} R'_j$ (respectively, $\prod_{j<\beta} L'_j$), and R_i's for distinct subscripts embed in disjoint subproducts.

Proof. To show (1), $Q[B]$ is a quotient ring of QB, which is a direct product of fields, thus so is $Q[B]$. Let $Q[B] = \prod_{i<\alpha} K_i$ and let v_i be the image of v in K_i. Further, $Q[B,x]$ is a quotient ring of $Q[B][X]/\langle X^t - v\rangle$, which is isomorphic to $\prod_{i<\alpha} K_i[X]/\langle X^t - v_i\rangle$. Since $v \in U(Z[B])$, each $v_i \neq 0$, hence $X^t - v_i$ has distinct roots over K_i and the quotient will be a product of fields. The same is true for $Q[B,x]$, thus (1) is shown since the image of B will be generated by a root of unity. Claim (2) is immediate, and for (3), we have a natural inclusion $QR \subseteq QR'$, the statement on embeddings following from the way idempotents must decompose since L'_j and R'_j are indecomposable rings. □

We continue with several lemmas on ranks of subgroups of units.

Lemma 2.2 *Let $R \subseteq \prod_{i<\alpha} R_i \subseteq \prod_{i<\alpha} L_i$ be as above, and let I_i denote the ring of algebraic integers in L_i. There exists $m \geq 1$ such that if u belongs to either $U(\prod_{i<\alpha} I_i)$ or $U(\prod_{i<\alpha} R_i)$, then $u^m \in U(R)$. Consequently, $\sum_{i<\alpha} rk\, U(I_i) \leq rk\, U(R) = \sum_{i<\alpha} rk\, U(R_i)$.*

Proof. Let S denote either $\prod_{i<\alpha} I_i$ or $\prod_{i<\alpha} R_i$. Since Q/Z is torsion and I_i is a finitely generated $\mathbb{Z}$-module, I_i is bounded modulo R_i, hence S is bounded modulo $\prod_{i<\alpha} R_i$ by some exponent $k_i > 0$. Clearly, $\prod_{i<\alpha} L_i$ is a quotient ring of $(QB)[X]$, hence we may choose pre-images of the idempotents in $\prod_{i<\alpha} L_i$. There exists $k_2 > 0$ which multiplies all of these pre-images into $(ZB)[X]$. For either choice of S, S/k_1k_2S is a finite ring, thus we may choose m to be the product of the orders of the unit groups of these two quotient rings. If $u \in U(S)$, then $u^m = 1 + k_1k_2s$ for some $s \in S$. The image of $(ZB)[X]$ in $\prod_{i<\alpha} R_i$ is R and contains k_2 multiples of the idempotents, thus the image contains $\prod_{i<\alpha} k_2R_i$. It follows that $k_1k_2s \in R$, hence $u^m \in R$. The second claim of the statement is now immediate since $U(R) \subseteq U(\prod_{i<\alpha} R_i)$. □

It will be necessary to consider subalgebras generated by finite groups of units.

Lemma 2.3 *Let $s \geq 1, h \geq 1$, and put $Z = \mathbb{Z}[s^{-1}, \zeta_{2^h}]$. Let E be a torsion-free Z-algebra and B a finite subgroup of $U(E)$ with $B \cong \prod_{1\leq i\leq n} C_{p_i^{e_i}}$.*

(1) There exist m_j $(1 \leq j \leq k)$, dividing the order of B, such that B embeds in $\prod_{1\leq j\leq k}\langle\zeta_{m_j}\rangle$, and $rkU(Z[B]) = \sum_{1\leq j\leq k} rk\, U(Z[\zeta_{m_j}])$.

(2) If $s = 1$ and $h \leq 2$, then $rk\,U(Z[B]) \geq \sum_{1\leq i\leq n} rk\,U(Z[\zeta_{p_i^{e_i}}])$.

Proof. Continuing the earlier notation with $x = 1$ and $R = Z[B]$, we have $Q[B] = \prod_{1\leq j\leq k} \mathbb{Q}[\zeta_{m_j}]$, where we may assume m_j is even, thus ζ_{m_j} will generate the torsion of $U(\mathbb{Q}[\zeta_{m_j}])$. Then B is embedded in $\prod_{1\leq j\leq k}\langle\zeta_{m_j}\rangle$, and Lemma 2.2 finishes the proof of (1) since $R_j = Z[\zeta_{m_j}]$. Now assume $s = 1$ and $h \leq 2$. Lemma 2.2 implies that $rk\,U(Z[B]) = \sum_{1\leq j\leq k} rk\,U(Z[\zeta_{m_j}])$. Since B embeds in $\prod_{1\leq j\leq k}\langle\zeta_{m_j}\rangle$, there exists $\sigma : \{1,\dots,n\} \to \{1,\dots,k\}$ such that $p_i^{e_i}|m_{\sigma(i)}$ for every i, and if $p_{i_1} = p_{i_2}$ for $i_1 \neq i_2$, then $\sigma(i_1) \neq \sigma(i_2)$. Therefore we may assume that $p_1,\dots,p_n$ are distinct, that $\prod_{1\leq i\leq n} p_i^{e_i}$ divides m, and it suffices to show that $rk\,U(Z[\zeta_m]) \geq \sum_{1\leq i\leq n} rk\,U(Z[\zeta_{p_i^{e_i}}])$.

Note that $2^h|m$ since $Q[B]$ is a Z-algebra. It suffices to show that if a and b are relatively prime natural numbers, then $rk\,U(Z[\zeta_{ab}]) \geq rk\,U(Z[\zeta_a]) + rk\,U(Z[\zeta_b])$. This is true if $Z[\zeta_a] = \mathbb{Z}$, therefore we may assume the cyclotomic fields involved have only complex embeddings. By the Dirichlet unit theorem, for such a $\mathbb{Q}(\zeta_c)$, we have $rk\,U(\mathbb{Z}[\zeta_c]) = \frac{1}{2}\varphi(c) - 1$. We may assume 2 divides one of a or b,therefore suppose $a = 2^\ell a_1, \ell > 0,\quad 2 \nmid a_1$. Then $2 \nmid b$. Put $M = \max\{h,\ell\}$. We must show that $\frac{1}{2}\varphi(2^M a_1 b) - 1 \geq \frac{1}{2}\varphi(2^M a_1) - 1 + \frac{1}{2}\varphi(2^h b) - 1$. By properties of the Euler function, this is equivalent to $2^{M-h}\varphi(a_1)(\varphi(b)-1) \geq \varphi(b) - 2^{2-h}$, which is true since $h \leq 2$.□

Corollary 2.4 *Let E be a torsion-free ring such that $U(E)$ has finite rank. If T is the torsion subgroup of $U(E)$, then T^{12} is finite, thus $U(E) = T \times F$ for some torsion-free group F of finite rank.*

Proof. Take $s = 1, h = 1$ in (2), and suppose $p_i^{e_i} > 4$ for every i. Then each $rk\,U(\mathbb{Z}[\zeta_{p_i^{e_i}}]) > 0$, and tends to infinity as $p_i^{e_i}$ does. The finite rank of $U(E)$ implies T^{12} must be finite. □

To show that certain abelian groups occur as automorphism groups, we need to construct indecomposable algebras. It is often useful if these algebras are augmented. We recall that a Z-algebra E is said to have an augmentation aug: $E \to Z$ if aug is a Z-algebra homomorphism.

Lemma 2.5 *Let $p_i^{e_i}$ be prime powers for $0 \leq i \leq k$, where $p_0 = 2$, and let $B \cong \prod_{0\leq i\leq k} C_{p_i^{e_i}}$. Let $Z = \mathbb{Z}[\zeta_{2^{e_0}}]$. Then there exists a torsion-free indecomposable augmented Z-algebra $Z[B]$ such that B is the torsion subgroup of $U(Z[B])$ and $rkU(Z[B]) = \Sigma_{0\leq i\leq k} rk\,(U(Z[\zeta_{p_i^{e_i}}]))$.*

Proof. Put $\theta_i = \zeta_{p_i^{e_i}}$. We will realize B inside the Z-algebra $\prod_{0\leq i\leq k} Z[\theta_i]$. Let B be the group of units in this algebra generated by $(\theta_0, \dots, \theta_0)$ and $(1, 1, \dots, \theta_i, \dots, 1)$ $(1 \leq i \leq k)$, where θ_i occurs as component i and 1 as other components. Clearly, $B \cong \prod_{0\leq i\leq k} C_{p_i^{e_i}}$, $Z[B]$ is torsion-free, and $Z[B]$ is augmented via the projection of the product onto component 0 since $Z[\theta_0] = Z$. The statement on rank follows from Lemma 2.2 since $Q[B] = \prod_{0\leq i\leq k} Q[\theta_i]$. Since $\theta_0 \in Z$, it is easy to see that $Z[B] = \Sigma_{r_1,\dots,r_k} Z \cdot (1, \theta_1^{r_1}, \dots, \theta_k^{r_k})$, thus an element in $Z[B]$ has form $(z, f_1(\theta_1), \dots, f_k(\theta_k))$, where $z \in Z$ and $f_i \in Z[X]$ such that $f_i(1) = z$ for $1 \leq i \leq k$.

We first show that every unit of finite order in $Z[B]$ lies in B. The important property of the representation above of elements of $Z[B]$ is that $\theta_i - 1$ divides $f_i(\theta_i) - f_i(1)$ in $Z[\theta_i]$, hence divides $f_i(\theta_i) - z$. If the element is a unit of finite order, certainly $z \in \langle\theta_0\rangle$, thus we may assume $z = 1$ by using a power of $(\theta_0, \dots, \theta_0) \in B$. Since $f_i(\theta_i)$ is a unit of finite order in $Z[\theta_i]$, we have $f_i(\theta_i) = \theta_0^s\theta_i^t$ for some s and t. Suppose that p_i is odd. For some ℓ we have $f_i(\theta_i)^\ell = \theta_0^s$. But $\theta_i - 1$ divides $f_i(\theta_i) - 1$ in $Z[\theta_i]$, thus divides $f_i(\theta_i)^\ell - 1 = \theta_0^s - 1$. If $\theta_0^s \neq 1$, then $\theta_0^s - 1$ factors in $Z[\theta_i]$ only into prime ideals dividing the rational prime 2, while $\theta_i - 1$ factors into prime ideals dividing p_i. This contradiction means that $\theta_0^s = 1$, thus $f_i(\theta_i) \in \langle\theta_i\rangle$. Now suppose that $p_i = 2$ and that $\theta_0^s \notin \langle\theta_i\rangle$. As above, for some ℓ we have $f_i(\theta_i)^\ell = \theta_0^s$ and $\theta_i - 1$ divides $\theta_0^s - 1$ in $Z[\theta_i] = Z \supsetneq \mathbb{Z}[\theta_i]$. Since $\theta_i - 1$ and $\theta_0^s - 1$ generate the totally ramified extensions of the prime 2 to $\mathbb{Z}[\theta_i]$ and $\mathbb{Z}[\theta_0^s]$, such a division is impossible since the extension is proper. Therefore, we have $f_i(\theta_i) \in \langle\theta_i\rangle$ and the unit of finite order must lie in B.

Finally, we must show that $Z[B]$ is indecomposable. If $Z[B]$ contains a nontrivial idempotent, we may assume that $z = 1$. Since $\theta_i - 1$ divides $f_i(\theta_i) - 1$, and $\theta_i - 1$ is not a unit, it is impossible that $f_i(\theta_i) = 0$. This means that a nontrivial idempotent cannot exist. □

If B is a finite abelian group with a factor isomorphic to C_{2^h}, we shall denote the algebra $Z[B]$ constructed above by $R(h, B)$, where $h = e_0$.

3 Abelian automorphism groups of finite rank

We are now prepared to start considering abelian automorphism groups of finite rank. The following lemma is easily proved.

Lemma 3.1 *Let A and B be abelian groups. Assume one of the following.*

(1) *There exist automorphisms $\alpha : A \to A, \beta : B \to B$ and a homomorphism $\gamma : A \to B$ such that $\gamma\alpha \neq \beta\gamma$.*

(2) There exist homomorphisms $\gamma_1 : A \to B$ and $\gamma_2 : B \to A$ such that $\gamma_2\gamma_1 \neq 0$.

Then Aut $(A \bigoplus B)$ is nonabelian.

Lemma 3.2 *Let Aut G be an abelian automorphism group of finite rank. There exists $t \geq 1$ such that $G \cong \mathbb{Z}_t \times G'$, where G' is torsion-free and Aut $G' = T \times F$, where T^{12} is finite and F is torsion-free of finite rank. If $t \not\equiv 2(\mathrm{mod}\ 4)$, put $s = t$. Then Aut $G \cong$ Aut $\mathbb{Z}_s \times$ Aut G'. If $t \equiv 2(\mathrm{mod}\ 4)$, put $s = t/2$. Then there is an elementary 2-group M such that Aut G $\cong M\times$ Aut $\mathbb{Z}_s\times$ Aut G'. In either case, End G' is a $\mathbb{Z}[s^{-1}]$-algebra.*

Proof. If G is torsion-free, let E be the subring of End G generated by Aut G and apply the Corollary to Lemma 2.3 to obtain Aut $G = T \times F$ with T^{12} finite. We may now assume that G has nontrivial torsion. Note that finite rank implies that the torsion of G is reduced. Lemma 3.1 implies that each primary component of torsion is cyclic .

Suppose that $p_1, \dots, p_n$ are distinct primes, and H is a summand of G such that multiplications by $p_1, \dots, p_n$ are automorphisms of H which are not free in Aut H. We claim H must be finite. For some $e_i \in \mathbb{Z}$, not all 0, we have $p_1^{e_1} \cdots p_n^{e_n} = 1_H$ in Aut H. Therefore, there exists $m > 0$ such that $m \cdot 1_H = 0$, hence H is bounded. Since each primary component of H must be cyclic, H is finite, as claimed.

First assume that G has no summand of order 2. Since G has torsion, $G = \mathbb{Z}_{p^e} \bigoplus H$ for some prime power $p^e \geq 3$, and H can have no p-torsion. If $H \neq pH$, then Lemma 3.1 is contradicted since $\mathbb{Z}_{p^e}$ has nontrivial automorphisms. Therefore, $H = pH$, Hom $(H, \mathbb{Z}_{p^e}) = 0$, Aut G=Aut $\mathbb{Z}_{p^e}\times$ Aut H and $p \in$ Aut H. If p has finite order in Aut H, then H is finite, hence $\mathbb{Z}_{p^e} \times H \cong \mathbb{Z}_s$ and Aut G= Aut $\mathbb{Z}_s$. Thus, we may assume that $G = (\bigoplus_{1 \leq i \leq n} \mathbb{Z}_{p_i^{e_i}}) \bigoplus H$, where $p_1, \dots, p_n$ are distinct primes giving free automorphisms of H, and n is maximal by the finite rank condition. If H is not torsion-free, then $H = \mathbb{Z}_{p^e} \bigoplus H'$, where $p \neq p_i$ for every i , and $p \in$ Aut H'. By the above, H' is finite, a contradiction since p_1 is free on H. Therefore H is torsion-free, Aut G=Aut $\mathbb{Z}_s\times$ Aut H, and $s = \prod_i p_i^{e_i}$. Take $G' = H$, and the torsion-free case above gives us the $T \times F$ decomposition.

Now assume that $G = \mathbb{Z}_2 \bigoplus H$, and let $M = \{\gamma + 1 : \gamma \in$ End $G, \gamma(\mathbb{Z}_2) = 0, \gamma(H) \subseteq \mathbb{Z}_2\}$. Then M is a subgroup of Aut G such that $A^2 = 1$, and Aut $G = M\times$ Aut H since Aut G is abelian. But H has no 2-torsion, therefore we are in the previous case and s will be the order of the torsion of H. □

Next, we need to break part of the endomorphism algebra down into indecomposable pieces. The reason for this will be apparent when constructing algebras with prescribed unit groups; we must ensure that a group algebra does not have too many units.

Lemma 3.3 *Let $s \geq 1, Z = \mathbb{Z}[s^{-1}]$, and let E be a torsion-free Z-algebra with $U(E)$ of finite rank. Then $E = \prod_{1 \leq i \leq n} E_i$, where E_i is a torsion-free Z-algebra, $U(E_i) = B_i \times A_i \times F_i$, F_i is torsion-free of finite rank, B_i is finite, and either $A_i = 1$, or else A_i is infinite with $A_i^{12} = 1$ and the 3-component of A_i is either 1 or infinite. Moreover, $Z[B_i]$ is indecomposable when $A_i = 1$, and if $s > 1$, then the p-torsion of B_i if cyclic for every prime p dividing s.*

Proof. By the Corollary to Lemma 2.3, $U(E) = B \times A \times F$, where B is finite, and either $A = 1$ or A is infinite and $A^{12} = 1$. Since $Z[B]$ is noetherian, we may write $1 = \Sigma_j \varepsilon_j$ as a finite sum of orthogonal primitive idempotents, thus $Z[B]\varepsilon_j$ is an indecomposable ring. Now consider $Z[B \times A]\varepsilon_j$. Since $E = \prod_j E\varepsilon_j$, the torsion units of $E\varepsilon_j$ are precisely $(B \times A)\varepsilon_j$. First suppose that this group is finite. Since $Z[B \times A]\varepsilon_j$ is noetherian, we may further decompose $\varepsilon_j = \sum_k \varepsilon_{jk}$ into orthogonal primitive idempotents of $Z[B \times A]\varepsilon_j$. Replace ε_j by the family $\{\varepsilon_{jk}\}$. In the case that the group $(B \times A)\varepsilon_j$ is infinite, leave ε_j alone. Now relabel the orthogonal idempotents that one has as $\{\varepsilon_i\}$, and put $E_i = E\varepsilon_i$, thus $E = \prod_i E_i$. If the torsion of $U(E_i)$ is finite, call it B_i and put $A_i = 1$. Then $U(E_i) = B_i \times F_i$ for a suitable torsion-free F_i, and $Z[B_i]$ is indecomposable since it equals $Z[B \times A]\varepsilon_{jk}$ for some ε_{jk}. If the torsion of $U(E_i)$ is finite, it equals $(B\varepsilon_j)(A\varepsilon_j)$, where $B\varepsilon_j$ is finite and $(A\varepsilon_j)^{12} = 1$. We may write it as $B_i \times A_i$, where B_i is finite, $A_i^{12} = 1$, and B_i contains all 3-torsion if that is finite.

Now let $s > 1$ and $p|s$. If B_i' is the p-torsion of B_i, then all idempotents of QB_i' lie in ZB_i' (see Lemma 3.2.2 in [5]), therefore $ZB_i' = \prod_j Z[\zeta_{m_j}]$. But $Z[B_i']$ is a torsion-free quotient of this ring and is indecomposable. Hence it is isomorphic to some $Z[\zeta_{m_j}]$, consequently B_i' is cyclic. □

The following two lemmas refine our information on the torsion units.

Lemma 3.4 *Let $s \geq 1, h \geq 1$, and put $Z = \mathbb{Z}[s^{-1}, \zeta_{2^h}]$. Let E be a torsion-free Z-algebra such that $U(E) = T \times F$ where T is bounded and F is torsion-free of finite rank.*

(1) If either $s > 1$, or else $s = 1$ and $h \geq 3$, then T is finite.

(2) If $s = 1$ and $h = 2$, then the 3-component of T is finite.

Proof. We shall apply Lemma 2.3 with various finite $B \subseteq T$ to contradict the finite rank of $U(E)$. Suppose that T is infinite. Since T is bounded, some p-socle must be infinite. By taking B to be large parts of this socle, we must have arbitrarily large numbers of summands $rk\, U(Z[\zeta_{m_j}])$ in (1) of Lemma 2.3. If $s > 1$, this would imply that $U(E)$ has infinite rank. If $h \geq 3$, the same is true since $8|m_j$ for every j. If T has infinite 3-component, take $p=3$ in this argument. If $h=2$, then $12|m_j$ for every j, once again implying $U(E)$ has infinite rank. □

Now we examine the torsion-free part of $U(E)$ and find that it has a large pure free subgroup.

Lemma 3.5 *Let $Z=\mathbb{Z}[s^{-1},\zeta_{2^h}]$ and let E be a torsion-free Z-algebra with $U(E)=T\times F$, where T is bounded and F is torsion-free of finite rank. Let B be a finite subgroup of $U(E)$. Then F has a pure free subgroup of rank equal to $rk\, U(Z[B])$.*

Proof. Put $P=U(Z[B])\subseteq U(E)$. Then PT/T has the same rank as P, as does its purification in $U(E)/T\cong F$. We shall assume that the purification is not free, and we will contradict F being of finite rank. By Corollary 7.2.8 in [6], the unit group of $Z[B]$ is finitely generated, hence PT/T is free and $U(E)/PT$ cannot have bounded torsion. Therefore $U(E)/PT$ has either elements of order p for infinitely many p, or else elements of order p^e for one p and every e. Apply Lemma 2.2 to $Z[B]$ and let m be as in the lemma. In the setting of Lemma 2.1, suppose we have $Q[B]=\prod_{i<\alpha}K_i, Q[B,x]=\prod_{j<\beta}L_j$, and let R_i denote the image of $Z[B]$ in K_i.

First suppose that the quotient has elements of order p for infinitely many p. We may assume that $p\nmid m$, that T has no p-torsion, and that $U(\prod_j L_j)$ has no p-torsion since the torsion is finite. Suppose that $x\in U(E), x\notin PT$, and $x^p\in PT$. We may assume that $x^p=v\in P$. Relative to the products of the K_i and L_j, we may write $v=(v_i)$ and $x=(x_j)$. Suppose that $v_i\in K_i^p$ for every i. Since $v_i\in R_i$ and R_i is integrally closed, we have $v_i\in U(R_i)^p$, thus $v\in U(\prod_i R_i)^p$. But p-th roots are unique by the way we have chosen p, hence $x\in U(\prod_i R_i)$. By the choice of m, we have $x^m\in U(Z[B])=P$, thus $x\in P$. This contradiction means that $v_i\notin K_i^p$ for some i. Some L_j will extend K_i and $L_j=K_i(x_j)$, where $x_j^p=v_i$. Thus $[L_j:K_i]=p$, implying that if I_j is the ring of algebraic integers in L_j, then $rk\, U(I_j)\geq\frac{1}{2}p-1$. Lemma 2.2 now implies that the rank of F is at least as big, contradicting the finite rank of F.

Now suppose that $U(E)/PT$ has elements of order p^e for all e. Let $m=p^\ell m', p\nmid m'$, and let p^k be an exponent for the p-torsion in T. Choose $e\geq 2+\ell+k$ and suppose $x^{p^e}=vt\,(v\in P, t\in T)$, and $x^{p^{e-1}}\notin PT$. We may assume $t^{p^k}=1$, thus $x^{p^{e+k}}=v^{p^k}$. Suppose that for every j, if L_j extends K_i, then $x_j^{p^{e-1-\ell}}\in K_i$. Then $x_j^{p^{e+k}}=v_i^{p^k}\in R_i$, hence R_i integrally closed implies $x_j^{p^{e-1-\ell}}\in U(R_i)$. Thus $x^{p^{e-1}}\in U(Z[B])=P$ since $m=p^\ell m'$. This contradiction allows us to conclude $x_j^{p^{e-1-\ell}}\notin K_i$ for some j. The p-torsion of $U(K_i(\zeta_p))$ is bounded by p^k since K_i is a quotient of

$Q[B]$. Referring to Lemma 1 in [9], the p-torsion of $U(K_i(\zeta_p))/U(K_i)$ is bounded by p^k (for $p = 2$, use ζ_4 and 2^{k+1}). Therefore $x^{p^{e-2-\ell-k}} \notin K_i(\zeta_p)$, and hence $[L_j(\zeta_p) : K_i(\zeta_p)] \geq p^{e-2-\ell-k}$. The degree of such L_j over $\mathbb{Q}$ may be arbitrarily large, contradicting the finite rank of F by Lemma 2.2. □

The preceding two lemmas dealt with algebras over $\mathbb{Z}[\zeta_{2^h}]$. We remark on how these appear. Let E be a ring such that $U(E)$ has bounded torsion and a cyclic summand of order $2^h, h \geq 1$, with h minimal. Then the 2-height of -1 in $U(E)$ is at least $h-1$, so we may select a corresponding root and consider the homomorphism of $\mathbb{Z}[X]$ into E sending X to this root. The kernel contains $X^{2^{h-1}} + 1$, but since this is the cyclotomic polynomial for ζ_{2^h}, we have E as a $\mathbb{Z}[\zeta_{2^h}]$-algebra.

Before proceeding, we need to define a sharp lower bound on the rank of $U(Z[B])$ in the case that $Z[B]$ is indecomposable. The next lemma guarantees the existence of such indecomposable algebras.

Lemma 3.6 *Let $s \geq 1$, let B be finite of even order, and put $Z = \mathbb{Z}[s^{-1}]$. If $s > 1$, for every prime p dividing s assume that the p-component of B is cyclic. Then there exists a torsion-free indecomposable Z-algebra $Z[B]$ such that B is the torsion subgroup of $U(Z[B])$.*

Proof. We may write $B = C_{2^h} \times B'$, where $h \geq 1$. An abelian group algebra over an integral domain is indecomposable if no prime is invertible for which there are group elements of that order. Moreover, units of finite order are so-called trivial units (see Corollary 3.3.13 and Theorem 5.1.5 in [5]). If $s = 1$, then $(\mathbb{Z}[\zeta_h])B'$ is torsion-free indecomposable, and the torsion of the unit group is $\langle \zeta_{2^h} \rangle \times B'$. If $s > 1$, write $B = C_{2^h} \times B_1 \times B_2$, where the p-components of B for $p|s$ comprise either B_1 or $C_{2^h} \times B_1$. In either case, $C_{2^h} \times B_1$ is cyclic with $|B_1| = m$ odd. We see that $(\mathbb{Z}[s^{-1}, \zeta_{2^h m}])B_2$ has the appropriate properties. □

We can now define a useful number in determining abelian automorphism groups of finite rank. Let $s \geq 1$ and let B be a finite group of even order such that the p-component of B is cyclic for every prime p dividing s. We define $r(s, B)$ to be the minimum rank of $U(Z[B])$, where $Z = \mathbb{Z}[s^{-1}]$ and $Z[B]$ is a torsion-free indecomposable Z-algebra such that B is the torsion subgroup of $U(Z[B])$. This number is usually smaller than the rank of the unit group of the algebra constructed in the lemma, however, we had to know that such algebras exist. In certain instances, the augmented algebra constructed in Lemma 2.5 will achieve the minimum rank if $s = 1$. For later use, we give this algebra a name. Let B be a finite group with a cyclic factor of order 2^h for $h \geq 1$. The augmented $\mathbb{Z}[\zeta_{2^h}]$-algebra constructed in Lemma 2.5 (with $e_0 = h$) will be denoted by $R(h, B)$.

Lemma 3.7 *Let B be finite with a cyclic factor of order 2^h, where $h = 1$ or 2 and is minimal. Let $Z = \mathbb{Z}[\zeta_{2^h}]$ and let $Z[B]$ be a torsion-free Z-algebra. Then $rk\, U(Z[B]) \geq rk\, U(R(h, B)) = r(1, B)$.*

Proof. Lemmas 2.3(2) and 2.5 show that $rkU(Z[B]) \geq rk\, U(R(h, B))$. By definition, there is a torsion-free indecomposable $\mathbb{Z}$-algebra R generated by B such that $r(1, B) = rk\, U(R) \leq rk\, U(R(h, B))$. The remark after Lemma 3.5 shows that R is a Z-algebra, thus what we have already shown implies $rk\, U(R(h, B)) \leq rk\, U(R)$, giving the equality. □

To state the first theorem, we shall identify a class of groups $\mathcal{A}_s$ for every $s \geq 1$. We say that $B \times A \times F$ *belongs to* $\mathcal{A}_s$ if B is finite of even order, F is torsion-free of finite rank with a pure free subgroup of rank $r(s, B)$, and:

(1) if $s > 1$, then $A = 1$ and the p-component of B is cyclic for every prime p dividing s;

(2) if $s = 1$, and if $h \geq 1$ is minimal such that B has a cyclic factor of order 2^h, then $A = 1$ if $h \geq 3$, $A^4 = 1$ if $h = 2$, and $A^{12} = 1$ if $h = 1$.

Theorem 3.8 *Let G be an abelian group such that $Aut\, G$ is abelian of finite torsion-free rank. The torsion subgroup of G is cyclic of order $t \geq 1$. If $t \not\equiv 2 (mod\ 4)$, put $s = t$. Then $Aut\, G \cong Aut\ \mathbb{Z}_s \times \prod_{1 \leq i \leq n} U_i$, where U_i belongs to $\mathcal{A}_s$ for every i. If $t \equiv 2\ (mod\ 4)$, put $s = t/2$.*
Then $Aut\, G \cong M \times Aut \mathbb{Z}_s \times \prod_{1 \leq i \leq n} U_i$, where U_i belongs to $\mathcal{A}_s$ for every i and M is an elementary 2-group.

Proof. Lemma 3.2 takes care of the torsion subgroup of G and the case $t \equiv 2$ (mod 4) once we have the other case (see the last paragraph of the proof). Continuing with the case $t \not\equiv 2$ (mod 4), Lemma 3.3 gives $\prod_{1 \leq i \leq n} U_i$, where $U_i = U(E_i)$. Since the torsion of U_i is bounded, the remark after Lemma 3.5 implies E_i is a $\mathbb{Z}[s^{-1}, \zeta_{2^h}]$-algebra, where $h \geq 1$ is minimal such that U_i has a cyclic factor of order 2^h. We can now apply Lemmas 3.4 and 3.5. Setting aside the cases $s = 1, h \leq 2$, we have that U_i has finite torsion B_i, thus $Z[B_i]$ is indecomposable and Lemma 3.5 guarantees that F_i has a pure free subgroup of rank $r(s, B_i)$. Now let $s = 1, h \leq 2$. The torsion of U_i may be written as $B_i \times A_i$, where $A_i^{12} = 1$, B_i is finite and contains a cyclic factor of order 2^k and also the 3-torsion if that is finite. Although $Z[B_i]$ may not be indecomposable, Lemmas 3.5 and 3.7 give the rank condition on F_i concerning $r(1, B_i)$. □

Next, we wish to show that if $t \not\equiv 2$(mod 4), then every possible group in Theorem 1 is realized as an automorphism group. To construct needed algebras, we shall utilize the pullback construction for rings, which is similar to the construction for groups. A brief lemma summarizes the properties we require.

Lemma 3.9 *Let $\varphi : S \to R$ and $\psi : T \to R$ be homomorphisms of rings.*

(1) If S and T are torsion-free and reduced, then so is $S \times_R T$.

(2) $U(S \times_R T) = U(S) \times_{U(R)} U(T)$.

(3) Suppose that $U(S) = V_1 \times V_2$, where $\varphi | V_1 : V_1 \to U(R)$ is an isomorphism. Then $U(S) \times_{U(R)} U(T) \cong V_2 \times U(T)$.

Proof. Both (1) and (2) are easy, working in $S \times T$. For (3), we may assume V_2 is the kernel of φ restricted to $U(S)$. If $v_i \in V_i$ and $t \in U(T)$ such that $\varphi(v_1, v_2) = \psi(t)$, then the map sending $((v_1, v_2), t)$ to (v_2, t) is an isomorphism. □

Theorem 3.10 *Let $s \not\equiv 2$ (mod 4) and let U_i belong to $\mathcal{A}_s$ for $1 \leq i \leq n$. Then there exists an abelian group G such that $Aut\, G \cong Aut\, \mathbb{Z}_s \times \prod_{1 \leq i \leq n} U_i$.*

Proof. Suppose that we can find a cortorsion-free $\mathbb{Z}[s^{-1}]$-algebra E_i such that $U(E_i) \cong U_i$ for $1 \leq i \leq n$. Then $E = \prod_{1<i<n} E_i$ is cotorsion-free and there exists G' such that Aut $G' \cong U(E) \cong \prod_i U_i$. If we put $G = \mathbb{Z}_s \times G'$ and note that G' is s-divisible, then Aut $G \cong$ Aut $\mathbb{Z}_s \times$ Aut G' and we are done.

Therefore we assume that $U = B \times A \times F$ belongs to $\mathcal{A}_s$, and we must find a cotorsion-free $\mathbb{Z}[s^{-1}]$-algebra E such that $U(E) \cong U$. Put $Z = \mathbb{Z}[s^{-1}]$. We have that there is an indecomposable Z-algebra $Z[B]$ such that the torsion of $U(Z[B])$ is B and $rkU(Z[B])$ is $r(s, B)$. In fact, from the embedding in Lemma 2.1, we see that $Z[B]$ has a free additive group, trivial nilradical, and $U(Z[B])$ is free modulo torsion by the Dirichlet unit theorem. Thus $U(Z[B]) = B \times F_0'$, where F_0' is free of rank $r(s, B)$. In the case that $s = 1$ and $h \leq 2$, Lemma 3.7 tells us that $Z[B]$ may be taken to be an augmented $\mathbb{Z}[\zeta_{2^h}]$-algebra.

Now let Z denote either $\mathbb{Z}[s^{-1}]$, or $\mathbb{Z}[\zeta_{2^h}]$ in case $s = 1$ and $h \leq 2$. We know that F contains a pure free subgroup F_0 of the same rank as F_0'. Let $ZF_0 \to ZF$ be inclusion and $ZF_0 \to Z[B]$ a Z-homomorphism identifying F_0 with F_0'. Put $E' = Z[B] \bigotimes_{ZF_0} ZF$. Since ZF is a free ZF_0-module, we conclude that E' is a free $Z[B]$-module, thus has free additive group. Consider $F_0 \leq \widetilde{F} \leq F$ such that $\widetilde{F}/F_0$ is finitely generated, hence $\widetilde{F} = F_0 \times F_1$ for some free group F_1 (F_0 is pure). Therefore, $Z[B] \bigotimes_{ZF_0} Z\widetilde{F} \cong Z[B] \bigotimes_{ZF_0} (ZF_0)F_1 \cong (Z[B])F_1$. Since $Z[B]$ is indecomposable and has trivial nilradical, an appeal to Proposition 2.2.1 in [5] shows that $U((Z[B])F_1) = U(Z[B]) \times F_1$, which is the pushout of $U(Z[B])$ and $\widetilde{F}$ by F_0. But there is a natural mapping of the pushout of $U(Z[B])$ and F by F_0 into $U(E')$. By what we have just seen, this mapping is both injective and surjective, thus $U(E')$ is isomorphic to the pushout of $U(Z[B])$ and F by F_0. Since $U(Z[B]) = B \times F_0$, we conclude by Lemma 3.9 that $U(E') \cong B \times F$. We may take $E = E'$ if $s > 1$, or if $s = 1$ and $h \geq 3$.

In case $s = 1$, we must still account for A if $h \leq 2$. As remarked earlier, we may assume E' is an augmented Z-algebra. Let $h = 2$. Then $A^4 = 1$. By the references cited in the proof of Lemma 3.6 and by Lemma 2.3, $U(ZA) = U(Z) \times A$. Since ZA is also an augmented Z-algebra, we may take $E = E' \times_Z ZA$, thus $U(E) = U(E') \times A$, as desired. Now let $h = 1$. Then $A^{12} = 1$, therefore $A = A_1 \times A_2$ where $A_1^4 = A_2^3 = 1$. Since $U(ZA_i) = U(Z) \times A_i$, we may take $E = (E' \times_Z ZA_1) \times_Z ZA_2$ to obtain $U(E) = U(E') \times A_1 \times A_2$. □

4 Abelian automorphism groups of rank $\aleph_0$

We first characterize the countable abelian automorphism groups of rank $\aleph_0$.

Theorem 4.1 *Let V be a countable abelian group of infinite torsion-free rank and possessing an element of order two. Then there exists a torsion-free abelian group G such that $Aut\, G \cong V$.*

It will suffice to produce a ring R with free additive group R^+ such that $U(R) \cong V$. In addition, if V can be written as $V = V_0 \times V'$, where V_0 is cyclic of order $2^h, 1 \leq h \leq 2$, then we want R to be an augmented $\mathbb{Z}[\zeta_{2^h}]$-algebra. We refer to this as the augmentation option. We may fix subgroups V_n $(n < \omega)$, such that $V_0 \subseteq V_1 \subseteq \dots, \bigcup_n V_n = V, V_0$ is cyclic of order 2, and V_{n+1}/V_n is cyclic of either infinite or prime order. We shall inductively construct $(R_n, S_n, \varphi_n, T_n)$, where $R_0 \subseteq R_1 \subseteq \dots$ are commutative rings where we put $U_n = U(R_n), S_n$ is a subring of R_n, $\varphi_n : V_n \to U_n$ is an injection, and $T_n \subseteq U_n$ is a subset with the elements of $\bigcup_n T_n$ indexed by a subset of $\mathbb{N}$ such that the index map $\bigcup_n T_n \to \mathbb{N}$ is injective. In the augmentation option, we have $V_n = V_0 \times V_n'$ for some V_n' $(n \geq 0)$, and we shall take $R_0 = \mathbb{Z}[\zeta_{2^h}]$ and inductively construct an R_0-algebra augmentation $\psi_n : R_n \to R_0$. We shall require the following conditions:

(1) R_n is a finitely generated ring, integral over S_n,and with no nontrivial nilpotent or idempotent elements. The additive group R_n^+ is free abelian, and R_{n+1}^+ is a free module over R_n;

(2) S_n is an integrally closed integral domain in which no prime p is a unit, and no element of S_n is a zero divisor on R_n;

(3) φ_{n+1} extends φ_n, and, if we are in the augmentation option, ψ_{n+1} extends ψ_n;

(4) U_n is the inner direct product of $\varphi_n(V_n)$ and a free abelian group generated by T_n;

(5) if we are in the augmentation option, then ψ_n takes $\varphi_n(V_n')\langle T_n\rangle$ to 1.

We start the induction with $R_0 = S_0 = \mathbb{Z}, \varphi_0 : V_0 \to \langle -1\rangle$, and $T_0 = \emptyset$, or else we take $R_0 = S_0 = \mathbb{Z}[\zeta_{2^h}]$ and $\varphi_0 : V_0 \to \langle \zeta_{2^h}\rangle$ if we are in the augmentation option. This is clearly a valid initial step. For the induction step, suppose we have R_n, S_n, φ_n, T_n, and possibly ψ_n. Let $a \in V_{n+1}$ be a generator for V_{n+1} modulo V_n. In the augmentation option, we may assume $a \in V_{n+1}'$. Then either $\langle a\rangle \cap V_n = 1$, or else $a \notin V_n$ and $a^p = b \in V_n$ for some prime p. We distinguish three types of induction steps.

Type 1: Suppose $\langle a\rangle \cap V_n = 1$ and $T_n = \emptyset$. We put $R_{n+1} = R_n[X, X^{-1}]$ and $S_{n+1} = S_n[X, X^{-1}]$. Since $V_{n+1} = V_n \times \langle a\rangle$, we may let φ_{n+1} extend φ_n by $\varphi_{n+1}(a) = X$. Take $T_{n+1} = \emptyset$. In the augmentation option, we let ψ_{n+1} extend ψ_n by $\psi_{n+1}(X) = 1$.

Type 2: Suppose $\langle a\rangle \cap V_n = 1$ and $T_n \neq \emptyset$. In this case, put $R_{n+1} = R_n$ and $S_{n+1} = S_n$. Using the indexing of elements of T_n, let t be the first element and put $T_{n+1} = T_n \backslash \{t\}$. Then $U_{n+1} = U_n = \varphi_n(V_n) \times \langle t\rangle \times \langle T_{n+1}\rangle$. Let φ_{n+1} extend φ_n by $\varphi_{n+1}(a) = t$ and, in the augmentation option, let $\psi_{n+1} = \psi_n$.

Type 3: Suppose that $a^p = b \in V_n$. Put $c = \varphi_n(b)$. Then $c \in U_n$. Put $R_{n+1} = R_n[X]/\langle X^p - c\rangle$ and let x denote the image of X in R_{n+1}. Since $X^p - c$ is monic, we have $R_{n+1} = \bigoplus_{0\le j\le p-1} R_n x^j$. We regard R_n as embedded in R_{n+1} and set $S_{n+1} = S_n$. Note that $x \in U_{n+1}$ since $x^p = c$. We may extend φ_n to an injection φ_{n+1} by $\varphi_{n+1}(a) = x$. Then $\varphi_{n+1}(A_{n+1}) = A_n\langle x\rangle$ and $U_n\langle x\rangle = (A_n\langle x\rangle) \times \langle T_n\rangle$. Suppose we can show, in this case, that U_{n+1} is the direct product of $U_n\langle x\rangle$ and a free abelian group of finite rank on a set T_n'. We then take $T_{n+1} = T_n \cup T_n'$. To index the elements of T_n', if there are k of them, we index with the natural numbers $2^n \cdot 3, \dots, 2^n \cdot 3^k$. It is clear the index map will be injective. In the augmentation option, we have $b \in V_n'$, thus $\psi_n(c) = 1$ and we may extend ψ_n by $\psi_{n+1}(x) = 1$. The augmentations of elements of T_n' are torsion elements, hence T_n' may be adjusted so that $\psi_{n+1}(T_n') = 1$.

If the induction is valid, we put $R = \bigcup_n R_n$. It is almost immediate that this is the desired ring R. Certainly R^+ is free by condition (1). We put $\varphi = \bigcup_n \varphi_n$ to obtain an injection $\varphi : V \to U(R)$. Only the surjectivity of φ must be checked. Any element of $U(R)$ lies in some U_k. Since $U_k = \varphi_k(V_k) \times \langle T_k\rangle$, it suffices to consider an element of T_k. Since the rank of A is infinite, steps of Types 1 or 2 occur infinitely many times. Thus, for some $n \ge k$, step n will be of Type 2 and the first element of T_n will be the given element, which will be in the image of φ_{n+1}. In the augmentation option, $\psi = \bigcup_n \psi_n$ will be the desired augmentation.

The remainder of the paper is devoted to validating the induction step. First, we check the easier items. It is clear that R_{n+1} is a finitely generated ring integral over S_{n+1}, and that R_{n+1}^+ is a free module over R_n, hence R_{n+1}^+ is free. In Types 1 and 2, there are no nontrivial nilpotent elements in R_{n+1} since there is none in R_n. Furthermore, in these two types, there are no nontrivial idempotent elements since a degree argument in Type 1 can use the fact on nilpotent elements.

Turning to S_n, it is clear that S_{n+1} is an integral domain which is integrally closed since that property is inherited by polynomial rings and localizations. To see that no element of S_{n+1} is a zero divisor on R_{n+1}, Type 2 is trivial, Type 1 follows from a degree argument, and Type 3 is because R_{n+1}^+ is a free R_n -module. The statement on primes being nonunits in S_{n+1} is evident. For the conditions on φ_n and T_n, in Type 1 we use Proposition 5.2.1 in [5] to conclude that $U_{n+1} = U_n \times \langle X \rangle = \varphi_{n+1}(V_{n+1})$. In Type 2, we have $U_{n+1} = U_n = \varphi_n(V_n) \times \langle T_n \rangle = \varphi_{n+1}(V_{n+1}) \times \langle T_{n+1} \rangle$. In the augmentation option, condition (5) is clear since $V'_{n+1} = V'_n\langle a \rangle$.

Everything is now verified except for three items involving Type 3. These are that R_{n+1} has no nontrivial nilpotent elements, no nontrivial idempotent elements, and that U_{n+1} is the direct product of $U_n\langle x \rangle$ and a free abelian group of finite rank. Since R_{n+1} is a finitely generated ring, U_{n+1} is finitely generated by Corollary 7.2.8 in [6]. Thus, for this third item, it suffices to show that $U_{n+1}/U_n\langle x \rangle$ is torsion-free. The three items will require some work to establish.

We will first need some lemmas dealing with roots of unity. If $L \supseteq K$ is a field extension of finite degree, let $T_{L/K}$ denote the trace mapping.

Lemma 4.2 *Let q be a prime, $r \geq 1$, and let ζ be a primitive q^r-th root of unity over $\mathbb{Q}$. Let E be a subfield of $\mathbb{Q}(\zeta)$ such that $[\mathbb{Q}(\zeta) : E]$ is relatively prime to q. Then there exists $\alpha \in \mathbb{Z}[\zeta]$ such that $T_{\mathbb{Q}(\zeta)/E}(\alpha) = -1$.*

Proof. Let F be the unique subfield of $\mathbb{Q}(\zeta)$ such that $[\mathbb{Q}(\zeta) : F] = q^{r-1}$. Then $F \subseteq E$. Let G be the Galois group of $\mathbb{Q}(\zeta)$ over F, and let H be the subgroup leaving E fixed. Let $\mathcal{R}$ be a set of coset representatives of H in G. If θ denotes a primitive q-th root of unity, then G is isomorphic by restriction to the Galois group of $\mathbb{Q}(\theta)$ over $\mathbb{Q}$. Put $\alpha = \sum_{\tau \in \mathcal{R}} \tau(\theta)$. Then $T_{\mathbb{Q}(\zeta)/E}(\alpha) = \sum_{\sigma \in H} \sigma(\alpha) = \sum_{\sigma,\tau} (\sigma\tau)(\theta) = T_{\mathbb{Q}(\zeta)/F}(\theta) = T_{\mathbb{Q}(\theta)/\mathbb{Q}}(\theta) = -1$ since the minimal polynomial of θ over $\mathbb{Q}$ is $X^{q-1} + X^{q-2} + \ldots + 1$. □

We now set up some generic notation to be used in the next several lemmas. Let D denote an integral domain of characteristic 0, which is integrally closed in its quotient field K. Let p be a prime number which is not a unit in D. Let $c \in U(D)$ and $f = X^p - c$. Then $K[X]f \cap D[X] =$

$D[X]f$ since f is monic. We may let x be the coset of X modulo the ideal generated by f in either polynomial ring, and regard $D[x] \subseteq K[x]$. Note that $x \in U(D)$. If $c \notin K^p$, then $X^p - c$ is irreducible over K and $K[x] = K(x)$ is a field.

Lemma 4.3 *Suppose $c \notin K^p$. Then every root of unity in $D[x]$ lies in $\langle x \rangle U(D)$.*

Proof. Let q^r be a prime power and ζ a primitive q^r-th root of unity lying in $D[x]$, but not in K. It suffices to show that $\zeta \in \langle x \rangle U(D)$. Since $[K(x) : K] = p$ and $\zeta \notin K$, we have $K(x) = K(\zeta)$. Put $E = K \cap \mathbb{Q}(\zeta)$. Then the Galois group of $K(x)$ over K is isomorphic to that of $\mathbb{Q}(\zeta)$ over E by restriction.

Suppose that $q \neq p$. Since $[\mathbb{Q}(\zeta) : E] = p$, we may apply Lemma 4.2 to obtain $\alpha \in \mathbb{Z}[\zeta]$ with $T_{\mathbb{Q}(\zeta)/E}(\alpha) = -1$. Since $\zeta \in D[x]$, we have $\alpha \in D[x]$, thus $\alpha = \sum_{0 \leq i \leq p-1} a_i x^i$ $(a_i \in D)$. For $1 \leq i \leq p-1, c^i \notin K^p$, thus $X^p - c^i$ is irreducible over K and x^i is a root. Therefore $K(x^i) = K(x)$, and we have that $T_{K(x)/K}(x^i) = 0$. Hence $T_{K(x)/K}(\alpha) = \sum_{0 \leq i \leq p-1} a_i T_{K(x)/K}(x^i) = pa_0$ since $T_{K(x)/K}(1) = p$. But the restriction isomorphism mentioned above implies $T_{K(x)/K}(\alpha) = T_{\mathbb{Q}(\zeta)/E}(\alpha) = -1$. This contradicts our hypothesis on $p \notin U(D)$, so this case cannot occur.

We now know $q = p$. Since $[\mathbb{Q}(\zeta) : E] = p$ and ζ is a primitive p^r-th root of unity, we have $\zeta^p \in E$ by the Galois Theory of $\mathbb{Q}(\zeta)$. Moreover, $\zeta^p \neq 1$ since the degree over $\mathbb{Q}$ of a primitive p-th root of unity is $p-1$. Thus K contains a primitive p-th root of unity and we may apply Kummer Theory to conclude that $\zeta = x^i a$ for some $1 \leq i \leq p-1$ and $a \in U(K)$. Then $a^{p^r} = x^{-i} p^r \in U(D)$. Since D is integrally closed, $a \in U(D)$ and hence $\zeta \in \langle x \rangle U(D)$. □

Lemma 4.4 *Suppose that $p = 2$ and $c \notin K^2$. Let ζ_4 denote a primitive 4-th root of unity, and assume that $K(x) = K(\zeta_4)$ and that $\sqrt{-2} \notin K$. Assume that $\theta \in K(x)$ is a primitive 2^r-th root of unity where $r \geq 2$, and put $v = 1 + \theta$ and $e = \theta + \theta^{-1} + 2$.*

(1) We have $v^2 = \theta e, e \in K$, and there is a unit $\mu \in \mathbb{Z}[\theta]$ such that $2 = \mu e^{2r-2}$.

(2) Assume that $\theta \in D[x]$. If $a \in K$ and $va \in D[x]$, then $a \in D$.

Proof. (1) The assumption on ζ_4 and $\sqrt{-2}$ imply that the nontrivial automorphism of $K(x)$ leaving K fixed takes θ to θ^{-1}. Consequently, $e \in K$. The primitive 2^r-th roots of unity are θ^i for i odd. Since $(1+\theta^i)(1+\theta)^{-1} = 1 - \theta + \ldots + \theta^{i-1} \in \mathbb{Z}[\theta]$, and by symmetry $(1+\theta)(1+\theta^i)^{-1} \in \mathbb{Z}[\theta]$, the quotient is a unit in $\mathbb{Z}[\theta]$ for odd i. The cyclotomic polynomial $X^{2^{r-1}}+1 = \prod_i (X-\theta^i)$, thus there is a unit $\mu \in \mathbb{Z}[\theta]$ such that $2 = \mu(1+\theta)^{2r-1}$. But $(1+\theta)^2 = \theta e$, hence $2 = \mu e^{2^{r-2}}$.

(2) By Kummer Theory applied to $K(x) = K(\zeta_4)$, there exists $b \in K$ such that $k = b\zeta_4$, therefore $c = x^2 = -b^2$. Since D is integrally closed, $b \in D$ and thus $D[x] \subseteq D[\zeta_4]$. But ζ_4 is a power of θ, thus $D[\theta] \subseteq D[x] \subseteq D[\theta]$, hence $D[x] = D[\theta]$. Note that $\theta + \theta^{-1} \in D[x] \cap K = D$, and $\theta^2 = -1+(\theta+\theta^{-1})\theta$. Consequently, $D[\theta] = D \oplus D\theta$. If $a + a\theta = va \in D[x] = D[\theta]$, then $a \in D$, as desired. □

The next two lemmas show that $D[x]$ has crucial properties needed to prove the remaining Type 3 items.

Lemma 4.5 *$D[x]$ has no nontrivial nilpotent or idempotent elements.*

Proof. First consider nilpotent elements. It suffices to show that $K[x]$ has trivial nilradical, which will be true if f has no repeated roots over K. No root of $f = X^p - c$ is a root of its derivative since $c \neq 0$ and K has characteristic 0, hence $K[x]$ has trivial nilradical.

Now we show that $D[x]$ is indecomposable. If f is irreducible over K, then $D[x]$ is an integral domain, thus indecomposable. If f is reducible, then $c = d^p$ for some $d \in K$. Since D is integrally closed, $d \in D$ and is a unit in D since c is one. Change variables $D[X] = D[Y]$ by $Y = d^{-1}X$. Thus $D[x] \cong D[Y]/\langle Y^p - 1\rangle$, showing that $D[x]$ is isomorphic to the group algebra over D of a cyclic group of order p. Since p is not a unit in $D, D[x]$ is indecomposable by Corollary 3.3.13 in [5]. □

Lemma 4.6 *$U(D[x])/\langle x\rangle U(D)$ is torsion-free.*

Proof. Since x is torsion modulo $U(D)$, we must show that every element of $U(D[x])$ that is torsion modulo $U(D)$ comes from $\langle x\rangle U(D)$. In the previous proof, we saw that $D[x]$ is either an integral domain or isomorphic to DC_p. In the group algebra case, the generator of the group corresponds to $d^{-1}x$. Let $u \in U(D[x])$ be torsion modulo $U(D)$. If $a \in D$ is the augmentation of u, then $a \in U(D)$ and $a^{-1}u$ has augmentation 1. If $u^m \in U(D)$, then $(a^{-1}u)^m \in U(D)$ and has augmentation 1, therefore $(a^{-1}u)^m = 1$. Since $p \notin U(D)$, by Theorem 5.1.5 in [5], we have that all units of finite order in $D[x]$ are "trivial units," thus $a^{-1}u \in \langle x\rangle U(D)$ and $u \in \langle x\rangle U(D)$.

We now consider the integral domain case. Then $D[x]$ is an integral domain with quotient field $K(x)$, where the irreducibility of f means $c \notin K^p$. Let $u \in U(D[x])$ be a torsion element modulo $U(D)$. We first claim that if $u = \zeta a$ for ζ a root of unity and $a \in U(K)$, then $\zeta \in D[x]$ and $a \in U(D)$.

Let $\zeta^m = 1$ and $u^m \in U(D)$. Then $a^m \in U(D)$, thus D integrally closed implies $a \in U(D)$. But then $\zeta \in D[x]$, as claimed.

Now suppose that u is arbitrary again. By Proposition 2 in [9], unless we are in the exceptional case indicated there, the elements of $U(K(x))$ which are torsion modulo $U(K)$ are those of $\langle x \rangle W U(K)$, where W is the group of roots of unity in $K(x)$. Thus, $u = x^i \zeta a$ for some root of unity ζ and $a \in U(K)$. But $ux^{-i} \in U(D[x])$ and the claim above shows that $a \in U(D)$ and $\zeta \in D[x]$. Lemma 4.3 applies to show that $\zeta \in \langle x \rangle U(D)$, hence $u \in \langle x \rangle U(D)$, as desired.

In the exceptional case, $p = 2, K(x) = K(\zeta_4)$, $\sqrt{-2} \notin K$, and there is a primitive 2^r-th root of unity $\theta \in K(x) (r \geq 2)$, such that the additional torsion modulo $\langle x \rangle W U(K)$ is of order 2 and generated by $v = 1 + \theta$ (see Lemma 1 in [8]). Moreover, it is pointed out in the reference that $K(x)$ contains no primitive 2^{r+1}-th root of unity. It will suffice to show that this additional torsion does not occur in $U(D[x])$. Therefore, assume we are in the exceptional case and that $u \in U(D[X])$ is torsion modulo $U(D)$. Since powers of x may be neglected, we may assume $u = \zeta v a$. Then $u^2 = \zeta^2 v^2 a^2 = \zeta^2 \theta e a^2$ by Lemma 4.4. Since $ea^2 \in U(K)$, by the claim in the paragraph above, we have $\zeta^2 \theta \in D[x]$ and $ea^2 \in U(D)$. Since $K(x)$ contains no primitive 2^{r+1}-th root of unity, we conclude that $\theta \in D[x]$, and hence $\zeta \in D[x]$. Thus $va \in U(D[X])$. By Lemma 4.4, $a \in D$. But $\theta \in D[x]$ implies $e \in D[x] \cap K = D$. Since we have already observed $ea^2 \in U(D)$, we now may conclude $e \in U(D)$. Lemma 4.4 now implies that 2 is a unit in $D[x]$. Since $D[x]$ is integral over D, 2 is a unit in D. This contradiction shows that the exceptional torsion does not occur in $U(D[x])$. □

We now return to R_n to verify the three outstanding Type 3 items. Since R_n is noetherian, there are finitely many minimal prime ideals, say $P_1, \dots, P_m$. Let K_i be the quotient field of R_n/P_i, and let D_i be the integral closure of R_n/P_i in K_i $(1 \leq i \leq m)$. Since R_n has trivial nilradical by induction, by using the natural embedding, we may regard $R_n \subseteq \prod_{1 \leq i \leq m} D_i \subseteq \prod_{1 \leq i \leq m} K_i$. Recall $c \in U_n$. If c_i denotes this image of c in D_i, then $c_i \in U(D_i)$. Since $f = X^p - c$ is monic, the restriction of the ideal f generates in $(\prod_i K_i)[X]$ to either $R_n[X]$ or $(\prod_i D_i)[X]$ is the ideal generated by f in the subring. Let $f_i = X^p - c_i$ and let x_i be the image of X in $D_i[X]/\langle f_i \rangle \subseteq K_i[X]/\langle f_i \rangle$. We then have

$$R_{n+1} = R_n[x] \subseteq \prod_{1 \leq i \leq m} D_i[x_i] \subseteq \prod_{1 \leq i \leq m} K_i[x_i].$$

To be able to apply Lemmas 4.5 and 4.6, we prove

Lemma 4.7 *No prime number is a unit in D_i. In particular, D_i has characteristic zero.*

Proof. Since P_i is minimal, the nonzero elements of P_i are zero divisors in R_n. Since no element of S_n is a zero divisor on R_n, we have $S_n \cap P_i = 0$. Thus S_n is naturally embedded in R_n/P_i, which is integral over the image, hence D_i is integral over the image. Since no prime number is a unit in S_n, it is easy to show the same holds for D_i. □

Lemma 4.8 *R_{n+1} has no nontrivial nilpotent or idempotent elements.*

Proof. First consider nilpotent elements. Lemma 4.5 says that $D_i[x_i]$ has trivial nilradical, thus the same is true for R_{n+1}. Now consider idempotent elements. By Lemma 4.5 again, the idempotents of $D_i[x_i]$ are trivial, thus the idempotents of $\prod_i D_i[x_i]$ all lie in $\prod_i D_i$. Since $R_{n+1} = \bigoplus_{0\leq j\leq p-1} R_n x^j \subseteq \bigoplus_{0\leq j\leq p-1}(\prod_i D_i)x^j$, we have $R_{n+1}\cap\prod_i D_i = R_n$, hence all idempotents of R_{n+1} lie in R_n. But R_n is indecomposable, therefore so is R_{n+1}. □

The final lemma establishes the last remaining item in the proof of Theorem 4.1.

Lemma 4.9 *$U_{n+1}/\langle x\rangle U_n$ is torsion-free.*

Proof. Since $U_{n+1} \subseteq \prod_i U(D_i[x_i])$ and $\prod_i(U(D_i[x_i])/\langle x_i\rangle U(D_i))$ is torsion-free by Lemma 3.2, it suffices to show that $U_{n+1} \cap \prod_i \langle x\rangle U(D_i) = \langle x\rangle U_n$. One containment is clear, so let $\alpha \in U_{n+1} \cap \prod_i \langle x\rangle U(D_i)$. Let the idempotent in $\prod_i D_i$ for the factor D_i be denoted by ε_i. In regarding $R_n[x] \subseteq (\prod_i D_i)[x] = \prod_i D_i[x_i]$, we have $x_i = \varepsilon_i x$. We may write $\alpha = \sum_i x_i^{k_i} d_i$, where $d_i \in U(D_i)$ and $0 \leq k_i \leq p-1$ for every i. For $0 \leq j \leq p-1$, put $I_j = \{i : k_i = j\}$. Then $\alpha = \sum_j(\sum_{i\in I_j} d_i\varepsilon_i)x^j$. Since $R_{n+1} = \bigoplus_j R_n x^j \subseteq \bigoplus_j(\prod_i D_i)x^j$, we have $\sum_{i\in I_j} d_i\varepsilon_i \in R_n$ for every j. But $\alpha^{-1} \in R_{n+1}$ also, and clearly $\alpha^{-1} = \sum_j(\sum_{i\in I_j} d_i^{-1}\varepsilon_i)x^{-j}$ by multiplying. Since x is a unit in R_{n+1}, $R_{n+1} = x^{-p+1}R_{n+1} = \bigoplus_j R_n x^{-j}$, hence $\sum_{i\in I_j} d_i^{-1}\varepsilon_i \in R_n$ for every j. Taking the product, we obtain $\sum_{i\in I_j}\varepsilon_i \in R_n$, thus $1 = \sum_j(\sum_{i\in I_j}\varepsilon_i)$ is a decomposition of 1 into orthogonal idempotents in R_n. Since R_n is indecomposable, there exists k such that $I_k = \{0,\dots,p-1\}$ and $I_j = \emptyset$ for $j \neq k$. Thus $\alpha = \sum_{0\leq i\leq p-1} d_i\varepsilon_i x^k$. But $\sum_{0\leq i\leq p-1} d_i\varepsilon_i \in R_n$, as we saw above, thus $\alpha \in \langle x\rangle U_n$. □

We next show that, with one excluded case, if the rank of G is infinite, then the torsion is no larger except for a bounded summand.

Lemma 4.10 *Assume that G has no summand of order two, that End G is $\mathbb{Z}[\zeta_{2^h}]$-algebra, and that $U=$ Aut G is abelian of infinite rank λ. Then*

$U = U' \times A$, where U' has cardinal λ, and $A = 1$ if $h \geq 3$, $A^4 = 1$ if $h = 2$, and $A^{12} = 1$ if $h = 1$.

Proof. Put $Z = \mathbb{Z}[\zeta_{2^h}]$ and $E =$End G. We first claim that the torsion of the additive group of E is countable. Since G has no summand of order two, Lemma 3.1 allows us to conclude that every p-component of G is cocyclic and that a complement for such a component must be p-divisible. The claim now follows since the p-torsion of E is finite.

Let $\overline{E} = Q \otimes_Z E$ and consider the natural map $E \to \overline{E}$. Its kernel is the torsion subgroup of E. Let T denote the torsion subgroup of $U = U(E)$. Let p^e $(e \geq 1)$, be a prime power subject to the following restrictions. If $h \geq 3$, then p^e may be any prime power; if $h = 2$, then $p^e \neq 2$ or 4; and if $h = 1$, then $p^e > 4$. We shall assume that T contains more than λ elements of order p^e, and shall derive a contradiction. The rank of $U(Z[T])$ is $\leq \lambda$, hence we may choose $B \subseteq T$ with $|B| \leq \lambda$ such that $U(Z[T])/U(Z[B])$ is torsion. Let $\overline{B}$ be the image of B in $\overline{E}$, and if $u \in T$, let $\overline{u}$ be its image in $\overline{E}$. By the claim above, the kernel of $U(E) \to U(\overline{E})$ is countable, thus we may choose $u \in T$ of order p^e such that $\overline{u}$ has order p^e and $\langle \overline{u} \rangle \cap U(Q[\overline{B}]) = 1$.

Before obtaining a contradiction, we claim that every element of $U(Z[\overline{u}])$ is torsion modulo $U(Z[\overline{B}])$. The group algebra $Q\langle u \rangle$ is a product of fields $Q\langle u \rangle = \prod_{k<\alpha} L_k$. The natural map $Q\langle u \rangle \to Q[\overline{u}]$ may be regarded as a projection $\prod_{k<\alpha} L_k \to \prod_{j<\beta} L_j$ for some $\beta \leq \alpha$. Letting R_k be the ring of algebraic integers in L_k, we have that the projection $\prod_{k<\alpha} R_k \to \prod_{j<\beta} R_j$ is surjective on units. Consider an element of $U(\prod R_k)$. By Lemma 2.2, some power lies in $U(Z\langle u \rangle)$. Taking the image of this in $U(Z[u])$ a power of this image lies in $U(Z[B])$. This shows the claim since $U(Z[\overline{u}]) \subseteq U(\prod R_j)$.

Since $Q[\overline{u}] \cap Q[\overline{B}]$ is a Q- subalgebra of $Q[\overline{u}]$, it may be regarded as a product $\prod_i K_i$ embedded in $Q[\overline{u}] = \prod L_j$. But $\langle \overline{u} \rangle \cap \prod K_i = 1$, thus some $L_j = Q[\zeta_{p^e}]$ extends some K_i such that either the extension is proper, or else there is $j' \neq j$ such that $L_{j'}$ extends K_i and $L_{j'} = L_j = K_i$. Appealing to Lemma 2.2, using the way p^e was taken, and noting $\zeta_{2^h} \in Q$, we conclude that the rank of $U(Z[\overline{u}])$ exceeds that of $U(Z[\overline{u}]) \cap U(Z[\overline{B}])$. This contradicts the torsion claim above, thus T contains $\leq \lambda$ elements of order p^e. To finish the proof, if $h \geq 3$, then $|T| \leq \lambda$, hence $|U| = \lambda$. For $h \leq 2$, we may use basic subgroups to obtain $U = U' \times A$ as required in the statement. □

Using Theorem 4.1, we apply the lemma to rank $\aleph_0$.

Theorem 4.11 *Consider abelian G with no summands of order two and with Aut G abelian of rank $\aleph_0$. Then U is isomorphic to some Aut G if and only if $U = U' \times A$, where $|U'| = \aleph_0, U'$ has an element of order two, and if h is minimal $(1 \leq h \leq \infty)$, such that U' has a direct factor isomorphic to $\mathbb{Z}_{2^h}$, then $A = 1$ if $h \geq 3$, $A^4 = 1$ if $h = 2$, and $A^{12} = 1$ if $h = 1$.*

Proof. For $U=$ Aut G, let h be minimal such that U has a factor isomorphic to $\mathbb{Z}_{2^h}$. Then the 2-height of -1 in U is $\geq h$, hence End G is a $\mathbb{Z}[\zeta_{2^h}]$-algebra, and Lemma 4.10 yields $U = U' \times A$. Conversely, if $U = U' \times A$ as above, then there is a cotorsion-free ring R' with $U(R') \cong U'$. If $h \leq 2$, then R' may be taken as an augmented $\mathbb{Z}[\zeta_{2^h}]$-algebra, thus we can utilize a pullback construction as at the end of the proof of Theorem 3.10 to produce a ring R with $U(R) \cong U' \times A$. There exists a torsion-free group G with Aut $G \cong U(R)$. □

References

[1] A.L.S. Corner, *Every countable reduced torsion-free ring is an endomorphism ring*, Proc. London Math. Soc. **13** (1963), 687–710.

[2] L. Fuchs, *Infinite Abelian Groups*, Vol. **II**, Academic Press, New York, (1973).

[3] J.T. Hallett and K.A. Hirsch, *Torsion-free groups having finite automorphism groups*, J. Algebra **2** (1965), 287–298.

[4] G. Higman, *The units of group rings*, Proc. London Math. Soc. **46** (1940), 231–248.

[5] G. Karpilovsky, *Commutative Group Algebras*, Marcel Dekker, New York, (1983).

[6] G. Karpilovsky, *Unit Groups of Classical Rings*, Oxford University Press, Oxford, (1988).

[7] S. Lang, *Algebraic Number Theory*, Addison-Wesley, Reading, (1970).

[8] W. May, *Multiplicative groups under field extension*, Can. J. Math **31** (1979), 436–440.

[9] W. May, *Fields with free multiplicative groups modulo torsion*, Rocky Mountain J. **10** (1980), 599–604.

Trends in Mathematics, © 1999 Birkhäuser Verlag Basel/Switzerland

Transitivity and full transitivity over subgroups of abelian p-groups

Georg Hennecke *

Abstract

A generalization of Kaplansky's notion of (full-)transitivity is introduced. We say a p-group G is (fully) transitive over a subgroup H, if for all $x \in G$ and $y \in H$ with $U(x) = U(y)$ $(U(x) \leq U(y))$ there exists an automorphism (endomorphism) of G which maps x to y. An adaption of a result by Corner shows that only the elements of infinite height in G are relevant for this definition. Therefore a group G is always transitive and fully transitive over any subgroup which contains only elements of finite height in G.

Furthermore we show that a group G is fully transitive over any subgroup with elements of height $< \lambda$, if λ is the least ordinal for which $G/p^{\lambda}G$ is not totally projective.

We also study subgroups which are maximal with respect to the above definition. These subgroups are dense in the p-adic topology, but not necessarily pure. With the restriction that the length of the group is $\omega + 1$ we can show that these maximal subgroups are indeed pure, and are even N-high subgroups, where N is a subgroup of the first Ulm subgroup. In this setting we also give examples of groups, which contain $2^{2^{\aleph_0}}$ non-isomorphic subgroups that are maximal in the above sense.

Introduction

The classical notions of transitivity and full transitivityfully transitive group were introduced by Kaplansky in his 1954 book [Ka]. Extensive classes of groups which are both transitive and fully transitive were found in [Gr] and [H2]. Groups with neither property were constructed in [M1] and for larger cardinalities in [CG]. The independence of both concepts was first obtained in [C1]. In a recent paper Files and Goldsmith [FG] proved that a group G is fully transitive if and only if its square $G \oplus G$ is transitive. This result indicates that there is a basic connection between these two notions.

*This work represents part of the author's Ph.D. thesis under the direction of Prof. Dr. R. Göbel.

The concept of transitivity over subgroups was first used by Files in [Fi], where he introduced the notion of transitivity over the torsion submodule of a mixed module.

All groups considered will be abelian groups. Notation follows the standard works of Fuchs [Fu] and Kaplansky [Ka] with the exception that maps are written on the right. If H is a direct summand of G, then we write $H \sqsubset G$.

1 Preliminaries

If G is a p-group, we define $p^\alpha G$ inductively to be $p^0 G = G$, $p^{\alpha+1}G = p(p^\alpha G)$ and $p^\beta G = \bigcap_{\alpha<\beta} p^\alpha G$ whenever β is a limit ordinal. The first ordinal λ for which $p^\lambda G = p^{\lambda+1}G$ is called the *length* of G. We say G is *reduced*, if there exists an ordinal λ with $p^\lambda G = 0$; all groups we consider are reduced unless stated otherwise.

The *height* $|g|_G$ in G of an element $g \in G$ is defined to be ∞ if $g \in p^\alpha G$ for all ordinals α and it is α if $g \in p^\alpha G \setminus p^{\alpha+1}G$, where infinity exceeds any ordinal. The *height sequence* of an element $g \in G$ is given by $U_G(g) = (\alpha_0, \alpha_1, \alpha_2, \dots)$ where $\alpha_i = |p^i g|_G$. The height sequences are partially ordered by agreeing that $U_G(x) \leq U_G(y)$ if $|p^i x|_G \leq |p^i y|_G$ for all $i \geq 0$.

We say a subgroup N of G is *nice* in G if the set $\{|x+y|_G : y \in N\}$ contains a maximal element for all $x \in G \setminus N$.

Let H and N be subgroups of G. We call H an N-high subgroup of G if H is maximal with respect to the property of being disjoint from N.

Now we introduce the concepts of Transitivity and Full Transitivity over subgroups of abelian p-groups. These concepts generalize Kaplansky's original notions of Transitivity and Full Transitivity [Ka].

Definition 1.1 Let G be a p-group and H a subgroup of G. We say

(i) G is *fully transitive over* H if for any $x \in G$ and $y \in H$ with $U_G(x) \leq U_G(y)$ there exists an endomorphism of G which maps x to y.

(ii) G is *transitive over* H if for any $x \in G$ and $y \in H$ with $U_G(x) = U_G(y)$ there exists an automorphism of G which maps x to y.

The following Proposition provides a useful criteria for a group to be transitive or fully transitive over a given subgroup.

Proposition 1.2 *Let G be a p-group and λ the least ordinal such that $G/p^\lambda G$ is not totally projective.*

(i) If $x, y \in G$ with $U_G(x) \leq U_G(y)$ and $\langle x\rangle \cap p^\lambda G = 0$ or $\langle y\rangle \cap p^\lambda G = 0$, then there exists an endomorphism ϕ of G such that $x\phi = y$.

(ii) If $x, y \in G$ with $U_G(x) = U_G(y)$ and $\langle x \rangle \cap p^\lambda G = \langle y \rangle \cap p^\lambda G = 0$, then there exists an automorphism ϕ of G such that $x\phi = y$.

Proof. We know that λ has to be a limit ordinal, therefore there exists a $\mu < \lambda$ such that $\langle x \rangle \cap p^\mu G = 0$ or $\langle y \rangle \cap p^\mu G = 0$ and $G/p^\mu G$ is totally projective.

We first prove (i). Let $\sigma : G \longrightarrow G/p^\mu G$ be the canonical epimorphism. We shall show that

$$U_{G/p^\mu G}(x\sigma) \leq U_G(y). \tag{1}$$

Let k be minimal with $p^k x \in p^\mu G$. For all $i < k$ we have $|p^i(x\sigma)|_{G/p^\mu G} = |p^i x|_G \leq |p^i y|_G$. Assume $\langle x \rangle \cap p^\mu G = 0$, i.e. $p^k x = 0$. Therefore $|p^k y|_G = \infty$, since $U_G(x) \leq U_G(y)$. On the other hand, if $\langle y \rangle \cap p^\mu G = 0$, we get $|p^k y|_G \geq |p^k x|_G \geq \mu$ and hence $p^k y = 0$. In any case (1) holds.

Now let $\phi_0 : \langle x\sigma \rangle \longrightarrow G$ be a homomorphism which maps $x\sigma$ to y. Because of (1) ϕ_0 does not decrease heights in G. Furthermore $\langle x\sigma \rangle$ is nice in $G/p^\mu G$ and $(G/p^\mu G)/\langle x\sigma \rangle$ is totally projective. From [Fu, Corollary 81.4] it follows that ϕ_0 can be extended to a homomorphism $\phi : G/p^\mu G \longrightarrow G$ and the composition of the canonical epimorphism σ and of ϕ leads to the wanted endomorphism which maps x to y.

To prove (ii) we extend the identity on $p^\mu G$ to a height-preserving isomorphism between $\langle p^\mu G, x \rangle$ and $\langle p^\mu G, y \rangle$ by mapping x to y. It follows from [H2, Theorem 1] that this isomorphism can be extended to an automorphism of G. □

Corollary 1.3 *Let G be a p-group and λ the least ordinal such that $G/p^\lambda G$ is not totally projective. Then G is transitive and fully transitive over any subgroup H of G with $H \cap p^\lambda G = 0$. Consequently G is transitive and fully transitive over any $p^\lambda G$-high subgroup.*

Note that if $G/p^\omega G$ is not totally projective, then ω is the least ordinal such that $G/p^\omega G$ is not totally projective. Therefore G is always transitive and fully transitive over any subgroup which contains only elements of finite height.

Recall that a C_Ω (or C_{ω_1}) group is a p-group G such that $G/p^\lambda G$ is totally projective for all $\lambda < \omega_1$ (see [M2]).

Corollary 1.4 *A C_Ω group is transitive and fully transitive over any subgroup which contains only elements of countable height.*

We now extend our definition of transitivity over subgroups in a way that corresponds to the generalized definition of ordinary transitivity by Corner [C1].

Definition 1.5 Given a (unital) subring Φ of the endomorphism ring of a p-group G, a Φ-invariant subgroup H of G, and a subgroup U of H, we say that

(i) Φ is fully transitive on H over U if, for any $x \in H$ and $y \in U$ with $U_G(x) \leq U_G(y)$, there is an element $\phi \in \Phi$ such that $x\phi = y$.

(ii) Φ is transitive on H over U if, for any $x \in H$ and $y \in U$ with $U_G(x) = U_G(y)$, there is a unit $\phi \in \Phi$ such that $x\phi = y$.

We can now adapt a fundamental observation by Corner [C1, Lemma 2.1] to our setting.

Proposition 1.6 *A p-group G is (fully) transitive over a subgroup H of G if and only if the endomorphism ring of G is (fully) transitive on $p^\omega G$ over $p^\omega G \cap H$.*

2 (Full-)Transitivity over maximal subgroups

Let G be a p-group. We denote by $\mathfrak{T}_G$ ($\mathfrak{F}_G$) the set of all subgroups H of G which are maximal with respect to the property that G is (fully) transitive over H.

Lemma 2.1 *Let G be a p-group and H a subgroup of G.*

(i) If G is (fully) transitive over H with $p^\omega G \cap H = 0$ and U is a subgroup of $p^\omega G$ such that G is (fully) transitive over U, then G is (fully) transitive over $\langle H, U\rangle$.

(ii) Either

(a) for all $H \in \mathfrak{T}_G$ ($H \in \mathfrak{F}_G$), $H \cap p^\omega G = 0$ (in this case all these H's are in fact $p^\omega G$-high subgroups of G)

or

(b) for all $H \in \mathfrak{T}_G$ ($H \in \mathfrak{F}_G$), $H \cap p^\omega G \neq 0$.

Proof. (i) In view of Proposition 1.6 it suffices to show that $\langle H, U\rangle \cap p^\omega G = U$. Trivially $U \subseteq \langle H, U\rangle \cap p^\omega G$. Let $x \in \langle H, U\rangle \cap p^\omega G$, then x can be written as $x = h + u$ with $h \in H$ and $u \in U$. We get $h = x - u \in p^\omega G \cap H = 0$ and therefore $x = u \in U$.

(ii) Let $H_1, H_2 \in \mathfrak{T}_G$ ($H_1, H_2 \in \mathfrak{F}_G$) with $p^\omega G \cap H_1 = 0$. From (i) it follows that G is (fully) transitive over $\langle H_1, p^\omega G \cap H_2\rangle \supseteq H_1$. Since H_1 is maximal we have $H_1 = \langle H_1, p^\omega G \cap H_2\rangle$. Thus $p^\omega G \cap H_2 = 0$. □

Theorem 2.2 *Let G be a p-group and $H \in \mathfrak{T}_G$ ($H \in \mathfrak{F}_G$). Then G/H is divisible.*

Proof. We assume that G/H is not divisible. Then there exists an $x \in G$ and a subgroup K of G such that

$$G/H = \langle x + H \rangle \oplus K/H.$$

Let p^k be the order of $x + H$ in G/H. Since $\langle x + H \rangle$ is a direct summand of G/H there exists an integer m such that

$$|p^{k-1}x + H|_{G/H} = m < \omega. \tag{2}$$

Now let $H' = \langle H, p^{k-1}x \rangle$. Since $p^{k-1}x \notin H$, H is a proper subgroup of H'. In view of Proposition 1.6 it is enough to show that $H \cap p^\omega G = H' \cap p^\omega G$ because this would mean that G is (fully) transitive over H', which is a contradiction to the maximality of H.

Let $z \in H' \cap p^\omega G$, then z can be written as $z = h + rp^{k-1}x$ with $h \in H$ and $r \in \mathbb{Z}$. Since $z \in p^\omega G$, there exists $g \in G$ such that $h + rp^{k-1}x = p^{m+1}g$. If r and p are relatively prime, it follows that $|rp^{k-1}x + H|_{G/H} \geq m+1 > m$, in contradiction to (2). Therefore p divides r and since $p^k x \in H$, we get $z \in H \cap p^\omega G$. □

After the next lemma we will give an example that shows that the subgroups in $\mathfrak{T}_G$ and $\mathfrak{F}_G$ need not to be pure.

Lemma 2.3 *Let G be a p-group and $x \in p^\omega G$ such that G is (fully) transitive over $\langle px \rangle$, but G is not (fully) transitive over $\langle x + \epsilon_\omega \rangle$ for all $\epsilon_\omega \in (p^\omega G)[p]$. Then there exists a nonpure subgroup $U \in \mathfrak{T}_G$ ($U \in \mathfrak{F}_G$).*

Proof. Let H be any $p^\omega G$-high subgroup of G. By Corollary 1.3, G is transitive and fully transitive over H and it follows from Lemma 2.1(i) that G is also (fully) transitive over $H' = \langle H, px \rangle$. By Zorn's Lemma there exists a $U \in \mathfrak{T}_G$ ($U \in \mathfrak{F}_G$) such that $H' \subseteq U$. To prove that U is not pure in G, we show that for all $\epsilon \in G[p]$, $x + \epsilon \notin U$.

Let $\epsilon \in G[p]$ and suppose $x + \epsilon \in U$. Using [Ir, Lemma 1(f)], we get

$$\epsilon = \epsilon_H + \epsilon_\omega \text{ with } \epsilon_H \in H[p] \text{ and } \epsilon_\omega \in (p^\omega G)[p].$$

Therefore $x + \epsilon - \epsilon_H = x + \epsilon_\omega \in H$. But this contradicts the assumption that G is not (fully) transitive over $\langle x + \epsilon_\omega \rangle$ for all $\epsilon_\omega \in (p^\omega G)[p]$. □

Example 2.4 Let $K = \langle a \rangle \oplus \langle b \rangle$, where a is of order p and b is of order p^2. Let Φ, Φ' be the subrings of the endomorphism ring of K generated by $\left(\begin{smallmatrix}1&0\\0&1\end{smallmatrix}\right)$ and $\left(\begin{smallmatrix}1&0\\0&1\end{smallmatrix}\right)$, $\left(\begin{smallmatrix}0&p\\0&0\end{smallmatrix}\right)$, respectively. By doing so we are identifying the endomorphism ring of K with the ring of 2×2 matrices in the usual way.

Φ is obviously transitive on K over $\langle pb\rangle$, but for all $x \in K$ with $x \notin \langle pb\rangle$, we have that Φ is not transitive on K over $\langle x\rangle$.

Furthermore, easy calculations show that Φ' is fully transitive on K over $\langle pb\rangle$ and that Φ' is not fully transitive on K over $\langle b+\epsilon\rangle$ for all $\epsilon \in K[p]$.

Theorem 6.1 of [C2] provides p-groups G, G' with $p^\omega G = p^\omega G' = K$ such that $E(G)$ acts on K as Φ and $E(G')$ acts on K as Φ'. Therefore G is transitive over $\langle pb\rangle$ and not transitive over $\langle b+\epsilon_\omega\rangle$ for all $\epsilon_\omega \in (p^\omega G)[p]$. Moreover, G' is fully transitive over $\langle pb\rangle$ and not fully transitive over $\langle b+\epsilon_\omega\rangle$ for all $\epsilon_\omega \in (p^\omega G')[p]$. Hence, by Lemma 2.3 there exist nonpure subgroups $H \in \mathfrak{T}_G$ and $H' \in \mathfrak{F}_{G'}$.

Though we have seen that the subgroups in $\mathfrak{T}_G$ and $\mathfrak{F}_G$ are not necessarily pure, the following Lemma provides a necessary and sufficient condition for these subgroups to be pure.

Lemma 2.5 *Let G be a p-group and $H \in \mathfrak{T}_G$ $(H \in \mathfrak{F}_G)$. Then H is pure in G if and only if $H \cap p^{\omega+1}G \subseteq p^\omega H$.*

Proof. If H is pure in G, then we have, for all $n \in \omega$, that $H \cap p^{\omega+1}G \subseteq H \cap p^n G = p^n H$ and therefore $H \cap p^{\omega+1}G \subseteq p^\omega H$.

Conversely, let $n \in \omega$ be minimal such that $H \cap p^n G \neq p^n H$. Then there exist $h \in H \setminus p^n H$ and $g \in G$ with $p^n g = h$ but $p^{n-1}g \notin H$. Let $H' = \langle H, p^{n-1}g\rangle$, then H is a proper subgroup of H' since $p^{n-1}g \notin H$. We show that $H \cap p^\omega G = H' \cap p^\omega G$. Then it follows that G is (fully) transitive over H' by Proposition 1.6, which contradicts the maximality of H.

Let $z \in H' \cap p^\omega G$, then $z = \bar{h} + rp^{n-1}g$ with $\bar{h} \in H$ and $r \in \mathbb{Z}$. Since $z \in p^\omega G$, there is a $g' \in G$ with $p^{n-1}g' = z$. It follows that

$$\bar{h} = p^{n-1}(g' - rg) \in p^{n-1}G \cap H = p^{n-1}H.$$

Thus there exists an $\bar{h}' \in H$ with $\bar{h} = p^{n-1}\bar{h}'$ and

$$pz = p\bar{h} + rp^n g = p^n\bar{h}' + rh \in H \cap p^{\omega+1}G \subseteq p^\omega H.$$

Hence, there is an $\bar{\bar{h}} \in H$ with $p^n\bar{\bar{h}} = p^n\bar{h}' + rh$. Suppose r is relatively prime to p, then it follows that $h \in p^n H$, which contradicts the assumption that $h \in H \setminus p^n H$. Therefore p divides r and $z \in H \cap p^\omega G$. □

As an immediate consequence of Lemma 2.5 all the subgroups in $\mathfrak{T}_G$ and $\mathfrak{F}_G$ of a p-group G of length $\omega + 1$ are indeed pure.

The next Lemma provides an often easier way to calculate the p-rank of the factor group G/H if $H \in \mathfrak{F}_G$, for this rank can be determined in $p^\omega G$. This enables us, for any cardinal α, to construct groups G such that the p-rank of G/H is α for an $H \in \mathfrak{F}_G$.

Lemma 2.6 *Let G be a p-group and suppose $H \in \mathfrak{F}_G$. Then $r_p(G/H) = r_p(p^\omega G/p^\omega G \cap H)$.*

Proof. Let B be a subgroup of G containing H such that $0 \neq B/H \subseteq G/H$. The maximality of H shows that $H \cap p^\omega G$ is a proper subgroup of $B \cap p^\omega G$, since, by Proposition 1.6, G would otherwise be fully transitive over B. Therefore $B/H \cap (H + p^\omega G)/H \neq 0$, i.e. $(H + p^\omega G)/H$ is an essential subgroup of G/H. Using the second isomorphism theorem we finally get $r_p(G/H) = r_p((p^\omega G + H)/H) = r_p(p^\omega G/p^\omega G \cap H)$. □

Theorem 2.7 *For any cardinal κ there exists a p-group G and a subgroup $H \in \mathfrak{F}_G$ such that $r_p(G/H) = \kappa$.*

Proof. Let $K = \bigoplus_{i<\kappa+1} \langle g_i \rangle$ where $o(g_i) = p$ (if κ is infinite, we can replace $\kappa + 1$ by κ). We denote by e_{ij} the mapping $e_{ij} : K \longrightarrow K$ that maps g_i to g_j and g_k to zero for all $k \neq i$. Let Φ be the subring of the endomorphism ring of K that is generated by all e_{ii} and e_{i1} for $i < \kappa + 1$. For all $i > 1$ it follows that g_1 cannot be mapped to g_i by any element of Φ though $U_K(g_1) \leq U_K(g_i)$. This means Φ is not fully transitive on K over $\langle g_i \rangle$ for all $i > 1$. On the other hand we have, for all $i < \kappa$, $g_i e_{i1} = g_1$, i.e. Φ is fully transitive on K over $\langle g_1 \rangle$. Let $x = \sum_{i<\kappa} k_i g_i \in K$. Then Φ is in fact fully transitive on K over $\langle x \rangle$ if and only if $k_i = 0$ for all $i > 1$. For if Φ is fully transitive on K over $\langle x \rangle$, then there exists an $\phi \in \Phi$ with $g_1\phi = x$ and therefore we have $(g_1\phi)e_{jj} = xe_{jj} = k_j g_j = 0$ for all $j > 1$ by the construction of Φ.

Theorem 2.1 of [Go] now provides a p-group G with $p^\omega G = K$ and the endomorphism ring of G acts on $p^\omega G$ as Φ. Furthermore there is an $H \in \mathfrak{F}_G$ with $g_1 \in H$ and

$$p^\omega G \cap H = p^\omega H = \langle g_1 \rangle \sqsubset p^\omega G.$$

Lemma 2.6 now yields the desired rank of G/H:

$$r_p(G/H) = r_p(p^\omega G/p^\omega G \cap H) = \kappa.$$

□

3 Groups of length $\omega + 1$

In this section we examine the subgroups in $\mathfrak{T}_G$ and $\mathfrak{F}_G$ of a p-group G whose length is $\omega + 1$. We show that, in this case, the subgroups in $\mathfrak{T}_G$ are exactly the $p^\omega G$-high subgroups of G and that the subgroups in $\mathfrak{F}_G$ are N-high subgroups of G where N is a subgroup of $p^\omega G$.

Lemma 3.1 *Let G be a non-transitive p-group of length $\omega+1$. Then $H \in \mathfrak{T}_G$ if and only if H is a $p^\omega G$-high subgroup of G.*

Proof. Let $H \in \mathfrak{T}_G$. It suffices to show that H has no elements of infinite height, since the maximality of H then provides that H is $p^\omega G$-high. We assume that there exists an element $0 \neq g \in p^\omega G \cap H$. Since all elements in $p^\omega G$ have the same height sequence, there exists for all $z \in p^\omega G$ an automorphism ϕ of G with $z\phi = g$. Now suppose $x, y \in p^\omega G$, then there are two automorphisms ϕ and ψ of G with $x\phi = g$ and $y\psi = g$. Therefore $(x\phi)\psi^{-1} = y$, where $\phi\psi^{-1}$ is an automorphism of G. Hence the endomorphism ring of G acts transitively on $p^\omega G$, i.e. G is transitive, a contradiction.

If, on the other hand, H is a $p^\omega G$-high subgroup of G, then G is by Corollary 1.3 transitive over H. Therefore the maximality of H yields that $H \in \mathfrak{T}_G$. □

This enables us to state the following Theorem, which is a collection of results by Irwin [Ir, Theorem 5 and 6] on $p^\omega G$-high subgroups.

Theorem 3.2 *Let G be a p-group of length $\omega+1$ and suppose $H, K \in \mathfrak{T}_G$. Then*

(i) H is pure in G

(ii) G/H is a divisible hull of $(p^\omega G \oplus H)/H \cong p^\omega G$

(iii) $G/H \cong G/K$

(iv) $p^n G/p^n H \cong G/H$ for all $n < \omega$

(v) H and K have the same Ulm invariants

(vi) If H and K are totally projective, then $H \cong K$.

The next theorem summarizes similar results we can obtain for subgroups in $\mathfrak{F}_G$.

Theorem 3.3 *Let G be a p-group of length $\omega+1$ and suppose $H, K \in \mathfrak{F}_G$. Then*

(i) H is pure in G

(ii) $p^\omega H = \{x \in p^\omega G : G$ fully transitive over $\langle x\rangle\}$

(iii) $G/H \cong G/K$

(iv) $p^n G/p^n H \cong G/H$ for all $n < \omega$

(v) Let N be a complementary summand of $p^\omega H$ in $p^\omega G$. Then H is an N-high subgroup of G.

(vi) Any $p^\omega H$-high subgroup of H is a $p^\omega G$-high subgroup of G.

(vii) H and K have the same Ulm invariants

(viii) If H and K are totally projective, then $H \cong K$.

Proof. (i) follows from Lemma 2.5.

(ii) It suffices to show that the set $V = \{x \in p^\omega G : G \text{ fully transitive over } \langle x \rangle\}$ is a subgroup of $p^\omega G$. Suppose $x, y \in V$. Then $U_G(x) = U_G(y) = U_G(x-y)$, since all elements in $p^\omega G$ have the same height sequences. Therefore there exist for all $g \in p^\omega G$ two endomorphisms ϕ and ψ of G with $g\phi = x$ and $g\psi = y$. Now $g(\phi - \psi) = x - y$, i.e. $x - y \in V$.

(iii) From (ii) we have $p^\omega G/p^\omega H = p^\omega G/p^\omega K$. By Lemma 2.6, we get $r_p(G/H) = r_p(p^\omega G/p^\omega H) = r_p(p^\omega G/p^\omega K) = r_p(G/K)$. Since G/H and G/K are divisible by Theorem 2.2, the claim follows.

(iv) The purity and density of H in G provide

$$G/H = (H + p^n G)/H \cong p^n G/(p^n G \cap H) = p^n G/p^n H.$$

(v) Our assumption yields $p^\omega G = p^\omega H \oplus N$ and therefore $H \cap N = 0$. To show that H is maximal with respect to this property let H' be a subgroup of G with $H' \supseteq H$ and $H' \cap N = 0$. Then it follows that $p^\omega G \cap H' = p^\omega G \cap H$ and therefore by Proposition 1.6, G is fully transitive over H'. Hence the maximality of H yields $H' = H$.

(vi) By (v), H is an N-high subgroup of G where $p^\omega G = p^\omega H \oplus N$. Then it follows by [Ir, Lemma 1(f)] that $G[p] = H[p] \oplus N$. Let U be a $p^\omega H$-high subgroup of H. Thus U is pure in H and [Ir, Lemma 1(f)] provides $H[p] = U[p] \oplus p^\omega H$. Since H is pure in G, U is pure in G and

$$G[p] = H[p] \oplus N = U[p] \oplus p^\omega H \oplus N = U[p] \oplus p^\omega G,$$

i.e. U is a $p^\omega G$-high subgroup of G by [IW, Lemma 1].

(vii) By (vi), H is an N-high subgroup of G, where N is a subgroup of $p^\omega G$. The claim now follows by [IW, Lemma 7(d)].

(viii) follows by (vii) and Ulm's theorem for totally projective groups (see [H1], [Wa]). □

Theorem 3.4 *Let G be a p-group of cardinality $2^{\aleph_0}$ and length $\omega+1$ with a countable basic subgroup. Suppose G is not transitive (not fully transitive) and $G[p] = p^\omega G \oplus S$ where the cardinality of S is $2^{\aleph_0}$. Then there exists a family of $2^{2^{\aleph_0}}$ pairwise nonisomorphic subgroups of G in $\mathfrak{T}_G$ ($\mathfrak{F}_G$).*

Proof. Since S only contains elements of finite height, there exists a $p^\omega G$-high subgroup H of G with socle S. H is pure and dense in G and therefore the basic subgroups of H are countable. Since S is not countable, it follows that $pH \cap S$ is of the power $2^{\aleph_0}$. This provides a set $L = \{h_i : i \in I\}$ of independent elements of order p^2 in H such that $|I| = 2^{\aleph_0}$. Since G is not (fully) transitive, there exists an element $g \in p^\omega G$ such that G is not (fully) transitive over $\langle g \rangle$. For each subset M of I we define the following set:

$$L_M = \{h_i + g : i \in M\} \cup \{h_i : i \in I \setminus M\}.$$

The socle of $\langle L_M \rangle$ is again contained in S and since S has only elements of finite height, we get $\langle L_M \rangle \cap p^\omega G = 0$. By Corollary 1.3 it follows that G is (fully) transitive over $\langle L_M \rangle$ and Zorn's Lemma provides a subgroup $H_M \in \mathfrak{T}_G$ ($H_M \in \mathfrak{F}_G$) with $\langle L_M \rangle \subseteq H_M$. If M and N are different subsets of I, then the corresponding subgroups H_M and H_N are distinct, since for no i may both h_i and $h_i + g$ belong to H_M or H_N. Therefore, G contains $2^{2^{\aleph_0}}$ different subgroups $H_M \in \mathfrak{T}_G$ ($H_m \in \mathfrak{F}_G$). The purity and density of the H_M's show that their basic subgroups B_M are also countable. Thus $\mathrm{Hom}(B_M, G)$ is of cardinality $(2^{\aleph_0})^{\aleph_0} = 2^{\aleph_0}$. From the short exact sequence

$$0 = \mathrm{Hom}(H_M/B_M, G) \longrightarrow \mathrm{Hom}(H_M, G) \longrightarrow \mathrm{Hom}(B_M, G),$$

there results an upper bound of $2^{\aleph_0}$ for $\mathrm{Hom}(H_M, G)$. This shows that the cardinality of an isomorphism class is $\leq 2^{\aleph_0}$, and hence the set of all nonisomorphic subgroups in $\mathfrak{T}_G$ ($\mathfrak{F}_G$) has power $2^{2^{\aleph_0}}$. □

Megibben's construction in [M1] of groups that are neither transitive nor fully transitive provides examples for this Theorem.

References

[CG] D. Carroll, B. Goldsmith, *On transitive and fully transitive abelian p-groups*, Proc. Royal Irish Academy **96A**(1) (1996), 33–41.

[C1] A.L.S. Corner, *The independence of Kaplansky's notions of transitivity and full transitivity*, Quart. J. Math. Oxford (2) **27** (1976), 15–20.

[C2] A.L.S. Corner, *On endomorphism rings of primary abelian groups* II, Quart. J. Math. Oxford (2) **27** (1976), 5–13.

[Fi] S. Files, *Transitivity and full transitivity for nontorsion modules*, J. Algebra **197** (1997), 468–478.

[FG] S. Files, B. Goldsmith, *Transitive and fully transitive groups*, Proc. Amer. Math. Soc. **126** (1998), 1605–1610.

[Fu] L. Fuchs, *Infinite Abelian Groups* – Vol. **I** and **II**, Academic Press, New York - London (1970, 1973).

[Go] B. Goldsmith, *On endomorphism rings of non-separable abelian p-groups*, J. Algebra **127** (1989), 73–79.

[Gr] P. Griffith, *Transitive and fully transitive primary abelian groups*, Pacific J. Math. **25** (1968), 249–254.

[H1] P. Hill, *On the classification of abelian groups*, (unpublished).

[H2] P. Hill, *On transitive and fully transitive primary groups*, Proc. Amer. Math. Soc. **22** (1969), 414–417.

[Ir] J. M. Irwin, *High subgroups of abelian groups*, Pacific J. Math. **11** (1961), 1375–1384.

[IW] J. M. Irwin, E. A. Walker, *On N-high subgroups of abelian torsion groups*, Pacific J. Math. **11** (1961), 1363–1374.

[Ka] I. Kaplansky, *Infinite Abelian Groups*, University of Michigan Press, Ann Arbor (1954, 1969).

[M1] C. Megibben, *Large subgroups and small homomorphisms*, Michigan Math. J. **13** (1966), 153–160.

[M2] C. Megibben, *A generalization of the classical theory of primary abelian groups*, Tôhoku Math. J. **22** (1970), 347–356.

[Wa] E.A. Walker, *Ulm's theorem for totally projective groups*, Proc. Amer. Math. Soc. **37** (1973), 387–392.

Trends in Mathematics, © 1999 Birkhäuser Verlag Basel/Switzerland

Subgroups of p^5-bounded groups

Fred Richman and Elbert A. Walker

Abstract

Each valuated module B with $B(5) = 0$ is a direct sum of simply presented valuated modules and copies of two valuated modules which come from (finite) hung trees. There are infinite-rank indecomposable valuated modules B with $B(6) = 0$.

1 Valuated modules

By a **module** we will mean a module over a fixed discrete valuation domain with prime p. The reader is assumed to be familiar with the notion of a **valuated module**, or **v-module**, which is a module B together with a filtration $B = B(0) \supset B(1) \supset B(2) \supset \cdots$ such that $p(B(n)) \subset B(n+1)$. We will not consider arbitrary ordinal values of n because we are interested in the case $B(5) = 0$. If $x \in B(n)$ and $x \notin B(n+1)$ we write $vx = n$ and say the **value** of x is n.

If B is a subgroup of a p^5-bounded group G, then B is naturally a module over the ring of integers localized at p, and the module B is filtered by setting $B(n) = B \cap p^n G$. Classifying such subgroups, up to isomorphism of G, is equivalent to classifying the associated v-modules, because bounded modules (with the filtration $B(n) = p^n B$) are injective in the category of v-modules [6, Theorem 9].

We classify v-modules by writing them as direct sums. These direct sums must **respect values**, that is, they must respect the filtration: $(A \oplus B)(n) = A(n) \oplus B(n)$. Recall that the n-th **Ulm invariant** of a v-module A is the dimension of $\{a \in A(n) : pa \in A(n+2)\}/A(n+1)$. If the n-th Ulm invariant of A is nonzero, then we say that n is an Ulm invariant of A.

Lemma 1.1 (respect value) *If A and H are submodules of a torsion v-module B, and $A \cap H = 0$, then $A \oplus H$ respects values provided*

$$(A \oplus H)(n+1) = A(n+1) \oplus H(n+1)$$

whenever n is an Ulm invariant of A.

Proof. Contrapositively, we will show that if the equation fails for some n, then it fails for some k that is an Ulm invariant of A. Suppose $a+h \in B(n+1)$ and $a \notin B(n+1)$. As $pa+ph \in B(n+2)$, we may assume that $pa \in B(n+2)$ by induction on the order of $a+h$. As $va = k \leq n$ and $vpa \geq n+2$, the number k is an Ulm invariant of A for which the equation fails. □

Note that this lemma and its proof are valid for n any ordinal. We will also be worried about respecting the height filtration. The next lemma is a modification of [4, Theorem 1.3].

Lemma 1.2 (respect height) *Let A be a reduced torsion submodule of a module B, and H a submodule of B. If $\mathrm{ht}_B(a+h) \leq \mathrm{ht}_A a$ whenever $a \in A[p]$ and $h \in H$, then*

$$(A \oplus H) \cap p^n B = p^n A \oplus (H \cap p^n B).$$

If, in addition, A is bounded, then $B = A \oplus K$ for some $K \supset H$.

Proof. Clearly the hypothesis implies that $A \cap H = 0$. Suppose $a+h \in (A \oplus H) \cap p^n B$. We want to show that $a \in p^n A$. By induction on the order of a, we have $pa \in p^{n+1}A$. Let $pa = p^{n+1}a'$, so $a - p^n a' \in A[p]$. Then

$$a - p^n a' + h \in (A[p] \oplus H) \cap p^n B$$

so $a - p^n a' \in p^n A$, by hypothesis. Thus $a \in p^n A$.

If A is bounded, then we can extend H to a complementary summand K of A because algebraically compact modules, such as A, are injective in the category of v-modules [6, Theorem 9]. □

We combine these two lemmas in one, which will be used frequently.

Lemma 1.3 *Let A be a bounded submodule of a v-module B, and H a submodule of B. Suppose*

- *$B(m+1) = A(m+1) \oplus H(m+1)$ whenever m is an Ulm invariant of the v-module A, and*
- *if $a \in A[p]$ and $h \in H$, then $\mathrm{ht}_B(a+h) \leq \mathrm{ht}_A a$.*

Then $B = A \oplus K$ for some $K \supset H$.

Proof. Lemma 1.2 says that we can find $K \supset H$ such that $A \cap K = 0$ and $A + K = B$. Lemma 1.1 says that $A \oplus K$ respects values. □

2 Finite valuated trees

By a **tree** we will mean a valuated tree with values in ω. To fix notation and terminology, consider the tree

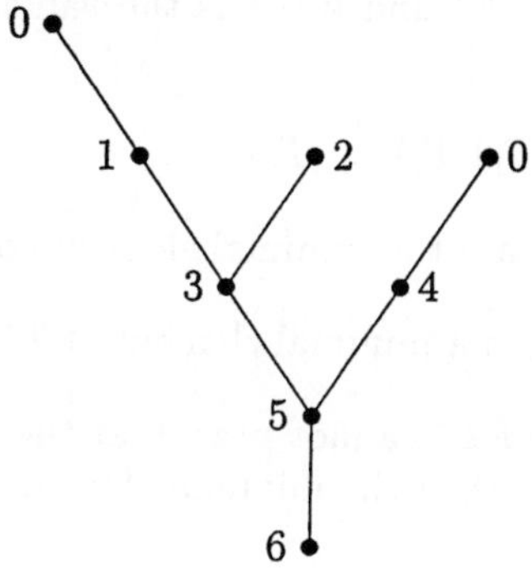

We denote this tree by 65(3(10)(2))(40). It is obtained by adjoining a node of value 6 as parent of the root of the tree 5(3(10)(2))(40). This tree, in turn is obtained by adjoining a node of value 5 as the parent of the roots of the trees 3(10)(2) and 40. A **forest** is a family of trees, like the two element family (3(10)(2))(40). Every tree is of the form nF where F is a forest, possibly empty. A **pole** is a tree with no branching.

If x is a node in a tree, then px denotes the parent of x, if any. If $p^n x$ is the root of the tree, then we say that the **level** of x is n, and write $\ell(x) = n$. A **map** of trees is a function f such that

- $f(px) = pf(x)$ whenever $\ell(x) > 0$,
- $vf(x) \geq vx$.

A map f of trees is **order preserving** if $\ell(f(x)) = \ell(x)$ for all x. The set of trees is pre-ordered by setting $T_1 \leq T_2$ if there is a map (which we can take to be order preserving) from T_1 to T_2. A finite tree is **irretractible** if the only idempotent map to itself is the identity. (Every infinite tree has a nontrivial retraction, [3, Theorem 11].) The irretractible trees form a partially ordered subset of the trees. In fact, the irretractible trees form a distributive lattice [5, Theorem 1.3]. The tree 0 is at the bottom of this lattice.

Each tree T gives rise to a simply presented v-module $\langle T \rangle$.

Let F be a forest. The forest F^* is defined to be the set of trees that cannot be mapped into (any tree of) F. The set F^* is an up-set in the set of all trees which is generated by a finite number of minimal elements.

We can compute F^* inductively. If F is empty, then F^* is all trees—the up-set generated by the tree 0. If

$$T = nF$$

is a tree, then T^* is the up-set generated by the tree $n+1$ together with all trees of the form kT' where $T' \in F^*$ and $k-1$ is the value of the root of T'. If

$$F = (T_1)(T_2)\cdots(T_k)$$

then the minimal elements of F^* are the minimal elements of the finite set

$$\{t_1 \vee t_2 \vee \cdots \vee t_k : t_i \text{ is a minimal element of } T_i^*\}.$$

Note that the minimal elements of T^* cannot branch at the root.

If T is the pole $n_1, n_2, \ldots, n_k$, then the minimal elements of T^* are the poles

$$n_j + j,\ n_j + j - 1, \ldots,\ n_j + 2,\ n_j + 1$$

if $j = 1$ or $n_{j-1} > n_j + 1$, and

$$k, k-1, \ldots, 1, 0$$

if $n_k > 0$. These are gapless poles. There is one for each Ulm invariant of the pole T, and one more if there is no node of value 0 (does this imply an Ulm invariant at -1?).

Some examples:

- If $T = (14, 9, 8, 7, 4, 1, 0)$, then T^* is generated by the poles (15), $(11, 10)$, $(9, 8, 7, 6, 5)$, and $(7, 6, 5, 4, 3, 2)$.
- If $T = 3(2)(10)$, then T^* is generated by 4 and 320.
- If $T = 4(310)(32)$, then T^* is generated by 5 and $43(10)(2)\}$.
- If $T = 6(430)(521)$, then T^* is generated by $\{7, 54(21)(3), 653, 43210\}$.

A v-module B with $\bigcap B(n) = 0$ is an honorary tree. The root of B is 0, has value ∞, and is its own parent. Technically, it's not a tree, but it's clear what a tree-map $T \to B$ is. If T is a tree, then $B(T)$ is defined to be the set of images of the root of T under tree-maps $T \to B$. If T is the gapless pole $(\alpha + n, \ldots, \alpha)$, then $B(T) = p^n(B(\alpha))$. If F is a forest, then $B(F)$ is the submodule of B generated by $\{B(T) : T \in F\}$.

The **v-height** of an element x of finite value in a v-module B is given by the equivalence class of the branch $\{b \in B : p^n b = x \text{ for some } n\}$ above x. There is a unique irretractible tree in this equivalence class, the smallest T such that $x \in B(T)$.

Call an irretractible tree **hangable** if it has distinct nodes t_0 and t_1 such that $vt_0 = vt_1$ and either $pt_0 = pt_1$ or $\min(vpt_0, vpt_1) > vt_0 + 1$. Poles are not hangable, nor is any tree in which distinct nodes have distinct values. The tree 4(310)(32) is the only hangable tree with all values less than 5.

Theorem 2.1 *Let B be a reduced v-module and T an irretractible subtree of B, with exactly two leaves, that is the v-height of its root. Then*

1. *T is a p-basis for the (unvaluated) submodule A that it generates.*
2. *If T is unhangable, then T is a p-basis for the v-module A.*

Proof. For 1, suppose contrapositively, that $\sum_i u_i t_i = 0$ where the u_i are units and the t_i are distinct nodes of T. We will construct a retraction of T. Consider the nodes t_i of minimum value. There must be at least two of them, lest the sum have value different from ∞, and, because T has only two leaves, there must be exactly two of them, t_0 and t_1. Clearly every other node that appears in the sum is a multiple of either pt_0 or pt_1. So we can write $t_0 = ut_1$ for some unit u. Therefore t_0 and t_1 have the same v-height, and $vp^i t_0 = vp^i t_1$ for each i. So there is a retraction of T that takes t_1 to t_0.

For 2, suppose contrapositively that $v(u_0t_0 + u_1t_1) > vt_0 = vt_1$. We may assume that this is the maximum value of $vt_0 = vt_1$ where this occurs. Either $pt_0 = pt_1$ or

$$vt_0 + 1 < v(u_0pt_0 + u_1pt_1) \leq \min(vpt_0, vpt_1).$$

In either case, T is hangable. □

3 Szele trees

For C a cyclic valuated p-group, the functor F_C was defined in [4, page 23] as

$$F_C(B) = \sum_{p^n(C(\alpha))=0} p^n(B(\alpha)).$$

If C comes from the pole T, then $p^n(C(\alpha)) = 0$ if and only if the gapless pole $(\alpha + n, \ldots, \alpha)$ is in T^*, so $F_C(B) = B(T^*)$. The following is a paraphrase of [4, Theorem 2.5].

Theorem 3.1 *Let B be a v-module, T a pole, and A a submodule of B which is a direct sum of copies of $\langle T \rangle$. Then*

1. *A is a summand of B if and only if* $A \cap B(T^*) = 0$.

2. *If* $A + H = B$ *and* $A \cap H = 0$ *and* $H \supset B(T^*)$, *then* $B = A \oplus H$.

For T the gapless pole $n \cdots 10$, and B an unvaluated module, Part 1 is the theorem of Szele [1, Prop. 27.1] stating that if A is a direct sum of cyclic groups of order p^n, and $A \cap p^n B = 0$, then A is a summand of B. Much of our work here consists of extending Part 1 to other trees.

A tree T that satisfies Part 1 of Theorem 3.1 will be called a **Szele tree**. So poles are Szele trees. The tree 3(10)(2) was shown to be a Szele tree in [4, Lemma 4.1]. Note that if A is a summand of B, then $A \cap B(T^*) = 0$ because $B(T^*)$ is an additive functor in B.

Part 2 of Theorem 3.1 does not carry over to Szele trees. Let $T = 3(10)(2)$ and $B = \langle T \rangle \oplus \langle 32 \rangle$. Then $B(T^*) = 0$. Let x be the node of T of value 1, and y the node of 32 of value 2. The submodule H generated by $x - y$ satisfies the hypothesis but not the conclusion of Part 2.

We can rephrase, and slightly strengthen, Theorem 3.1.

Corollary 3.2 *Let B be a v-module, T a pole, and A a submodule of B which is a direct sum of copies of* $\langle T \rangle$. *If K is a submodule of B such that* $K \cap A = 0$ *and* $B(T^*) \subset K$, *then* $B = A \oplus H$ *for some* $H \supset K$.

Proof. We want a submodule H containing K such that $A + H = B$ and $A \cap H = 0$. Then Theorem 3.1 Part 2 finishes the job. As A is bounded, it suffices to show that $\mathrm{ht}_B(a + k) \leq \mathrm{ht}_A a$ for each $a \in A[p]$ and $k \in K$ (Lemma 1.2). But if $\mathrm{ht}_B(b) > \mathrm{ht}_A a$ for $a \in A[p]$, then $b \in B(T^*) \subset K$. □

If T is a tree, then the T-th **valuated Ulm invariant** of a valuated module G is defined in [5] to be

$$U_T G = \frac{G(T)[p]}{G(T^*)[p] \cap G(T)[p]}.$$

Theorem 3.3 *Let B be a v-module and T an unhangable Szele tree. Then* $B = A \oplus K$ *where A is a direct sum of copies of* $\langle T \rangle$, *and* $U_T K = 0$.

Proof. Consider families of submodules S_i of B with the properties that S_i is isomorphic to $\langle T \rangle$ for each i, the sum $\sum S_i$ is direct, and $B(T^*) \cap \sum S_i = 0$. Zorn's lemma applies, so there is a maximal such family S_i. Let $A = \sum S_i$. As T is a Szele tree, we can write $B = A \oplus K$. It remains to show that $U_T K = 0$.

Suppose $U_T K \neq 0$. Let c be an element of $K(T)[p]$ that is not in $K(T^*) = B(T^*)$. As T is unhangable, Theorem 2.1 says that c is contained in a submodule of K isomorphic to $\langle T \rangle$. As T is a Szele tree, this submodule is a summand of K, contradicting the maximality of the family S_i. □

No doubt any Szele tree is unhangable, but we don't need that.

For the purpose of showing that T is a Szele tree, we may assume that $B(n+1) = 0$, where n is the value of the root of T. Indeed, $B = A \oplus K$ follows easily from

$$\frac{B}{B(n+1)} = A \oplus \frac{K}{B(n+1)}.$$

The smallest tree, 3(10)(2), has two Ulm invariants, 1 and 3. The next theorem is effectively a generalization of [4, Lemma 4.1] from the smallest tree to any tree with two Ulm invariants.

Theorem 3.4 *If T is an irretractible tree with exactly two Ulm invariants, then T is a Szele tree.*

Proof. Let B be a v-module and A a submodule which is a direct sum of copies of $\langle T \rangle$. Suppose $A \cap B(T^*) = 0$. We must show that A is a summand of B.

Let the Ulm invariants of T be $k < n$, so

$$T = n \ldots m(m-1 \ldots i)(k \ldots j)$$

where $k < m-1$ and $k - j > m - 1 - i$, the dots indicate no gaps. Note that T^* consists of the poles $n+1$ and $k+n-m+2 \ldots k+1, m-1-i \ldots 0$, and, if $j > 0$, the pole $n - m + k - j + 2 \ldots 0$.

Let $T_0 = T(k+1)$. Then T_0 is the gapless pole $n \ldots s$ where $s = \max(k+1, i)$, so T_0^* consists of the poles $n+1$ and $n - s + 1 \ldots 0$. For any v-module K,

$$K(T_0^*) = K(n+1) + p^{n-s+1}K.$$

Thus

$$B(k+1)(T_0^*) = B(n+1) + p^{n-s+1}B(k+1) \subset B(T^*),$$

because $p^{n-s+1}B(k+1) \subset B(n-s+k+2) \subset B(n-m+k+2)$, so

$$B(k+1)(T_0^*) \subset B(T^*)(k+1).$$

As $A \cap B(T^*) = 0$, we can write

$$B(k+1) = A(k+1) \oplus H_{k+1}$$

where $H_{k+1} \supset B(T^*)(k+1)$, by Corollary 3.2.

To write $B = A \oplus H$ with $H \supset H_{k+1}$, it suffices, by Lemma 1.3, to show that if a in $A[p]$ and $h_{k+1} \in H_{k+1}$, then $\mathrm{ht}_B(a + h_{k+1}) \leq \mathrm{ht}_A\, a$. Suppose $a + h_{k+1} \in p^t B$. If $va = n$ and $t > \mathrm{ht}_B\, a$, then $t > \mathrm{ht}_A\, a$ so $p^t B \subset B(T^*)$. Thus $a + h_{k+1} \in p^t B \cap B(k+1) \subset H_{k+1}$, and $a \in H_{k+1}$, a contradiction.

So suppose $va = k$ and $a + h_{k+1} \in p^{m-i}B$. Then $a = a_{m-1} - a_k$ where $va_{m-1} = m - 1$ and $a_k \in p^{m-i}A$ has value k. Note that $\mathrm{ht}_A\, a \geq m - i - 1$. Then $a_{m-1} + h_{k+1} \in p^{m-i}B \cap B(k+1)$ so

$$p^{n-m+1}(a_{m-1} + h_{k+1}) \in B(T^*)(k+1) \subset H_{k+1},$$

whence $p^{n-m+1}a_{m-1} \in H_{k+1}$, a contradiction. □

4 Unhangable trees in $\mathcal{T}_4$ are Szele trees

Let $\mathcal{T}_4$ denote the set of trees with values less than 5 The only hangable tree in $\mathcal{T}_4$ is 4(32)(310). It is not a Szele tree, as we shall see (5.3 Example). In this section we will prove that the other trees in $\mathcal{T}_4$ are Szele trees. Before starting, we note that the restriction to $\mathcal{T}_4$ is essential. Consider the tree $T = 5(41)(32)$. Let $A = \langle T \rangle$ and let x and y be the leaves of T. Adjoin z to A with $pz = x - y$ to get B. Then A is not pure in B, but $B(T^*) = 0$.

Aside from the poles, which we know are Szele trees, the eleven unhangable trees in $\mathcal{T}_4$ are

43(2)(10)	4(210)(32)	4(210)(31)	4(210)(30)	4(210)(3)
4(30)(21)	4(21)(3)	4(20)(3)	4(10)(3)	4(10)(2)

and 3(10)(2). A computer count of $\mathcal{T}_4$ came up with 43 trees [2]. There are $2^5 - 1$ poles in $\mathcal{T}_4$, so it looks like we've listed all the rest here.

Theorem 3.4 takes care of all but these five

4(210)(31)	4(210)(30)	4(30)(21)
4(20)(3)	4(10)(2)	

4.0.1 The tree 4(30)(21)

The star is generated by 5, 431 and 3210. We may assume $B(5) = 0$, so $B(T^*) \subset (p^2 B)(3)$. Note that $A(31) = A(4)$ is a direct sum of copies of $\langle 4 \rangle$, and $(4)^*$ is generated by (5) and (10). So $B(31)((4)^*) \subset B(T^*) \subset B(31)$, whence Corollary 3.2 says that we can write

$$B(31) = A(31) \oplus H_{31}$$

with $H_{31} \supset B(T^*)$.

As $A(3)$ is a direct sum of copies of $\langle 43\rangle$, and $p^2(B(3)) = 0$, and $A(3) \cap H_{31} = 0$, Corollary 3.2 says that we can write

$$B(3) = A(3) \oplus H_3$$

with $H_3 \supset H_{31}$.

We want to write

$$B(1) = A(1) \oplus H_1$$

with $H_1 \supset H_3$. It suffices, by Lemma 1.3, to show that if $a \in A(1)[p]$, and $h_3 \in H_3$, then $\text{ht}_{B(1)}(a + h_3) \leq \text{ht}_{A(1)}\, a$. Write $a = a_{30} + a_{21}$ where $a_{30} \in A(30)$ and $a_{21} \in A(21)$. If $a_{30} + a_{21} + h_3 \in p(B(1))$, then $a_{30} + h_3 \in p(B(1))$, so

$$a_{30} + h_3 \in B(31) = A(31) \oplus H_{31} \subset A(31) \oplus H_3.$$

Therefore $a_{30} \in A(31)$, so $a \in A(4) \subset p^2(A(1))$.

4.0.2 The tree $4(210)(31)$

The star is generated by 5, 432, and 4310, so $B(T^*) \subset B(4)$. First write

$$(p^2B)(4) = A(4) \oplus K_4$$

where $K_4 \supset B(T^*)$. This is just a vector space argument. Note that

$$(p^2B)(3)(10) = (p^2B)(43) = B(4310) \subset B(T^*).$$

As $A(4)$ is a direct sum of copies of $\langle 4\rangle$, and $A(4) \cap K_4 = 0$, Corollary 3.2 says that we can write

$$(p^2B)(3) = A(4) \oplus K_3$$

where $K_3 \supset K_4$. As $A(3)$ is a direct sum of copies of $\langle 43\rangle$, and $B(3)(210) = B(543) \subset B(T^*) \subset K_3$, Corollary 3.2 says that we can write

$$B(3) = A(3) \oplus H_3$$

with $H_3 \supset K_3$.

Now we want to show that if $a \in A(2)[p]$, and $h_3 \in H_3$, then $\text{ht}_{B(2)}(a + h_3) \leq \text{ht}_{A(2)}\, a$. In particular, $A(2) \cap H_3 = 0$. Every element of $A(2)[p]$ can be written as $a_3 + a_2$ with $a_3 \in A(3)$ and $a_2 \in p^2A$. Suppose

$$a_3 + a_2 + h_3 \in p(B(2)).$$

Then $a_2 \in B(3)$, so $a_2 \in (p^2B)(3) = A(4) \oplus K_3$, so $a_2 \in A(4)$ because $K_3 \cap A = 0$. So $a_3 + a_2 \in A(4) \subset p(A(2))$. If

$$a_3 + a_2 + h_3 \in p^2(B(2)) \subset B(T^*) \subset H_3$$

then, as before, $a_3 + a_2 \in A(4)$. But $a_3 + a_2 \in H_3$, so $a_3 + a_2 = 0$.

So Lemma 1.3 says that we can write

$$B(2) = A(2) \oplus H_2$$

where $H_2 \supset H_3$. We want to show that this respects heights in B. If $a_3 + a_2 \in p^2 B$, then $a_3 \in p^2 B$, so $a_3 \in A(4)$, so $a_3 + a_2 \in A(4) \subset p^2 A$.

So Lemma 1.3 says that we can write

$$B = A \oplus H$$

with $H \supset H_2$.

4.0.3 The tree 4(210)(30)

The star is generated by 5 and 431, so $B(T^*) \subset B(4)$. As $B(31)(40) = B(431) \subset B(T^*)$, and $A(4)$ is a direct sum of copies of $\langle 4 \rangle$, Corollary 3.2 says that we can write

$$B(31) = A(4) \oplus K_3$$

with $K_3 \supset B(431)$.

We want to write

$$B(3) = A(3) \oplus H_3$$

with $H_3 \supset K_3$. It suffices, by Lemma 1.3, to show that if $a \in A(3)[p]$, and $k_3 \in K_3$, then $\mathrm{ht}_{B(3)}(a + k_3) \leq \mathrm{ht}_{A(3)}\, a$. But if $a \neq 0$, then $\mathrm{ht}_{A(3)}\, a = 1$, while $p^2(B(3)) \subset K_3$, so $\mathrm{ht}_{B(3)}(a + k_3) \leq 1$.

Now we want to write

$$B(21) = A(21) \oplus H_{21}$$

where $H_{21} \supset H_3(31)$. We can do this as before because each nonzero element of $A(21)[p]$ has height 1 in $A(21)$, and $p^2(B(21)) \subset B(431) \subset H_3(31)$.

Now we want to write

$$B(1) = A(1) \oplus H_1$$

with $H_1 \supset H_3 + H_{21}$. We can do this, by Lemma 1.3, if we can show that, for $a \in A(1)[p]$ and $h_3 \in H_3$ and $h_{21} \in H_{21}$, that $\mathrm{ht}_{B(1)}(a + h_3 + h_{21}) \leq \mathrm{ht}_{A(1)}\, a$.

We can write $a = a_3 - a_2$ where $a_3 \in A(3)$ and $a_2 \in p^2 A$. Suppose $a_3 - a_2 + h_3 + h_{21} \in p(B(1))$, as otherwise we're done. Then $a_3 + h_3 \in p(B(1))$, so $a_3 + h_3 \in B(31) = A(4) \oplus K_3$, which says that $a_3 \in A(4) + H_3$, hence $a_3 \in A(4)$ and so $a \in A(4)$. Then $\mathrm{ht}_{A(1)}\, a \geq 2$ while $p^3(B(1)) \subset B(431) \subset H_3$, so if $\mathrm{ht}_{B(1)}(a + h_3 + h_{21}) > 2$, then $a \in H_3 + H_{21}$, so $a \in H_3$, so $a = 0$.

Finally, to apply Lemma 1.3 one more time, we want to show that, for $a \in A[p]$ and $h_1 \in H_1$, that $\mathrm{ht}_B(a + h_1) \leq \mathrm{ht}_A a$. Suppose, in the notation of the preceding paragraph, that $a_3 - a_2 + h_1 \in p^2B$, so $a_3 + h_1 \in p^2B \subset B(21) = A(21) \oplus H_{21}$. So $a_3 + h_1 + h_{21} \in A(21)$. But that implies $h_1 + h_{21} \in A$, hence $h_1 + h_{21} = 0$, so $a_3 \in A(21)$, so $a \in A(4)$. Thus $\mathrm{ht}_A a \geq 3$ while $p^4B \subset B(5) \subset H_1$, so if $\mathrm{ht}_B(a + h_1) > 3$, then $a \in H_1$, so $a = 0$.

4.0.4 The trees 4(20)(3) and 4(10)(2)

We can handle all the four-element trees at once.

Theorem 4.1 *The tree $T = n(mj)(i)$ with $j < m < i < n$ is a Szele tree.*

Proof. Suppose $A \cap B(T^*) = 0$. We may assume $B(n + 1) = 0$. Note that T^* is generated by

$$(n + 1),\ (i + 2, i + 1),\ (m + 2, m + 1, 0),\ (j + 3, j + 2, j + 1),\ 3210$$

where some of these poles may be redundant. Also, $(ni)^*$ is generated by

$$(n + 1),\ (i + 2, i + 1),\ 210$$

so

$$B(T^*)(m + 1) \supset B(m + 1)((ni)^*).$$

First we want to write

$$B(m + 1) = A(m + 1) \oplus H_{m+1}$$

where $H_{m+1} \supset B(T^*)(m+1)$. As $A(m+1)$ is a direct sum of copies of $\langle ni \rangle$, Corollary 3.2 says that we can do that.

Next we want to write

$$pB = pA \oplus K$$

with $K \supset (H_{m+1} + B(T^*)) \cap pB = (H_{m+1} \cap pB) + B(T^*)$. Now pA is a direct sum of copies of $\langle nm \rangle$, and $(nm)^*$ is generated by $(n+1)$ and $(m+2, m+1)$ and (210). So $(pB)((nm)^*) \subset B(T^*)$. Moreover $pA \cap (H_{m+1} + B(T^*)) = 0$ because if $a = h_{m+1} + b^* \in (pA)[p]$, then $b^* \in B(T^*)(m + 1) \subset H_{m+1}$, so $a \in H_{m+1}$ whence $a = 0$. So Corollary 3.2 says that we can write $pB = pA \oplus K$.

Next we want to write

$$B(j + 1) = A(j + 1) \oplus H_{j+1}$$

with $H_{j+1} \supset H_{m+1}+K(j+1)+B(T^*)(j+1)$. The Ulm invariants of $A(j+1)$ are all at least m so, by Lemma 1.1 it suffices that $H_{j+1} \cap A(j+1) = 0$ and $H_{j+1} + A(j+1) = B(j+1)$. By Lemma 1.2 we can do this if,

$$\mathrm{ht}_{B(j+1)}(a + h_{m+1} + k_{j+1} + b^*) \leq \mathrm{ht}_{A(j+1)}\, a$$

for each $a \in A(j+1)[p]$. We can write $a = a_i - a_m$ where $a_i \in A(i)$ and $a_m \in p(A(j))$. Suppose

$$a_i - a_m + h_{m+1} + k_{j+1} + b^* \in p(B(j+1))$$

then

$$a_i + h_{m+1} \in pB$$

so

$$pa_i + ph_{m+1} \in B((m+2, m+1, 0),$$

so $pa_i \in H_{m+1}$, whence $pa_i = 0$, so $a \in A(n) \subset p(A(j+1))$. Suppose, in addition, that

$$a + h_{m+1} + k_{j+1} + b^* \in p^2(B(j+1)) \subset B(T^*) \subset K.$$

Then $a + h_{m+1} \in K(m+1)$. As $K \subset pB$, we have $h_{m+1} \in H_{m+1} \cap pB \subset K$, so $a \in K$, whence $a = 0$.

Finally, $p^tB = p^tA \oplus p^{t-1}K$ for $t > 0$. We will show that

$$(A \oplus H_{j+1}) \cap p^tB \subset p^tA \oplus H_{j+1}.$$

If $a+h_{j+1} = p^ta'+k$, then $pa+ph_{j+1} = p^{t+1}a'+pk$, so $pk \in K(j+1) \subset H_{j+1}$ whence $pa = p^{t+1}a'$. This can only happen if $a \in A(j+1)$. Therefore $k \in K(j+1)$, so $a = p^ta'$. □

5 The structure of B when $B(5) = 0$

Suppose B is a v-module with $B(5) = 0$. The plan is to remove summands that are direct sums of copies of $\langle T \rangle$, for various trees T, until $B(T) = 0$ for all trees T, hence $B = 0$.

By Theorem 3.3 we may assume that $U_TB = 0$ for any unhangable Szele tree T. So, for such a tree, if $B(T^*) = 0$, then $B(T) = 0$. Thus $B(T) = 0$ for $T = 43210$, so $p^4B = 0$. Similarly $B(T) = 0$ for T successively equal to 4321, 4320, and $43(2)(10) = T_0$. Note that T_0^* is generated by 5 and 4320.

The next tree in line, 4(310)(32), comes in two forms: one, T_1, with a hang of 4 across the 3's, and the other, T_2, plain. Note that T_2^* is generated by 5 and 43(2)(10).

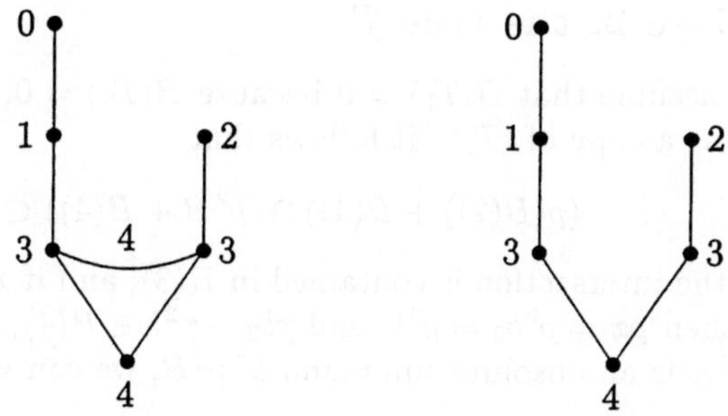

Case 1 Case 2

It's clear what we mean by $B(T_1)$, and $\langle T_1 \rangle$, even though T_1 has a hang. We will treat Case 1 first, and in Case 2 we will be able to assume that $B(T_1) = 0$.

Suppose $B(5) = p^4 B = B(T_0) = 0$. Let A be a direct sum of copies of $\langle T \rangle$, with $T = T_1$ or $T = T_2$.

5.1 Case 1, the tree T_1

As $p^2 A$ is pure in $p^2 B$ (because $p^4 B = 0$), we can write

$$p^2 B = p^2 A + H'' \text{ with } p^2 A \cap H'' = 0.$$

Then

$$A \cap H'' \subset A \cap p^2 B = p^2 A$$

so $A \cap H'' = 0$.

We want to write

$$B(2) = A(2) \oplus H'$$

where $H' \supset H''$. Note that 4 is the only Ulm invariant of $A(2)$ (because of the hang), and $B(5) = 0$. So, by Lemma 1.3, it suffices to show that if $a_2 \in A(2)[p] = A[p]$, and $h'' \in H''$, then $\mathrm{ht}_{B(2)}(a_2 + h'') \leq \mathrm{ht}_{A(2)}\, a_2$.

There is $a_3 \in p^2 A$ such that $a_3 \in a_2 + p(A(2))$, so if $a_2 + h'' \in p(B(2))$, then $a_3 + h'' \in p(B(2))$. But

$$p(p(B(2)) \cap p^2 B) = B(T_0) = 0,$$

so $pa_3 = 0$. Therefore $a_3 \in A(4310) = p^2(A(2))$, whence $a_2 \in p(A(2))$, so $a_2 \in A(4310) = p^2(A(2))$.

The only Ulm invariants of A are 1 and 4. To finish the proof, it suffices, by Lemma 1.3, to show that if $z \in A[p]$, and $h' \in H'$, then $\mathrm{ht}_B(z + h') \leq \mathrm{ht}_A\, z$. If $z + h' \in p^2 B$, then $z + h' = p^2 a + h''$ so $z = p^2 a$.

5.2 Case 2, the tree T_2

We may assume that $B(T_1) = 0$ because $B(T_0) = 0$, so if $B(T_1) \neq 0$, then B contains a copy of $\langle T_1 \rangle$. It follows that

$$(p(B(2)) + B(4)) \cap (p^2 B + B(4)) \subset B[p] \qquad (*)$$

because the intersection is contained in $B(3)$, and if $x = pb_2 + b_4 = p^2 b + b_4'$ is in it, then $px = p^2 b_2 = p^3 b$, and $pb_2 - p^2 b \in B(4)$, so $px \in B(T_1) = 0$.

As $p^2 A$ is an absolute summand of $p^2 B$, we can write

$$p^2 B = p^2 A + H''' \text{ with } p^2 A \cap H''' = 0$$

and

$$(p^2 B)(4) = A(4) \oplus H'''(4).$$

Because $B(5) = 0$, we can write

$$B(4) = A(4) \oplus K_4$$

with $K_4 \supset H'''(4)$. Let $H'' = H''' + K_4$. Note that $H''(4) = K_4$. Then

$$A \cap H''' \subset A \cap p^2 B = p^2 A$$

so $A \cap H''' = 0$. We will show that $A \cap H'' = 0$.

If $a = h''' + k_4$, then $pa \in A \cap H''' = 0$. So we can find $a_1 \in p(A(2))$ and $a_2 \in p^2 A$ such that $a = a_1 - a_2$. So

$$a_1 - k_4 = a_2 + h''' \in p^2 B.$$

From $(*)$ it follows that $p(a_1 - k_4) = 0$, so $pa_1 = 0$, whence $a \in A(4)$. Thus $a = 0$ because $A(4) \cap H'' = A(4) \cap K_4 = 0$. We have shown that $A(2) \cap H'' = 0$.

We want to write

$$B(2) = A(2) \oplus H'$$

with $H' \supset H'' = H''' + K_4$. The Ulm invariants of $A(2)$ are 3 and 4, so by Lemma 1.3 it suffices to show that if $a \in A[p]$, then $\mathrm{ht}_{B(2)}(a + h''' + k_4) \leq \mathrm{ht}_{A(2)}\, a$. Suppose $a + h''' + k_4 \in p(B(2))$. By adding an element in $p(A(2))$ to a, we can get an element a_3 in $(p^2 A)(3)$ such that $a_3 + h''' + k_4 \in p(B(2))$. From $(*)$ it follows that $p(a_3 + h''') = 0$, so $pa_3 = 0$. This means $a_3 \in A(4)$, so $a \in A(4) \subset p^2(A(2))$.

We want to write

$$B = A \oplus H$$

with $H \supset H'$. By Lemma 1.3 it suffices to show that if $a \in A[p]$, then $\mathrm{ht}_B(a + h') \leq \mathrm{ht}_A\, a$. If $a + h' \in p^2 B = p^2 A \oplus H'''$, then $a + h' = p^2 a' + h'''$ so $a = p^2 a'$. That completes the proof of case 2.

5.3 Example

Here is an example showing that the hung tree T_1 must be eliminated before eliminating the unhung tree T_2. Alternatively, that T_2 is not a Szele tree unless T_1 is included in its star. Consider $B = \langle 32 \rangle \oplus \langle T_1 \rangle$.

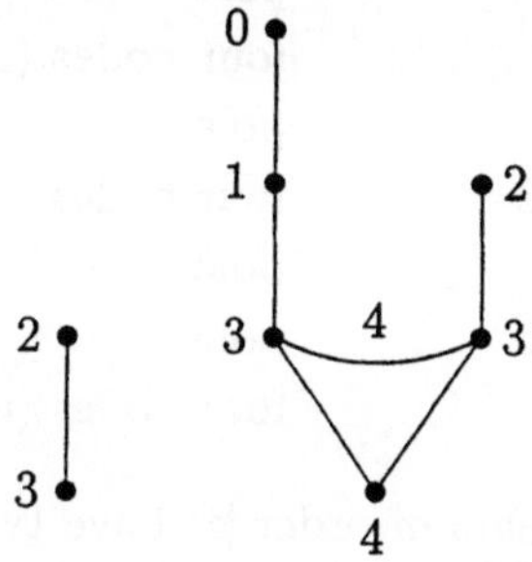

Let x generate the first summand $\langle 32 \rangle$ of B, and y and z generate $\langle 4310 \rangle$ and $\langle 432 \rangle$ respectively of $\langle T_1 \rangle$. Let A be the subgroup of B generated by $z - x$ and y. Then $B(T_2^*) = 0$, and $A = \langle T_2 \rangle$ with the v-height of 4 equal to T_2, but A is not a summand.

5.4 The schedule for removal

Here are the trees, in order of removal, together with the relevant generators of their stars. After step T, we have $B(T) = 0$. Why? For unhangable trees we know that $B(T)[p] \subset B(T^*)[p]$. If there's at most one pole left in $B(T^*)$, it will be easy to see that $B(T) \cap B(T^*) = 0$. The hangable tree gets a separate treatment.

43210		pole
4321	43210	pole
4320	4321	pole ·
43(2)(10)	4320	two Ulm invariants
4(310)(32)		the hangable tree: hung, then unhung
4310	432	pole
4(210)(32)	4310	two Ulm invariants
432	3210	pole
4(210)(31)	432 4310	
431	432 3210	pole
4(210)(30)	431	
4(30)(21)	431 3210	
430	321	pole
4(210)(3)	430	two Ulm invariants
4210	43	pole
3210	4	pole. At this point $p^3 B = 0$.
4(21)(3)	430	four nodes (and two Ulm invariants)

421	43	pole
321	4	pole
4(20)(3)	430 321	four nodes
420	43 321	pole
320	321 4	pole
4(10)(3)	320	four nodes (and two Ulm invariants)
43	210	pole
4(10)(2)	320 43	four nodes
410	32	pole
42	43 210	pole
3(10)(2)	320 4	four nodes (and two Ulm invariants)

At this point all elements of order p^3 have type 310. Elements of type 210 miss $B(3)$, so they split out.

5.4.1 The hung forest (32)(310)

We may now assume that $B(T) = 0$ unless T is a pole, that $(p^2B)[p] = B(310)$, and that $B(42) = B(410) = 0$. Write

$$B(4) = (B(4) \cap (B(310) + B(32))) \oplus K_4.$$

Let X be a direct sum of copies of $\langle 310 \rangle$ such that $X[p] = B(310)$. Let Y be a direct sum of copies of $\langle 32 \rangle$ such that $Y[p] = B(32)$. Then $X + Y$ is an absolute direct summand of the unvaluated module B, so we can write

$$B = X + Y + Z$$

an unvaluated direct sum, where $K_4 \subset Z$. Then

$$B = (X + Y) \oplus Z$$

because $X+Y$ has Ulm invariants only at 1, 3 and 4. So we need only check the filtration at 2 and 4. For 2, note that $B(2) = (X + Z)(2) + Y$, and if $x + z \in B(2)$, then $px + pz \in B(32) = Y[p]$, so $px = pz = 0$, so $x \in B(3)$. For 4 we have arranged that $B(4) = (X + Y)(4) \oplus K_4$.

Let $T = 32$. As $Y[p] = B(T)$, and $U_TB = 0$, it follows that $Y[p] \subset B(T^*) = X[p] + B(4)$. Similarly $X[p] \subset Y[p] + B(4)$. Because $B(4) \cap X = B(4) \cap Y = 0$, there is a natural isomorphism $f : X[p] \to Y[p]$ such that $x - f(x) \in B(4)$ for all $x \in X[p]$. Let E be a basis for $X[p]$. Then E and $f(E)$ support bases for X and Y, showing that $X + Y$ is a direct sum of copies of $\langle F \rangle$, where F is the hung forest

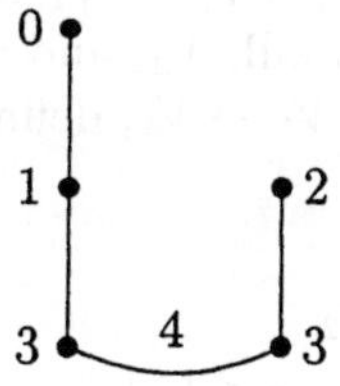

The complementary summand Z is a p^2-bounded v-module with finitely many values, hence a direct sum of cyclics [4, Theorem 3.2]. Alternatively, we can easily show that Z is a direct sum of cyclics by continuing the process of eliminating poles.

6 Uniqueness

So each v-module B with $B(5) = 0$ is the direct sum of a simply presented v-module with a direct sum of copies of $\langle T_1 \rangle$, and a direct sum of copies of $\langle F \rangle$, where T_1 is the hung tree 4(32)(310), and F is the hung forest (32)(310) (so $T_1 = 4F$). If we extend our notion of Ulm invariant slightly, to cover T_1 and F, then the number of copies of each indecomposable $\langle T \rangle$ is equal to the dimension of $U_T B$, the T-th Ulm invariant, hence is an invariant of B.

We have already used the submodule $B(T_1)$, which has the obvious meaning. For T_2, the unhung tree 4(32)(310), we must extend T_2^* to include T_1, while T_1^* is simply the old T_2^*. Then the definitions of U_{T_1} and U_{T_2} are formally the same as for any other Ulm invariant. Finally, we define

$$B(F) = B(4) \cap (B(32) + B(310))$$

and let F^* be generated by 5, 42, and 3(10)(2). These definitions are all natural—they could be formulated in a general context of certain kinds of hung forests—and do the trick.

7 Indecomposable pairs bounded by p^6

We present a simplification of the categorical equivalence of [4, Cor. 5.3]. Let k be a field, and $\mathcal{C}_1$ the category of modules over $k[X]$ (not a discrete valuation domain). The category $\mathcal{C}_2$ consists of vector spaces V over k, together with a (labeled) family of four distinguished subspaces V_1, V_2, V_3 and V_4 such that

$$V = V_1 \oplus V_2 = V_2 \oplus V_3 = V_1 \oplus V_3 = V_4 \oplus V_2.$$

This implies that $V_1 \cong V_2 \cong V_3 \cong V_4$. Given the object (V, V_1, V_2, V_3, V_4) in $\mathcal{C}_2$, we get a linear transformation $f : V_1 \to V_1$ by setting $fx = \pi_2\pi_4 x$, where π_4 is the projection on V_4 that kills V_2, and π_2 is the projection on V_2 that kills V_1. Conversely, given $f : V_1 \to V_1$, define

$$\begin{aligned} V_2 &= V_1 \\ V &= V_1 \oplus V_2 \\ V_3 &= \{(x, x) : x \in V_1\} \\ V_4 &= \{(x, fx) : x \in V_1\}. \end{aligned}$$

But $f : V_1 \to V_1$ is simply a $k[X]$-module on V_1, where f gives the action of X.

There are indecomposable modules in $\mathcal{C}_1$ of every finite dimension over k, and we know that there are a ton of infinite-dimensional ones. Let k be the residue class field of our discrete valuation domain. For each indecomposable object (V, V_1, V_2, V_3, V_4) in $\mathcal{C}_2$, we will construct an indecomposable v-group B, with $B(6) = 0$, such that if $C = B/B(5)$, then (V, V_1, V_2, V_3, V_4) is isomorphic to

$$C[p],\ C(32),\ C(310),\ B[p]/B(5),\ B(4)/B(5).$$

The dimension of V is $2m$, where m could be infinite. Let B be the direct sum of m copies of $\langle 5(32)(310)\rangle$, and $C = B/B(5)$. In $C[p]$, let

$$\begin{aligned} W_1 &= C(32) = p(C(2)) \\ W_2 &= C(310) = p^2 C \\ W_3 &= B[p]/B(5). \end{aligned}$$

Choose the subspace W_4 of $C[p] = C(3)$ that makes (V, V_1, V_2, V_3, V_4) isomorphic to $(C[p], W_1, W_2, W_3, W_4)$, and redefine $B(4)$ to be the preimage if W_4. This doesn't affect W_1, W_2 or W_3. Then C is an indecomposable v-module. If $B = B' \oplus B''$, then either B' or B'' is contained in $B(5)$ because C is indecomposable. But $B(5) \subset pB$, so such a summand must be zero.

References

[1] L. Fuchs, *Infinite Abelian Groups*, Academic Press (1970).

[2] D. Beers, R. Hunter, F. Richman and E. A. Walker, *Computing valuated trees*, Abelian Group Theory (Oberwolfach 1986), Gordon and Breach, 65–88.

[3] R. Hunter, F. Richman and E. A. Walker, *Simply presented valuated p-groups*, J. Algebra **49** (1977), 125–133.

[4] R. Hunter, F. Richman and E. A. Walker, *Subgroups of bounded abelian groups*, Abelian Groups and Modules (Udine 1984), CISM Courses and Lectures **287**, Springer-Verlag, 17–35.

[5] R. Hunter, F. Richman and E. A. Walker, *Ulm's theorem for simply presented valuated p-groups*, Abelian Group Theory (Oberwolfach 1986), Gordon and Breach, 33–64.

[6] F. Richman and E. A. Walker, *Valuated groups*, J. Algebra **56** (1979), 145–167.

Trends in Mathematics, © 1999 Birkhäuser Verlag Basel/Switzerland

Groups acting on modules

Phillip Schultz

Abstract

When a group of automorphisms Δ acts on a module M, each invariant submodule H of M determines two normal subgroups of Δ: the centraliser of H and the centraliser of M/H. Similarly, each normal subgroup Γ of Δ determines two invariant submodules of M: the fixed module of Γ and the residual of Γ. These connections determine Galois correspondences between the lattice of normal subgroups of Δ and lattice of invariant submodules of M.

1 Introduction

Let R be a unital ring and M a unital R–module. Let Δ be a group of R–automorphisms of M. The aim of this paper is to use the tools of group actions and Galois theory to extract information about the module M from the group Δ and vice versa.

As well as the action of Δ on M, it is important to consider the action of Δ on the lattice of submodules and on the lattice of factor modules of M. These actions induce a pair of Galois Correspondences between lattices of normal subgroups of Δ and certain lattices of submodules and factor modules of M. The results have applications to structure theorems in several categories of modules which will be dealt with in subsequent publications.

The notation not specifically defined comes from the standard texts [1] for module theory, [6] for group theory and [3] for abelian groups. Latin upper case letters are used for modules and lower case for their elements; Greek upper case letters for groups and lower case for their elements. For example, $N \leq M$ means N is a submodule of M, $\Gamma \leq \Delta$ means Γ is a subgroup of Δ and $\Gamma \trianglelefteq \Delta$ means Γ is a normal subgroup of Δ.

2 Properties of group actions on modules

Let Δ be a group of R–automorphisms of an R–module M. If $N \leq M$, the **normaliser** of N in Δ is $\Delta_N = \{\delta \in \Delta : N\delta = N\} \leq \Delta$. If $\delta \in \Delta_N$ then δ induces automorphisms of N and of M/N. The **centraliser of** N **in** Δ is

$$N' = \{\delta \in \Delta_N : \delta|_N = 1\}$$

and the **centralizer of M/N in Δ** is

$$N^* = \{\delta \in \Delta_N : \delta|_{M/N} = 1\}.$$

N is called **Δ-invariant** if $\Delta_N = \Delta$, **Δ-fixed** if $N' = \Delta$ and **Δ-residual** if $N^* = \Delta$. If $\Gamma \leq \Delta$, the **fixed module** of Γ is $\Gamma' = \{m \in M : m\Gamma = \{m\}\}$ and the **residual** of Γ, $\Gamma^* = \langle m\gamma - m : m \in M,\ \gamma \in \Gamma\rangle$.

Lemma 2.1 *Let $N \leq M$ and let $\Gamma \leq \Delta$.*

1. *$N' \leq \Delta$ and $N^* \leq \Delta$.*
2. *If N is Δ-invariant then $N' \trianglelefteq \Delta$ and $N^* \trianglelefteq \Delta$.*
3. *$\Gamma' \leq M$ and $\Gamma^* \leq M$.*
4. *If $\Gamma \trianglelefteq \Delta$, then Γ' is Δ-invariant and Γ^* is Δ-invariant.*

Proof. 1. It is routine to check that both N' and N^* are closed under products and inverses.
2. Since N' is the kernel of the homomorphism $\delta \to \delta|_N$ of Δ into $\mathrm{Aut}(N)$, and N^* is the kernel of the homomorphism $\delta \to \delta|_{M/N}$ of Δ into $\mathrm{Aut}(M/N)$, they are both normal subgroups of Δ.
3. Let $m,\ n \in \Gamma'$ and $r,\ s \in R$. Then for all $\gamma \in \Gamma$, $(rm + sn)\gamma = rm + sn$, so $rm + sn \in \Gamma'$ and $\Gamma' \leq M$.

Similarly, any linear combination of elements of M of the form $\sum_{i=1}^n m_i\gamma_i - m_i$ has the same form, so $\Gamma^* \leq M$.
4. Suppose that $\Gamma \trianglelefteq \Delta$, and let $m \in \Gamma'$ and $\delta \in \Delta$. Then for all $\gamma \in \Gamma$, $\delta\gamma\delta^{-1} \in \Gamma$, so $m\delta\gamma\delta^{-1} = m$, i.e. $m\delta\gamma = m\delta$, so $m\delta \in \Gamma'$.

Finally, let $\sum_{i=1}^n m_i\gamma_i - m_i \in \Gamma^*$. Then for all $\delta \in \Delta$, $(\sum_{i=1}^n m_i\gamma_i - m_i)\delta = \sum_{i=1}^n (m_i\delta)\delta^{-1}\gamma_i\delta - m_i\delta \in \Gamma^*$. □

Let $\mathcal{S}$ be the set of submodules and $\mathcal{F}$ the set of factor modules of M. $\mathcal{S}$ is a complete lattice under the relation of inclusion and $\mathcal{F}$ is a complete lattice under the relation $M/N \preccurlyeq M/K$ if $K \leq N$. The group Δ acts on $\mathcal{S}$ by $\alpha : N \mapsto N\alpha$ and on $\mathcal{F}$ by $\alpha : M/N \mapsto M/N\alpha$. Denote these group actions by $\Delta^{\mathcal{S}}$ and $\Delta^{\mathcal{F}}$. These actions respect the lattice structures in the sense that $N \leq L$ if and only if $N\delta \leq L\delta$ and $M/N \preccurlyeq M/L$ if and only if $M/N\delta \preccurlyeq M/L\delta$ for all $\delta \in \Delta$.

The following Proposition is stated without proof, since it is simply a matter of sorting out the definitions.

Proposition 2.2 *The map $\Delta^{\mathcal{S}} \longrightarrow \Delta^{\mathcal{F}}$ which associates each $N \in \mathcal{S}$ with $M/N \in \mathcal{F}$ and $\alpha \in \Delta$ with itself is an equivalence of group actions and an anti-isomorphism of lattices.* □

3 Invariant submodules

It is routine to check that the set of Δ–invariant submodules of M forms a sublattice $\mathcal{I}$ of $\mathcal{S}$. Its opposite lattice $\mathcal{I}^{\mathrm{op}}$ is isomorphic to the lattice of Δ–invariant factor modules under the order described in Section 2.

Let $H \in \mathcal{I}$ and denote by Δ^H the group of automorphisms of H induced by Δ. Thus the centralizer H' of H in Δ is the kernel of the action Δ^H. Denote by $\Delta^{M/H}$ the group of automorphisms of M/H induced by Δ so that the centralizer H^* of M/H is the kernel of the action $\Delta^{M/H}$.

Thus $H' = \{\alpha \in \Delta : x\alpha = x \text{ for all } x \in H\} \trianglelefteq \Delta$ and $H^* = \{\alpha \in \Delta : x\alpha - x \in H \text{ for all } x \in M\} \trianglelefteq \Delta$. Since the operations $'$ and * reverse orders, these centralizers form lattices of normal subgroups of Δ.

Proposition 3.1 *Let $H \in \mathcal{I}$. There is an exact commutative diagram of groups and group homomorphisms:*

$$\begin{array}{ccccc}
H' \cap H^* & \xrightarrow{h} & H' & \xrightarrow{i} & \dfrac{H' \cdot H^*}{H^*} \\
j\big\downarrow & & k\big\downarrow & & \ell\big\downarrow \\
H^* & \xrightarrow{m} & \Delta & \xrightarrow{n} & \Delta^{M/H} \\
o\big\downarrow & & p\big\downarrow & & q\big\downarrow \\
\dfrac{H' \cdot H^*}{H'} & \xrightarrow{r} & \Delta^H & \xrightarrow{s} & \dfrac{\Delta}{H' \cdot H^*}
\end{array}$$

In this diagram, all rows and columns are short exact sequences, so the top left corner is a pullback and the bottom right a pushout.

Proof. It is enough to define the homomorphisms; the rest follows by straightforward diagram–chasing.

Let h, j, k and m be the inclusion maps.

Let i be the natural epimorphism of H' onto $\dfrac{H'}{H' \cap H^*}$ followed by the natural isomorphism of $\dfrac{H'}{H' \cap H^*}$ onto $\dfrac{H' \cdot H^*}{H^*}$.

For $\alpha \in \Delta$, let $\overline{\alpha}$ be the induced automorphism of M/H. Let $\ell : \alpha \cdot H^* \mapsto \overline{\alpha}$ and let $o : \alpha \mapsto \alpha \cdot H'$. Let $n : \alpha \mapsto \overline{\alpha}$ and let $p : \alpha \mapsto \alpha|_H$. Let r be the natural embedding of $\dfrac{H' \cdot H^*}{H'}$ into Δ^H and let $s : \alpha \mapsto \alpha \cdot (H' \cdot H^*)$. Finally, let $q : \overline{\alpha} \mapsto \alpha \cdot (H' \cdot H^*)$. □

To characterise the group $H' \cap H^*$ we need for the first time to introduce the ring $\mathcal{E}_R(M)$ of R–endomorphisms of the module M and the additive group $\mathrm{Hom}_R(M/H, H)$ of R-homomorphisms of M/H into H.

Lemma 3.2 *Let $H \in \mathcal{I}$. There is a natural embedding of* $\mathrm{Hom}_R(M/H, H)$ *into $\mathcal{E}_R(M)$ such that, for all $f \in \mathrm{Hom}_R(M/H, H)$, $1 + f \in \mathrm{Aut}_R(M)$ and $1 + f$ fixes H and M/H.*

Proof. The natural epimorphism $M \longrightarrow M/H$ induces an embedding of $\mathrm{Hom}_R(M/H, H)$ into $\mathrm{Hom}_R(M, H)$ considered as a subgroup of $\mathcal{E}_R(M)$.

It is convenient to consider each $f \in \mathrm{Hom}_R(M/H, H)$ as an endomorphism of M which annihilates H and maps M into H. With this identification, if $f \in \mathrm{Hom}_R(M/H, H)$, then $f^2 = 0$ so $1 + f$ is an automorphism which is the identity on H and induces the identity on M/H. □

In view of Lemma 3.2, let $\mathrm{Hom}_R^{\Delta}(M/H, H)$ denote $\{f \in \mathrm{Hom}_R(M/H, H) : 1 + f \in \Delta\}$.

Proposition 3.3 *Let $H \in \mathcal{I}$.*

1. *The group $H' \cap H^*$ is an abelian normal subgroup of Δ.*

2. *The set $\mathrm{Hom}_R^{\Delta}(M/H, H)$ is a zero subring of $\mathcal{E}_R(M)$.*

3. *The map*

$$\alpha \mapsto \alpha - 1 : H' \cap H^* \longrightarrow \mathrm{Hom}_R^{\Delta}(M/H, H)$$

is an isomorphism of abelian groups.

Proof. 1. By Lemma 2.1, $H' \cap H^* \trianglelefteq \Delta$. Let α, $\beta \in H' \cap H^*$. For all $m \in M$, $m\alpha - m \in H$, so $m\alpha\beta - m\beta = m\alpha - m$. Similarly, $m\beta\alpha - m\alpha = m\beta - m$. Hence $m\alpha\beta = m\beta\alpha$, so $H' \cap H^*$ is abelian.
2. Let f, $g \in \mathrm{Hom}_R^{\Delta}(M/H, H)$. Then $fg = 0$ so $(1 + f)(1 - g) = 1 + f - g$. Hence $\mathrm{Hom}_R^{\Delta}(M/H, H)$ is closed under subtraction.
3. The map is certainly a bijection from $H' \cap H^*$ onto $\mathrm{Hom}_R^{\Delta}(M/H, H)$, both abelian groups. Since $(1 + f)(1 + g) = 1 + f + g$, the map is a homomorphism. □

Corollary 3.4 *Let H be a Δ-invariant submodule of M. If $\mathrm{Hom}_R^{\Delta}(M/H, H) = 0$ then Δ is a subdirect union* [4] *of Δ^H and $\Delta^{M/H}$ with kernels H' and H^*.* □

The following Lemma shows that every submodule is contained in a minimal invariant submodule:

Lemma 3.5 *If $H \leq K$ with $K \in \mathcal{I}$, then $\langle H\Delta \rangle \leq K$ and $\langle H\Delta \rangle \in \mathcal{I}$.*

Proof. Let $a \in \langle H\Delta \rangle$. Then a is a finite sum $\sum_i a_i$ with $a_i \in H\delta_i$ for some $\delta_i \in \Delta$. Hence for each $\alpha \in \Delta$, $a\alpha = \sum_i a\delta_i\alpha \in \langle H\Delta \rangle$. Since $K \in \mathcal{I}$, K must contain each $H\delta$, $\delta \in \Delta$. □

It follows that M is generated by its Δ–invariant submodules. On the other hand, the subgroup of Δ generated by the subgroups of Proposition 3.1 is a useful invariant.

Lemma 3.6 *Let* $\mathrm{Elem}(\Delta) = \langle \{H' \cap H^* : H \in \mathcal{I}\} \rangle$. *Then* $\mathrm{Elem}(\Delta) \trianglelefteq \Delta$.

Proof. A subgroup generated by normal subgroups is necessarily normal.□

4 Galois connections

A **Galois connection** [2] between complete lattices $\mathcal{F}$ and $\mathcal{G}$ is a pair of functions $\Theta : \mathcal{F} \longrightarrow \mathcal{G}$ and $\Psi : \mathcal{G} \longrightarrow \mathcal{F}$ which satisfy

1. for all $x,\ y \in \mathcal{F},\ x \leq y$ implies $x\Theta \geq y\Theta$
2. for all $a,\ b \in \mathcal{G},\ a \leq b$ implies $a\Psi \geq b\Psi$
3. for all $x \in \mathcal{F}$ and $a \in \mathcal{G},\ x \leq x\Theta\Psi$ and $a \leq a\Psi\Theta$.

Elements x of $\mathcal{F}$ and a of $\mathcal{G}$ are called **closed** if $x = x\Theta\Psi$ and $a = a\Psi\Theta$. It is routine to check (see for example [2], Theorem 20) that elements of $\mathcal{F}$ in the range of Ψ, and elements of $\mathcal{G}$ in the range of Θ are closed, and that the closed elements of each lattice form complete sublattices $\overline{\mathcal{F}}$ and $\overline{\mathcal{G}}$. Furthermore, the functions Θ and Ψ restricted to these sublattices are inverse lattice isomorphisms. The structure $\langle \overline{\mathcal{F}},\ \overline{\mathcal{G}},\ \Theta \rangle$ is called a **Galois correspondence.**

Galois correspondences were used in [8] to show that an abelian p–group is determined by its torsion radical, the lattices in question being the ideals of the torsion radical and the invariant subgroups of the abelian p–group.

Let $\mathcal{N}$ be the lattice of normal subgroups of Δ. The aim of this section is to define two Galois connections between the lattices $\mathcal{I}$ and $\mathcal{I}^{\mathrm{op}}$ of Δ–invariant submodules and the corresponding factor modules of a module M, and the lattice $\mathcal{N}$.

We have already defined in Lemma 2.1 functions $H \mapsto H'$ and $H \mapsto H^*$ from $\mathcal{I}$ to $\mathcal{N}$, and in the same Lemma functions $\Gamma \mapsto \Gamma'$ and $\Gamma \mapsto \Gamma^*$ from $\mathcal{N}$ to $\mathcal{I}$. These functions induce two Galois connections.

1. The **left Δ–connection for** M: It is routine to check that the functions $H \mapsto H' : \mathcal{I} \longrightarrow \mathcal{N}$ and $\Gamma \mapsto \Gamma' : \mathcal{N} \longrightarrow \mathcal{I}$ determine a Galois connection between $\mathcal{I}$ and $\mathcal{N}$. The **left closed** Δ–invariant submodules H of M are those for which every element of M which is fixed by every automorphism in Δ which fixes H is already in H. The **left closed** normal subgroups Γ of Δ are those for which every element of Δ which fixes every element of M fixed by Γ is already in Γ. Let $\overline{\mathcal{I}}$ be the lattice of left closed Δ–invariant submodules of M and $\overline{\mathcal{N}'}$ the lattice of left closed normal subgroups of Δ. Then the maps $\Psi' : H \mapsto H' : \overline{\mathcal{I}} \longrightarrow \overline{\mathcal{N}'}$ and $\Theta' : \Gamma \mapsto \Gamma' : \overline{\mathcal{N}'} \longrightarrow \overline{\mathcal{I}}$ are inverse lattice isomorphisms and the triple $\langle \overline{\mathcal{I}}, \overline{\mathcal{N}'}, \Psi' \rangle$ is the **left correspondence between M and Δ.**

2. The **right Δ–connection for M**: We have seen that $H^* \in \mathcal{N}$ and that $\Gamma^* \in \mathcal{I}^{\mathrm{op}}$. Since the functions * reverse orders, the functions $H \mapsto H^* : \mathcal{I}^{\mathrm{op}} \to \mathcal{N}$ and $\Gamma \mapsto \Gamma^* : \mathcal{N} \to \mathcal{I}^{\mathrm{op}}$ determine a Galois connection between $\mathcal{I}^{\mathrm{op}}$ and $\mathcal{N}$. The **right closed** Δ–invariant submodules H are those for which every element of H has the form $m\alpha - m$ for some $\alpha \in \Delta$ whose residual is contained in H, and the **right closed** normal subgroups Γ are those for which every element of Δ fixing every coset which is fixed by Γ is already in Γ. Let $\overline{\mathcal{I}^{\mathrm{op}}}$ be the lattice of right closed Δ–invariant submodules of M and $\overline{\mathcal{N}^*}$ the lattice of right closed normal subgroups of Δ. Then the maps $\Psi^* : H \mapsto H^* : \overline{\mathcal{I}^{\mathrm{op}}} \to \overline{\mathcal{N}^*}$ and $\Theta^* : \Gamma \mapsto \Gamma^* : \overline{\mathcal{N}^*} \to \overline{\mathcal{I}^{\mathrm{op}}}$ are Galois connections and the triple $\langle \overline{\mathcal{I}^{\mathrm{op}}}, \overline{\mathcal{N}^*}, \Psi^* \rangle$ is the **right correspondence between $\mathcal{I}^{\mathrm{op}}$ and $\mathcal{N}$.**

In general, it is difficult to find necessary and sufficient conditions for closure of arbitrary invariant submodules with respect to the two connections. The following Lemma, whose proof is straightforward, presents some weak necessary conditions for closure. In some important cases, these conditions are also sufficient.

Lemma 4.1 *1. Let $H \in \mathcal{I}$ be left closed. Then $\Delta' \le H \le M$.*

2. Let $H \in \mathcal{I}$ be right closed. Then $\{0\} \le H \le \Delta^*$. □

For the closure of normal subgroups, recall from Proposition 3.1 that if H is a Δ–invariant submodule of M, then the map $p : \Delta \to \Delta^H$ defined by $p : \alpha \mapsto \alpha|_H$ is a group epimorphism with kernel H' and the map $n : \Delta \to \Delta^{M/H}$ defined by $\alpha \mapsto \overline{\alpha}$ is a group epimorphism with kernel H^*. Furthermore, if $\Gamma \trianglelefteq \Delta$, then Γ is contained in the kernel of the map $p : \Delta \to \Delta^{\Gamma'}$ and also in the kernel of the map $n : \Delta \to \Delta^{M/\Gamma^*}$. Hence the action of Δ induces actions of Δ/Γ on Γ' and on M/Γ^* and we have the following characterizations of closed normal subgroups:

Theorem 4.2 *Let $\Gamma \in \mathcal{N}$.*

1. *The following are equivalent:*

   (a) *Γ is left closed;*

   (b) *Γ is the kernel of $p : \Delta \to \Delta^{\Gamma'}$;*

(c) *Δ/Γ is faithful on Γ'.*

2. *The following are equivalent:*

(a) *Γ is right closed;*

(b) *Γ is the kernel of $n : \Delta \to \Delta^{M/\Gamma^*}$;*

(c) *Δ/Γ is faithful on M/Γ^*.*

Proof. 1. (a) $\Rightarrow$ (b) Since Γ is contained in the kernel of p, we must show that if $\alpha \in \Delta$ fixes every element of M that is fixed by Γ then $\alpha \in \Gamma$. This follows from (a).

(b)$\Rightarrow$ (c) The action of Δ/Γ on Γ' is defined by $m(\delta\Gamma) = m\delta$. So if $m(\delta\Gamma) = m$ for all $m \in \Gamma'$ then δ is in the kernel of p. Hence $\delta \in \Gamma$ so the action is faithful.

(c)$\Rightarrow$ (a) Let $\delta \in \Delta$ fix every element of Γ' which is fixed by every element of Γ. Since the action is faithful, $\delta \in \Gamma$. Hence Γ is closed.

2. The proof is clause by clause similar to the proof of 1. □

Corollary 4.3 *Let $\Gamma \in \mathcal{N}$. If Γ is left closed, then there exists a right closed $\Omega \in \mathcal{N}$ such that $\Gamma \cap \Omega$ is abelian and there is an exact commutative diagram*

$$\begin{array}{ccccc}
\Gamma \cap \Omega & \longrightarrow & \Gamma & \longrightarrow & \dfrac{\Gamma \cdot \Omega}{\Omega} \\
\downarrow & & \downarrow & & \downarrow \\
\Omega & \longrightarrow & \Delta & \longrightarrow & \dfrac{\Delta}{\Omega} \\
\downarrow & & \downarrow & & \downarrow \\
\dfrac{\Gamma \cdot \Omega}{\Gamma} & \longrightarrow & \dfrac{\Delta}{\Gamma} & \longrightarrow & \dfrac{\Delta}{\Gamma \cdot \Omega}
\end{array}$$

in which all the rows and columns are short exact sequences.

Proof. Let $\Omega = (\Gamma')^*$. Now apply Proposition 3.1 with $H' = \Gamma$ and Theorem 4.2. □

A similar criterion exists for right closure, with the rôles of Γ and Ω reversed.

The next result extends the relationship of Lemma 3.2 to lattice intervals. Recall that for H and $K \in \mathcal{I}$, if $K \leq H$, then $H' \leq K'$ and $K^* \leq H^*$. Let $K \leq H$ and let $\mathrm{Hom}_R^\Delta(M/H, K)$ denote $\{f \in \mathrm{Hom}_R(M/H, K) : 1 + f \in \Delta\}$. From the definitions it follows that $\mathrm{Hom}_R^\Delta(M/H, K) \leq \mathrm{Hom}_R^\Delta(M/H, H)$ and $\mathrm{Hom}_R^\Delta(M/H, K) \leq \mathrm{Hom}_R^\Delta(M/K, K)$. Denote the sublattice $\{L \in \mathcal{I} : K \leq L \leq H\}$ by $[K, H]$. Then we have:

Lemma 4.4 *Let $K \leq H \in \mathcal{I}$. Then for all $L \in [K, H]$,*

1. *$L' \cap K^*$ is an abelian normal subgroup of Δ isomorphic to the additive group* $\mathrm{Hom}_R^\Delta(M/L, K)$.

2. *$H' \cap L^*$ is an abelian normal subgroup of Δ isomorphic to the additive group* $\mathrm{Hom}_R^\Delta(M/H, L)$.

Proof. Certainly $L' \cap K^*$ and $H' \cap L^*$ are in $\mathcal{N}$, so to complete the proofs of 1. and 2., it suffices to show that for any $K \leq H \in \mathcal{I}$, $H' \cap K^* \cong \mathrm{Hom}_R^\Delta(M/H, K)$.

Define $\theta : H' \cap K^* \longrightarrow \mathrm{Hom}_R^\Delta(M/H, K)$ by $\theta : \alpha \mapsto \alpha - 1$. Note that since $\alpha \in H' \cap K^*$, $\alpha - 1 \in \mathrm{Hom}_R^\Delta(M/H, K)$. If $\alpha, \beta \in H' \cap K^*$, then $\alpha\beta - 1 = (\alpha - 1)(\beta - 1) + (\alpha - 1) + (\beta - 1) = (\alpha - 1) + (\beta - 1)$, since $(\alpha - 1)(\beta - 1) = 0$ in $\mathrm{Hom}_R^\Delta(M/H, K)$.

Thus θ is a homomorphism of groups. Its kernel is clearly $\{1\}$ and it is surjective by definition of $\mathrm{Hom}_R^\Delta(M/H, K)$. □

Corollary 4.5 *$\{H' \cap L^* : L \in [K, H]\}$ and $\{L' \cap K^* : L \in [K, H]\}$ are sublattices of $\mathcal{N}$ whose elements are abelian normal subgroups of Δ.* □

5 Groups of normalised units

The results of Section 4 suggest that we should consider groups of **normalised units** in $\mathrm{End}_R M$, that is, groups Δ of the form $\Delta = 1 + \mathbf{t} = \{1 + f : f \in \mathbf{t}\}$, where $\mathbf{t}$ is a nil ideal in $\mathrm{End}_R M$. Clearly, any such Δ is a normal subgroup of $\mathrm{Aut}_R M$.

For example, if M is an abelian p–group and $\mathbf{t}$ is the torsion radical, it was shown in [5] that $\Delta = 1+\mathbf{t}$ is the maximal normal p–subgroup of $\mathrm{Aut}\, M$. Groups of normalised units were also used in [7] to analyse the structure of the automorphism group of an almost completely decomposable group.

The following Theorem shows how to characterise the left and right closed subgroups of a group of normalised units.

Theorem 5.1 *Let Δ be a group of normalised units.*

1. *A submodule H of M is Δ–invariant if and only if H is $\mathbf{t}$–invariant, i.e. for all $f \in \mathbf{t}$, $Hf \subseteq H$.*

2. *For all ideals I of $\mathbf{t}$, $\Omega = 1 + I$ is a normal subgroup of* $\mathrm{Aut}\, M$.

3. *For all left closed or right closed normal subgroups Γ of Δ, $\Gamma - 1 = \{\alpha - 1 : \alpha \in \Gamma\}$ is an ideal of $\mathbf{t}$.*

4. *Let* $\Gamma \trianglelefteq \Delta$ *and let* $I = \Gamma - 1$. *Then* Γ *is left closed if and only if there is an ideal* J *of* $\mathbf{t}$ *such that* I *is the right annihilator of* J.

5. *Let* $\Omega \trianglelefteq \Delta$ *and let* $J = \Omega - 1$. *Then* Ω *is right closed if and only if there is an ideal* K *of* $\mathbf{t}$ *such that* J *is the left annihilator of* K.

Proof. 1. Let $f \in \mathbf{t}$. Then $\alpha = 1 + f \in \Delta$ so f maps H into H if and only if α does.
2. Let $1+f, 1+g \in \Omega$, and let g have quasi-inverse h. Then $(1+f)(1+g)^{-1} = (1+f)(1+h) = 1+f+h+fh \in \Omega$, so Ω is a subgroup of Δ. Let $\alpha = 1+r \in \Delta$ and let r have quasi-inverse s. Then $\alpha^{-1}(1+f)\alpha = (1+s)(1+f)(1+r) \in \Omega$, since I is an ideal in $\mathbf{t}$. Thus $\Omega \trianglelefteq \Delta$.
3. Let Γ be a left closed normal subgroup of Δ. Then $\Gamma = H'$ for some $H \in \mathcal{I}$. Hence $I = \Gamma - 1 = \{f \in \mathbf{t} : Hf = 0\}$ which is an ideal of $\mathbf{t}$. If on the other hand Γ is right closed, then $\Gamma = K^*$ for some $K \in \mathcal{I}$. Hence $I = \{f \in \mathbf{t} : Mf \subseteq K\}$, which is again an ideal.
4. If $\Gamma = H'$ is left closed, then $\Gamma - 1$ is an ideal by 3. Let $J = H^* - 1 = \{f \in \mathbf{t} : Mf \subseteq H\}$. Then $JI = 0$. Furthermore, if $g \in \mathbf{t}$ satisfies $gI = 0$, then Mg is annihilated by I and hence fixed by Γ. Since Γ is left closed, $Mg \leq H$ so $g \in J$. Hence I is the right annihilator of J.

Conversely, suppose I is the right annihilator of an ideal J and let $H = MJ$. Then $HI = 0$ and furthermore, if $g \in \mathbf{t}$ satisfies $gI = 0$, then $g \in J$ so $Mg \leq H$. Hence $I = \{f \in \mathbf{t} : Hf = 0\}$, so $\Gamma = H'$ and hence Γ is left closed.
5. If $\Omega = H^*$ is right closed then J is the left annihilator of K where $1 + K = H'$. Conversely, if $J = \Omega - 1$ is the left annihilator of K, let $H = \{m \in M : mK = 0\}$. Then $J = \{f \in \mathbf{t} : Mf \subseteq H\}$ so Ω is right closed. □

We can now use the properties of $\mathbf{t}$ to extract useful information about Δ. The **Loewy sequence** of $\mathbf{t}$ is the descending sequence of ideals

$$\mathbf{t} = \mathbf{t}^1 \sqsupseteq \mathbf{t}^2 \sqsupseteq \cdots \sqsupseteq \mathbf{t}^n \sqsupseteq \cdots$$

where $\mathbf{t}^n$ is the ideal of $\mathbf{t}$ additively generated by products of n elements.

Hence there is an ascending sequence

$$0 \leq M_1 \leq M_2 \leq \cdots \leq M_n \leq \cdots$$

of $\mathbf{t}$-invariant submodules, where $M_n = \{m \in M : m\mathbf{t}^n = 0\}$ and a descending sequence

$$M \geq M^1 \geq M^2 \geq \cdots M^n \geq \cdots$$

of $\mathbf{t}$–invariant submodules of M.

Since each $\mathbf{t}^n$ is an ideal of $\operatorname{End}_R M$ contained in the Jacobson radical of $\operatorname{End}_R M$, the Loewy sequence of $\mathbf{t}$ induces a descending chain of normal subgroups of Δ:

$$\Delta = \Delta^{(1)} \trianglerighteq \Delta^{(2)} \trianglerighteq \cdots \trianglerighteq \Delta^{(n)} \trianglerighteq \cdots$$

where $\Delta^{(n)} = 1 + \mathbf{t}^n$. Without risk of confusion we may call this sequence the Loewy sequence of Δ. The submodule M_n described above is clearly the fixed module, and M^n is the residual, of $\Delta^{(n)}$.

Lemma 5.2 *For all $n \in \mathbb{N}$, M^n is closed for the right Galois connection.*

Proof. To show that M^n is right closed, it suffices to show that each $a \in M^n$ can be expressed as $a = m\alpha - m$ for some $\alpha \in \Delta$ whose residual is contained in M^n. But $a = mf$ for some m in M and some $f \in \mathbf{t}^n$. Hence if $a = m\alpha - m$ for some $\alpha = 1 + f$ having residual in M^n, then $a \in M^n$. □

Since $\mathbf{t}^r\mathbf{t}^s \subseteq \mathbf{t}^{r+s}$ for all $r,\ s \geq 1$, the Loewy sequence of $\mathbf{t}$ is actually a filtration and so has a corresponding graded ring $T = \oplus_{n\in\mathbb{N}}\mathbf{t}^n/\mathbf{t}^{n+1}$, where multiplication is defined on the homogeneous elements by

$$(x + \mathbf{t}^{r+1})(y + \mathbf{t}^{s+1}) = xy + \mathbf{t}^{r+s+1} \in \mathbf{t}^{r+s}/\mathbf{t}^{r+s+1}.$$

It will now be shown that the Loewy sequence of Δ is in fact a central series. The notation used for commutators is: if $a,\ b \in \mathbf{t}$, then $(a,b) = ab - ba$. If $\alpha,\ \beta \in \Delta$, then $[\alpha,\beta] = \alpha^{-1}\beta^{-1}\alpha\beta$.

The ring and group theoretic Loewy sequences are related as follows:

Lemma 5.3 *With the notation above, let $\alpha = a + 1 \in \Delta^{(r)}$ and$\beta = b + 1 \in \Delta^{(s)}$. Then*

1. $\beta^{-1} - 1 \equiv -b \pmod{\mathbf{t}^{2s}}$;

2. $\alpha\beta - 1 \equiv a + b \pmod{\mathbf{t}^{s+r}}$;

3. $[\alpha,\beta] - 1 \equiv (a,b) \pmod{\mathbf{t}^{r+s+1}}$.

Proof. 1. Let $\beta^{-1} = 1 + c$. Then b and c are quasi–inverses in $\mathbf{t}$ with $c \in \mathbf{t}^s$, so $c = -b + bc$ with $bc \in \mathbf{t}^{2s}$.
2. $\alpha\beta - 1 = a + b + ab$ with $ab \in \mathbf{t}^{r+s}$.
3. A routine calculation shows that $[\alpha,\beta] = 1 + \alpha^{-1}\beta^{-1}(a,b) = 1 + (1 - a + a')(1 - b + b')(a,b)$ for some $a' \in \mathbf{t}^{r+1}$ and some $b' \in \mathbf{t}^{s+1}$. Hence $[\alpha,\beta] - 1 = (a,b) + c$ for some $c \in \mathbf{t}^{r+s+1}$. □

Theorem 5.4 1. *The Loewy sequence $\{\Delta^{(n)}\}$ is central.*

2. *The factors of the Loewy sequence generate a graded abelian group* $\Lambda = \oplus_{n\in\mathbb{N}}\Delta^{(n)}/\Delta^{(n+1)}$.

3. *The map* $\Theta : T \longrightarrow \Lambda$ *defined on homogeneous generators by* $a + \mathbf{t}^{n+1} \mapsto \alpha\Delta^{(n+1)}$ *is an isomorphism of graded abelian groups.*

Proof. 1. and 2. It must be checked that $[\Delta^{(n)}, \Delta] \leq \Delta^{(n+1)}$ for all n. This follows from Lemma 5.3, 3.
3. The map Θ is well-defined, since $c \in \mathbf{t}^{n+1}$ implies $\gamma = c + 1 \in \Delta^{(n+1)}$. It is a morphism by Lemma 5.3 1. and 2., and 1–1 and onto by the definitions of $\mathbf{t}^n$ and $\Delta^{(n)}$. □

The graded abelian groups T and Λ carry a Lie ring structure by means of the products (a, b) and $[\alpha, \beta]$. This structure makes Θ a graded Lie ring isomorphism:

Theorem 5.5 *The isomorphism* Θ *of Theorem 5.4 is a graded Lie ring isomorphism.*

Proof. It must be checked that Θ preserves the induced Lie ring structures on T and Λ. Let $\alpha = 1 + a \in \Delta^{(n)}$ and $\beta = 1 + b \in \Delta^{(m)}$. Then $[\alpha, \beta] \in \Delta^{(n+m)}$ and $((a, b) + \mathbf{t}^{n+m+1})\Theta = [\alpha, \beta]\Delta^{(n+m+1)}$ as required. □

References

[1] F. W. Anderson and K. R. Fuller, *Rings and Categories of Modules*, Springer–Verlag, (1973).

[2] G. Birkhoff, *Lattice Theory*, American Mathematical Society Colloquium Publications, Vol. **25** (1967).

[3] L. Fuchs, *Infinite Abelian Groups*, Vols. **I** and **II**, Academic Press (1970–1973).

[4] L. Fuchs, *On Subdirect Unions*, Acta Math. Acad. Sci. Hung. **3** (1952), 103–119.

[5] J. Hausen and P. Schultz, *The maximal normal p-subgroup of the automorphism group of an abelian p-group*, Proc. Amer. Math. Soc **216** (1998), 2525–2533.

[6] B. Huppert and N. Blackburn, *Finite Groups*, Vols. **I, II** and **III**, Springer–Verlag (1962–1973).

[7] A. Mader and P. Schultz, *Endomorphism rings and automorphism groups of almost completely decomposable groups*, Comm. in Algebra, (to appear).

[8] P. Schultz, *When is an abelian p-group determined by the Jacobson radical of its endomorphism ring?*, Abelian Group Theory and Related Topics: Conference on Abelian Groups August 1–7, 1993 Oberwolfach, Contemporary Mathematics Vol. **171** (1994), 385–396.

Trends in Mathematics, © 1999 Birkhäuser Verlag Basel/Switzerland

Some mixed abelian groups as modules over the ring of pseudo–rational numbers*

Alexander A. Fomin

Abstract

An element of a direct product of rings of p–adic integers over all prime numbers p belongs to *the* subring R if almost all of its components are equal to one rational number. Finitely generated R–modules are closely connected with popular classes of abelian groups such as the class of self-small groups $\mathcal{G}$, the classes of mixed and torsion–free quotient divisible groups, the class of locally free groups and the class of torsion–free finite–rank groups. Two new characterizations of the class $\mathcal{G}$ are given as groups with distinguished endomorphisms and as R–modules.

1 The class $\mathcal{M}$

Throughout this paper "group" will mean abelian group. Z, Q and $\widehat{Z}_p$ are the rings of integers, rationals and p–adic integers respectively. $T(G)$ and $T_p(G)$ are the torsion part of a group G and its p–primary component. P is the set of all prime numbers. A type τ is called **locally free** if it contains a characteristic (m_p) with each $m_p < \infty$. Whenever a notion is not defined, it is adopted from [10] as well as the notation.

Definition 1.1 Let $\chi = (m_p)$ be a characteristic of a locally free type. The class $\mathcal{M}_\chi$ consists of all the groups G satisfying the following two conditions:

$\mathcal{M}_1$ For every prime number p there exists an endomorphism $\nu_p : G \to G$ such that $p^{m_p}\nu_p = 0$.

$\mathcal{M}_2$ Both groups, $\bigcap_p \ker \nu_p$ and $G/\sum_p \mathrm{Im}\ \nu_p$, are torsion–free divisible. In particular, they may be equal to 0.

The system of endomorphisms ν_p, $p \in P$, is called **a relevant system** of endomorphisms of a group $G \in \mathcal{M}_\chi$.

Theorem 1.2 *Let $G \in \mathcal{M}_\chi$ have a relevant system of endomorphisms ν_p, $p \in P$, where $\chi = (m_p)$ is a characteristic of a locally free type. Then:*

---

*This research is supported by the grant RFFI 97-01-00169 and by "La Universidad Antonio Narino".

1. *For every prime number p, the p–primary component $T_p(G)$ of the torsion part of the group G coincides with* $\mathrm{Im}\ \nu_p$. *Hence* $p^{m_p}T_p(G) = 0$.

2. *The restriction of ν_p to the subgroup $T_p(G) = \mathrm{Im}\ \nu_p$ is an automorphism.*

3. *An element $g \in G$ is divisible by any power of a prime number p if and only if $\nu_p(g) = 0$.*

Proof. By $\mathcal{M}_1$, it follows that $\mathrm{Im}\ \nu_p \subset T_p(G)$ for each p, therefore, $\sum_p \mathrm{Im}\ \nu_p \subset \sum_p T_p(G) = T(G)$. On the other hand, $\mathcal{M}_2$ implies that $T(G) \subset \sum_p \mathrm{Im}\ \nu_p$. Hence, the torsion subgroups $T(G)$ and $\sum_p \mathrm{Im}\ \nu_p$ coincide and their p–primary components coincide as well, $T_p(G) = \mathrm{Im}\ \nu_p$.

If $h = \nu_p(g) \in T_p(G) = \mathrm{Im}\ \nu_p$, then $\nu_q(h) = 0$ for every prime $q \neq p$. If additionally $\nu_p(h) = 0$, then $h \in \bigcap_p \ker \nu_p$ and $h = 0$ as a torsion element of a torsion–free group. Therefore, ν_p is injective on the group $T_p(G) = \mathrm{Im}\ \nu_p$.

It follows from $\mathcal{M}_2$ that for every natural n every element $g \in G$ may be presented in the form $g = p^n g_1 + t$, where $g_1 \in G$ and $t \in \sum_p \mathrm{Im}\ \nu_p$ is a torsion element. The element t may be presented as $t = t_p + t_1$, where $t_p \in T_p(G) = \mathrm{Im}\ \nu_p$ and the order of t_1 is relatively prime to p. Applying ν_p to the element

$$g = p^n g_1 + t_p + t_1 \tag{1}$$

we obtain for $n \geq m_p$

$$\nu_p(g) = \nu_p(t_p). \tag{2}$$

If the element g is divisible by any power of p, then for $n \geq m_p$ the element t_p is divisible by p^{m_p} by (1). Therefore $t_p = 0$ and $\nu_p(g) = 0$ by (2). Conversely, if $\nu_p(g) = 0$, then $\nu_p(t_p) = 0$ by (2). Since ν_p is injective on $T_p(G)$ the element t_p has to be equal to 0, and by (1) again, g is divisible by p^n. Thus, the third item of the theorem is proved. The formula (2) means in particular that the mapping $\nu_p : T_p(G) \to T_p(G)$ is surjective. This accomplishes the proof. □

Definition 1.3 The class $\mathcal{M}$ is the union of all the classes $\mathcal{M}_\chi$ over all the characteristics χ of locally free types, $\mathcal{M} = \cup_\chi \mathcal{M}_\chi$.

Theorem 1.4 *For every group $G \in \mathcal{M}$ there exists a unique relevant system of endomorphisms π_p, $p \in P$, such that $\pi_p^2 = \pi_p$ for all primes p.*

Proof. Let ν_p, $p \in P$, be any relevant system of endomorphisms of a group $G \in \mathcal{M}$. Since for every p the restriction of the homomorphism $\mu_p = \nu_p \mid_{T_p(G)}: T_p(G) \to T_p(G)$ is an isomorphism by Theorem 1.2(2), we can define a system of endomorphisms of the group G by $\pi_p = \mu_p^{-1}\nu_p$, $p \in P$. Plainly these form a relevant system of endomorphisms of G. Moreover

$$\pi_p(\pi_p(g)) = \mu_p^{-1}(\nu_p\mu_p^{-1})\nu_p(g) = \mu_p^{-1}\nu_p(g) = \pi_p(g)$$

for every element $g \in G$. If π'_p, $p \in P$, is another relevant system of endomorphisms satisfying $(\pi'_p)^2 = \pi'_p$, then for any element $g \in G$, the element $g - \pi'_p(g) \in \ker \pi'_p$ and therefore it is divisible by any power of p. Hence $\pi_p(g - \pi'_p(g)) = 0$ and, since π_p acts identically on p–torsion elements, $\pi_p(g) = \pi_p(\pi'_p(g)) = \pi'_p(g)$. Therefore $\pi_p = \pi'_p$ for all prime numbers p. □

Definition 1.5 A relevant system of endomorphisms π_p, $p \in P$, satisfying $\pi_p^2 = \pi_p$ is called the system of **projections** of a group $G \in \mathcal{M}$.

We use the same notation π_p, $p \in P$, for projections of different groups of the class $\mathcal{M}$ when no confusion will result. Note the following functorial property of projections.

Theorem 1.6 *Let $f : G \to H$ be a homomorphism of groups in $\mathcal{M}$. Then $\pi_p f = f\pi_p$ for all prime numbers p.*

Proof. Let $g \in G$. Then the element $g - \pi_p(g) \in \ker \pi_p$ is divisible by any power of p due to Theorem 1.2(3). Hence so is the element $f(g - \pi_p(g)) = f(g) - (f\pi_p)(g)$. By Theorem 1.2(3) again, $\pi_p(f(g) - (f\pi_p)(g)) = 0$ and $\pi_p(f(g)) = \pi_p((f\pi_p)(g))$. Since π_p is the identity on $T_p(H)$ and $(f\pi_p)(g) \in T_p(H)$ we obtain eventually $(\pi_p f)(g) = (f\pi_p)(g)$ for all elements $g \in G$. □

Theorem 1.6 in particular means that the projection π_p is a functor from the category $\mathcal{M}$ to the category of bounded p–groups.

Theorem 1.7 *A group G is in the class $\mathcal{M}$ if and only if it is of the form $G = D \oplus A$, where D is divisible torsion-free and A is a reduced group with torsion subgroup $T = \bigoplus_p T_p$ satisfying:*

1. *each T_p is bounded and*

2. *A is (may be embedded as) a pure subgroup of $\prod_p T_p$.*

Proof. Let $G \in \mathcal{M}$. Homomorphisms $\pi_p : G \to T_p(G)$ induce a homomorphism $G \to \prod_p T_p(G)$ $(g \mapsto (\pi_p(g)))$. The kernel of this homomorphism is $D = \bigcap_p \ker \pi_p$. By $\mathcal{M}_2$, D is a divisible torsion–free subgroup and, therefore, a direct summand, $G = D \oplus A$. The restriction of the homomorphism $G \to \prod_p T_p(G)$ to the subgroup A is an embedding $\varphi : A \to \prod_p T_p(G)$. Denote $T_p = T_p(A) = T_p(G)$. By the property $\mathcal{M}_1$ and Theorem 1.2(1), all the groups T_p are bounded. The divisibility of the group $G/T(G)$ provides the divisibility of the groups $A/T(A)$ and $\varphi(A)/\bigoplus_p T_p$, which in turn provides the purity of the subgroup $\varphi(A)$ in the group $\prod_p T_p$.

Conversely, if $\varphi : A \to \prod_p T_p$ is an embedding satisfying the conditions of the theorem, then it may be extended to a homomorphism $\psi : D \oplus A \to \prod_p T_p$, where $d + a \mapsto \varphi(a)$. The composition of ψ and the natural projection $\prod_p T_p \to T_p \subset A \subset D \oplus A = G$ is an endomorphism $\pi_p : G \to G$. It is easy to check that the system of endomorphisms π_p, $p \in P$, is relevant and $G \in \mathcal{M}$. □

2 Projection condition and the class $\mathcal{G}$

A subgroup F of a group G is called **full** if G/F is torsion.

Definition 2.1 A group $G \in \mathcal{M}$ satisfies the **strong projection condition (the projection condition)** if it satisfies in addition the following condition $\mathcal{M}_3$ ($\mathcal{M}_3'$).

$\mathcal{M}_3$ There exists a free finite-rank full subgroup F of the group G such that $\nu_p(F) = \nu_p(G)$ for all prime numbers p.

$\mathcal{M}_3'$ There exists a free finite-rank full subgroup F of the group G such that $\nu_p(F) = \nu_p(G)$ for **almost** all prime numbers p.

We use the following definition for the class $\mathcal{G}$ of self–small groups G with $G/T(G)$ finite–rank divisible.

Definition 2.2 ([8]) The class $\mathcal{G}$ consists of all groups of the form $V \oplus G$, where V is finite–rank torsion–free divisible and G is a reduced group with torsion subgroup $T = \bigoplus_p T_p$ satisfying:

- each T_p is finite and
- G is (can be embedded as) a pure subgroup of $\prod_p T_p$ and
- G/T is finite–rank (necessarily divisible) and
- (the projection condition) if F is a full free subgroup of G then F projects onto T_p for almost all primes p.

All the groups in $\mathcal{G}$ have finite torsion-free rank as well as groups satisfying the (strong) projection condition, see examples in [8]. Since finitely generated bounded groups are finite the following theorem is a straightforward consequence of Theorem 1.7.

Theorem 2.3 *Let $\mathcal{G}'$ and $\mathcal{G}''$ be the classes of groups $G \in \mathcal{M}$ satisfying the strong projection condition and the projection condition respectively. Then:*

1. $\mathcal{G}' \subset \mathcal{G} \subset \mathcal{G}''$.
2. *A group $G \in \mathcal{G}$ if and only if $G = H \oplus A$, where $H \in \mathcal{G}'$ and A is finite.*
3. *A group $G \in \mathcal{G}''$ if and only if $G = H \oplus A$, where $H \in \mathcal{G}'$ and A is bounded.* □

This theorem shows that the classes $\mathcal{G}'$, $\mathcal{G}$ and $\mathcal{G}''$ differ insignificantly. If we consider the categories whose objects are groups of these classes and whose morphisms are quasi–homomorphisms, denoted by $\mathcal{QG}'$, $\mathcal{QG}$ and $\mathcal{QG}''$ respectively, then they are all equivalent.

The class $\mathcal{G}$ introduced in [11] is now a focus of research. A number of papers have dealt with various properties of the groups in $\mathcal{G}$ and of the category $\mathcal{QG}$ (see [1-9, 11, 12, 14]). For every locally free type $\tau = [\chi]$, W. Wickless [14] has introduced an equivalence E_τ from the category $\mathcal{QG}_\tau$ (objects are groups of the class $\mathcal{M}_\chi$ satisfying the projection condition) to the category of locally free torsion free groups of finite rank with outer type less than or equal to τ and with quasi–homomorphisms as morphisms. A duality d from the whole category $\mathcal{QG}$ to the whole category of locally free torsion free groups of finite rank with quasi–homomorphisms as morphisms is obtained in [7]. It is shown there that the equivalence E_τ is the composition of the duality d and the Warfield duality $\mathrm{Hom}(-, A)$ in the class of locally free groups [13], where A is a torsion–free rank 1 group of type τ. The duality d is extended in [9] to a duality between the category of quotient divisible (mixed) groups and the category of all torsion–free finite–rank groups with quasi–homomorphisms. It suggests to consider classes that are larger than $\mathcal{G}$ yet share the good properties of the latter. This is the motivation for the present paper.

3 The ring of pseudo–rational numbers

Definition 3.1 Let $\chi = (m_p)$ be any characteristic and $K_p = Z/p^{m_p}Z$ or $K_p = \hat{Z}_p$ depending on whether $m_p < \infty$ or $m_p = \infty$.

1. If the characteristic $\chi = (m_p)$ contains infinitely many non-zero p-components we define a subring R_χ of the ring $\prod_p K_p$ as follows. An element

$(\alpha_p) \in \prod_p K_p$ belongs to R_χ if there exists a rational number $\frac{m}{n} \in Q$ such that $n\alpha_p = m \cdot 1_{K_p}$ for almost all prime numbers p. For an element $r \in R_\chi$, this rational number $\frac{m}{n}$ will be denoted as $|r|$.

In particular, if $\chi = (\infty, \infty, \dots)$ is the characteristic of type Q, then we set $R = R_\chi$ for simplicity and call it the ring of **pseudo–rational** numbers.

2. If all the p–components of χ are equal to zero except for $p_1, p_2, \dots, p_k$, then let $R_\chi = (K_{p_1} \oplus \cdots \oplus K_{p_k}) \oplus Q$ and, for an element $r \in R_\chi$, $|r|$ is its **rational component**.

The following properties of the rings R_χ are more or less obvious.

1. R_χ is a ring with unit $1 = (1, 1, \dots)$. Every element $r \in R_\chi$ determines uniquely a rational number $|r|$; $|1| = 1$, $|0| = 0$.

2. Denote by ε_p the element of R_χ such that its p–component is equal to 1 and all other components, including the rational one if it exists, are equal to 0. Then ε_p is an idempotent and the ideal $\varepsilon_p R_\chi = K_p$ is a direct summand of the ring R_χ, $|\varepsilon_p| = 0$. Note that ε_p may be equal to 0, namely, if the ring K_p is zero.

3. Elements of the form

$$\varepsilon = \varepsilon_{p_1} + \cdots + \varepsilon_{p_k}, \tag{a}$$

where the prime numbers $p_1, \dots, p_k$ are distinct, $0 < k \in Z$, and elements of the form

$$1 - \varepsilon, \varepsilon = \varepsilon_{p_1} + \cdots + \varepsilon_{p_k}, \tag{b}$$

and also 0 and 1 form the set of all the idempotents of the ring R_χ.

4. For every two elements $r_1, r_2 \in R_\chi$,

$$|r_1 \pm r_2| = |r_1| \pm |r_2|, |r_1 r_2| = |r_1| \, |r_2|.$$

Since for every idempotent $\varepsilon = \varepsilon_{p_1} + \cdots + \varepsilon_{p_k}$ we have a direct decomposition $R_\chi = K_{p_1} \oplus \cdots \oplus K_{p_k} \oplus (1 - \varepsilon) R_\chi$, it follows that every element $r \in R_\chi$ can be presented in the form

$$r = \varepsilon r + (1 - \varepsilon)|r|,$$

where ε is an idempotent of the form $\varepsilon = \varepsilon_{p_1} + \cdots + \varepsilon_{p_k}$. Here all the prime divisors of the denominator of $|r|$ are in the set $\{p_1, \dots, p_k\}$ and the multiplication by $|r|$ is uniquely determined in $(1 - \varepsilon) R_\chi$.

5. An ideal of the ring R_χ is called **substantial** if it contains an element r with $|r| \neq 0$. Every substantial ideal contains an idempotent $1 - \varepsilon$ of the form (b) and, therefore, an ideal $(1 - \varepsilon) R_\chi$.

6. An ideal of the ring R_χ is called a **finitely component ideal** if it is contained in an ideal εR_χ, where ε is an idempotent of the form (a). Substantial ideals containing an idempotent $1 - \varepsilon$ are in one–to–one correspondence with those finitely component ideals that are in $\varepsilon R_\chi = K_{p_1} \oplus \cdots \oplus K_{p_k}$. The last two propeties imply the next one.

7. Every substantial ideal of the ring R_χ is principal. Every finitely component ideal of the ring R_χ is principal. Every principal ideal of the ring R_χ is either substantial or finitely component.

8. Every ideal of the ring R_χ either is substantial or is contained in the ideal generated by all the idempotents ε_p, that is in the **main** ideal $T = \bigoplus_p K_p$. If the characteristic χ has infinitely many non–zero components, then the ring R_χ contains non–principal ideals, the main ideal T itself for example.

9. **The homomorphisms of descent**. Let $\chi = (m_p) \geq \kappa = (k_p)$ be arbitrary characteristics. Define the maps

$$\xi^\chi_\kappa : R_\chi \to R_\kappa \tag{3}$$

in the following way. Let $r = (\alpha_p) \in R_\chi$. Then $\xi^\chi_\kappa(r) = (\beta_p) \in R_\kappa$, where $\beta_p = \alpha_p$ if $k_p = m_p$. If $k_p < m_p$ then $\beta_p \equiv \alpha_p \pmod{p^{k_p}}$. If the ring R_κ contains a rational component, that is, if almost all p–components of the characteristic κ are equal to zero, then the rational component of the element $\xi^\chi_\kappa(r)$ is equal to $|r|$. It is easy to see that all the maps ξ^χ_κ are surjective homomorphisms of rings for every pair of characteristics $\chi \geq \kappa$ and that $|\xi^\chi_\kappa(r)| = |r|$.

10. $\xi^\chi_\chi = \mathrm{id}$, $\xi^\kappa_\lambda \xi^\chi_\kappa = \xi^\chi_\lambda$, for every three characteristics $\chi \geq \kappa \geq \lambda$. All the homomorphisms (3) simultaneously form a direct and an inverse system.

Theorem 3.2 *The direct limit of the system* (3) *is the ring of rational numbers Q. The inverse limit of the system* (3) *is the ring of pseudo-rational numbers R.*

Proof. The direct limit coincides with the final object of this direct system, that is, Q. The inverse limit coincides with the initial object, that is, R. □

Theorem 3.3 *The following rings and only they are homomorphic images of the ring R_χ, where $\chi = (m_p)$.*

1. *Rings R_κ, where $\chi \geq \kappa$.*
2. *Finite direct sums of the rings $L_{p_1} \oplus \cdots \oplus L_{p_n}$, where L_{p_i} is either the ring of the classes of residues modulo k_{p_i} if $k_{p_i} \leq m_{p_i} < \infty$, or the ring of p_i–adic integers if $m_{p_i} = \infty$.*

Proof. If an ideal I of the ring $R_\chi \subset \prod_p K_p$ is not substantial, then it is of the form $I = \bigoplus_p p^{k_p} K_p$, where $k_p \leq m_p$ for every prime p and $p^{k_p} = 0$ if $k_p = \infty$. Then $R_\chi / I \cong R_\kappa$, where $\kappa = (k_p)$. In particular, if $\kappa = (0, 0, \ldots)$, then $I = T$ is the main ideal and $R_\kappa = R_\chi / T \cong Q$. The homomorphism

$R_\chi \to R_\kappa$ coincides with a homomorphism of descent ξ^χ_κ. If the ideal I is substantial, then it contains an ideal $J = (1-\varepsilon)R_\chi$, where $1-\varepsilon$ is an idempotent of the form (b). The ring $R_\chi/J \cong K_{p_1} \oplus \cdots \oplus K_{p_n}$ is a finite direct sum of rings of residue classes or rings of p-adic integers by one for each prime number p. The ring R_χ/I must have the same structure as a homomorphic image of the ring R_χ/J. □

Corollary 3.4 *The following ideals of the ring of pseudo-rational numbers R and only they are maximal: the main ideal $T = \bigoplus_p \hat{Z}_p$ and the ideals $(1+(p-1)\varepsilon_p)R$. Besides these, only the ideals $(1-\varepsilon_p)R$ are prime.*

4 Modules over the ring of pseudo–rational numbers

Due to the homomorphisms of descent $\xi^\chi_\kappa : R_\chi \to R_\kappa$, $\chi \geq \kappa$, every R_κ–module is an R_χ–module as well. In particular, all the R_χ–modules may be considered as modules over the ring of pseudo–rational numbers R. Also note that every module over the ring of universal integers $\prod_p \hat{Z}_p$ is an R–module due to the natural embedding $R \to \prod_p \hat{Z}_p$. Finally, every $\hat{Z}_p$–module is an R–module due to the projection $\pi_p : R \to \hat{Z}_p$.

Definition 4.1 Let χ be any characteristic. An R_χ–module D is called **divisible** if its additive group is divisible torsion free and $rd = |r|d$ for all elements $d \in D$ and $r \in R$. If an R_χ–module does not contain non–zero divisible submodules, it is called **reduced.**

Proposition 4.2 *Let χ be any characteristic. Then:*

1. *Every R_χ-module M contains a largest divisible submodule* $\operatorname{div} M$.

2. $\operatorname{div} M = \{m \in M \mid Tm = 0\}$ *where $T = \bigoplus_p K_p$ is the main ideal of the ring R_χ.*

Proof. It is sufficient to prove that the submodule $D = \{m \in M \mid Tm = 0\}$ is the greatest divisible submodule. Let D contain an element m of order p. Since $1-\varepsilon_p$ is divisible by p in R_χ, we obtain $m - \varepsilon_p m = (1-\varepsilon_p)m = 0$. Therefore $m = \varepsilon_p m$ and since $\varepsilon_p \in T$ and $m \in D$, $m = \varepsilon_p m = 0$. It follows that the additive group of D is torsion free. For any element $d \in D$ and for every prime p, by the same reason, $d = (1-\varepsilon_p)d$ is divisible by p in D, that is the additive group of D is divisible. By Property 4 of Section 3, every element $r \in R_\chi$ may be presented in the form $r = \varepsilon r + (1-\varepsilon)|r|$. Since $\varepsilon \in T$, we have $rd = \varepsilon rd + (1-\varepsilon)|r|d = |r|d$ for every element $d \in D$, hence D is a divisible R_χ–module. Let c be any element of any divisible submodule C. Then $\varepsilon_p c = |\varepsilon_p| c = 0$ for every prime number p, hence $Tc = 0$ and $c \in D$. Therefore, $C \subset D$ and D is greatest. □

Theorem 4.3 *Let χ be any characteristic. Every divisible submodule of an R_χ–module M is its direct summand. In particular, $M = \operatorname{div} M \oplus A$, where A is a reduced R_χ–module.*

Proof. Let D be any divisible submodule of the R_χ–module M. By the Lemma of Zorn, there exists a maximal R_χ–submodule $B \subset M$ such that $D \cap B = 0$. Let $M \neq D \oplus B$. Suppose that $x \in M$ and $x \notin D \oplus B$. Consider the ideal $I = \{r \in R_\chi \mid rx \in D \oplus B\}$.

If I is substantial then it is principal, that is $I = r_0 R_\chi$ for an element $r_0 \in R_\chi$, $|r_0| \neq 0$. Then $r_0 x = d + b = |r_0| d' + b = r_0 d' + b$, where $d, d' \in D, b \in B$. Hence $r_0\,(x - d') = b$. We prove that the intersection of the module generated by B and the element $x - d'$ with the module D is equal to 0, $D \cap \langle B, x - d' \rangle_{R_\chi} = 0$. In fact, if $d_1 = b_1 + r(x - d')$, then $r = s r_0$ and $d_1 = b_1 + sb = 0, s \in R_\chi$. Since $x - d' \notin B$ we obtain a contradiction to the choice of B.

If the ideal I is not substantial then already $D \cap \langle B, x \rangle_{R_\chi} = 0$ and we obtain the same contradiction to the choice of B again. In fact, an equality

$$d_1 = b_1 + rx$$

implies that $|r| = 0$ and $r = \varepsilon r$ for an idempotent $\varepsilon = \varepsilon_{p_1} + \cdots + \varepsilon_{p_k}$. Subtracting from this equality the same equality multiplied by ε, we finally obtain $d_1 = b_1 - \varepsilon b_1$. Therefore, this element belongs to $D \cap B$ and must be equal to 0. The first statement of the theorem is proved.

In particular, $M = \operatorname{div} M \oplus A$. If A contains a non–zero divisible submodule D, then the module $\operatorname{div} M \oplus D$ is divisible contradicting the definition of $\operatorname{div} M$. Therefore, A is reduced. □

The additive group of a reduced R–module is not necessarily reduced as it is shown in the following two examples of reduced R–modules.

Example 4.4 The abelian group Q/Z is a module over its ring of endomorphisms $\prod_p \hat{Z}_p$ and, therefore, an R–module.

Example 4.5 The field of p–adic numbers is a $\hat{Z}_p$–module and, therefore, an R–module due to the projection $\pi_p : R \to \hat{Z}_p$.

Definition 4.6 Let $A \subset B$ be R–modules. A is called a **pure** submodule of B, if the solvability of every equation $rx = a$, $r \in R$, $a \in A$, in the module B implies its solvability in the module A. In particular, the purity of R–modules implies the purity of their additive groups.

For every R–module M, its submodule $M_p = \varepsilon_p M$ is a module over the ring of p–adic integers $\hat{Z}_p = \varepsilon_p R$. The projections $M \to M_p$, where $m \longmapsto m_p = \varepsilon_p m$, induce a homomorphism $\gamma : M \to \prod_p M_p$, where $m \mapsto (m_p) \in \prod_p M_p$. By Proposition 4.2, $\ker \gamma = \operatorname{div} M$. Since the homomorphism γ is an embedding for reduced R–modules, we can identify M with $\gamma(M)$ and assume that every reduced R–module M is embedded in the R–module $\prod_p M_p$ such that $\bigoplus_p M_p \subset M \subset \prod_p M_p$.

Theorem 4.7 *For every reduced R–module M, both embeddings $\bigoplus_p M_p \subset M \subset \prod_p M_p$ are pure.*

Proof. Let $a \in \bigoplus_p M_p$ and $rb = a$ for $r \in R, b \in M$. There exists an idempotent $\varepsilon = \varepsilon_{p_1} + \cdots + \varepsilon_{p_k}$ such that $\varepsilon a = a$. Then $r(\varepsilon b) = a$ and $\varepsilon b \in \bigoplus_p M_p$. Therefore, the embedding $\bigoplus_p M_p \subset M$ is pure.

Let $a \in M$ and $rb = a$ for $r \in R, b \in \prod_p M_p$. If $|r| = 0$ the equation has a solution in $\bigoplus_p M_p \subset M$ as it is shown above. Otherwise, the element $r \in R$ may be presented in a form $r = \varepsilon r + (1-\varepsilon)|r|$, where $|r| \neq 0$. Then $r(\varepsilon b) + |r|((1-\varepsilon)b) = a$. The elements εb and $r(\varepsilon b)$ both belong to $\bigoplus_p M_p \subset M$. Hence $|r|((1-\varepsilon)b) \in M$. Since the R–module $M/\bigoplus_p M_p$ is divisible, we have $(1-\varepsilon)b + x \in M$ for an element $x \in \bigoplus_p M_p$. Hence $(1-\varepsilon)b \in M$ and $b = \varepsilon b + (1-\varepsilon)b \in M$ and the embedding $M \subset \prod_p M_p$ is pure as well. □

Theorem 4.8 *Let χ be a characteristic of a locally free type. The additive group of every R_χ–module belongs to the class $\mathcal{M}_\chi$. Every group G of the class $\mathcal{M}_\chi$ admits uniquely a multiplication by elements of the ring R_χ such that G becomes an R_χ–module.*

Proof. The multiplications by ε_p on an R_χ–module determine a system of projections on its additive group which satisfies the conditions $\mathcal{M}_1$ and $\mathcal{M}_2$.

On the other hand, let $G \in \mathcal{M}_\chi$ with a system of projections $\pi_p, p \in P$. First, prove the uniqueness of the multiplication by elements of the ring R_χ. Assume that G has the structure of an R_χ–module. Then the multiplications by ε_p on G form a system of projections which is unique by Theorem 1.4. Hence $\varepsilon_p x = \pi_p(x)$ for every $x \in G$. Let $\chi = (m_p)$. The multiplication by elements of $Z/p^{m_p}Z$ can be defined uniquely on the p–group $\pi_p(G) = T_p(G)$. Since each element $r = (\alpha_p) \in R_\chi$ may be presented in a form $r = \varepsilon r + (1-\varepsilon)|r|$, where $\varepsilon = \varepsilon_{p_1} + \cdots + \varepsilon_{p_k}$, it follows that

$$rx = \alpha_{p_1}\pi_{p_1}(x) + \cdots + \alpha_{p_k}\pi_{p_k}(x) + \frac{m}{n}(x - (\pi_{p_1}(x) + \cdots + \pi_{p_k}(x))), \tag{4}$$

where $\frac{m}{n} = |r|$. Since all the prime divisors of the integer n are in the set $\{p_1, \dots, p_k\}$, the division by n is uniquely determined on the direct summand $(\mathrm{id} - (\pi_{p_1} + \cdots + \pi_{p_k}))G$ of the group G. Therefore, the multiplication by elements $r \in R_\chi$ must be defined by (4) that really determines the structure of an R_χ–module on the group G as is easily seen. □

By means of this theorem, we can identify a group of the class $\mathcal{M}_\chi$ with an R_χ–module which is uniquely determined. In particular, all the groups of the class $\mathcal{M}$ are modules over the ring of pseudo–rational numbers R.

Theorem 4.9 *Let* $G_1, G_2 \in \mathcal{M}$. *Then* $\mathrm{Hom}_R(G_1, G_2) = \mathrm{Hom}_Z(G_1, G_2)$.

Proof. Obviously, $\mathrm{Hom}_R(G_1, G_2) \subset \mathrm{Hom}_Z(G_1, G_2)$. Let $f : G_1 \to G_2$ be a homomorphism of the abelian groups. It follows from Theorem 1.6 that $f(\varepsilon_p x) = \varepsilon_p f(x)$ for every $x \in G_1$ and for every prime p. Hence $f(r\varepsilon_p x) = r\varepsilon_p f(x)$ for every $r \in R$. Therefore $f(r\varepsilon x) = r\varepsilon f(x)$ for every $x \in G_1$, for every $r \in R$, and for every idempotent of the form $\varepsilon = \varepsilon_{p_1} + \cdots \varepsilon_{p_k}$. On the other hand, if all the prime divisors of an integer n are in the set $\{p_1, \dots, p_k\}$, then the division by n is uniquely determined in the subgroups $(1-\varepsilon)G_1 \subset G_1$ and $(1-\varepsilon)G_2 \subset G_2$, moreover $f((1-\varepsilon)G_1) \subset (1-\varepsilon)G_2$. Therefore $f(\frac{m}{n}(1-\varepsilon)x) = \frac{m}{n}f(x - \varepsilon x) = \frac{m}{n}(f(x) - \varepsilon f(x)) = \frac{m}{n}(1-\varepsilon)f(x)$. By Property 4., Section 3, an arbitrary element $r \in R$ may be presented in the form $r = r\varepsilon + |r|(1-\varepsilon)$. Then $f(rx) = f(r\varepsilon x + |r|(1-\varepsilon)x) = f(r\varepsilon x) + f(|r|(1-\varepsilon)x) = r\varepsilon f(x) + |r|(1-\varepsilon)f(x) = rf(x)$. Hence $f \in \mathrm{Hom}_R(G_1, G_2)$. □

The last two theorems mean the following surprising fact. The "forgetful" functor, which associates with each R_χ–module its additive group, does not forget anything actually.

5 Finitely generated modules over the ring of pseudo–rational numbers

Let M be a finitely generated R–module with a system $x_1, \dots, x_n$ of generators. Then, obviously, the $\widehat{Z}_p$–module $M_p = \varepsilon_p M$ is generated by the elements $\varepsilon_p x_1, \dots, \varepsilon_p x_n$. The finitely generated p–adic module M_p may be presented as a direct sum of cyclic $\widehat{Z}_p$–modules.

$$M_p = \langle a_1 \rangle_{\widehat{Z}_p} \oplus \cdots \oplus \langle a_n \rangle_{\widehat{Z}_p} . \tag{5}$$

Some direct summands may be equal to 0. Every cyclic $\widehat{Z}_p$–module is isomorphic either to $Z/p^{k_{pi}}Z$, where $0 \leq k_{pi} \in Z$, or to $\widehat{Z}_p$.

Definition 5.1 The R–module M is called R**–torsion** if all the $\widehat{Z}_p$–modules $M_p = \varepsilon_p M$ are torsion.

Taking into account Theorems 4.8 and 4.9, we can formulate the next theorem in the following way.

Theorem 5.2 *The category of groups $\mathcal{G}$ coincides with the category of R–torsion finitely generated R–modules.*

Proof. Let $G \in \mathcal{G}$. Then by Theorem 2.3, $G \in \mathcal{M}$ and it has a unique R–module structure by Theorem 4.8. Furthermore, by Theorem 2.3, $G = H \oplus A$, where the group H satisfies the strong projection condition and A is a finite group. Let $F = \langle x_1, \ldots , x_n \rangle$ be a full free subgroup of the group H such that $\pi_p(F) = \pi_p(H)$ (or equivalently $\varepsilon_p F = \varepsilon_p H$) for each prime number p. We will prove that $H = \langle x_1, \ldots , x_n \rangle_R$. Let $x \in H$ be an arbitrary element. Since F is full,

$$mx = m_1 x_1 + \cdots + m_n x_n$$

for suitable integer coefficients, $m \neq 0$. Let $p_1, \ldots , p_k$ be all the prime divisors of m. Consider an idempotent $\varepsilon = \varepsilon_{p_1} + \cdots + \varepsilon_{p_k}$ of the form (a) (see Section 3.3). Then $H = \varepsilon H \bigoplus (1 - \varepsilon)H$. In the group $(1 - \varepsilon)H$ as well as in the ideal $(1 - \varepsilon)R$, every element is uniquely divisible by m. Therefore

$$(1 - \varepsilon)x = r_1 x_1 + \cdots + r_n x_n, \tag{6}$$

where $r_i = \frac{m_i}{m}(1 - \varepsilon) \in R$, $i = 1, \ldots , n$. On the other hand,

$$\varepsilon x \in \varepsilon H = T_{p_1}(H) \oplus \cdots \oplus T_{p_k}(H),$$

where $T_p(H)$ is the p-primary component of H. By the strong projection condition

$$\varepsilon x = s_1 x_1 + \cdots + s_n x_n \tag{7}$$

for suitable $s_1, \ldots , s_n \in \varepsilon R$. It follows from (6) and (7) that

$$x = \varepsilon x + (1 - \varepsilon)x = (r_1 + s_1)x_1 + \cdots + (r_n + s_n)x_n .$$

That is, $x_1, \ldots, x_n$ is a system of generators of the R–module H. Adding to them a finite number of elements of the group A, we obtain a finite system of generators of the R–module G.

Conversely, let G be a finitely generated R–torsion R–module and let $x_1, \ldots, x_n$ be a system of generators. Then all the direct summands in the decomposition (5) of the module $\varepsilon_p G$ are finite for every prime number p. Hence $p^{m_p}(\varepsilon_p G) = 0$ for a suitable integer $m_p \geq 0$. Therefore, G is an R_χ–module for the characteristic $\chi = (m_p)$ and its additive group belongs to the class $\mathcal{M}_\chi \subset \mathcal{M}$ by Theorem 4.8. Consider the finitely generated subgroup $\langle x_1, \ldots, x_n \rangle$ of the additive group G. A group $m\langle x_1, \ldots, x_n \rangle$ is either free or equal to 0 for a sufficiently great integer m. But in the last case we obviously get that G is finite and so $G \in \mathcal{G}$. Let $F = m\langle x_1, \ldots, x_n \rangle$ be a free group. Let $r \in R, x \in G$. It follows from Property 4, Section 3, that $krx = lx$ for a sufficiently great integer k and some $l \in Z$. Now it is easy to see that F is a full subgroup, that is G/F is torsion. Besides, $\varepsilon_p F = \varepsilon_p G$ for all prime numbers such that $m\varepsilon_p G = \varepsilon_p G$, that is for almost all prime numbers. Therefore, G satisfies the projection condition. Since all the p–primary components of G are finite, we have $G \in \mathcal{G}$ by Theorem 2.3. Thus the class of groups $\mathcal{G}$ coincides with the class of R–torsion finitely generated modules over the ring of pseudo–rational numbers R.

Since by Theorem 4.9 every group homomorphism in the class $\mathcal{G}$ is an R–module homomorphism, we can say that these two categories coincide. □

References

[1] U. Albrecht, *A-projective resolutions and an Azumaya theorem for a class of mixed abelian groups*, to appear.

[2] U. Albrecht, H.P. Goeters and W. Wickless, *The flat dimension of mixed abelian groups as E–modules*, Rocky Mt. J. Math. **25** (1995), 569–590.

[3] U. Albrecht and J. Hausen, *Mixed abelian groups with the summand intersection property*, Lecture Notes in Pure and Applied Math. **182**, Marcel Dekker, New York (1996), 123-132.

[4] U. Albrecht and J.–W. Jeong, *Homomorphisms between A-projective abelian groups and left Kasch rings*, to appear.

[5] S. Files and W. Wickless, *The Baer–Kaplansky theorem for a class of global mixed abelian groups*, Rocky Mt. J. Math. **26** (1996), 593–613.

[6] S. Files and W. Wickless, *Direct sums of self–small mixed groups*, to appear.

[7] A.A. Fomin and W. Wickless, *Categories of mixed and torsion-free finite rank abelian groups*, in Abelian Groups and Modules (A. Facchini and C. Menini, eds) Kluwer Acad. Publ. (1995), 185–192.

[8] A.A. Fomin and W. Wickless, *Self-small mixed abelian groups G with G/T(G) finite rank divisible*, Comm. in Algebra **26** (1998), 3563–3580.

[9] A.A. Fomin and W. Wickless, *Quotient divisible abelian groups*, Proc. Amer. Math. Soc. **126** (1998), 45–52.

[10] L. Fuchs, *Infinite Abelian Groups*, Vols. **I** and **II**, Academic Press (1971, 1973).

[11] S. Glaz and W. Wickless, *Regular and principal projective endomorphism rings of mixed abelian groups*, Comm. in Algebra **22** (1994), 1161–1176.

[12] C. Vinsonhaler and W. Wickless, *Realizations of finite dimensional algebras over the rationals*, Rocky Mt. J. Math. **24** (1994), 1553–1565.

[13] R. Warfield, *Homomorphisms and duality for torsion-free groups*, Math. Z. **107** (1968), 189–200.

[14] W. Wickless, *A functor from mixed groups to torsion-free groups*, Contemp. Math. **171** (1995), 407–419.

Trends in Mathematics, © 1999 Birkhäuser Verlag Basel/Switzerland

The Baer-Kaplansky theorem for direct sums of self-small mixed groups

William J. Wickless

Abstract

We consider Baer-Kaplansky type theorems for the class $\sum \mathcal{G}$. The class consists of arbitrary direct sums of the form $G = \bigoplus_{i \in I} G_i$, where each G_i is a reduced group in the class $\mathcal{G}$ of self-small mixed group of finite torsion-free rank. Without loss, the summands G_i can be taken to be either essentially indecomposable mixed groups or finite cyclic groups. Let $G' = \bigoplus_{i \in I} G'_i$ be another such group with $E(G') \cong E(G)$. Then $G' \cong G$ if the essentially indecomposable summands of both groups G, G' are A_0-cyclic or if the essentially indecomposable summands of G are A_0-cyclic and satisfy certain projection conditions on torsion-free elements. A reduced group $H \in \mathcal{G}$ is called A_0-cyclic if $Q \bigotimes H$ is a cyclic A_0 module for some commutative subalgebra $A_0 \subset Q \bigotimes E(H)$.

All groups in the paper are abelian and, unless otherwise indicated, all homomorphisms and tensor products are taken over the integers.

The class $\mathcal{G}$ consists of self-small mixed groups G such that G/T is finite rank divisible. Recall that a group G is **self-small** if $Hom(G, \bigoplus_{i \in I} G_i) \cong \bigoplus_{i \in I} Hom(G, G_i)$ for any indexed set of groups $\{G_i : i \in I\}$ with each $G_i \cong G$. An alternate description of the groups in $\mathcal{G}$ is that they are of the form $V \bigoplus G$, where V is finite rank torsion-free divisible and G is a reduced mixed group with torsion subgroup $T = \bigoplus T_p$ satisfying: ($\mathcal{G}$-1) each T_p is finite; ($\mathcal{G}$-2) G is (can be embedded as) a pure subgroup of $\prod_p T_p$; ($\mathcal{G}$-3) G/T is finite-rank (necessarily divisible); and ($\mathcal{G}$-4) if F is a maximal rank free subgroup of G, then F projects onto T_p for almost all primes p. To see that the above two descriptions of $\mathcal{G}$ coincide, see [AGW]. Note that in the definition $G = T$ is allowed. In this case $F = 0$ and conditions ($\mathcal{G}$-1) and ($\mathcal{G}$-4) require that G be a finite group. Thus, reduced groups in $\mathcal{G}$ are either honest mixed groups (neither torsion nor torsion-free) or are finite. The category $\mathcal{QG}$ has objects the groups in $\mathcal{G}$ and maps quasi-homomorphisms; that is $Hom_{\mathcal{QG}}(G, G') = Q \bigotimes Hom(G, G')$. Write $G \sim G'$ to indicate that two groups $G, G' \in \mathcal{G}$ are quasi-isomorphic, equivalently are isomorphic in $\mathcal{QG}$. We can characterize quasi-isomorphism for groups in $\mathcal{G}$ as follows: For $G, G' \in \mathcal{G}$, $G \sim G'$ if and only if $G = G_0 \bigoplus B, G' = G'_0 \bigoplus C$ with $G_0 \cong G'_0$ as abelian groups and B, C finite.

In [W] it is shown that $\mathcal{QG}$ is a Krull-Schmidt category with indecomposables of arbitrary finite rank. The mixed indecomposables in $\mathcal{QG}$ are the **essentially indecomposable groups:** honest mixed groups G such that if $G = H \bigoplus K$ then one of the summands is finite. We note that nontrivial $\mathcal{QG}$ direct sum decompositions of a group $G \in \mathcal{G}$ correspond to proper idempotents in $Hom_{\mathcal{QG}}(G, G)$. Hence $G \in \mathcal{G}$ is essentially indecomposable if and only if the finite dimensional rational algebra $Q \bigotimes End(G)$ is local. A number of papers have dealt with various properties of the groups in $\mathcal{G}$ and the category $\mathcal{QG}$ (see [A], [AGW], [AH], [AJ], [FiW-1,2], [FoW1-3], [GW], [VW], [W]).

In [A], Albrecht iniated the study of arbitrary direct sums of groups, $\bigoplus_{i \in I} G_i$, where each G_i is a reduced group in $\mathcal{G}$. The main theorem of [A] was a strong Azumaya-type result: a summand of such a group is of the form $\bigoplus_{j \in J} H_j$, where for each $j \in J$ there exists $i = i(j) \in I$ such that H_j is isomorphic to a summand of G_i. Albrecht's result indicates that this class of groups, which we denote by $\sum \mathcal{G}$, is a good candidate for a reasonably tractable class of mixed groups. In particular, it follows directly from the definition that the class $\mathcal{G}$ is closed under taking direct summands. Hence, Albrecht's theorem implies that the class $\sum \mathcal{G}$ is also closed under taking direct summands.

In this paper we prove some versions of a Baer-Kaplansky type theorem for groups in $\sum \mathcal{G}$. Since each reduced group in $\mathcal{G}$ is either finite or is a finite direct sum of essentially indecomposable mixed groups in $\mathcal{G}$, we henceforth assume that each $G \in \sum \mathcal{G}$ is written in the form $G = \bigoplus_{i \in I} G_i$ with each G_i either essentially indecomposable or of prime power order. Let $\mathcal{H}$ be a set consisting of one representative from each quasi-isomorphism class of essentially indecomposable groups in $\mathcal{G}$. For $H \in \mathcal{H}$ and $G = \bigoplus_{i \in I} G_i$ in $\sum \mathcal{G}$, let $\sigma_H(G)$ be the cardinality of the set $I_H(G) = \{i \in I : G_i \sim H\}$. In [FiW-2] it was shown that the $\sigma_H(G)$'s are invariants of G, that is are independent of the particular decomposition $G = \bigoplus_{i \in I} G_i$ chosen. Moreover, the main theorem of that paper proved that these invariants, together with the Ulm invariants, form a complete set of isomorphism invariants for groups in $\sum \mathcal{G}$. On the other hand, as an application of the Main Theorem of [MT], any pair of reduced abelian groups with isomorphic endomorphism rings will have isomorphic torsion subgroups, hence equal Ulm invariants. Since all groups in $\sum \mathcal{G}$ are reduced, the following theorem is immediate.

Theorem 0.1 *Let G, G' be in $\sum \mathcal{G}$ with $E = End(G)$ ring isomorphic to $E' = End(G')$. Then $G \cong G'$ if and only if $\sigma_H(G) = \sigma_H(G')$ for each $H \in \mathcal{H}$.*

Henceforth, $G = \bigoplus_{i \in I} G_i, G' = \bigoplus_{j \in J} G'_j$ will be groups in $\sum \mathcal{G}$ with isomorphic endomorphism rings. Our goal now is to find conditions under which $\sigma_H(G) = \sigma_H(G')$. As a first step in this direction we have the following lemma.

Lemma 0.2 *For all $H \in \mathcal{H}$ with $I_H(G)$ nonempty, there exists a summand of G' of the form $\bigoplus_{i \in I_H(G)} G'_i$, where each G'_i is essentially indecomposable and $G'_i \sim G'_k$ for all $i, k \in I_H(G)$. The groups G'_i can be chosen such that $End(G'_i) \cong End(G_i)$ for all $i \in I_H(G)$.*

Proof. Let $\theta : E \cong E'$ be a ring isomorphism and fix $H \in \mathcal{H}$. Put $I_H = I_H(G)$. For each $i \in I_H$, let e_i be projection of G onto G_i. The collection $\{e_i : i \in I_H\}$ is a set of orthogonal idempotents in E so that, if $e'_i = \theta e_i$, then $\{e'_i : i \in I_H\}$ is a set of orthogonal idempotents in E'. Hence, if $G'_i = e'_i G'$, we have that each G'_i is a summand of G' and $\sum_{i \in I_H} G'_i = \bigoplus_{i \in I_H} G'_i \subset G'$.

Let $i \in I_H$. As is well known, $End(G'_i) \cong e'_i E' e'_i \cong e_i E e_i \cong End(G_i)$. Since $G_i \sim H$ is essentially indecomposable, $Q \bigotimes End(G_i)$ is local. Hence the isomorphic rational algebra $Q \bigotimes End(G'_i)$ is local as well, so that G'_i is essentially indecomposable.

Fix $k \in I_H$. We show that, for each $i \in I_H$, $G'_i \sim G'_k$. For each $i \in I_H$, $G_i \sim G_k$ since both groups are quasi-isomorphic to H. Thus, we can construct $u_i, v_i \in E$ such that $v_i u_i = t_i e_k$, $u_i v_i = t_i e_i$ for some positive integer t_i. Let $u'_i = \theta u_i, v'_i = \theta v_i$. Then $v'_i u'_i = t_i e'_k$, $u'_i v'_i = t_i e'_i$. It follows that $G'_i \sim G'_k$.

Let $e \in E$ be projection of $\bigoplus_{i \in I} G_i$ onto $\bigoplus_{i \in I_H} G_i$ and $e' = \theta e$. We claim that $\bigoplus_{i \in I_H} G'_i = e'G'$, proving that $\bigoplus_{i \in I_H} G'_i$ is a summand of G'. Since $e'e'_i = e'_i$ it follows that $\bigoplus_{i \in I_H} G'_i \subset e'G'$. Call an idempotent $\rho \in E$ an **essentially indecomposable idempotent** if ρG is an essentially indecomposable summand of G. By Albrecht's theorem, ρG is isomorphic to a summand of G_i for some i. Since the class $\mathcal{G}$ is closed under taking summands both ρG and $e\rho G$ are in $\mathcal{G}$. By properties ($\mathcal{G}$-1) and ($\mathcal{G}$-4) of the alternate definition of groups in $\mathcal{G}$, there exists a finite subset $I_\rho \subset I_H$ with $e\rho G = (\sum_{i \in I_\rho} e_i)\rho G$. Hence, each essentially indecomposable idempotent $\rho \in E$ satisfies the **finite support property** (defined with respect to e and $\{e_i : i \in I_H\}$): $e\rho = (\sum_{i \in I_\rho} e_i)\rho$ for some finite subset $I_\rho \subset I_H$. Trivially, indecomposable idempotents in E (corresponding to $Z(p^k)$ summands of G) also satisfy this finite support property.

Note that an idempotent $\rho \in E$ is essentially indecomposable if and only if $Q \bigotimes End(\rho G) \cong Q \bigotimes \rho E \rho$ is local. Hence $\rho \to \theta \rho$ is a bijection between the essentially indecomposable idempotents in E and E'. It follows that the finite support property, defined analogously in E' with respect to e' and $\{e'_i : i \in I_H\}$, holds for indecomposable or essentially indecomposable indempotents in E'. Thus, if f'_j is projection of $G' = \bigoplus_{j \in J} G'_j$ onto one of its indecomposable or essentially indecomposable summands G'_j, then $e'f'_j$ is a finite sum $e'f'_j = (\sum e'_i)f'_j$. Hence, for $x \in G'$, $e'f'_j x \in \bigoplus_{i \in I_H} G'_i$.

Since any $x \in G'$ is a finite sum $x = \sum f'_j x$, then, for $x \in e'G'$ we have $x = e'x = \sum e' f_j x \in \bigoplus_{i \in I_H} G'_i$. Thus, $\bigoplus_{i \in I_H} G'_i = e'G'$ and the proof is complete. □

Our final task to prove a Baer-Kaplansky theorem for $\sum \mathcal{G}$ is to find some decent conditions under which essentially indecomposable groups $H, H' \in \mathcal{G}$ with isomorphic endomorphism rings will be quasi-isomorphic. (Incidently, in view of Theorem 1 and the fact that $\mathcal{QG}$ is a Krull-Schmidt category, quasi-isomorphic groups in $\mathcal{G}$ with isomorphic endomorphism rings must actually be isomorphic.) We need a notion from [FiW-1].

Let $H \in \mathcal{G}$ be reduced and $A = Q \bigotimes E(H)$ be the $\mathcal{QG}$ endomorphism ring of H. The algebra A acts on the finite dimensional rational vector space $V = H/T(H)$ (recall property $\mathcal{G}$-3 in the alternate definition of $\mathcal{G}$) in the natural way. A group $H \in \mathcal{G}$ is **A_0-cyclic** if there is a commutative subalgebra $A_0 \subset A$ such that V is a cyclic A_0 module. The theme of [FiW1] was to find various conditions under which if $H, H' \in \mathcal{G}$ are such that: (1) $E(H) \cong E(H')$, (2) rank H = rank H', and (3) H is A_0-cyclic, then $H \cong H'$. (By the rank of a mixed group we mean the torsion-free rank.) The justification for imposing assumption (3) on H at the outset is that, although rank one groups in $\mathcal{G}$ are trivially determined by their endomorphism rings, if $H \in \mathcal{G}$ is any group of rank two that is not A_0-cyclic, then there is an $H' \in \mathcal{G}$ of rank two with $E(H) \cong E(H')$ but such that H' is not quasi-isomorphic to H.

Lemma 0.3 *Let H (respectively H') be an A_0-cyclic (respectively A'_0-cyclic) group in $\mathcal{G}$. Suppose that H and H' are essentially indecomposable and that $E(H) \cong E(H')$. Then $H \sim H'$.*

Proof. Let $V' = H'/T(H')$ and suppose that $A_0 v = V$ and $A'_0 v' = V'$ for $v \in V, v' \in V'$ and commutative subalgebras $A_0 \subset A, A'_0 \subset A' = Q \bigotimes E(H')$. In the first part of the proof of Lemma 2 in [FiW1], it is shown that, if H is any A_0-cyclic group in $\mathcal{G}$ with $A_0 v = V$ and H' in $\mathcal{G}$ has $E(H) \cong E(H')$, then for any $v' \in V'$ we can construct $\Phi \in Q \bigotimes Hom(H, H')$ with $\Phi(v) = v'$. (Here we regard $Q \bigotimes Hom(H, H')$ as contained in the natural way in $Hom(V, V')$). Hence, under our assumptions, there exist $\Phi \in Q \bigotimes Hom(H, H'), \Phi' \in Q \bigotimes Hom(H', H)$ such that $\Phi(v) = v', \Phi'(v') = v$. The element $\Phi'\Phi \in A$ fixes v and, consequently, is not nilpotent. Since $A = End_{\mathcal{QG}}(H)$ is local, there exists an automorphism $\Psi \in A$ with $\Psi\Phi'\Phi = 1_V$. Thus, Φ is monic. By symmetry, there exists an automorphism $\Psi' \in A'$ with $\Psi'\Phi\Phi' = 1_{V'}$. Thus, Φ is epic. It follows that $H \sim H'$. □

We have all the ingredients for our first Baer-Kaplansky theorem for $\sum \mathcal{G}$. We write that G_i is $A_0(i)$-cyclic or that G'_j is $A'_0(j)$-cyclic to make clear the fact that the commutative subalgebras $A_0(i) \subset Q \bigotimes End(G_i)$ (resp. $A'_0(j) \subset Q \bigotimes End(G'_j)$) with $V_i = G_i/T(G_i)$ a cyclic $A_0(i)$ module (resp. $V'_j = G'_j/T(G'_j)$ a cyclic $A'_0(j)$ module) can vary with the indices i, j.

Theorem 0.4 *Let $G = \bigoplus_{i\in I} G_i, G' = \bigoplus_{j\in J} G'_j$ be groups in $\sum \mathcal{G}$ with $E(G) \cong E(G')$. Suppose that we can write $G = \bigoplus_{i\in I} G_i, G' = \bigoplus_{j\in J} G'_j$ such that each essentially indecomposable G_i is $A_0(i)$-cyclic and each essentially indecomposable G'_j is $A'_0(j)$-cyclic. Then $G \cong G'$.*

Proof. Since the hypotheses of our theorem are symmetric with respect to G, G', it suffices to prove that $\sigma_H(G) \leq \sigma_H(G')$ for each $H \in \mathcal{H}$. Then we can cite Theorem 1 to complete the proof. Say that $\sigma_H(G) > 0$. Using Lemma 2, we have $G' = (\bigoplus_{i\in I_H} G'_i) \bigoplus X$ where all the G'_i essentially indecomposable and $E(G'_i) \cong E(G_i)$ for each i. By Albrecht's theorem and the fact that all the summands G'_j are either finite or essentially indecomposable, for each $i \in I_H$ there exists $j \in J$ with $G'_i \sim G'_j$. Since the A_0-cyclic property is invariant under quasi-isomorphism G'_i is $A'_0(j)$-cyclic. Now we can apply Lemma 3 to conclude that $G'_i \cong G_i$ for $i \in I_H$ so that $\sigma_H(G) \leq \sigma_H(G')$. □

A group G in $\sum \mathcal{G}$ is **mixed completely decomposable** if G can be written $G = \bigoplus G_i$ with rank $G_i = 1$ for all i.

Corollary 0.5 *Let G, G' be mixed completely decomposable groups with $E(G) \cong E(G')$. Then $G \cong G'$.*

Proof. Any rank one mixed group is A_0-cyclic. □

Let $H \subset \prod T_p(H)$ be a reduced group in $\mathcal{G}$ and, as in Lemma 3, for $x \in H$ and a fixed prime p, let x_p be the natural projection of x into $T_p(H)$. We say that an infinite set of primes P is a **weak projective set (strong projective set)** for H if each nonzero element $x \in H$ of infinite order has $x_p \neq 0$ ($x_p \notin pT_p(H)$) for almost all $p \in P$. It is enough to check either projection property on a (finite) Z-basis for any maximal rank free subgroup of H. This fact makes it easy to construct a large variety of groups in $\mathcal{G}$ with strong or weak projective sets. The following results are proved in [FiW1].

Theorem 0.6 *Let H, H' be reduced groups in $\mathcal{G}$ with $E(H) \cong E(H')$, rank H = rank H' and suppose that H is A_0-cyclic. (a) If H has a strong* *projective set then $H \cong H'$. (b) If almost all the p-torsion subgroups $T_p(H)$ are cyclic and H has a weak projective set then $H \cong H'$.*

We note that the property of having a strong or weak projective set is closed under quasi-isomorphism. Arguing as in the proof of Theorem 4, using Theorem 6 in place of Lemma 3, yields our second Baer-Kaplansky theorem for $\sum \mathcal{G}$.

Theorem 0.7 *Let $G = \bigoplus G_i, G' = \bigoplus G'_j$ be groups in $\sum \mathcal{G}$ with $E(G) \cong E(G')$ and let n be a fixed positive integer such that each essentially indecomposable summand G_i of G, G'_j of G' has rank n . Suppose that each essentially indecomposable G_i is $A_0(i)$-cyclic. (a) If each essentially indecomposable G_i has a strong projective set P_i then $G \cong G'$. (b) If each essentially indecomposable G_i has a weak projective set P_i and has almost all of its p-torsion subgroups cyclic, then $G \cong G'$.*

References

[A] U. Albrecht, *A-projective resolutions and an Azumaya theorem for a class of mixed abelian groups*, to appear.

[AGW] U. Albrecht, H. P. Goeters and W. Wickless, *The flat dimension of mixed abelian groups as E-modules*, Rocky Mt. J. Math. **25**(2) (1995), 569–590.

[AH] U. Albrecht and J. Hausen, *Mixed abelian groups with the summand intersection property*, Lecture Notes in Pure and Applied Math. **182** (1996), 123–132.

[AJ] U. Albrecht and J.–W. Jeong, *Homomorphisms between A-projective abelian groups and left Kasch rings*, to appear.

[FiW1] S. Files and W. Wickless, *The Baer-Kaplansky theorem for a class of global mixed abelian groups*, Rocky Mt. J. Math. **26**(2) (1996), 593–613.

[FiW2] S. Files and W. Wickless, *Direct sums of self-small mixed groups*, to appear.

[FoW1] A. Fomin and W. Wickless, *Categories of mixed and torsion-free abelian groups*, Abelian Groups and Modules, Kluwer, Boston, (1995), 185–192.

[FoW2] A. Fomin and W. Wickless, *Self-small mixed abelian groups G with $G/T(G)$ finite rank divisible*, to appear in Comm. in Algebra.

[FoW3] A. Fomin and W. Wickless, *Quotient divisible groups*, to appear in AMS Proceedings.

[GW] S. Glaz and W. Wickless, *Regular and principal projective endomorphism rings of mixed abelian groups*, Comm. in Algebra **22**(4) (1994), 1161–1176.

[MT] W. May and E. Toubassi, *Endomorphisms of abelian groups and the theorem of Baer and Kaplansky*, J. Algebra **43** (1976), 1–13.

[VW] C. Vinsonhaler and W. Wickless, *Realizations of finite dimensional algebras over the rationals*, Rocky Mt. J. Math. **24**(4) (1994), 1553–1565.

[W] W. Wickless, *A functor from mixed groups to torsion-free groups*, Contemp. Math. **171** (1995), 407–419.

Trends in Mathematics, © 1999 Birkhäuser Verlag Basel/Switzerland

Finite rank Butler groups with small typesets

David M. Arnold* and Manfred Dugas*

The class of finite rank Butler groups is known to be complicated. Nevertheless, the structure of an indecomposable group G can sometimes be determined by restricting the typeset T of G. For example, if T is linearly ordered, then G has rank 1, while if T has branching order at most 2, then G is an indecomposable almost completely decomposable group [Butler 65].

In this note, attention is directed towards finite rank Butler groups with typesets of small cardinality. Example 8 consists of explicit non-trivial examples of indecomposable Butler groups and almost completely decomposable groups with small typeset and arbitrarily large finite rank.

A finite rank *Butler group* is a pure subgroup G of a finite direct sum of torsion-free abelian groups of rank 1. The *typeset* of G is the set of isomorphism classes of pure rank-1 subgroups, a finite subset of the distributive lattice of all types. Since G has finite rank, the typeset of G is closed under finite meets [Arnold 82]. Included in the class of Butler groups are *almost completely decomposable (acd) groups*, those torsion-free abelian groups G containing a finite direct sum of rank-1 groups C as a subgroup of finite index. The *critical typeset* of an acd group G is the set of isomorphism classes of rank-1 summands of C.

Let G be a finite rank indecomposable Butler group with typeset T. If $\mid T \mid \leq 3$, then either T is linearly ordered or else $T = T_2 = \{\tau_0, \tau_1, \tau_2\}$ with τ_1 and τ_2 incomparable and $\tau_0 = \tau_1 \cap \tau_2$, recalling that T is meet closed. In the latter case, G is an acd group with rank $G \leq 2$. Groups with typeset equal to T_2 are investigated in [Cruddis 70], [Lewis 92], [Arnold, Dugas 93], and [Files, Goebel 96A]. Infinite rank Butler groups with typeset contained in T_2 are also considered in the last two references.

The case that $\mid T \mid = 4$ is more complicated. First suppose that $T = T_3 = \{\tau_0, \tau_1, \tau_2, \tau_3\}$ with τ_1, τ_2, and τ_3 incomparable and $\tau_0 = \tau_1 \cap \tau_2 = \tau_1 \cap \tau_3$. Strongly indecomposable groups with typeset contained in T_3 have rank ≤ 2 [Arnold 89]. If $\tau_1 + \tau_2 + \tau_3 =$ type Q, then each indecomposable acd group with typeset contained in T_3 has rank ≤ 3 but there are indecomposable Butler groups that are not acd groups [Arnold, Dugas 95B].

Next assume that T_3 is p-locally free for some prime p, as defined below. In this case $\tau_1 + \tau_2 + \tau_3 \neq$ type Q. The category of finite rank Butler groups G with typeset $G \leq T_3$ is *wild modulo p*, i.e. if R is a finite dimensional

*Research supported, in part, by the Baylor University Summer Sabbatical Program

Z/pZ-algebra with $R = R'/pR'$ for some Z_p-algebra R' that is finitely generated and free as a Z_p-module then there is a G with R a factor ring of $\text{End}(G)_p$, the localization of the endomorphism ring of G at the prime p. In this case, there is no hope of being able to classify indecomposable Butler groups with typeset contained in T_3 due to the diversity of endomorphism rings. Even more, if $j \geq 3$, then the category of acd groups G with $p^j G \leq G(\tau_1) \oplus G(\tau_2) \oplus G(\tau_3)$ and typeset $G \leq T_3$ is wild modulo p [Arnold, Dugas 98]. Special classes of infinite rank Butler groups with typeset contained in T_3 are considered in [Files, Goebel 96B].

Another possibility with $| T | = 4$ is $T = \{\tau_0, \tau_1, \tau_2, \tau_3\}$ with τ_1 and τ_2 incomparable, $\tau_3 \geq \tau_1 \cup \tau_2$, and $\tau_0 = \tau_1 \cap \tau_2$. It is not difficult to see that, for indecomposable groups, this case reduces to the case that $T = T_2$.

The remaining case with $| T | = 4$, the typeset of primary concern in this paper, is $T = \{\tau_0, \tau_1, \tau_2 > \tau_3\}$ with τ_1 incomparable to both τ_2 and τ_3 and $\tau_0 = \tau_1 \cap \tau_2 = \tau_1 \cap \tau_3$. Let L denote the lattice of types generated by T. This notation for T and L is fixed for the remainder of the paper.

Strongly indecomposable finite rank Butler groups with typeset contained in L have rank ≤ 2 [Arnold 89]. Since T is meet closed and the lattice of all types is distributive, L is the join closure of T and $T \backslash \{\tau_0\} = JI(L)$, the poset of join irreducible elements of the lattice L. We stipulate that τ_0, the least element of T, is not join irreducible. Notice that $JI(L) = \{\tau_1, \tau_2 > \tau_3\}$ and if G is a Butler group with typeset $G \leq L$, then typeset $G \leq T$ if and only if $G(\tau_1) \cap G(\tau_3) = 0 = G(\tau_1) \cap G(\tau_2)$. This is because $G(\tau_1 \cup \tau_3) = G(\tau_1) \cap G(\tau_3)$ and $G(\tau_1 \cup \tau_2) = G(\tau_1) \cap G(\tau_2)$.

Isomorphism at p is the local version of near-isomorphism for finite rank torsion-free abelian groups, first defined in [Lady 75]; also see [Arnold 82] and [Arnold, Dugas 97]. An *isomorphism at p* category of torsion-free abelian groups, p a prime, means that morphisms in the category are $\text{Hom}_Z(G, H)_p$ the localization of $\text{Hom}_Z(G, H)$ at p for groups G and H in the category.

Let p be a prime and define $B(L, j)_p$ to be the isomorphism at p category of finite rank Butler groups G such that typeset $G \leq L$ and $p^j G \leq \Sigma\{G(\tau) : \tau \in JI(L)\}$. Restriction to $B(L, j)_p$ is not a serious loss of generality since if G is a finite rank Butler group with typeset contained in L, then G is isomorphic at p to a group in $B(L, j)_p$ for some j. Furthermore, if G is indecomposable in $B(L, j)_p$ for some p, then G is indecomposable as a group.

Define $C_{crit}(L, j)_p$ to be the full subcategory of $B(L, j)_p$ consisting of all acd groups G in $B(L, j)_p$ with critical typeset contained in $JI(L)$ and $UC_{crit}(L, j)_p$ the full subcategory of $C_{crit}(L, j)_p$ consisting of all acd groups G that are *uniform*, i.e. $G/(\Sigma\{G(\tau) : \tau \in JI(L)\})$ is a free $Z/p^j Z$-module.

An additive category with finite direct sums has *finite representation type* if there are only finitely many isomorphism classes of indecomposable objects. The set of types T is *p-locally free* if X_p, the localization of X

at p, is isomorphic to Z_p for each rank-1 group X with type $X \in T$. The following theorem summarizes all, to our knowledge, that is known about finite rank Butler groups with typeset contained in T.

Theorem 1 *Assume that* $T = \{\tau_0, \tau_1, \tau_2 > \tau_3\}$ *is p-locally free for some* p *with* $\tau_0 = \tau_1 \cap \tau_2 = \tau_1 \cap \tau_3$ *and* L *is the lattice of types generated by* T.

(a) [Arnold, Dugas 98] The category $UC_{crit}(L, j)_p$ *has finite representation type and each indecomposable has rank* ≤ 3.

(b) [Arnold, Dugas 97] The category $B(L, 0)_p$ *has infinite representation type, but each indecomposable has rank* ≤ 2.

Part (b) of the previous theorem demonstrates that there is a distinction between finite representation type and *rank-finite representation type*, i.e. the ranks of indecomposables are bounded.

The main tools of this paper are Z_p-representations of a finite poset S. Define $rep_{fpure}(S, Z_p, m)$ to be the category with objects consisting of representations $U = (U_0, U_i : i \in S)$ such that U_0 is a finitely generated free Z_p-module, each U_i is a pure submodule hence a summand of $U_0, U_i \leq U_j$ if $i \leq j$ in S, and $p^m U_0 \leq \Sigma\{U_i : i \in S\}$. Morphisms from representations U to U' are those Z_p-homomorphisms: $U_0 \to U_0'$ with $f(U_i) \leq U_i'$ for each i in S. A representation U is *decomposable* provided that $U_0 = V_0 \oplus W_0$ with $U_i = (V_0 \cap U_i) \oplus (W_0 \cap U_i)$ for each $i \in S$.

For a poset S, S^{op} is the set S with the opposite order.

Lemma 1 [Richman 94] *Assume* T *is p-locally free for some* p *and* L *is the lattice of types generated by* T. *Then* $B(L, j)_p$ *is equivalent to the category* $rep_{fpure}(S, Z_p, j)$ *where* $S = JI(L)^{op}$. *The equivalence is given by* $G \to (G_p, G(\tau)_p : \tau \in JI(L))$.

As noted above, if $T = \{\tau_0, \tau_1, \tau_2 > \tau_3\}$, then $JI(L) = \{\tau_1, \tau_2 > \tau_3)$, hence $JI(L)^{op}$ can be identified with the poset $(1, 2) = \{1, 2 < 3)$.

The category of representations of an antichain S_n with n elements over a field, in this case $\text{rep}(S_n, Z/pZ)$, is well understood from the point of view of representation type, see [Arnold 89]. The next theorem, together with Lemma 1, relates $\text{rep}(S_{n+2}, Z/pZ)$ to $B(L, 2n+1)_p$ via $\text{rep}_{fpure}(S, Z_p, 2n+1)$. This provides a systematic construction of indecomposable groups with typeset contained in L from indecomposable representations in $\text{rep}(S_{n+2}, Z/pZ)$, as demonstrated in Example 8.

Define $E(n, k)$ to be the subcategory of $\text{rep}(S_{n+2}, k)$, consisting of representations $V = (V_0, V_i : 1 \leq i \leq n+2)$ with $V_0 = V_1 \oplus \cdots \oplus V_n$, there is a fixed s with $V_i = k^s$ for each $i \leq n, V_{n+1} \oplus V_{n+2} \leq V_0$, and $V_{n+1} = (1 + 1 + \cdots + 1)k^s = \{(x, x, \ldots, x) : x \in k^s\}$.

Theorem 2 *Let $S = (1, 2 < 3)$, p a prime, and $n \geq 0$. There is a correspondence $\phi\colon E(n, Z/pZ) \to rep_{fpure}(S, Z_p, 2n+1)$ such that if $V \in E(n, Z/pZ)$ is indecomposable, then $\phi(V)$ is indecomposable.*

Proof. Let $k = Z/pZ$ and write $V_0 = (V_1 \oplus \ldots \oplus V_{n+2})/K$, where K is the row space of a k-matrix

$$M' = \begin{pmatrix} I & I & \cdots & I & I & 0 \\ M_1' & M_2' & \cdots & M_n' & 0 & I \end{pmatrix}$$

with each M_i' an $s \times s$ k-matrix, s the k-dimension of V_i for each $i \leq n$. Choose $s \times s$ Z_p-matrices M_i with $M_i(\text{mod } p) = M_i'$ for $1 \leq i \leq n$. Here $M(\text{mod } p)$ denotes the k-matrix obtained by reducing each element of M modulo pZ_p.

Let W and X be free Z_p-modules with rank $2s$ and Y and Z free Z_p-modules with rank $(n+1)s$. Define

$$\phi(V) = U = (U_0, U_1, U_2 \leq U_3) \in \text{rep}_{fpure}(S, Z_p, 2n+1)$$
$$\text{by } U_1 = W \oplus Z, U_3 = X \oplus Y \oplus Z, U_0 =$$
$$W \oplus \ X \oplus Y \oplus Z + (1/p^{2n+1})(1+1+0+0)Z_p^{2s} \text{ and } U_2 \leq X \oplus Y \oplus Z$$

is the row space of the matrix $M_U = (E_1, E_2, E_3 = I) =$

$$\left(\begin{array}{cc|ccccc|cccccc} pM_1 & p^2I & p^3I & 0 & \cdots & 0 & 0 & I & 0 & \cdots & 0 & 0 \\ p^2M_2 & p^3I & 0 & p^5I & \cdots & 0 & 0 & 0 & I & \cdots & 0 & 0 \\ & & & & \cdots & & & & & & & \\ p^nM_n & p^{n+1}I & 0 & 0 & \cdots & p^{2n+1}I & 0 & 0 & 0 & \cdots & I & 0 \\ p^{n+1}I & 0 & 0 & 0 & \cdots & 0 & p^{2n+3}I & 0 & 0 & \cdots & 0 & I \end{array}\right)$$

The columns of M_U are indexed by X, Y, and Z respectively, the I's stand for $s \times s$ identity matrices, the 0's stand for $s \times s$ matrices of all zeros, and $(1+1+0+0)Z_p^{2s} = \{w \oplus x \oplus 0 \oplus 0 \in U_0 : w \in W, x \in X, w = x\}$. Note that U_2 is a summand of U_3 since $E_3 = I$.

Let f be an endomorphism of U. Then $f\colon U_1 = W \oplus Z \to U_1 = W \oplus Z$, $f : U_3 = X \oplus Y \oplus Z \to U_3 = X \oplus Y \oplus Z, f : Z \to Z$ as $U_1 \cap U_3 = Z$, and $f : U_2 \to U_2$, Thus, f can be represented as an endomorphism of $W \oplus X \oplus Y \oplus Z$ with matrix

$$f = \begin{pmatrix} f_{11} & 0 & 0 & f_{14} \\ 0 & f_{22} & f_{23} & f_{24} \\ 0 & f_{32} & f_{33} & f_{34} \\ 0 & 0 & 0 & f_{44} \end{pmatrix}$$

where $f_{11} : W \to W, f_{14} : W \to Z, f_{22} : X \to X, f_{23} : X \to Y, f_{24} : X \to Z, f_{32} : Y \to X, f_{33} : Y \to Y, f_{34} : Y \to Z, f_{44} : Z \to Z$, and $f : U_2 \to U_2$.

We adopt the notation that if A is a free Z_p-module, then $A(\text{mod } p)$ stands for A/pA, while if h is an endomorphism of A, then $h(\text{mod } p)$ stands for the coset $h + p\text{End } A$. Since $f : U_0 \to U_0$, $f : (1+1+0+0)Z_p^{2s}(\text{mod } p^{2n+1}) \to (1+1+0+0)Z_p^{2s}(\text{mod } p^{2n+1})$. Hence, $f_{11} \equiv f_{22}(\text{mod } p^{2n+1})$ and $f_{14} \equiv f_{23} \equiv f_{24} \equiv 0 \ (\text{mod } p^{2n+1})$. Consequently,

$$f \equiv \begin{pmatrix} f_{11} & 0 & 0 & 0 \\ 0 & f_{11} & 0 & 0 \\ 0 & f_{32} & f_{33} & f_{34} \\ 0 & 0 & 0 & f_{44} \end{pmatrix} \qquad (\text{mod } p^{2n+1})$$

Since $f : U_2 \to U_2$, we have $M_U f \ |_{U(3)} \ (\text{mod } p^{2n+1}) \equiv (E_1 f_{11} + E_2 f_{32} \mid E_2 f_{33} \mid E_2 f_{34} + f_{44}) \equiv (gE_1 \mid gE_2 \mid gI)(\text{mod } p^{2n+1})$ for some matrix g. Hence,

$$\begin{aligned} E_2 f_{34} + f_{44} &\equiv gI(\text{mod } p^{2n+1}), \\ E_1 f_{11} + E_2 f_{32} &\equiv gE_1 \equiv (E_2 f_{34} + f_{44} I)E_1(\text{mod } p^{2n+1}), \text{ and} \\ E_2 f_{33} &\equiv gE_2 \equiv (E_2 f_{34} + f_{44} I)E_2(\text{mod } p^{2n+1}). \end{aligned}$$

Write $f_{33} = (a_{ij})_{(n+1)\times(n+1)}$ as an endomorphism of $Y = (Z_p^s)^{n+1}$ and $f_{44} = (b_{ij})_{(n+1)\times(n+1)}$ as an endomorphism of $Z = (Z_p^s)^{n+1}$, each a_{ij} and b_{ij} an $s \times s$ Z_p-matrix representing an endomorphism of Z_p^s. Then

$$\begin{aligned} E_2 f_{33} \equiv (p^{2i+1} a_{ij})(\text{mod } p^{2n+1}), f_{44} E_2 \equiv (p^{2j+1} b_{ij})(\text{mod } p^{2n+1}), \\ E_2 f_{33} \equiv E_2 f_{34} E_2 + f_{44} E_2 (\text{mod } p^{2n+1}). \end{aligned}$$

Hence, $p^{2i+1} a_{ij} \equiv p^{2j+2+2i} x_{ij} + p^{2j+1} b_{ij}(\text{mod } p^{2n+1})$ for $1 \le i, j \le n+1$ with $(x_{ij}) = f_{34}$. Therefore, $p^{2i+1} a_{ij} \equiv p^{2j+1} b_{ij}(\text{mod } p^{max(2i+1,2j+1)})$.

If $i < j$, then $a_{ij} \equiv 0(\text{mod } p)$ while if $i > j$, then $b_{ij} \equiv 0(\text{mod } p)$. Hence, $f_{33}(\text{mod } p)$ is lower triangular, $f_{44}(\text{mod } p)$ is upper triangular, and $a_{ii} \equiv b_{ii}(\text{mod } p)$ for each $1 \le i \le n+1$ by the congruence above.

Write $f_{11} = (c_{ij})_{2\times 2}$ as an endomorphism of $X = (Z_p^s)^2$ with c_{ij} representing an endomorphism of Z_p^s. Recall that $E_1 f_{11} + E_2 f_{32} \equiv (E_2 f_{34} E_1 + f_{44} E_1)(\text{mod } p^{2n+1})$. From the last row of this equation,

$$c_{11} \equiv b_{n+1,n+1}(\text{mod } p), \text{ and } c_{12} \equiv 0(\text{mod } p^2).$$

The first row yields

$$c_{11} \equiv b_{11}(\text{mod } p) \equiv c_{22}(\text{mod } p).$$

From the first column, $M_i c_{11} \equiv b_{ii} M_i(\text{mod } p)$ for each $i \le n$. Finally, the second column of this equation gives

$$c_{22} \equiv b_{ii}(\text{mod } p) \text{ for each } i \leq n.$$

Let $a = c_{11}(\text{mod } p) \in End_k(k^s)$. By the results of the preceding two paragraphs, $a \equiv c_{22}(\text{mod } p) \equiv b_{ii}(\text{mod } p) \equiv a_{ii}(\text{mod } p)$ for each $i \leq n+1$ and $aM_i \equiv M_i a(\text{mod } p)$ for each $i \leq n$. In particular, $aM' = M'a$, recalling the definition of M' and the fact that that $M'_i = M_i(\text{mod } p)$. Consequently, $aI_{(n+2)s\times(n+2)s}$ is a representation endomorphism of V, being an endomorphism of $V_1 \oplus \cdots \oplus V_{n+2}$ with $a(K) \leq K$ as K is the row space of M'.

We now have that $f(\text{mod } p) = aI + N$, where

$$N = \begin{pmatrix} N_1 & 0 & 0 & 0 \\ 0 & N_2 & 0 & 0 \\ 0 & N_{32} & N_3 & N_{34} \\ 0 & 0 & 0 & N_4 \end{pmatrix}$$

N is a nilpotent matrix with zeros on the diagonal and N_i is of the form

$$N_1 = \begin{pmatrix} 0 & 0 \\ * & 0 \end{pmatrix} N_2 = \begin{pmatrix} 0 & 0 \\ * & 0 \end{pmatrix} N_3 = \begin{pmatrix} 0 & 0 \\ * & 0 \end{pmatrix} N_4 = \begin{pmatrix} 0 & * \\ 0 & 0 \end{pmatrix}$$

corresponding to $f_{11}(\text{mod } p), f_{22}(\text{mod } p), f_{33}(\text{mod } p)$, and $f_{44}(\text{mod } p)$ respectively.

Assume that V is indecomposable and that $f^2 = f \in \text{End } U$. Then $(f(\text{mod } p))^2 = (aI+N)^2 = f(\text{mod } p) = aI + N$. Since N is a nilpotent matrix with 0's on the diagonal, $a^2 = a$. Thus, $aI_{(n+2)s\times(n+2)s}$ is an idempotent endomorphism of V. But V is indecomposable, so $a = 0$ or 1. If $a = 0$, then $f(\text{mod } p) = N$ is both nilpotent and idempotent, hence $f(\text{mod } p) = 0$. Thus, $f \in p\text{End } U$ with $f^2 = f$, whence $f = 0$. On the other hand if $a = 1$, then $1 - f = 0$ and $f = 1$. This shows that U is indecomposable. □

Corollary 3 *Suppose that $T = \{\tau_0, \tau_1, \tau_2 > \tau_3\}$ is p-locally free with $\tau_0 = \tau_1 \cap \tau_2 = \tau_1 \cap \tau_3$ and L is the lattice of types generated by T.*

(a) $B(L, j)_p$ has rank-infinite representation type for $j \geq 5$.

(b) $B(L, j)_p$ is wild modulo p for $j \geq 7$.

Proof. (a) Apply Theorem 2 and Lemma 1 with $n = 2$ and representations in $\text{rep}(S_4, k)$ of the form

$$V = (V_0, V_1, V_2, V_3, V_4) = (k^s \oplus k^s, k^s \oplus 0, 0 \oplus k^s, (1+1)k^s, (1+A)k^s)$$

for some indecomposable $s \times s$ k-matrix A. In this case,

$$M' = \begin{pmatrix} I & I & I & 0 \\ I & A & 0 & I \end{pmatrix}$$

Note that $V \in E(2,k)$ as $V_3 + V_4 \leq V_1 + V_2$. Moreover, End $V \approx k[x]/ < f(x)^e >$, where $f(x)$ is an irreducible polynomial and $f(x)^e$ is the minimal polynomial of the matrix A. Thus, V is indecomposable. By Theorem 2 and Lemma 1 there is an indecomposable G in $B(L,5)_p$ (this group is defined explicitly in Example 8.a). Since there is an indecomposable A for each s, namely a Jordan block matrix, $B(T,j)_p$ has indecomposables of arbitrarily large finite rank if $j \geq 2n+1 = 5$.

(b) Let R be a finite dimensional Z/pZ-algebra and R' a Z_p-algebra that is finitely generated and free as a Z_p-module with $R'(\text{mod } p) = R$. The same proof as in [Brenner 74] for fields shows that $R' = C(A,B)$, the centralizer of A and B, for some $s \times s$ Z_p-matrices A and B with $R = C(A(\text{mod } p), B(\text{mod } p))$. Define $V \in \text{rep}(S_5, k)$, where

$$V = (V_0, V_1, V_2, V_3, V_4, V_5) =$$
$$(k^s \oplus k^s \oplus k^s, k^s \oplus 0 \oplus 0, 0 \oplus k^s \oplus 0, 0 \oplus 0 \oplus k^s, (1+1+1)k^s, (1+A'+B')k^s)$$

for $s \times s$ k-matrices $A' = A(\text{mod } p)$ and $B' = B(\text{mod } p)$. In this case,

$$M' = \begin{pmatrix} I & I & I & I & 0 \\ I & A & B & 0 & I \end{pmatrix}$$

and End $V = C(A', B') = R$. Now let $U = \phi(V) \in \text{rep}_{fpure}(S, Z_p, 7)$ as defined in Theorem 2. Then $R' = C(A,B) \leq \text{End } U$. Finally, define a ring homomorphism $\pi\colon \text{End } U/p\text{End } U \to \text{End } V$ by $\pi(f) = aI_{(n+2)s \times (n+2)s}$, with notation as in the proof of Theorem 2. Since $\pi(R') = R$, it follows that π is an epimorphism. Now apply Lemma 1 to see that $B(L,7)$ is wild modulo p. □

Corollary 3 does not address the question of finding finite rank Butler groups G with typeset $\leq T$. This is because the representation $U = (U_0, U_1, U_2 < U_3)$ constructed in the proof of Theorem 2 has $U_1 \cap U_3 = Z \neq 0$. But U_i corresponds to $G(\tau_i)_p$ in Lemma 1. Hence, $G(\tau_1) \cap G(\tau_3) = G(\tau_1 \cup \tau_3) \neq 0$ and so $\tau_1 \cup \tau_3 \in$ typeset G but not T.

The next proposition shows that finite rank Butler groups with typeset contained in T are actually acd groups.

Proposition 4 *If $G \in B(L,j)_p$ is indecomposable with typeset $G \leq T = \{\tau_0, \tau_1, \tau_2 > \tau_3\}, \tau_0 = \tau_1 \cap \tau_2 = \tau_1 \cap \tau_3$, and L is the lattice of types generated by T, then G is an acd group with critical typeset contained in $\{\tau_1, \tau_2 > \tau_3\}$ and $G(\tau_1) \cap G(\tau_2) = G(\tau_1) \cap G(\tau_3) = 0$.*

Proof. As noted above, if typeset $G \leq T = \{\tau_0, \tau_1, \tau_2 > \tau_3\}$, then $G(\tau_2) \leq G(\tau_3)$ and $G(\tau_1) \cap G(\tau_3) = 0$. Moreover, $p^j G \leq G(\tau_1) \oplus G(\tau_3)$ since $G \in B(L,j)_p$ and $JI(L) = T$. But $G(\tau_1)$ and $G(\tau_3)$ are finite rank Butler groups with linearly ordered typeset, hence completely decomposable. Finally, note that $JI(L) = T \geq$ typeset $G \geq$ critical typeset G so that G is an acd group in $C_{crit}(L, j)_p$. □

There is also a category of Z_p-representations related to $C_{crit}(L, j)_p$. Define $Crep_{fpure}(S, Z_p, j)$ to be the full subcategory of $\text{rep}_{fpure}(S, Z_p, j)$ consisting of representations (U_0, U_i) with rank $U_0 = \Sigma\{rank\ U_i/U_i^{\#} : i \in S\}$, where $U_i^{\#}$ is the purification of $\Sigma\{U_s : i < s \in S\}$ in U_i. The next lemma is a special case of Lemma 1.

Lemma 5 [Arnold, Dugas 98] *Suppose that $T = \{\tau_0, \tau_1, \tau_2 < \tau_3\}$ is p-locally free with $\tau_0 = \tau_1 \cap \tau_2 = \tau_1 \cap \tau_3$ and L the lattice of types generated by T. Then $C_{crit}(L, j)$ is equivalent to $Crep_{fpure}(S, Z_p, j)$.*

Theorem 6 *Let $S = (1, 2 \leq 3)$, $k = Z/pZ$ a field, and $n \geq 0$. There is a correspondence $\phi: rep(S_n, k) \to Crep_{fpure}(S, Z_p, 2n)$ such that if $V \in rep(S_n, k)$ is indecomposable, then $\phi(V)$ is indecomposable.*

Proof. It suffices to define $\phi(V)$ for $V = (V_0, V_i : 1 \leq i \leq n+2) \in \text{rep}(S_n, k)$ with no summands of dimension 1. In this case, $V_0 = V_1 + \cdots + V_n$, otherwise the decomposition $V_0 = (V_1 + \cdots + V_n) \oplus W$ induces a representation decomposition of V and W is a direct sum of dimension 1 representations. Write $V_0 = (V_1 \oplus \cdots \oplus V_n)/K$, where K is the row space of a k-matrix $M' = (M_1' \mid M_2' \mid \cdots \mid M_n')$ for some $s \times s_i$ matrix M_i'. As V has no summands of dimension $1, s \geq s_i$ for each i. For notational convenience, we assume that $t = s_i$ for each i, the proof of the general case is the same.

Choose $s \times t$ Z_p-matrices M_i with $M_i(\text{mod } p) = M_i'$ for $1 \leq i \leq n$. Let W and X be free Z_p modules with rank tn and $W = W_1 \oplus \cdots \oplus W_n, X = X_1 \oplus \cdots \oplus X_n$, where W_i, X_i, and Y_i are free Z_p-modules of rank t. Let Y be a free Z_p-module of rank s. Define

$$\phi(V) = U = (U_0, U_1, U_2 \leq U_3) \text{ by } U_1 = W,\ U_3 = X \oplus Y,$$
$$U_0 = (W \oplus X \oplus Y) + U_4$$

where $U_4 = \Sigma\{Z(x_i + x_i)/p^{2(n-i)+2} : 1 \leq i \leq n,\ x_i \in X_i = Z_p = W_i\}$ and

$$U_2 \leq X \oplus Y$$

is the row space of a $(tn) \times (tn + s)$ matrix

$$M = \begin{pmatrix} & p^n M_1^{tr} \\ I_{tnxtn} & \ddots \\ & pM_n^{tr} \end{pmatrix}$$

with each M_i^{tr}, the transpose of M_i, a $t \times s$ Z_p-matrix. The columns of M are labelled by X and Y. In particular, U_2 is generated by elements of the form $x_i \oplus x_i p^{n-i+1} M_i^{tr} \in X \oplus Y$ for $x_i \in X_i = Z_p^t$.

Now $U \in Crep_{fpure}(S, Z_p, 2n)$, noting that $p^{2n}U_0 \leq U_1 \oplus U_3 \leq U_0$, U_2 is a summand of U_3 due to the presence of the identity matrix in M, and rank $U_0 =$ rank $U_1 +$ rank $U_3 =$ rank U_1+ rank U_2+ rank (U_3/U_2).

Let f be a representation endomorphism of U. Then $f: U_1 = W \to U_1 = W$, $f: U_3 = X \oplus Y \to U_3 = X \oplus Y$, and $f(\text{mod } p^2n): U_4(\text{mod } p^2n) \to U_4(\text{mod } p^{2n})$. Thus, f can be represented as an endomorphism of $W \oplus X \oplus Y$ with matrix

$$f = \begin{pmatrix} f_{11} & 0 & 0 \\ 0 & f_{22} & f_{23} \\ 0 & f_{32} & f_{33} \end{pmatrix}$$

where $f_{11}: W \to W$, $f_{22}: X \to X$, $f_{23}: X \to Y$, $f_{32}: Y \to X$, $f_{33}: Y \to Y$ and $f: U_2 \to U_2$. Write $f_{11} = (\alpha_{ij})_{nxn}$, $f_{22} = (\beta_{ij})_{nxn}$, where each α_{ij} and β_{ij} is a $t \times t$ Z_p-matrix.

Now $\alpha_{ii} \equiv \beta_{ii}(\text{mod } p^{2(n-i)+2})$ for each $1 \leq j \leq n$ and $\alpha_{ii} \equiv \beta_{ii} \equiv 0(\text{mod } p^{2(j-i)})$ whenever $j > i$, since $f: U_4(\text{mod } p^{2n}) \to U_4(\text{mod } p^{2n})$. To see this, let $x_i \in X_i$, $w_i \in W_i$ and write $f((x_i - w_i)/p^{2(n-i)+2)}) = \Sigma\{z_j(x'_j + w'_j)/p^{2(n-j)+2} : 1 \leq j \leq n\}(\text{mod } p^{2n})$ for some $z_j \in Z$, $x'_j \in X_j$, and $w'_j \in W_j$. Then, $f(x_i - w_i) = (x_i)f_{11} - (w_i)f_{22} = \Sigma\{z_j(x'_j + w'_j)/p^{2(i-j)+2} : 1 \leq j \leq n\}(\text{mod } p^{2n}) \in W \oplus X$. It follows from this equation that $\alpha_{ii} \equiv \beta_{ii}(\text{mod } p^{2(n-i)+2})$ and $\alpha_{ij} \equiv \beta_{ij} \equiv 0(\text{mod } p^{2(j-i)})$ if $j > i$. In particular,

$$f_{11} \equiv f_{22}(\text{mod } p) \text{ is upper triangular.}$$

Now write $f_{23} = (\gamma_{ij})_{n\times 1}$ and $f_{32} = (\delta_{ij})_{1\times n}$ with γ_{ij} a $t \times t$ Z_p-matrix and δ_{ij} an $s \times s$ Z_p-matrix. Since $f: U_2 \to U_2$,

$$\begin{gathered} f(x_i + x_i p^{n-i+1} M_i^{tr}) = \\ ((x_i)f_{22} + (x_i p^{n-i+1} M_i^{tr})f_{32}) \oplus ((x_i)f_{23} + (x_i p^{n-i+1} M_i^{tr})f_{33}) = \\ (\Sigma\{x_i\beta_{ij} + x_i p^{n-i+1} M_i^{tr}\delta_{ij} : 1 \leq j \leq n\}) \oplus (x_i\gamma_{i1} + x_i p^{n-i+1} M_i^{tr} f_{33}) = \\ \Sigma\{x'_j \oplus x'_j p^{n-j+1} M_j^{tr}\} : j \leq n\} \in U_2 \leq X \oplus Y. \end{gathered}$$

for some $x'_j \in X_j$. Equating coefficients and substituting yields $\gamma_{ij} \equiv 0(\text{mod } p^{2(n-i)+2})$ for each i. In particular,

$$f_{23} \equiv 0(\text{mod } p).$$

An examination of the above equations also yields

$$p^{n-j+1} M_i^{tr} f_{33} \equiv \beta_{ii} p^{n-j+1} M_i^{tr} (\text{mod } p^{n-j+2}) \text{ for each } 1 \leq j \leq n.$$

Hence,

$$g = \begin{pmatrix} \beta_{11}^{tr}(\text{mod } p) & 0 & \dots & 0 & 0 \\ 0 & \beta_{22}^{tr}(\text{mod } p) & \dots & 0 & 0 \\ & & \dots & & \\ 0 & 0 & \dots & \beta_{nn}^{tr}(\text{mod } p) & 0 \\ 0 & 0 & \dots & 0 & f_{33}^{tr}(\text{mod } p) \end{pmatrix}$$

is an endomorphism of V, since $g : V_1 \oplus \cdots \oplus V_n \to V_1 \oplus \cdots \oplus V_n$ with $g(K) \leq K$, as K is the row space of M', $M_i(\text{mod } p) = M_i'$ for each i, and $V_0 = (V_1 \oplus \cdots \oplus V_n)/K$. Moreover, $f\mid_{U(3)} (\text{mod } p) = g + N$, N a nilpotent matrix with 0's on the diagonal.

Assume that V is indecomposable and $f^2 = f \in \text{End } U$. Then $f\mid_{U(3)} (\text{mod } p)^2 = (g+N)^2 = g + N = f\mid_{U(3)} (\text{mod } p)$, whence $g^2 = g$ as g is a diagonal matrix. Since $g \in \text{End } V$ and V is indecomposable, $g = 0$ or 1. If $g = 0$, then $f\mid_{U(3)} (\text{mod } p) = 0$, being both idempotent and nilpotent. Since $f_{11} \equiv f_{22}(\text{mod } p)$, it follows that $f \in p\text{End } U$. But f is idempotent so that $f = 0$. Similarly, if $g = 1$, then $1 - f = 0$. This shows that U is indecomposable. □

Corollary 7 *Suppose that $T = \{\tau_0, \tau_1, \tau_2 < \tau_3\}$ is p-locally free with $\tau_0 = \tau_1 \cap \tau_2 = \tau_1 \cap \tau_3$.*

(a) $C_{crit}(L, j)_p$ has rank infinite representation type for $j \geq 8$.

(b) $C_{crit}(L, j)_p$ is wild modulo p for $j \geq 10$.

Proof. (a) Apply Theorem 6 and Lemma 5 with $n = 4$ and representations in $\text{rep}(S_4, k)$ of the form

$$V = (k^s \oplus k^s, k^s \oplus 0, 0 \oplus k^s, (I + I)k^s, (I + A)k^s)$$

for some indecomposable $m \times m$ k-matrix A. In this case,

$$M' = \begin{pmatrix} I & I & I & 0 \\ I & A & 0 & I \end{pmatrix}$$

(b) Use a proof analogous to that of Corollary 3 and then apply Theorem 6 and Lemma 5 with $n = 5$ and representations in $\text{rep}(S_5, k)$ of the form $V = (k^s \oplus k^s \oplus k^s, k^s \oplus 0 \oplus 0, 0 \oplus k^s \oplus 0, 0 \oplus 0 \oplus k^s, (1+1+1)k^s, (1+A'+B')k^s)$ for $s \times s$ k-matrices $A' = A(\text{mod } p)$ and $B' = B(\text{mod } p)$ and A and B $s \times s$ Z_p-matrices. In this case,

$$M' = \begin{pmatrix} I & I & I & I & 0 \\ I & A' & B' & 0 & I \end{pmatrix}$$

□

Example 8 Following are explicit constructions of Butler groups and acd groups G illustrating Corollaries 3 and 7. Let p be a prime and $T = \{\tau_0, \tau_1, \tau_2 > \tau_3\}$ be a set of types with $\tau_0 = Z$, $\tau_1 = type(X_1)$, and $\tau_3 = type(X_3) < \tau_2 = type(X_2)$, where each X_i is a subgroup of Q containing 1 with $(X_i)_p = Z_p$ and $\tau_0 = \tau_1 \cap \tau_2 = \tau_1 \cap \tau_3$. Let L be the lattice of types generated by T.

(a) If m is a positive integer, then there is an indecomposable $G \in B(L,5)$ with rank $\geq m$ and typeset $G = \{\tau_0, \tau_1, \tau_2 > \tau_3, \tau_1 \cup \tau_3\}$.

Define a subgroup of $Q^{2s} \oplus Q^{2s} \oplus Q^{3s} \oplus Q^{3s}$ by

$$G = (X_1^{2s} \oplus X_3^{2s} \oplus X_3^{3s} \oplus (X_1 + X_3)^{3s}) + X_2^{3s}N + (1/p^5)Z^{4s}(1+1+0+0)$$

where

$$N = \begin{pmatrix} pI & p^2I & | & p^3I & 0 & 0 & | & I & 0 & 0 \\ p^2A & p^3I & | & 0 & p^5I & 0 & | & 0 & I & 0 \\ p^3I & 0 & | & 0 & 0 & p^7I & | & 0 & 0 & I \end{pmatrix}$$

the row space of N is contained in $0 \oplus Z^{2s} \oplus Z^{3s} \oplus Z^{3s}$, I's are $s \times s$ identity matrices, 0's are $s \times s$ zero matrices, and A is an $s \times s$ Jordan block Z-matrix with 1's on the diagonal. Then rank $G = 10s, G(\tau_1) = X_1^{2s} \oplus 0 \oplus 0 \oplus X_1^{3s}$, $G(\tau_3) = 0 \oplus X_3^{2s} \oplus X_3^{3s} \oplus X_3^{3s}$, $G(\tau_2) = X_2^{3s}N \leq G(\tau_3)$, and G is indecomposable with typeset $G = \{\tau_0, \tau_1, \tau_2 > \tau_3, \tau_1 \cup \tau_3\}$.

(b) If R is a finite dimensional Z/pZ-algebra, there is an indecomposable $G \in B(L,7)$ with R an epimorphic image of End G and typeset $G = \{\tau_0, \tau_1, \tau_2 > \tau_3, \tau_1 \cup \tau_3\}$.

Choose $s \times s$ Z-matrices A and B so that $R = C(A(\text{mod } p), B(\text{mod } p))$. Define a subgroup of $Q^{2s} \oplus Q^{2s} \oplus Q^{4s} \oplus Q^{4s}$ by

$$G = (X_1^{2s} \oplus X_3^{2s} \oplus X_3^{4s} \oplus (X_1 + X_3)^{4s}) + X_2^{4s}N + (1/p^7)(1+1+0+0)Z^{2s},$$

$$\text{where } N = \begin{pmatrix} pI & p^2I & | & p^3I & 0 & 0 & 0 & | & I & 0 & 0 & 0 \\ p^2A & p^3I & | & 0 & p^5I & 0 & 0 & | & 0 & I & 0 & 0 \\ p^3B & p^4I & | & 0 & 0 & p^7I & 0 & | & 0 & 0 & I & 0 \\ p^4I & 0 & | & 0 & 0 & 0 & p^9I & | & 0 & 0 & 0 & I \end{pmatrix}$$

Then rank $G = 12s$, $G(\tau_1) = X_1^{2s} \oplus 0 \oplus 0 \oplus X_1^{4s}$, $G(\tau_3) = 0 \oplus X_3^{2s} \oplus X_3^{4s} \oplus X_3^{4s}$, $G(\tau_2) = X_2^{4s}N \leq G(\tau_3)$, and $C(A(\text{mod } p), B(\text{mod } p)) \approx R$ is an endomorphic image of $\text{End}(G)_p$.

(c) If m is a positive integer, then there is an indecomposable acd group $G \in B(L,8)$ with typeset $G = T$ and rank $G \geq m$;

Define a subgroup of $Q^{4s} \oplus Q^{4s} \oplus Q^{2s}$ by

$$G = (X_1^s)^4 \oplus (X_3^s)^4 \oplus (X_3^s)^2 + X_2^{4s}N + (1/p^8)Y, \text{ where}$$
$$Y = \Sigma\{Z(x_i \oplus x_i \oplus 0)/p^{2(n-i)+2} : i \leq 4,\ x_i \in Z_p\}$$

where $X_2^{4s}N \leq 0 \oplus (X_3^s)^4 \oplus (X_3^s)^2$,

$$N = \begin{pmatrix} I & 0 & 0 & 0 & | & p^4I & p^4I \\ 0 & I & 0 & 0 & | & p^3I & p^3A \\ 0 & 0 & I & 0 & | & p^2I & 0 \\ 0 & 0 & 0 & I & | & 0 & pI \end{pmatrix}$$

and A is an $s \times s$ Jordan block Z-matrix with 1's on the diagonal. Then rank $G = 10s$, $G(\tau_1) = X_1^{4s} \oplus 0 \oplus 0$, $G(\tau_3) = 0 \oplus X_3^{4s} \oplus X_3^{2s}$, and $G(\tau_2) = X_2^{4s}N \leq G(\tau_3)$. Hence, G is an indecomposable acd group with typeset $G = \{\tau_0, \tau_1, \tau_2 > \tau_3\}$.

(d) If R is a finite dimensional Z/pZ-algebra, there is an acd group $G \in B(L, 10)$ with typeset $G = T$ and R an epimorphic image of $E(G)$.

Choose $s \times s$ Z-matrices A and B so that $R = C(A(\text{mod } p), B(\text{mod } p))$. Define a subgroup of $Q^{5s} \oplus Q^{5s} \oplus Q^{2s}$ by

$$G = (X_1^s)^5 \oplus (X_3^s)^5 \oplus (X_3^s)^2) + X_2^{5s}N + (1/p^{10})Y,$$
$$\text{where } Y = \Sigma\{Z(x_i \oplus x_i \oplus 0)/p^{2(n-i)+2} : i \leq 5,\ x_i \in Z^s\}$$

with $X_2^{5s}N \leq (X_3^s)^5 \oplus (X_3^s)^2$,

$$N = \left(\begin{array}{ccccc|cc} I & 0 & 0 & 0 & 0 & p^5I & p^5I \\ 0 & I & 0 & 0 & 0 & p^4I & p^4A \\ 0 & 0 & I & 0 & 0 & p^3I & p^3B \\ 0 & 0 & I & 0 & 0 & p^2I & 0 \\ 0 & 0 & 0 & I & 0 & 0 & pI \end{array}\right)$$

and A is an $s \times s$ Jordan block Z-matrix with 1's on the diagonal. Then rank $G = 12s$, $G(\tau_1) = X_1^{5s} \oplus 0 \oplus 0$, and $G(\tau_3) = 0 \oplus X_3^{5s} \oplus X_3^{2s} \geq G(\tau_2) = X_2^{5s}N$. Hence, G is an indecomposable acd group with typeset $G = \{\tau_0, \tau_1, \tau_2 > \tau_3\}$. □

Several questions remain unresolved. The first is what are the best bounds for Corollaries 3 and 7? The bound for Corollary 3.a has been lowered to $j \geq 4$ and the bound for Corollary 7.a has been lowered to $j \geq 7$ by the authors. Since the constructions are adhoc, they are not included. Secondly, can indecomposable representations be classified for the values of j that give rise to rank-infinite representation type but not wild representation type? Finally, analogs of Corollaries 3 and 7 are not known in case T is not p-locally free for some p as the techniques of this paper do not apply.

References

[Arnold 82] D. M. Arnold, *Finite Rank Torsion-Free Abelian Groups and Rings*, Lect. Notes in Math. **931**, Springer-Verlag, New York, (1982).

[Arnold 89] D. M. Arnold, *Representations of partially ordered sets and abelian groups*, Contemporary Math. **87** (1989), 91–109.

[Arnold, Dugas 93] D. M. Arnold and M. Dugas, *Butler groups with finite typesets and free groups with distinguished subgroups*, Comm. Algebra **21** (1993), 1947–1982.

[Arnold, Dugas 95A] D. M. Arnold and M. Dugas, *Locally free finite rank Butler groups and near isomorphism*, Abelian Groups and Modules, Kluwer, Boston, (1995), 41–48.

[Arnold, Dugas 95B] D. M. Arnold and M. Dugas, *Representations of finite posets and near-isomorphism of finite rank Butler groups*, Rocky Mountain J. Math. **25** (1995), 591–609.

[Arnold, Dugas 97] D. M. Arnold and M. Dugas, *Representation type of finite rank Butler groups*, Colloq. Math. **74** (1997), 299–320.

[Arnold, Dugas 98] D. M. Arnold and M. Dugas, *Representation type of finite rank almost completely decomposable groups*, Forum Math, to appear.

[Brenner 74] S. Brenner, *Decomposition properties of some small diagrams of modules*, Symposia Mathamatica **XIII** (1974), 127–141.

[Butler 65] M. C. R. Butler, *A class of torsion-free abelian groups of finite rank*, Proc. London Math. Soc. (3) **15** (1965), 680–698.

[Cruddis 70] T. B. Cruddis, *On a class of torsion-free abelian groups*, Proc. London Math. Soc (3) **21** (1970), 243–276.

[Dugas, Oxford 93] M. Dugas and E. Oxford, *Near isomorphism invariants for a class of almost completely decomposable groups*, Abelian Groups, Lect. Notes in Pure and Applied Math., Marcel Dekker, New York, (1993), 129–150.

[Files, Goebel 96A] S. Files and R. Göbel, *Gauß' theorem for two submodules*, Math. Zeit. **228** (1998), 511–536.

[Files, Goebel 96B] S. Files and R. Göbel, *Representations over PID's with three distinguished submodules*, Trans. AMS, to appear.

[Fuchs 73] L. Fuchs, *Infinite Abelian Groups* Vol. **II**, Academic Press, New York, (1973).

[Lady 75] E. L. Lady, *Nearly isomorphic torsion-free abelian groups*, J. Algebra **35** (1975), 235–238.

[Lewis 92] W. Lewis, *Almost completely decomposable groups with two critical types*, Comm. in Algebra **21** (1992), 423–427.

[Richman 94] F. Richman, *Isomorphism of Butler groups at a prime*, Proc. of 1993 Oberwolfach Conference on Abelian Groups, Contemporary Math. **171** (1994), 333–337.

Trends in Mathematics, © 1999 Birkhäuser Verlag Basel/Switzerland

Normal forms of matrices with applications to almost completely decomposable groups

Otto Mutzbauer

Abstract

We will find normal forms of matrices which describe almost completely decomposable groups in the class of uniform groups defined by Dugas and Oxford [2] up to near-isomorphism. This problem is equivalent to the diagonal equivalence of matrices over $\mathbb{Z}/p^e\mathbb{Z}$.

1 Preliminaries

An almost completely decomposable group X is the extension of the completely decomposable regulator $R = R(X)$ and a finite group X/R, the regulator quotient. If $\exp(X/R) = h$, denote the canonical epimorphism by

$$- : R \to \overline{R} = h^{-1}R/R,\ x \mapsto \overline{x} = h^{-1}x + R.$$

Let $R = \bigoplus_{j=1}^{n} \langle x_j \rangle_*^R$ with an *h-decomposition basis* $(x_1, \ldots, x_n)$, i. e. $\mathrm{hgt}_p^R(x_j) \in \{0, \infty\}$ for all primes p dividing h, $j = 1, \ldots, n$, and let $\overline{\mathbf{a}} = (\overline{a}_1, \ldots, \overline{a}_r)$ be a basis of X/R. The basis $\overline{\mathbf{x}} = (\overline{x}_1, \ldots, \overline{x}_n)$ of the $\mathbb{Z}/h\mathbb{Z}$-module $\overline{R}$ is called *induced decomposition basis.* Then the basis elements $\overline{a}_i$ may be written as linear combinations of the induced decomposition basis

$$\overline{a}_i = \sum_{j=1}^{n} \alpha_{ij}\overline{x}_j, \quad \text{for} \quad i = 1, \ldots, r,$$

where $\alpha_{ij} \in \mathbb{Z}/h\mathbb{Z}$. The matrix $A = (\alpha_{ij}) \in M^{r \times n}(\mathbb{Z}/h\mathbb{Z})$ is called *representing matrix* of X over R relative to $\overline{\mathbf{a}}$ and $\overline{\mathbf{x}}$.

A representing matrix describes a group X as extension of R by X/R relative to an h-decomposition basis of R and a basis of X/R. In [4] is shown that two groups X and Y with $R = R(X) = R(Y) \subset X, Y \subset h^{-1}R$ are isomorphic if and only if X/R and Y/R are in the same orbit of subgroups of $\overline{R}$ under the operation induced by the automorphism group of R.

Recall, cf. [6], that the near-isomorphisms or type isomorphisms are

$$\mathrm{TypAut}\,\overline{R} = \{\xi \in \mathrm{Aut}\,\overline{R} \mid \forall_{\tau \in \mathrm{T}_{\mathrm{cr}}(R)}\ \xi\overline{R(\tau)} = \overline{R(\tau)}\}.$$

Then the groups X and Y are near-isomorphic if and only if there exists a $\xi \in \operatorname{TypAut} \overline{R}$ such that $\xi(X/R) = Y/R$.

Hence the classification problems for almost completely decomposable groups up to isomorphism and up to near-isomorphism can be rephrased as equivalence problems for the representing matrices. Namely, the two representing matrices A and B describe isomorphic or near-isomorphic groups if and only if there are invertible matrices U and V such that $B = UAV$, where U is an automorphism of X/R and V is an automorphism of $\overline{R}$ which is induced by an automorphism of R or is a type automorphism of $\overline{R}$. These equivalence problems are different from those in Linear Algebra since the matrix entries are not in a field and the transforming matrices U, V are from subgroups of the general linear groups over $\mathbb{Z}/h\,\mathbb{Z}$.

Note that the classification up to near-isomorphism, compared with the classification up to isomorphism, avoids two serious difficulties, namely the global problem, cf. [5], and the included number theory problem, cf. [1] and [4]. Moreover, we consider uniform groups, i. e. rigid groups with homocyclic primary regulator quotient. These two restrictions simplify again and considerably the equivalence problem, since the matrix U can be any invertible matrix over $\mathbb{Z}/p^e\,\mathbb{Z}$ and need not be taken from some subgroup, and the matrix V is a diagonal matrix.

To decide if two uniform groups are near-isomorphic it is necessary first to fix a common ordering of the critical typeset. This ordering is a priori arbitrary. So Dugas and Oxford arranged the critical typeset with a so-called pivot set to obtain representing matrices in Hermite Normal Form (E_r, A), where E_r is the r-rowed unit matrix. Near-isomorphism of groups is then equivalent to diagonal equivalence of the rest matrices A, a well known and difficult combinatorial problem. This equivalence problem for matrices is solved in terms of graph theory by Saunders and Schneider in cf. [8], and using a geometric description by Dugas and Oxford in [2]. We present a third solution here in terms of matrices. We hope to convince the reader that this solution is much more adequate to almost completely decomposable groups.

Permutations of the pivot set and its complement separately result in a symmetric equivalence of the rest matrix A, i. e., instead of the diagonal equivalence we deal with symmetric by diagonal equivalence. Since the equivalence classes are bigger we get simpler normal forms of the rest matrices A, so called *straight* matrices, cf. Corollary 3.1, avoiding graph theory and the difficult loop invariants of Dugas and Oxford. The corresponding orderings of the critical typeset are called *admissible* and those orderings are near-isomorphism invariants which have even a direct group theoretic meaning, cf. Theorem 5.2. Eventually we obtain, cf. Theorem 4.3, that two groups given by representing matrices in Hermite normed form are near-isomorphic if and only if the normed rest matrices are modified diagonal similar.

2 Equivalence of matrices

Let S be a commutative ring with 1, let r, n be natural numbers. Let the set of $(r \times n)$-matrices over S be denoted by $M^{r \times n}(S)$. A matrix B which is obtained by striking out rows and columns of a matrix A is called a *submatrix of* A. The maximal natural number k such that there is an invertible k-rowed submatrix of A is called *determinantal rank of* A. Note that a square matrix over S is invertible if and only if its determinant is a unit in S. Now let $\mathfrak{M}$ be the subset of matrices in $M^{r \times n}(S)$ of determinantal rank r. The matrices $A, B \in M^{r \times n}(S)$ are said to be *equivalent* if there are invertible matrices $U \in \mathrm{GL}_r(S)$ and $V \in \mathrm{GL}_n(S)$ such that

$$B = UAV. \tag{1}$$

We need some special equivalences where U and V are taken from of certain subgroups of the whole linear groups, respectively. If V is specialized to be a permutation matrix this is called *$g|s$-equivalence* (g = general, s = symmetric). If V is specialized to be a diagonal matrix this is called *$g|d$-equivalence.* If U and V both are specialized to be permutation matrices this is called *symmetric equivalence.* If U and V both are specialized to be diagonal matrices this is called *diagonal equivalence.* If V is specialized to be a product of a permutation matrix and a diagonal matrix this is called *$g|sd$-equivalence.* Note that the subgroup W of invertible matrices generated by the permutation matrices and the diagonal matrices is a semi-direct product where the diagonal matrices form a normal subgroup, since for all permutation matrices P and all diagonal matrices D there is another diagonal matrix D' such that $DP = PD'$. Hence the matrices V run through the group W. If U and V are in the respective groups W this is called *symmetric by diagonal equivalence.* Two square matrices A and B over S are called *diagonal similar* if there is an invertible diagonal matrix D with $B = D^{-1}AD$.

Now we specialize to the ring $S = \mathbb{Z}/p^e\mathbb{Z}$. In fact we use only that $\mathbb{Z}/p^e\mathbb{Z}$ is a local ring and that all ideals are principal. Let $p^\delta\,\mathbb{Z}/p^e\mathbb{Z}$ be the ideal of $\mathbb{Z}/p^e\mathbb{Z}$ generated by the set of all entries α_{ij} of A. The exponent δ is called the *stripping exponent of* A. The matrix A^{st} with entries in $\mathbb{Z}/p^e\mathbb{Z}$ is called *stripped form of* A if $p^\delta A^{st} = A$. The determinantal rank of A^{st} is called *lower determinantal rank of* A. Note that the entries of A^{st} are uniquely determined up to $p^{e-\delta}\,\mathbb{Z}/p^e\mathbb{Z}$.

Let $A = (\alpha_{ij}) \neq 0$ be an $(r \times n)$-matrix over $\mathbb{Z}/p^e\mathbb{Z}$. The tuple $(i_1, \ldots, i_f)$ determines a block structure on the matrix A if $1 \leq i_1 < \ldots < i_f = \min(r, n)$ as follows. Let $i_0 = 0$. The square $(i_l - i_{l-1})$-rowed submatrices $A_l = (\alpha_{ij})_{i_{l-1}<i,j\leq i_l}$ of A, $1 \leq l \leq f$, are the *diagonal blocks of* A. Let $A'_l = (\alpha_{ij})_{i_{l-1}<i,j}$ denote the *rest block* of the diagonal block A_l. Let δ_l be the stripping exponent of A'_l and let $C'_l = (\gamma_{i,j})$ be the stripped form

of A'_l. The matrix A is called *straight with block structure* $(i_1, \dots, i_f)$ and *stripping sequence* $(\delta_1, \dots, \delta_f)$, if

(1) $i_l - i_{l-1}$ is the lower determinantal rank of the rest block A'_l, $1 \leq l \leq f$.

(2) All main submatrices $(\gamma_{ij})_{i_{l-1}<i,j\leq i_{l-1}+m}$ of all stripped diagonal blocks $\neq 0$ of A, separately, are invertible.

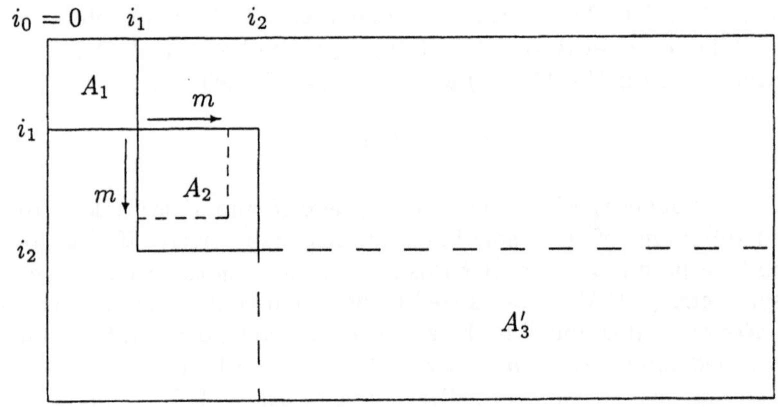

Note that $0 \leq \delta_1 \leq \delta_2 \leq \dots \leq \delta_f \leq e$ and that $\delta_l = e$ is equivalent to $A'_l = 0$, i. e. in particular, $A_l = 0$. Moreover, the properties of straight only relate to the maximal main square submatrix of a matrix.

A straight matrix is said to be *normed* if all the main submatrices of the stripped diagonal blocks $C_l \neq 0$ have determinant 1.

Proposition 2.1 *Every matrix is symmetric equivalent to a straight matrix, and every straight matrix is diagonal equivalent to a normed matrix.*

More precise, for a matrix A there are permutation matrices P and Q such that PAQ is straight, and for every straight matrix B there are invertible diagonal matrices D and H such that DBH is normed.

Proof. A matrix $A \neq 0$ of lower determinantal rank i_1 has a square i_1-rowed submatrix whose stripped form is invertible. This submatrix can be permuted to the left upper corner of A. It is straightforward to permute the rows and columns of this i_1-rowed main submatrix such that all determinants of the stripped form of the main i-rowed submatrices, $i \leq i_1$, are units. Hence, by induction on the number of blocks, each matrix is symmetric equivalent to a straight matrix.

Now let $B \in M^{r\times n}(\mathbb{Z}/p^e\,\mathbb{Z})$ be straight. If $r \leq n$ and $\lambda_1, \dots, \lambda_r$ is the sequence of the determinants of the main submatrices of the stripped diagonal blocks of B, i. e. the λ_i are units, then $D = \operatorname{diag}(\rho_1, \dots, \rho_r) \cdot B$ is normed if $\rho_i = \lambda_{i-1}\lambda_i^{-1}$ for all $i \leq r$. The case $r > n$ is similar. □

Two diagonal equivalent normed matrices with the same block structure $(i_1, \dots, i_f)$ and the same stripping sequence $(\delta_1, \dots, \delta_f)$ are called *modified diagonal similar* if the stripped forms of the diagonal blocks $A_l \neq 0$, $1 \leq l \leq f$, are diagonal similar modulo $p^{e-\delta_l}\,\mathbb{Z}/p^e\,\mathbb{Z}$, respectively. Note that modified diagonal similarity is defined for non-square matrices, too, and that for square normed matrices with $\delta_1 = \dots = \delta_f = 0$, modified diagonal similarity is exactly diagonal similarity.

Proposition 2.2 *The stripping exponents of corresponding submatrices of diagonal equivalent matrices are equal. The determinants of the stripped forms of corresponding square submatrices of diagonal equivalent matrices are in the same ideal of* $\mathbb{Z}/p^e\,\mathbb{Z}$.

If a diagonal equivalence class contains a straight matrix then all matrices in this class are straight and have the same block structure and the same stripping sequence. Moreover, there is a normed matrix in this class and two normed matrices in this class are modified diagonal similar.

Proof. Under diagonal equivalence the entries of matrices will be multiplied by units thus the stripping exponent of any entry is left unchanged. This inherits to submatrices. Moreover, under diagonal equivalence the rows and the columns of submatrices are multiplied by units thus the determinants of the stripped forms of square submatrices change only by units, i. e. these determinants stay in the same ideal of $\mathbb{Z}/p^e\,\mathbb{Z}$. So only straight matrices are diagonal equivalent to straight matrices. Moreover, the block structure and the stripping sequence are transferred by diagonal equivalence.

By Proposition 2.1 straight matrices are diagonal equivalent to normed matrices, and since the main submatrices of the stripped diagonal blocks of normed matrices have determinants 1 or 0, diagonal equivalence specializes to diagonal similarity of the stripped form of the diagonal blocks modulo $p^{e-\delta_l}\,\mathbb{Z}/p^e\,\mathbb{Z}$, respectively, i. e. to modified diagonal similarity. □

Our solution of the diagonal equivalence problem for matrices is now formulated in the following theorem.

Theorem 2.3 *Two* $(r \times n)$*-matrices over* $\mathbb{Z}/p^e\,\mathbb{Z}$ *are diagonal equivalent if and only if there is a common permutation of the rows and a common permutation of the columns to straight matrices such that the respective diagonal equivalent normed matrices are modified diagonal similar.*

Proof. Let A and B be $(r \times n)$-matrices over $\mathbb{Z}/p^e\,\mathbb{Z}$. By Proposition 2.1 there are there are permutation matrices P and Q such that PAQ is straight.

Now let A and B be diagonal equivalent. Thus there are invertible diagonal matrices D and H such that $B = DAH$. Since the subgroup W of invertible matrices generated by the permutation matrices and the diagonal matrices is a semi-direct product with normal subgroup formed by the diagonal matrices, there are invertible diagonal matrices D' and H' such that $PD = D'P$ and $HQ = QH'$. Hence

$$PBQ = PDAHQ = D'PAQH'.$$

Thus A is diagonal equivalent to B if and only if PAQ is diagonal equivalent to PBQ. Since PAQ is straight, PBQ is also straight by Propositions 2.2. Furthermore, by Proposition 2.1, PAQ and PBQ are diagonal equivalent to normed matrices. Thus we have by Proposition 2.2 that these corresponding normed matrices are modified diagonal similar as claimed. □

To decide if two square matrices A and B over $\mathbb{Z}/p^e\mathbb{Z}$ are diagonal similar is a finite problem since there are only finitely many matrices to check with. But there is a faster algorithm that allows to decide whether or not exists an invertible diagonal matrix D such that $B = D^{-1}AD$. The argument for this is that we get from $DB = AD$ that the first entry of D multiplies the first line of the matrix B and we obtain conditions for all entries of D since right multiplication of A by D multiplies the columns of A.

Example. In the set of all invertible 2-rowed normed matrices over $\mathbb{Z}/p^e\mathbb{Z}$ we want to select a representative in each diagonal similarity class. The 2-rowed matrix A is invertible and normed if and only if

$$A = \begin{pmatrix} 1 & \alpha \\ \beta & 1+\alpha\beta \end{pmatrix}.$$

Thus there are precisely the following representatives of diagonal similarity classes

$$\begin{pmatrix} 1 & \eta p^m \\ p^l & 1+p^{l+m}\eta \end{pmatrix} \quad \text{and} \quad \begin{pmatrix} 1 & p^m \\ \eta p^l & 1+p^{l+m}\eta \end{pmatrix}, \tag{2}$$

where $\eta \in \mathbb{Z}/p^e\mathbb{Z}$ is a unit, and $0 \le l \le m \le e$, and $0 \le m < l \le e$, respectively. The normal form is the same for all η in the coset modulo $p^{e-m}\mathbb{Z}/p^e\mathbb{Z}$ or modulo $p^{e-l}\mathbb{Z}/p^e\mathbb{Z}$, respectively. Note that there are $e+1$ matrices of the form $\begin{pmatrix} 1 & 0 \\ p^l & 1 \end{pmatrix}$, where $0 \le l \le e$, and that there are e matrices of the form $\begin{pmatrix} 1 & p^m \\ 0 & 1 \end{pmatrix}$, where $0 \le m \le e-1$. Hence the number N of diagonal similarity classes of normed invertible (2×2)-matrices over $\mathbb{Z}/p^e\mathbb{Z}$

is

$$\begin{aligned} N &= e+1+\sum_{0\le l\le m<e}(p-1)p^{e-m-1}+e+\sum_{0\le m<l<e}(p-1)p^{e-l-1} \\ &= 2e+1+(p-1)[\sum_{0\le m<e}(m+1)p^{e-m-1}+\sum_{0<l<e}lp^{e-l-1}] \\ &= 2e+1+(p-1)\sum_{i=0}^{e-1}(2i+1)p^{e-i-1} \\ &= p^e+2\sum_{i=0}^{e-1}p^i=\frac{p^{e+1}+p^e-2}{p-1}. \end{aligned}$$

3 Matrices in Hermite normal form

Now let $r<n$ and let E_r be the r-rowed unit matrix. In the set of $(r\times n)$-matrices (E_r,A) over $\mathbb{Z}/p^e\mathbb{Z}$ in Hermite Normal Form we consider symmetric by diagonal equivalence, and we may apply our solution for diagonal equivalence of matrices. Such a matrix (E_r,A) is said to be in *Hermite straight form* if A is straight, and in *Hermite normed form* if A is normed. The block structure $(i_1,\dots,i_f)$ and the stripping sequence $(\delta_1,\dots,\delta_f)$ of A are said *to correspond* to the Hermite straight form (E_r,A). Note that $i_f=\min(r,n-r)$. Recall that $\mathfrak{M}$ is the subset of matrices in $M^{r\times n}(\mathbb{Z}/p^e\mathbb{Z})$ of determinantal rank r.

Corollary 3.1 *Every $g|s$-equivalence class of matrices in $\mathfrak{M}$ contains a matrix in Hermite straight form and every $g|sd$-equivalence class of matrices in $\mathfrak{M}$ contains a matrix in Hermite normed form.*

Proof. A matrix $A\in\mathfrak{M}$ has an invertible square r-rowed submatrix which can be permuted to the front by a right multiplication with a permutation matrix. Multiplying with the inverse of this invertible submatrix in front we obtain A in Hermite Normal Form (E_r,B). Thus, by Proposition 2.1, there is an r-rowed permutation matrix P and an $(n-r)$-rowed permutation matrix Q such that

$$(E_r,C)=P\cdot(E_r,B)\cdot\begin{pmatrix}P^{-1}&0\\0&Q\end{pmatrix}$$

with $C=PBQ$ straight. Because of

$$(E_r,D^{-1}CD')=D^{-1}\cdot(E_r,C)\cdot\begin{pmatrix}D&0\\0&D'\end{pmatrix}, \tag{3}$$

where D and D' are invertible diagonal matrices, the straight matrix C is diagonal equivalent to $D^{-1}CD'$. By Proposition 2.1 we obtain a Hermite normed form. □

Corollary 3.2 *Let (E_r, A) be in Hermite straight form. If (E_r, B) is diagonal equivalent to (E_r, A) then B is straight and has the same block structure and the same stripping sequence as A.*

If (E_r, A) and (E_r, B) are diagonal equivalent and both in Hermite normed form, then A and B are modified diagonal similar.

Proof. By (3) diagonal equivalence of matrices in Hermite straight form inherits to the rest blocks A and B. Thus by Proposition 2.2 the matrices A and B are both straight and have the same block structure and the same stripping sequences. Moreover, A and B are diagonal equivalent to normed matrices, respectively, which are then modified diagonal similar, again by Proposition 2.2. □

4 Representing matrices

Let X be an almost completely decomposable group of rank n with regulator R and a homocyclic p-group of rank r, $r < n$, as regulator quotient. Dugas and Oxford called such groups *uniform* and in the class $C(T, p, e, r)$ if they are additionally p-reduced, rigid, with critical typeset T and if $\exp(X/R) = p^e$. A decomposition basis $(x_1, \ldots, x_n)$ of R is called a *p-decomposition basis* if $\mathrm{hgt}_p^R(x_j) = 0$ for all $j = 1, \ldots, n$. Let $(\overline{x}_1, \ldots, \overline{x}_n)$ be the induced decomposition basis of $\overline{R}$.

Lemma 4.1 ([2, 2.4], [7, 6.11 and 6.12]) *Let R be a completely decomposable p-reduced group with p-decomposition basis $(x_1, \ldots, x_n)$. Let e be a natural number and let $(\alpha_{ij}) \in M^{r \times n}(\mathbb{Z}/p^e\mathbb{Z})$. Let $a_i = \sum_{j=1}^{n} \alpha'_{ij} x_j$, $1 \le i \le r$, where $\alpha'_{ij} \in \mathbb{Z}$, $\alpha'_{ij} + p^e\mathbb{Z} = \alpha_{ij}$ for all i, j. Let $X = R + \sum_{i=1}^{r} \mathbb{Z} p^{-e} a_i \subset \mathbb{Q} R$. Then $X/R \cong (\mathbb{Z}/p^e\mathbb{Z})^r$ if and only if the matrix (α_{ij}) over $\mathbb{Z}/p^e\mathbb{Z}$ has determinantal rank r.*

In particular, representing matrices of uniform groups in the class $C(T, p, e, r)$ relative to induced p-decomposition bases have determinantal rank r, i. e. they are in $\mathfrak{M}$. Conversely for all matrices $(\alpha_{ij}) \in \mathfrak{M}$ the corresponding group X is in the class $C(T, p, e, r)$.

Proof. We have

$$X/R \cong p^e X/p^e R = \langle a_i + p^e R \mid 1 \le i \le r \rangle$$
$$\subset R/p^e R = \bigoplus_{j=1}^{n} (\mathbb{Z}/p^e\,\mathbb{Z})(x_j + p^e R) \cong (\mathbb{Z}/p^e\,\mathbb{Z})^n,$$

since $(x_1, \ldots, x_n)$ is a p-decomposition basis and R is p-reduced. Hence $X/R \cong (\mathbb{Z}/p^e\,\mathbb{Z})^r$ if and only if the set $\{a_i + p^e R \mid 1 \le i \le r\}$ is p-independent, cf. [3, 32.1].

Now we show that p-independence of the rows of an $(r \times n)$-matrix A is equivalent to the fact that A has determinantal rank r. Note that a square matrix over $\mathbb{Z}/p^e\,\mathbb{Z}$ has p-independent rows if and only if it describes an automorphism of a homocyclic group, i. e. if and only if it is invertible. If the $(r \times n)$-matrix A has determinantal rank r, then it has an invertible square r-rowed submatrix, whose rows are p-independent. But then the rows of A are p-independent.

Conversely if A has p-independent rows then A is a submatrix of some square n-rowed matrix B whose n rows are p-independent, since a p-independent set is contained in a maximal p-independent set. Thus B is invertible and A contains an invertible r-rowed submatrix by Laplace's expansion of determinants. The additional statements are direct consequences. □

Note that if a uniform group X is given by a representing matrix relative to an induced p-decomposition basis of R in Hermite Normal Form (E_r, A), then each row of A contains a unit. Thus if (E_r, A) is $g|s$-equivalent to (E_r, B), where B is straight, the stripping sequence of B starts with $\delta_1 = 0$. Moreover, if the group is clipped, i. e. it has no rational direct summand, then A has no zero column.

Since uniform groups are rigid, we may choose the critical typeset $T = \{\tau_1, \ldots, \tau_n\}$ as the index set of the columns of the representing matrix M via $j \leftrightarrow \tau_j = \mathrm{tp}(x_j)$.

Lemma 4.2 *Let X be a uniform group. There is an ordering of the critical typeset, a p-decomposition basis and a basis of X/R such that the corresponding representing matrix is in Hermite straight form.*

There is a group near-isomorphic to X, which relative to the same ordering of the critical typeset, has a representing matrix in Hermite normed form.

Proof. By Lemma 4.1 a representing matrix of X is in $\mathfrak{M}$. Rearranging of the critical typeset and changing the basis of X/R means $g|s$-equivalence of the representing matrices. Thus by Corollary 3.1 there is an ordering of the critical typeset and a representing matrix in Hermite straight form.

If there is no change of the ordering of the critical typeset then near-isomorphism is given by a right multiplication of the representing matrix by an invertible diagonal matrix. If the representing matrix is in Hermite straight form and we ask for a Hermite straight form of the resulting representing matrix then as in (3) we obtain a Hermite normed form. □

Recall that if a representing matrix of a uniform group is in Hermite Normal Form, then the ordering of the critical typeset starts with a *pivot set* $\{\tau_1, \dots, \tau_r\}$ in the sense of Dugas and Oxford. In view of Lemma 4.2 we call an indexing of the critical typeset *admissible for* X, if for this indexing, X has a representing matrix in Hermite straight form. Thus there are a *corresponding* block structure and a stripping sequence. Note that the critical typeset of a group can have several admissible orderings.

Theorem 4.3 *The admissible orderings, with corresponding block structure and stripping sequence, of the critical typeset of a uniform group are near-isomorphism invariants.*

Two uniform groups in $C(T,p,e,r)$ *described by representing matrices relative to a (common) admissible ordering of the critical typeset in Hermite normed form* (E_r, A) *and* (E_r, B)*, respectively, are near-isomorphic if and only if* A *and* B *are modified diagonal similar.*

Let T *be an antichain of types of length* n *with fixed ordering. There is a bijection between the modified diagonal similarity classes of normed* $[r \times (n-r)]$*-matrices over* $\mathbb{Z}/p^e\,\mathbb{Z}$ *and the near-isomorphism classes of uniform groups in* $C(T,p,e,r)$*, where the given ordering of* T *is admissible.*

Proof. Let the group X be given by a representing matrix (E_r, A) in Hermite straight form relative to a fixed admissible ordering of the critical typeset. Let Y be near-isomorphic to X. Let (D, B) be the representing matrix of Y relative to the same ordering of the critical typeset. Near-isomorphism relative to the fixed ordering of the critical typeset means $g|d$-equivalence of the representing matrices. Thus D is invertible and Y has a representing matrix (E_r, C) in Hermite Normal Form. The $g|d$-equivalence specifies to diagonal equivalence of A and C. Hence, by Corollary 3.2, C is straight with the same block structure and the same stripping sequence as A. Then the fixed ordering of the critical typeset is also admissible for Y, i. e. it is a near-isomorphism invariant.

If two groups are given by representing matrices in Hermite normed form (E_r, A) and (E_r, B), respectively, relative to the same ordering of the critical typeset, then near-isomorphism means diagonal equivalence of the representing matrices. Thus by Corollary 3.2 the matrices A and B are modified diagonal similar.

By Lemma 4.1 all matrices (E_r, A) in Hermite normed form are realized in the class $X \in C(T,p,e,r)$ where an admissible ordering of the critical typeset is fixed. By the above, near-isomorphism is equivalent to modified diagonal similarity of the rest blocks A. This is the desired bijection. □

5 Near-isomorphism invariants

We will show that some of the ingredients of the representing matrices of uniform groups have an immediate group theoretic meaning.

Notations. Let $X \in \mathcal{C}(T,p,e,r)$ be a uniform group with regulator R and ordered critical typeset $T = (\tau_1, \ldots, \tau_n)$. We use the notations in the Preliminaries and in Section 4 to deal with representing matrices. Let (E_r, A) be a representing matrix in Hermite Normal Form relative to the bases $\overline{\mathbf{x}}$ and $\overline{\mathbf{a}}$. We will establish a one-to-one correspondence between the submatrices of A and certain quotients of X. Let

$$R(L) = \bigoplus_{j \in L} X(\tau_j), \quad \text{where} \quad L \subset [1,n],$$

and, for $I \subset [1,r]$, $J \subset (r,n] = \{r+1, \ldots, n\}$ and $J' = (r,n] \setminus J$ let

$$X(I,J) = \frac{R(I \cup (r,n])_*^X}{R(I \cup J')_*^X}.$$

Identify $R(J)$ with

$$\frac{R(J) \oplus R(I \cup J')_*^X}{R(I \cup J')_*^X}.$$

Let $A(I,J) = (\alpha_{i,j})_{i \in I, j \in J}$ be the submatrix of $A = (\alpha_{ij})$ with row index in I and column index in J.

Lemma 5.1 *$X(I,J)$ is an almost completely decomposable group with completely decomposable subgroup $R(J)$ of finite index. If $\overline{\mathbf{x}}$ and $\overline{\mathbf{a}}$ are bases of $\overline{R}$ and X/R, respectively, then for $i \in I$ and $j \in J$ the elements*

$$x_j^* = \overline{x}_j + (R + R(I \cup J')_*^X)/R \quad \text{and} \quad a_i^* = \overline{a}_i + (R + R(I \cup J')_*^X)/R$$

form bases $\overline{\mathbf{x}}^*$ and $\overline{\mathbf{a}}^*$ of $\overline{R(J)}$ and of $X(I,J)/R(J)$, respectively. The corresponding representation matrix is $A(I,J)$. This correspondence between the submatrices $A(I,J)$ of A and the quotients $X(I,J)/R(J)$ is a bijection.

Proof. Since

$$X(I,J) \cong \frac{R(I\cup(r,n])_*^X + p^{-e}R(I\cup J')}{p^{-e}R(I\cup J')}$$

the bases $\overline{\mathbf{x}}$ and $\overline{\mathbf{a}}$ have the indicated canonical analogues $\overline{\mathbf{x}}^*$ and $\overline{\mathbf{a}}^*$. Since $a_i^* = \sum_{j\in J} \alpha_{ij} x_j^*$, $i \in I$, the representing matrix of $X(I,J)$ over $R(J)$ relative to these bases is the submatrix $A(I,J)$. This correspondence is the desired bijection between the submatrices $A(I,J)$ of A and the quotients $X(I,J)/R(J)$. □

Let (E_r, A) be in Hermite straight form with block structure $(i_1,\dots,i_f)$ where $i_f = \min(r, n-r)$. Let $I_l = (i_{l-1}, i_l]$, $J_{l,m} = (r+i_{l-1}, r+i_{l-1}+m]$ for $1 \le m \le i_l - i_{l-1}$, let

$$Y_{l,m} = \frac{X(I_l, J_{l,m})}{R(J_{l,m})}, \quad 1 \le m \le i_l - i_{l-1},$$

and

$$Z_l = \frac{X\big((i_{l-1}, r], (r+i_{l-1}, n]\big)}{R\big((r+i_{l-1}, n]\big)}$$

for $1 \le l \le f$, agreeing that $i_0 = 0$.

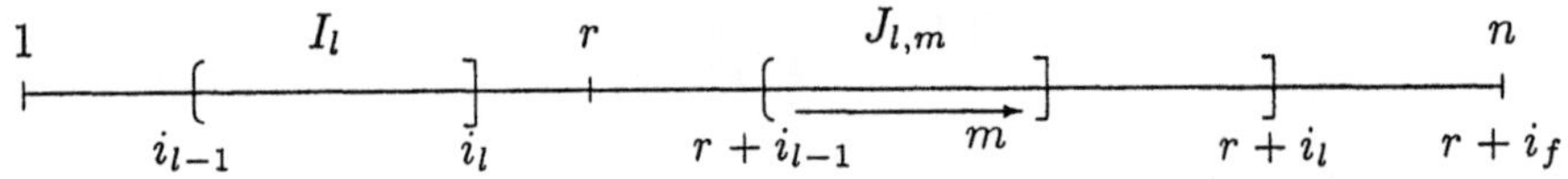

If a representing matrix is given for a uniform group relative to any ordering of the critical typeset then all admissible orderings of the critical typeset can be found by considering certain canonical quotients of the group.

Theorem 5.2 *Let X be a uniform group and assume the notations above. The ordering $T = (\tau_1,\dots,\tau_n)$ is admissible for X with corresponding block structure $(i_1,\dots,i_f)$ and stripping sequence $(\delta_1,\dots,\delta_f)$ if and only if the following hold:*

(1) $\dfrac{X}{R + R\big((r,n]\big)_*^X} \cong X/R$ *is homocyclic of exponent p^e and of rank r;*

(2) *Z_l is homocyclic of exponent $p^{e-\delta_l}$ and of rank $i_l - i_{l-1}$ for all $l = 1,\dots,f$;*

(3) *$Y_{l,m}$ is homocyclic of exponent $p^{e-\delta_l}$ and of rank m for all l and $1 \le m \le i_l - i_{l-1}$.*

Proof. The quotient in (1) corresponds to the r-rowed square submatrix in front of a representing matrix. Thus this submatrix is invertible if and only if this quotient is homocyclic of full rank r and condition (1) is equivalent to the possibility of a representing matrix (E_r, A) in Hermite Normal Form, relative to the given ordering of the critical typeset.

By Lemma 5.1 the quotients in (2) and (3) correspond to submatrices of a representing matrix of X. The quotient Z_l corresponds to the rest block A_l' of A relative to the given block structure. Condition (2) is equivalent to the fact that the block structure and the stripping sequence are the same as for straight matrices, cf. the condition (1) in the definition of straight matrices.

The quotients $Y_{l,m}$ correspond to the main submatrices of the diagonal blocks A_l. Condition (3) is equivalent to the fact that all main submatrices of the stripped diagonal blocks of A are invertible, cf. the condition (2) in the definition of straight matrices. □

Observe that the Hermite straight form of representing matrices of a uniform group can be read off the factors in Theorem 5.2, but not the Hermite normed forms.

Example. Now we determine the normal forms of representing matrices in the class $\mathcal{C}(T, p, 1, 2)$ where $T = (\tau_1, \tau_2, \tau_3, \tau_4)$ is an antichain of length 4 (i. e. the groups are of rank 4) and none of the types is p-divisible, and where in addition this ordering of T is admissible for the groups. The representing matrices are in Hermite normed form, i. e. (E_2, A) with $A \in M^{2\times 2}(\mathbb{Z}/p\mathbb{Z})$ normed. The example in Section 2 gives a list of $p+2$ normal forms of invertible (2×2)-matrices over $\mathbb{Z}/p\mathbb{Z}$ such that each matrix belongs to precisely one near-isomorphism class. Since R is the regulator there is at least one unit in each row of the rest matrix A, and there remain the following normed matrices, which are not invertible,

$$\begin{pmatrix} 1 & 0 \\ 1 & 0 \end{pmatrix} \quad \text{and} \quad \begin{pmatrix} 1 & 1 \\ 1 & 1 \end{pmatrix},$$

as representatives of the different similarity classes. Thus we have all $p+4$ normal forms (E_2, A) where A is of the form

$$\begin{pmatrix} 1 & 0 \\ 1 & 0 \end{pmatrix}, \begin{pmatrix} 1 & 0 \\ 0 & 1 \end{pmatrix}, \begin{pmatrix} 1 & 1 \\ 1 & 1 \end{pmatrix}, \begin{pmatrix} 1 & 1 \\ 0 & 1 \end{pmatrix}, \begin{pmatrix} 1 & \eta \\ 1 & 1+\eta \end{pmatrix},$$

where $\eta = 0, \dots, p-1$. Note that these are all normal forms relative to the fixed admissible ordering of T. Only the first two cases correspond to decomposable groups.

References

[1] R. Burkhardt and O. Mutzbauer, *Decompositions of torsion-free abelian groups of rank* 3, Arch. Math. **35** (1980), 501–504.

[2] M. Dugas and E. Oxford, *Near isomorphism invariants for a class of almost completely decomposable groups*, Abelian Groups, Proceedings of the 1991 Curaçao Conference, Marcel Dekker, Inc., (1993), 129–150.

[3] L. Fuchs, *Infinite Abelian Groups* **I, II**, Academic Press, (1970, 1973).

[4] K.-J. Krapf and O. Mutzbauer, *Classification of almost completely decomposable groups*, Abelian Groups and Modules, Proceedings of the Udine Conference 1984, CISM Courses and Lecture Notes **287**, Springer Verlag (1984), 151–161.

[5] A. Mader, O. Mutzbauer and C. Vinsonhaler, *Local-global relations for almost completely decomposable groups*, Rocky Mountain J. (1998).

[6] A. Mader and C. Vinsonhaler, *Classifying almost completely decomposable groups*, J. Algebra **170** (1994), 754–780.

[7] M. Nahler, *Near-isomorphism invariants of almost completely decomposable groups*, Diplomarbeit (Würzburg 1998).

[8] B. D. Saunders and H. Schneider, *Flows on graphs applied to diagonal similarity and diagonal equivalence for matrices*, Discrete Mathematics **24** (1978), 205–220.

Trends in Mathematics, © 1999 Birkhäuser Verlag Basel/Switzerland

Admissible matrices as base changes of $\mathbf{B}^{(1)}$-groups: a realizing algorithm

Clorinda De Vivo and Claudia Metelli

Abstract

In the study of Abelian groups a relevant role is played by realization of rings as endomorphism rings; this entails building from the ring a group on which the ring will act. We solve here a similar problem: given a $\mathbb{Q}$-matrix $\mathbf{M}$ in the suitable class, build from it the Butler $\mathbf{B}^{(1)}$-groups for which $\mathbf{M}$ is a base change. The building procedure we give is an algorithm, easy to perform by hand in one-digit ranks.

Introduction

Let V be an m-dimensional vector space over the field $\mathbb{Q}$ of rationals (m a natural number). Say a base change for V is an ordered couple $((x_1, \ldots, x_m), (y_1, \ldots, y_m))$ of bases of V. Writing $y_j = \sum_{i=1}^m a_{ij} x_i$, the matrix $\mathbf{M} = (a_{ij}) \in \mathrm{GL}_{\mathbb{Q}}(m)$ performs the base change. In this setting, base changes are nothing else but automorphisms of V. If, instead of a vector space, we consider a completely decomposable torsionfree Abelian group of rank m: $X = R_1 x_1 \oplus \cdots \oplus R_m x_m \leq V$ ($\mathbb{Z} \leq R_i \leq \mathbb{Q}$ for each i), the two concepts differ. For instance the automorphism determined by the matrix $\left(\begin{smallmatrix} 0 & 1 \\ 1 & 0 \end{smallmatrix}\right)$ on $V = \mathbb{Q}x_1 \oplus \mathbb{Q}x_2$ does not restrict to an automorphism on $X = \mathbb{Z}x_1 \oplus \mathbb{Q}x_2$, although the couple $((x_1, x_2), (x_2, x_1))$ deserves to be considered as a base change for X, since $X = \mathbb{Q}x_2 \oplus \mathbb{Z}x_1$. Determining all base changes becomes relevant for groups which are defined via their realization on a 'base'; there, 'knowing' the group means knowing its realizations, that is knowing its base changes. This can be the case for Butler groups, which are (by one of their possible definitions) torsionfree quotients of completely decomposable groups of finite rank. Our focus will be on $\mathbf{B}^{(1)}$-groups, those Butler groups where the quotient is taken over a subgroup of rank 1. Both the general and the particular case have been amply studied; we refer to [AV] for the basic bibliography. The equivalence we use is quasi-isomorphism, that is isomorphism up to finite index.

We will solve here the following problem: given an $m \times m$ matrix over $\mathbb{Q}$, determine the $\mathbf{B}^{(1)}$-groups of rank $m-1$ on which the matrix performs a base change. An on-going example will illustrate the algorithmic solution.

Let us sketch the problem for completely decomposable groups. If, as above, $X = R_1x_1 \oplus \cdots \oplus R_mx_m$, we say an m-tuple $(y_1, \ldots, y_m)$ of elements of X is a base for X if $X = R'_1y_1 \oplus \cdots \oplus R'_my_m$, where the R'_i are subgroups of $\mathbb{Q}$ containing $\mathbb{Z}$. Setting $y_j = \sum_{i=1}^m a_{ij}x_i$, the matrix $\mathbf{M} = (a_{ij})$ is then invertible. For each $i \in I = \{1, \ldots, m\}$ let t_i be the type of x_i in X (i.e. the isomorphy class of the pure subgroup $\langle x_i\rangle_* = R_ix_i \cong R_i$); u_i the type of y_i. Then for each $j \in I$ we have $u_j = \bigwedge\{t_i \mid a_{ij} \neq 0\}$, and, letting $\mathbf{M}^{-1} = (b_{ij})$, analogously $t_j = \bigwedge\{u_i \mid b_{ij} \neq 0\}$. Note that, in order to spell out these necessary conditions, we need an order-theoretical setting besides the linear condition of invertibility of the matrix. We may in fact re-define a *base of* X to be a couple $((x_1, \ldots, x_m), (t_1, \ldots, t_m))$, where $(x_1, \ldots, x_m)$ is a base of $V = X \otimes \mathbb{Q}$, contained in X; and the types $t_i = \text{type}_X(x_i)$ are elements of the lattice $\mathbb{T}$ of types. (By using the types t_i instead of the actual coefficient groups R_i we identify our groups up to finite index; this will be required later with quasi-isomorphisms, and we adopt it here to simplify the description). A matrix $\mathbf{M}$ performing a base change transforms $(x_1, \ldots, x_m)$ into $(y_1, \ldots, y_m)$ linearly by its columns; and transforms $(t_1, \ldots, t_m)$ into $(u_1, \ldots, u_m)$ order-theoretically by the *supports* of its columns: setting

$$\mu_j = \begin{pmatrix} a_{1j} \\ \vdots \\ a_{mj} \end{pmatrix},$$

$$y_j = \sum_{i=1}^m a_{ij}x_i, \qquad u_j = \bigwedge\{t_i \mid a_{ij} \neq 0\} = \bigwedge\{t_i \mid i \in \operatorname{supp} \mu_j\}.$$

In other words, just as the elements of a base are 'valuated', by their type, into the lattice of types, so the columns of the matrix are 'valuated', by their support, into the lattice $\mathcal{P}(m)$ of subsets of I. The matrix $\mathbf{M}$ performs a base change on X only if both its linear form acts invertibly on the elements and its 'valuated form' acts invertibly on the types. It is easy to realize that, while linear invertibility is a property of the matrix, the invertibility of its order-theoretical action depends on the group X: it is enough to note that any invertible matrix is a base change for a homogeneous X (i.e. when $t_1 = \cdots = t_m$); while the matrix $\left(\begin{smallmatrix} 1 & 1 \\ 1 & 0 \end{smallmatrix}\right)$ irreparably diminishes $X = \mathbb{Z}x_1 \oplus \mathbb{Q}x_2$ to $Y = \mathbb{Z}x_1 \oplus \mathbb{Z}x_2$, for its inverse won't bring Y back up to X again. Therefore it makes sense, given an invertible matrix, to determine those completely decomposable groups on which it acts as a base change; the solution of this case follows the lines of the forthcoming proof for $\mathbf{B}^{(1)}$-groups, where invertibility will be replaced by admissibility.

For a $\mathbf{B}^{(1)}$-group $G = X/H$ we will use two simplifications. The first is useful for the linear setting. From [FM], we may suppose G to be *regular*, which means there is no loss of generality in supposing the rank 1 subgroup H to be $R(x_1 + \cdots + x_m)$, where $R = \bigcap\{R_i \mid i \in I\}$. Thus we can write $G = R_1g_1 + \cdots + R_mg_m$, where the elements $g_i = x_i + H$ are subject to the (only) condition $g_1 + \cdots + g_m = 0$. Since in this setting G is determined once X is given by a base (the denominator H is determined by the base of X), we will call a *base* for G the image of the base of X, that is the couple $((g_1, \dots, g_m), (t_1, \dots, t_m))$, where $g_1 + \cdots + g_m = 0$ is the only relation among the g_i, and $t_i = \text{type}_G(g_i) = \text{type}_X(x_i)$ (the last equality is a consequence of regularity, see [FM]). Since it may happen that $G \cong Y/K$ with Y completely decomposable, K rank 1 of the required form, but Y non isomorphic to X (ex. 4.1 in [FM]), the problem of determining base changes for $\mathbf{B}^{(1)}$-groups does not reduce to the analog for completely decomposable groups. Note also that base changes describe all representations of G as a $\mathbf{B}^{(1)}$-group (up to quasi-isomorphism).

In the second simplification, given in [DVM2], we may suppose the occurring types to be reduced, that is to consist only of 0's and ∞'s (equivalently, the R_i may be taken to be subrings of $\mathbb{Q}$). This will turn out to be crucial for the order-theoretical setting.

1 Admissibility

Let us start by determining the linear condition a matrix must satisfy to be a base change for a $\mathbf{B}^{(1)}$-group G. Note that $G \otimes \mathbb{Q} = \mathbb{Q}g_1 + \cdots + \mathbb{Q}g_m = V/\mathbb{Q}(x_1 + \cdots + x_m)$; we will call this quotient space V', and say *a* $\mathbf{B}^{(1)}$*-base for the vector space* V' is a base for V', viewed as a $\mathbf{B}^{(1)}$-group (here $t_1 = \cdots = t_m = (\infty, \infty, \infty, \dots)$ is the type of $\mathbb{Q}$). Thus a $\mathbf{B}^{(1)}$-base for V' is an m-tuple $(g_1, \dots, g_m)$ of elements of V' such that

a) $V' = \mathbb{Q}g_1 + \cdots + \mathbb{Q}g_m$

b) $g_1 + \cdots + g_m = 0$ is the only relation the m-tuple satisfies in V'.

If $((h_1, \dots, h_m), (u_1, \dots, u_m))$ is a base for G, we have in particular $G \otimes \mathbb{Q} = \mathbb{Q}h_1 + \cdots + \mathbb{Q}h_m = V'$; hence the same $m \times m$ matrix $\mathbf{M}$ over $\mathbb{Q}$ operating the base change for G will perform a $\mathbf{B}^{(1)}$-base-change for V'. The linear transformation operated by $\mathbf{M}$ on V must then induce an automorphism on V'. A matrix satisfying such a condition will be called admissible, and so will be called the associated linear transformation of V.

Let $M(m, K)$ denote the K-algebra of $m \times m$ matrices over a field K; $V = K^m$ the vector space whose canonical base vectors we denote by $\delta_1, \dots, \delta_m$. The elements of V will be viewed as columns:

$$v = k_1\delta_1 + \cdots + k_m\delta_m = \begin{pmatrix} k_1 \\ \vdots \\ k_m \end{pmatrix} \qquad (k_i \in K);$$

in particular we will denote by $\delta_E = \sum_{i \in E} \delta_i$ the characteristic vector of the subset E of I (e.g. $\delta_{\{i\}} = \delta_i$, $\delta_I = \delta_1 + \cdots + \delta_m$). We can view a matrix as the m-tuple of its column vectors (with square brackets: e.g. the identity matrix $\mathbf{I} = [\delta_1, \ldots, \delta_m]$).

Definition 1.1 Denote by $\mathrm{Stab}(K\delta_I)$ the subalgebra of $\mathrm{End}_K(V)$ consisting of the endomorphisms mapping the diagonal subspace $K\delta_I$ into itself. A linear transformation $\alpha \in \mathrm{Stab}(K\delta_I)$ will be called *admissible* if it induces an automorphism on $K^m/K\delta_I = V'$. When α is admissible, the matrix $\mathbf{M}(\alpha) = [\alpha(\delta_1), \ldots, \alpha(\delta_m)]$ that realizes α with respect to the canonical base will also be called *admissible.* □

Clearly, the map

$$\phi\colon \mathrm{Stab}(K\delta_I) \to \mathrm{End}_K(V'),$$
$$\alpha \to \phi(\alpha)\colon v + K\delta_I \to \alpha(v) + K\delta_I$$

is a K-algebra-epimorphism. In fact, the set $\mathrm{Adm}(m, K)$ of admissible linear transformations is nothing but $\phi^{-1}(\mathrm{GL}_K(V'))$, a multiplicative subsemigroup of $\mathrm{End}_K(V)$.

When for two elements α, β of $\mathrm{Stab}(K\delta_I)$ we have $\phi(\alpha) = \phi(\beta)$ we will write $\alpha \equiv \beta$ and say α is *equivalent* to β. We have $\alpha \equiv \beta$ iff, for each $i \in I$, $\beta(\delta_i) = \alpha(\delta_i) + k_i\delta_I$ with suitable $k_i \in K$; that is, iff the matrix $\mathbf{M}(\alpha)$ differs from $\mathbf{M}(\beta)$ by adding, for each $i \in I$, to each entry of the i-th column the same scalar k_i: $\mathbf{M}(\beta) = \mathbf{M}(\alpha) + [k_1\delta_I, \ldots, k_m\delta_I]$. When this is the case we will write $\mathbf{M}(\alpha) \equiv \mathbf{M}(\beta)$. We have the easy consequence

Corollary 1.2 *If $\alpha \in \mathrm{Stab}(K\delta_I)$, there is a matrix equivalent to $\mathbf{M}(\alpha)$ with an assigned row of zeroes; and a matrix equivalent to $\mathbf{M}(\alpha)$ with an assigned row of all zeroes but for a* 1. □

We show now that this definition of admissibility is essentially equivalent to the definition given in [FM, 4]. We start with an easy computation.

Lemma 1.3 *Let $\mathbf{M}_i$ be the matrix obtained from $\mathbf{M}$ by replacing the i-th column with δ_I (a column of 1's). If $\mathbf{M} \equiv \mathbf{N}$ then* $\det(\mathbf{M}_i) = \det(\mathbf{N}_i)$.

Proof. Directly, or following the lines of [FM, 4.4]. □

Let now $W = \langle \delta_1, \ldots, \delta_{m-1} \rangle \leq V$, $\alpha \in \mathrm{Stab}(K\delta_I)$, and choose a matrix $\mathbf{N}$ equivalent to $\mathbf{M}(\alpha)$ with the last row consisting of all zeroes but for 1 in the m-th place; we have $\mathbf{N}(V) = \mathbf{N}(W) \oplus K\delta_I$. If $\alpha \in \mathrm{Adm}(m, K)$ then $W \cong \mathbf{N}(W)$, thus the upper left $(m-1) \times (m-1)$ determinant of $\mathbf{N}$ is nonzero; hence $\det(\mathbf{N}_m) \neq 0$, and by the above Lemma $\det(\mathbf{M}_m) \neq 0$. This can be repeated to yield $\det(\mathbf{M}_i) \neq 0$ for all $i \in I$. We thus have the necessary part of the following

Proposition 1.4 *If* $\alpha \in \mathrm{Stab}(K\delta_I)$, $\mathbf{M} = \mathbf{M}(\alpha)$ *is admissible if and only if for each* $i \in I$ *the matrix* $\mathbf{M}_i$ *obtained from* $\mathbf{M}$ *by replacing the i-th column with* δ_I *has nonzero determinant.*

Proof. We need to show sufficiency; for this purpose, choose $\mathbf{N}$ equivalent to $\mathbf{M}$ with the last row of all zeroes but for 1 in the m-th place. Since $\det(\mathbf{N}_m) = \det(\mathbf{N})$ this last is nonzero; then the endomorphism β determined by $\mathbf{N}$ is an automorphism of V, belongs to $\mathrm{Stab}(K\delta_I)$, hence is admissible. Thus so is α, being equivalent to β. □

Example 1.5 To build an admissible 6×6 matrix $\mathbf{M}$ over $\mathbb{Q}$ (w.l.o.g. invertible) we start by an invertible 5×5 matrix

$$\mathbf{M}' = \begin{pmatrix} 1 & -1 & 1 & 0 & -1 \\ 1 & 0 & 0 & 1 & -1 \\ 1 & -1 & 0 & 1 & 0 \\ 0 & -1 & 1 & 1 & 0 \\ 0 & 0 & 1 & 0 & 0 \end{pmatrix}, (\mathbf{M}')^{-1} = \begin{pmatrix} 0 & 0 & 1 & -1 & 1 \\ -1/2 & 1/2 & 0 & -1/2 & 1 \\ 0 & 0 & 0 & 0 & 1 \\ -1/2 & 1/2 & 0 & 1/2 & 0 \\ -1/2 & -1/2 & 1 & -1/2 & 1 \end{pmatrix},$$

and line both by a last row 0 0 0 0 0 1 and a last column that makes all row sums = 1:

$$\mathbf{M} = \begin{pmatrix} 1 & -1 & 1 & 0 & -1 & 1 \\ 1 & 0 & 0 & 1 & -1 & 0 \\ 1 & -1 & 0 & 1 & 0 & 0 \\ 0 & -1 & 1 & 1 & 0 & 0 \\ 0 & 0 & 1 & 0 & 0 & 0 \\ 0 & 0 & 0 & 0 & 0 & 1 \end{pmatrix}, \mathbf{M}^{-1} = \begin{pmatrix} 0 & 0 & 1 & -1 & 1 & 0 \\ -1/2 & 1/2 & 0 & -1/2 & 1 & 1/2 \\ 0 & 0 & 0 & 0 & 1 & 0 \\ -1/2 & 1/2 & 0 & 1/2 & 0 & 1/2 \\ -1/2 & -1/2 & 1 & -1/2 & 1 & 1/2 \\ 0 & 0 & 0 & 0 & 0 & 1 \end{pmatrix}.$$

□

2 Double action of a matrix as a base change

From now on $K = \mathbb{Q}$, and all our matrices will be admissible. To find a necessary condition for base changes on the order-theoretical side, we must first recall (say from [FM]) how to compute the type of an element g of G.

Partitions $\mathcal{A} = \{A_1, \ldots, A_k\}$ of $I = \{1, \ldots, m\}$ (that is, $I = A_1 \dot\cup \cdots \dot\cup A_k$ where $\dot\cup$ stands for disjoint union of the nonempty blocks $A_i \subseteq I$) form a (non distributive) lattice $\mathbb{P}(m)$ under the ordering "greater = coarser".

Associate to each $g = \sum_{i=1}^m r_i g_i$ the partition $\mathcal{A} = \{A_1, \ldots, A_k\}$ of the index set I into subsets of equal coefficients: a nonempty subset A_j of I consists of all indices n such that the coefficients r_n have the same value a_j. Writing $g_{A_j} = \sum_{i \in A_j} g_i$, we have then $g = \sum_{j=1}^k a_j g_{A_j}$ with $a_j = r_n$ for all $n \in A_j$, and $a_j \neq a_{j'}$ if and only if $j \neq j'$ $(j, j' \in I)$. We have from [FM, 2.3] and the distributivity of the lattice $\mathbb{T}$ of types that the type of g in G

depends only on the partition $\mathcal{A}$:

$$\mathcal{A}(t_1,\dots,t_m) = \Big(\bigwedge_{i\in I\setminus A_1} t_1\Big) \vee\cdots\vee \Big(\bigwedge_{i\in I\setminus A_k} t_k\Big) = \operatorname{type}_G(g). \tag{1}$$

So, just as the element $g = \sum_{i=1}^{m} r_i g_i$ is obtained from the base elements g_i by the coefficients vector $\rho = \begin{pmatrix} r_1 \\ \vdots \\ r_m \end{pmatrix}$, the type of g is obtained from the base types t_i by the partition of I associated to ρ, $\mathcal{A} = p\rho$. Partitions (elements of the lattice $\mathbb{P}(m)$) play for $\mathbf{B}^{(1)}$-groups the role held by supports (elements of the lattice $\mathcal{P}(m)$) in completely decomposable groups. (A challenging problem: what will play that role for $\mathbf{B}^{(2)}$-groups?)

As a consequence, we may consider the matrix $\mathbf{M}$ performing a base change as consisting of two m-tuples: an m-tuple $[\mu_1,\dots,\mu_m]$ of columns (vectors of $\mathbb{Q}^m$) acting on the base elements linearly, and an m-tuple $p\mathbf{M} = (\mathcal{A}_1,\dots,\mathcal{A}_m)$ of partitions (the *partition matrix* of $\mathbf{M}$), acting on the base types order-theoretically; here and in the following, $\mathcal{A}_i = p\mu_i$ denotes the partition of the index set I into equal coefficients for the column vector μ_i. Hence

$$(u_1,\dots,u_m) = p\mathbf{M}(t_1,\dots,t_m) = (\mathcal{A}_1(t_1,\dots,t_m),\dots,\mathcal{A}_m(t_1,\dots,t_m)).$$

Example (cont.) 2.1 Let $\min = \{\{1\},\{2\},\{3\},\{4\},\{5\},\{6\}\}$.

For our 6×6 matrices we have

$$\begin{aligned} p\mathbf{M} = (\mathcal{A}_1,\dots,\mathcal{A}_6) = (&\{\{1,2,3\},\{4,5,6\}\},\{\{1,3,4\},\{2,5,6\}\},\\ &\{\{1,4,5\},\{2,3,6\}\},\{\{1,5,6\},\{2,3,4\}\},\\ &\{\{1,2\},\{3,4,5,6\}\},\{\{1,6\},\{2,3,4,5\}\});\\ p\mathbf{M}^{-1} = (\mathcal{B}_1,\dots,\mathcal{B}_6) = (&\{\{1,3,6\},\{2,4,5\}\},\{\{1,3,6\},\{2,4\},\{5\}\},\\ &\{\{1,5\},\{2,3,4,6\}\},\{\{1\},\{2,5\},\{3,6\},\{4\}\},\\ &\{\{1,2,3,5\},\{4,6\}\},\{\{1,3\},\{2,4,5\},\{6\}\}). \end{aligned}$$

For later use, we compute (see (6))

$$\begin{aligned} \bigwedge_{i\in\{2,4,5\}} \mathcal{A}_i &= \{\{1,3,4\},\{2,5,6\}\}\wedge\{\{1,5,6\},\{2,3,4\}\}\wedge\\ &\quad\wedge\{\{1,2\},\{3,4,5,6\}\} = \{\{1\},\{2\},\{3,4\},\{5,6\}\} = \mathcal{C}_{1,2},\\ \bigwedge_{i\in\{1,3,5,6\}} \mathcal{A}_i &= \{\{1\},\{2\},\{3\},\{4,5\},\{6\}\} = \mathcal{C}_{2,2},\\ \bigwedge_{i\in\{1,2,3,4,6\}} \mathcal{A}_i &= \min = \mathcal{C}_{3,2}. \end{aligned}$$

□

Let us now use the fact that the types are sequences of 0 and ∞:

$$\begin{array}{lcllllll} t_1 & = & 0 & \infty & 0 & 0 & \infty & 0 & \dots \\ t_2 & = & \infty & \infty & 0 & 0 & 0 & \infty & \dots \\ \dots & & & & & & & & \\ t_m & = & 0 & 0 & 0 & \infty & 0 & 0 & \dots \end{array}$$

The simplification given in [DVM2] allows us also to suppose that there is no repetition of columns. Then the m-tuple of types shapes up as a finite table called *tent*[1], whose rows are the (base-) types and whose columns, which we call the *primes*, are (characteristic vectors of) subsets of I; in fact, the tent can be viewed (column-wise) as an ordered subset of $\mathcal{P}(m)$. The only restriction is that by regularity subsets of cardinality $m-1$ are not allowed (see [DVM2]): each column $\neq I$ must have at least two 0s. We make now a crucial observation.

Lemma 2.2 *In the action* (1) *of a partition* $\mathcal{A}$ *on* $(t_1, \dots, t_m)$, $\mathcal{A}$ *operates column-by-column.*

Proof. Let $t_1 = \alpha_1\ \alpha_2\ \dots\ \alpha_h$, $t_2 = \beta_1\ \beta_2\ \dots\ \beta_h$, with $\alpha_i, \beta_j \in \{0, \infty\}$. Then $t_1 \wedge t_2 = \alpha_1 \wedge \beta_1\ \alpha_2 \wedge \beta_2\ \dots\ \alpha_m \wedge \beta_m$; analogously for $\vee$.

Finite induction completes the proof. □

We thus have a simple way to determine how the partition matrix $p\mathbf{M}$ operates on the tent of G. We transform in turn each prime first by $p\mathbf{M}$, then by $p\mathbf{M}^{-1}$. If every time we re-obtain the initial prime, then $\mathbf{M}$ (if admissible) is a base change for G; if even one prime does not revert to itself, then $\mathbf{M}$ doesn't work on G.

Example (cont.) 2.3 Let G have the tent

$$\begin{array}{lcll} t_1 & = & \infty & 0 \\ t_2 & = & \infty & 0 \\ t_3 & = & \infty & \infty \\ t_4 & = & \infty & \infty \\ t_5 & = & 0 & \infty \\ t_6 & = & 0 & \infty \end{array}$$

consisting of the primes $A = \begin{pmatrix} \infty \\ \infty \\ \infty \\ \infty \\ 0 \\ 0 \end{pmatrix}$, $A' = \begin{pmatrix} 0 \\ 0 \\ \infty \\ \infty \\ \infty \\ \infty \end{pmatrix}$. Applying the first partition $\mathcal{A}_1 = \{\{1,2,3\},\{4,5,6\}\}$ of $\mathbf{M}$ to A we get $(\infty \wedge \infty \wedge \infty) \vee (\infty \wedge 0 \wedge 0) = \infty$, which is the first entry of $p\mathbf{M}(A)$. Proceeding with $\mathcal{A}_2, \dots, \mathcal{A}_6$ we get

[1] A more complex, but basically equivalent, definition of tent was given in [DVM2].

$p\mathbf{M}(A) = \begin{pmatrix} \infty \\ \infty \\ 0 \\ \infty \\ \infty \\ 0 \end{pmatrix}$; similarly, $p\mathbf{M}(A') = \begin{pmatrix} \infty \\ 0 \\ 0 \\ 0 \\ \infty \\ 0 \end{pmatrix}$; $p\mathbf{M}^{-1}(p\mathbf{M}(A)) = \begin{pmatrix} \infty \\ \infty \\ 0 \\ \infty \\ \infty \\ 0 \end{pmatrix} =$

A, $p\mathbf{M}^{-1}(p\mathbf{M}(A')) = \begin{pmatrix} 0 \\ 0 \\ \infty \\ 0 \\ 0 \\ 0 \end{pmatrix} \neq A'$; hence $\mathbf{M}$ is not a base change for G. □

Rather than doing this on the tent of a given G, let us check $\mathbf{M}$ once and for all on each possible prime, i.e. on all subsets of I of cardinality $\neq m-1$. Call $D(\mathbf{M})$, the *domain* of $\mathbf{M}$, the set of primes on which $\mathbf{M}$ acts invertibly. For instance, $D(\mathbf{M})$ always contains the prime consisting of all 0s (the empty subset of I) and the prime with all ∞s (I itself). We can summarize

Theorem 2.4 *The admissible matrix $\mathbf{M}$ is a base change for a $\mathbf{B}^{(1)}$-group G if and only if the primes of G are contained in $D(\mathbf{M})$.* □

Even if being a base change depends on the group, there is an interesting property that depends on the matrix itself. Recall from [FM] that a $\mathbf{B}^{(1)}$-group G is strongly indecomposable (indecomposable together with its subgroups of finite index) if and only if

(∗) no base type t_i of the tent of G is topped or equalled by a type $\mathcal{A}(t_1, \ldots, t_m)$, for any partition $\mathcal{A} \in \mathbb{P}(m)$, $\mathcal{A} \neq \{\{i\}, I \setminus \{i\}\}, \{I\}$.

If G is not strongly indecomposable, there is a base type t_i and a partition $\mathcal{A} \neq \{\{i\}, I \setminus \{i\}\}, \{I\}$ such that $t_i \leq \mathcal{A}(t_1, \ldots, t_m)$. This relation remains true on each prime: if $\begin{matrix} \alpha_{1j} \\ \vdots \\ \alpha_{mj} \end{matrix}$ is the j-th prime of the tent (j-th column of the types table), we have $\alpha_{ij} \leq \mathcal{A}(\alpha_{1j}, \ldots, \alpha_{mj})$; in fact (∗) is a property of the (unordered) set of primes of the tent. Say a set of primes is indecomposable if it satisfies (∗).

Proposition 2.5 **([FM])** *An admissible matrix $\mathbf{M}$ is the base change for a strongly indecomposable $\mathbf{B}^{(1)}$-group if and only if $D(\mathbf{M})$ is indecomposable.*

Proof. If $D(\mathbf{M})$ is indecomposable, a $\mathbf{B}^{(1)}$-group G whose tent has exactly $D(\mathbf{M})$ as a set of primes is strongly indecomposable. Conversely, let all the primes of $D(\mathbf{M})$ satisfy $\alpha_{ij} \leq \mathcal{A}(\alpha_{1j}, \ldots, \alpha_{mj})$ for an $i \in I$ and a partition $\mathcal{A}$. Then if all the primes of a $\mathbf{B}^{(1)}$-group H occur among the primes of $D(\mathbf{M})$, the type t_i of H will satisfy the same relation, hence H will not be strongly indecomposable. □

3 Building the domain of an admissible matrix

Given an admissible matrix $\mathbf{M}$ and a computer, it is not difficult to determine its domain $D(\mathbf{M})$ by checking $\mathbf{M}$ on all allowed subsets of I. What we want to do here is to describe a (much faster) realizing procedure, that allows to build the domain directly from $\mathbf{M}$, and can be performed easily by hand for 1-digit ranks, thus yielding plenty of tractable examples of $\mathbf{B}^{(1)}$-groups.

In order to determine $D(\mathbf{M})$, we must find all primes A such that $p\mathbf{M}^{-1}(p\mathbf{M}(A)) = A$. For economy of space we will write from now on the primes as rows rather than as columns. Start by observing that since $p\mathbf{M} = (\mathcal{A}_1, \ldots, \mathcal{A}_m)$ is an m-tuple of elements of a lattice, we can operate on it by a partition $\mathcal{B} = (B_1, \ldots, B_h)$ following (1):

$$\mathcal{B}(\mathcal{A}_1, \ldots, \mathcal{A}_m) = \left(\bigwedge_{i \in I \setminus B_1} \mathcal{A}_i\right) \vee \cdots \vee \left(\bigwedge_{i \in I \setminus B_h} \mathcal{A}_i\right) \tag{2}$$

Therefore we can define a product of partition matrices by

$$p\mathbf{M}p\mathbf{N} = p\mathbf{N}(p\mathbf{M}) = (\mathcal{B}_1(\mathcal{A}_1, \ldots, \mathcal{A}_m), \ldots, \mathcal{B}_m(\mathcal{A}_1, \ldots, \mathcal{A}_m)), \tag{3}$$

if $p\mathbf{N} = (\mathcal{B}_1, \ldots, \mathcal{B}_m)$. Unfortunately in general $p\mathbf{M}p\mathbf{N} \neq p(\mathbf{MN})$. The best we can do for a generic prime A is

Proposition 3.1 $p\mathbf{N}(p\mathbf{M}(A)) \leq (p\mathbf{M}p\mathbf{N})(A) \leq p(\mathbf{MN})(A)$.

Proof. It is simple to verify that for two partitions $\mathcal{A}$, $\mathcal{B}$ and a prime A we have

$$\begin{aligned} &(\mathcal{A} \wedge \mathcal{B})(A) = \mathcal{A}(A) \wedge \mathcal{B}(A) \\ &\mathcal{A}(A) \vee \mathcal{B}(A) \leq (\mathcal{A} \vee \mathcal{B})(A) \\ &\mathcal{A} \leq \mathcal{B} \text{ implies } \mathcal{A}(A) \leq \mathcal{B}(A). \end{aligned} \tag{4}$$

These extend easily to a finite number of partitions. Let us use the first two to prove that $A' = p\mathbf{N}(p\mathbf{M}(A)) \leq (p\mathbf{M}p\mathbf{N})(A)$. We have

$$\begin{aligned} [\mathcal{B}(\mathcal{A}_1, \ldots, \mathcal{A}_m)](A) &= \left[\left(\bigwedge_{i \in I \setminus B_1} \mathcal{A}_i\right) \vee \cdots \vee \left(\bigwedge_{i \in I \setminus B_h} \mathcal{A}_i\right)\right](A) \geq \\ &\geq \left(\bigwedge_{i \in I \setminus B_1} \mathcal{A}_i\right)(A) \vee \cdots \vee \left(\bigwedge_{i \in I \setminus B_h} \mathcal{A}_i\right)(A) = \\ &= \bigwedge_{i \in I \setminus B_1} \mathcal{A}_i(A) \vee \cdots \vee \bigwedge_{i \in I \setminus B_h} \mathcal{A}_i(A) = \\ &= \mathcal{B}(\mathcal{A}_1(A), \ldots, \mathcal{A}_m(A)) = \mathcal{B}[(\mathcal{A}_1, \ldots, \mathcal{A}_m)(A)]. \end{aligned}$$

Reading this with $\mathcal{B} = \mathcal{B}_i$ for all i, we have the desired inequality.

To get $(p\mathbf{M}p\mathbf{N})(A) \leq p(\mathbf{MN})(A)$ let us prove first

$$p\mathbf{M}p\mathbf{N} \leq p(\mathbf{MN}); \tag{5}$$

that is, if $\nu_i = b_1\delta_{B_1} + \cdots + b_h\delta_{B_h}$ is the i-th column of $\mathbf{N}$, and $\mathcal{D}_i$ the partition associated to $\nu_i(\mathbf{M}) = \nu_i(\mu_1, \ldots, \mu_m) = b_1 \sum_{r\in B_1} \mu_r + \cdots + b_h \sum_{r\in B_h} \mu_r$, we show $\mathcal{D}_i \geq \mathcal{B}_i(\mathcal{A}_1, \ldots, \mathcal{A}_m)$. For this we need the identity, valid in $\mathbb{P}(m)$ even if it is not a distributive lattice [DVM2]:

$$\mathcal{B}_i(\mathcal{A}_1, \ldots, \mathcal{A}_m) = \Big(\bigwedge_{i\in I\setminus B_1} \mathcal{A}_i\Big) \vee \cdots \vee \Big(\bigwedge_{i\in I\setminus B_h} \mathcal{A}_i\Big) =$$

$$= \Big[\Big(\bigwedge_{i\in B_1} \mathcal{A}_i\Big) \vee \Big(\bigwedge_{i\in I\setminus B_1} \mathcal{A}_i\Big)\Big] \wedge \cdots \wedge \Big[\Big(\bigwedge_{i\in B_h} \mathcal{A}_i\Big) \vee \Big(\bigwedge_{i\in I\setminus B_h} \mathcal{A}_i\Big)\Big].$$

Observe that the partition associated to a sum of vectors is greater than or equal to the infimum of the partitions associated to the vectors; therefore the partition associated to $\sum_{r\in B_j} \mu_r = \mathbf{M}(\delta_I) - \sum_{r\in I\setminus B_j} \mu_r$ is greater or equal to $(\bigwedge_{r\in B_j} \mathcal{A}_r) \vee (\bigwedge_{r\in I\setminus B_j} \mathcal{A}_r)$. This same observation now leads to $\mathcal{D}_i \geq \mathcal{B}_i(\mathcal{A}_1, \ldots, \mathcal{A}_m)$. Applying the third of (4) we get the desired conclusion. □

From (5) there follows immediately

Corollary 3.2 *We have* $p\mathbf{M}p\mathbf{M}^{-1} \leq p\mathbf{I} = (\{\{1\}, I\setminus\{1\}\}, \ldots, \{\{m\}, I\setminus\{m\}\})$. □

In particular, by setting, for $j \in I$, $r = 1, \ldots, h_j$,

$$p\mathbf{M}^{-1} = p\mathbf{N} = (\mathcal{B}_1, \ldots, \mathcal{B}_m),$$

where $\mathcal{B}_j = (B_{1,j}, \ldots, B_{h_j,j})$, and

$$\mathcal{C}_{r,j} = \bigwedge_{i\in I\setminus B_{r,j}} \mathcal{A}_i, \tag{6}$$

we have

$$\mathcal{B}_j(\mathcal{A}_1, \ldots, \mathcal{A}_m) = \mathcal{C}_{1,j} \vee \cdots \vee \mathcal{C}_{h_j,j} \leq \{\{j\}, I\setminus\{j\}\};$$

this means that $\{j\}$ is a block of each $\mathcal{C}_{r,j}$.

Example (cont.) 3.3 In our example,

$$\mathcal{B}_2 = \{B_{1,2}, B_{2,2}, B_{3,2}\} = \{\{1,3,6\},\{2,4\},\{5\}\};$$

$$\mathcal{B}_2(\mathcal{A}_1, \ldots, \mathcal{A}_6) = \mathcal{C}_{1,2} \vee \mathcal{C}_{2,2} \vee \mathcal{C}_{3,2} = \{\{1\},\{2\},\{3,4\},\{5,6\}\} \vee \vee \{\{1\},\{2\},\{3\},\{4,5\},\{6\}\} \vee \min = = \{\{1\},\{2\},\{3,4,5,6\}\}.$$ □

We are now ready to give the rule to build $D(\mathbf{M})$.

Let $\mathbf{M}$ be an admissible $m \times m$ matrix over $\mathbb{Q}$, w.l.o.g. $\mathbf{M}$ invertible, with $p\mathbf{M} = (\mathcal{A}_1, \dots, \mathcal{A}_m)$, $p\mathbf{M}^{-1} = (\mathcal{B}_1, \dots, \mathcal{B}_m)$, $\mathcal{B}_j = (B_{1,j}, \dots, B_{h_j,j})$ for each $j \in I$. Let

$$\begin{aligned} p\mathbf{M}^{-1}(p\mathbf{M}(A)) &= p\mathbf{M}^{-1}[(\mathcal{A}_1, \dots, \mathcal{A}_m)(A)] = p\mathbf{M}^{-1}(\mathcal{A}_1(A), \dots, \mathcal{A}_m(A)) \\ &= (\mathcal{B}_1(\mathcal{A}_1(A), \dots, \mathcal{A}_m(A)), \dots, \mathcal{B}_m(\mathcal{A}_1(A), \dots, \mathcal{A}_m(A))), \end{aligned}$$

with

$$\begin{aligned} \mathcal{B}_j(\mathcal{A}_1(A), \dots, \mathcal{A}_m(A)) &= \Big(\bigwedge_{i \in I \setminus B_{1,j}} (\mathcal{A}_i(A)) \Big) \vee \dots \vee \Big(\bigwedge_{i \in I \setminus B_{h_j,j}} (\mathcal{A}_i(A)) \Big) \\ &= \Big(\bigwedge_{i \in I \setminus B_{1,j}} \mathcal{A}_i \Big)(A) \vee \dots \vee \Big(\bigwedge_{i \in I \setminus B_{h_j,j}} \mathcal{A}_i \Big)(A). \end{aligned}$$

Recall that $A = I$ belongs to $D(\mathbf{M})$ for every m. If $A \neq I$, we have $A \in D(\mathbf{M})$ if and only if $p\mathbf{M}^{-1}(\mathcal{A}_1(A), \dots, \mathcal{A}_m(A)) = A$; that is if and only if $[\mathcal{B}_j(\mathcal{A}_1(A), \dots, \mathcal{A}_m(A)) = \infty$ if and only if $j \in A]$; i.e. if and only if $[[\mathcal{C}_{r,j}(A) = \infty$ for at least an $r = 1, \dots, h_j]$ if and only if $j \in A]$; equivalently, if and only if $[[A$ contains a coblock of $\mathcal{C}_{r,j}$ for at least an $r = 1, \dots, h_j]$ if and only if $j \in A]$. This means checking a) that A contains a coblock of $\mathcal{C}_{r,j}$ for at least an $r = 1, \dots, h_j$ for all $j \in A$, and b) that A does not contain coblocks of any $\mathcal{C}_{r,j}$ if $j \notin A$.

Now if $j \notin A$, a coblock of $\mathcal{C}_{r,j}$ contained in A cannot contain j; by Corollary 3.2, the only coblock of $\mathcal{C}_{r,j}$ not containing j is $I \setminus \{j\}$; but by regularity if A contains $I \setminus \{j\}$ then $A = I$, which was excluded before. Thus b) needn't be checked; and we have the simple rule

Rule 3.4 $A \in D(\mathbf{M})$ if and only if either $A = I$, $A = \emptyset$, or A contains a coblock of some $\mathcal{C}_{r,j}$ for each $j \in A$; equivalently (better for computations)

(**) $A \in D(\mathbf{M})$ if and only if either $A = I$, $A = \emptyset$, or $I \setminus A$ is contained in a block of some $\mathcal{C}_{r,j}$ for each $j \in A$. □

We list $\mathcal{C}_{1,1}, \dots, \mathcal{C}_{h_1,1}$; $\mathcal{C}_{1,2}, \dots, \mathcal{C}_{h_2,2}$; $\dots$; $\mathcal{C}_{1,m}, \dots, \mathcal{C}_{h_m,m}$. Start with the first block of $\mathcal{C}_{1,1}$; let C be a subset of at least two elements. If C is contained in some $\mathcal{C}_{r,j}$ for each $j \notin C$, then C is the set of 0s for a prime of $D(\mathbf{M})$; otherwise it isn't. Proceeding through the subsets of the blocks of the $\mathcal{C}_{r,j}$ we complete $D(\mathbf{M})$. Observe that if a prime A of $D(\mathbf{M})$ has ∞ as the j-th entry, $I \setminus A$ must belong to a block of some $\mathcal{C}_{r,j}$. Thus, once we finish checking the subsets of the blocks of the $\mathcal{C}_{r,j}$, the only new sets $I \setminus A$ we need to check must contain j.

Example (cont.) 3.5 We have for our 6×6 matrix $\mathbf{M}$

$$\mathcal{C}_{1,1} = \{\{1\},\{2\},\{3,4\},\{5,6\}\}; \quad \mathcal{C}_{2,1} = \{\{1\},\{2,3\},\{4,5\},\{6\}\}.$$
$$\mathcal{C}_{1,2} = \{\{2\},\{1\},\{3,4\},\{5,6\}\}; \quad \mathcal{C}_{2,2} = \{\{2\},\{1\},\{3\},\{4,5\},\{6\}\};$$
$$\mathcal{C}_{3,2} = \min .$$
$$\mathcal{C}_{1,3} = \min; \quad \mathcal{C}_{2,3} = \{\{3\},\{1,2\},\{4,5,6\}\}.$$
$$\mathcal{C}_{1,4} = \min; \quad \mathcal{C}_{2,4} = \{\{4\},\{1\},\{2,3\},\{5\},\{6\}\};$$
$$\mathcal{C}_{3,4} = \{\{4\},\{1\},\{2\},\{3\},\{5,6\}\}; \quad \mathcal{C}_{4,4} = \min .$$
$$\mathcal{C}_{1,5} = \{\{5\},\{1,6\},\{2,3,4\}\}; \quad \mathcal{C}_{2,5} = \min .$$
$$\mathcal{C}_{1,6} = \{\{6\},\{1\},\{2\},\{3,4\},\{5\}\}; \quad \mathcal{C}_{2,6} = \{\{6\},\{1\},\{2,3\},\{4,5\}\};$$
$$\mathcal{C}_{3,6} = \min .$$

Now starting with $I \setminus A = \{3,4\}$ from $\mathcal{C}_{1,1}$ we note that it is contained in $\mathcal{C}_{1,2}$, $\mathcal{C}_{1,5}$ and $\mathcal{C}_{1,6}$, that is in some $\mathcal{C}_{r,j}$ for each $j = 1,2,5,6$. Hence $A = \begin{pmatrix} \infty \\ \infty \\ 0 \\ 0 \\ \infty \\ \infty \end{pmatrix} \in D(\mathbf{M})$. Instead, $I \setminus A' = \{1,2\}$ (which we find in $\mathcal{C}_{2,3}$) is not contained in a block of any $\mathcal{C}_{r,4}$, hence $A' \notin D(\mathbf{M})$. We don't need to check $\{4,5,6\}$ from $\mathcal{C}_{2,3}$: since it doesn't contain 1, if we didn't find it already while checking the first row (as happened for $\{3,4\}$) it doesn't belong to the tent. Finally we have

$$D(\mathbf{M}) = \begin{array}{lcllllll} t_1 & = & \infty & \infty & \infty & \infty & 0 & \infty \\ t_2 & = & \infty & \infty & 0 & \infty & 0 & \infty \\ t_3 & = & 0 & \infty & 0 & \infty & 0 & \infty \\ t_4 & = & 0 & \infty & \infty & 0 & 0 & \infty \\ t_5 & = & \infty & 0 & \infty & 0 & 0 & \infty \\ t_6 & = & \infty & 0 & \infty & \infty & 0 & \infty \end{array}$$

Here $t_1 \geq t_2$, so from [FM] we get that t_1 is extractible (that is $R_1 g_1$ is a direct summand of G). Hence any $\mathbf{B}^{(1)}$-group with primes in $D(\mathbf{M})$ is decomposable. In fact, they turn out to be completely decomposable. □

Observe finally that the partitions $\mathcal{C}_j = \mathcal{C}_{1,j} \vee \cdots \vee \mathcal{C}_{h_j,j}$ are the components of $p\mathbf{M}p\mathbf{M}^{-1}$. In the final part of our example we show that the inequalities of Proposition 3.1 and (5) do not, in general, reduce to equalities.

Example (end) 3.6 We have

$$p\mathbf{M}p\mathbf{M}^{-1} = (\mathcal{C}_1, \ldots, \mathcal{C}_m) = (\{\{1\},\{2,3,4,5,6\}\},\{\{2\},\{1\},\{3,4,5,6\}\},$$
$$\{\{3\},\{1,2\},\{4,5,6\}\},\{\{4\},\{1\},\{2,3\},\{5,6\}\},\{\{5\},\{1,6\},$$
$$\{2,3,4\}\},\{\{6\},\{1\},\{2,3,4,5\}\}) < p(\mathbf{M}\mathbf{M}^{-1}) = p\mathbf{I}.$$

For $A = \begin{pmatrix} \infty \\ \infty \\ \infty \\ 0 \\ 0 \\ 0 \end{pmatrix}$, $A' = \begin{pmatrix} 0 \\ 0 \\ \infty \\ \infty \\ \infty \\ \infty \end{pmatrix}$, we have

$$A = (p\mathbf{M}p\mathbf{M}^{-1})(A) > p\mathbf{M}^{-1}(p\mathbf{M}(A)) = \begin{pmatrix} 0 \\ 0 \\ \infty \\ 0 \\ 0 \\ 0 \end{pmatrix};$$

$$A' = p\mathbf{I}(A') > (p\mathbf{M}p\mathbf{M}^{-1})(A') = \begin{pmatrix} 0 \\ 0 \\ \infty \\ 0 \\ 0 \\ 0 \end{pmatrix}.$$

□

References

[FM] L. Fuchs and C. Metelli, *On a class of Butler groups*, Manuscr. Math. **71** (1991), 1–28.

[AV] D. M. Arnold and C. Vinsonhaler, *Finite rank Butler groups. A survey of recent results*, Abelian Groups, Lecture Notes in Pure and Applied Math. **146** (Marcel Dekker 1991), 17–41.

[DVM1] C. De Vivo and C. Metelli, $\mathbf{B}^{(1)}$*-groups: some counterexamples*, Abelian Groups and Modules, Lecture Notes in Pure and Applied Math. **182** (Marcel Dekker 1996), 227–232.

[DVM2] C. De Vivo and C. Metelli, *Finite partition lattices and Butler groups*, to appear in Comm. Algebra.

Trends in Mathematics, © 1999 Birkhäuser Verlag Basel/Switzerland

Butler modules over 1-dimensional Noetherian domains

H. Pat Goeters

Abstract

Let R be a noetherian domain of Krull dimension 1. A completely decomposable module is a direct sum of rank one modules, and a Butler module is any torsion-free image of a finite rank, completely decomposable module. Butler modules are closed under quasi-isomorphism. We characterize when the class of Butler modules is closed under the formation of pure submodules. Additionally, when R is analytically unramified, we show that the Butler modules coincide with the class of pure submodules of finite rank completely decomposable modules if and only if for each maximal ideal P of R, there is a unique maximal ideal in the integral closure of R lying over P.

1 Introduction

One of the most widely studied classes of torsion-free abelian groups is the class of Butler groups. The study of these objects was initiated by Butler in [8]. Subsequent developments were given in [5], [12], [11], and in [3] and [4] where Butler groups were popularized. Many problems concerning torsion-free abelian groups involve the study of torsion-free modules over subrings of algebraic number fields, and the building blocks for torsion-free modules over an integral domain are the rank one modules, and hence the completely decomposable modules (i.e., direct sums of rank one modules).

We define a *Butler* module over a noetherian domain R of Krull dimension one, to be any torsion-free image of a finite rank completely decomposable module. This is not obviously equivalent and most likely not generally equivalent to the definition of a Butler module given in [6]. In section 4 we show that our Butler modules are closed under quasi-isomorphism thereby equating our Butler modules with the modules receiving primary attention in [7], justifying our use of the term Butler module. We extend the work of [7] and determine precisely when a noetherian domain of Krull dimension one has its lattice of types distributive, which results in a characterization of when the class of Butler modules is closed under the formation of pure submodules.

Perhaps the most significant feature of Butler groups, which is the case $R = \mathbb{Z}$, is that such groups coincide with the class of pure subgroups of finite rank completely decomposable groups. Butler points out in [7] that this result is due to the fact that the lattice of subgroups of $\mathbb{Q}$ containing $\mathbb{Z}$ is a distributive lattice. We show that for an analytically unramified R, that the class of Butler modules coincides with the class of pure submodules of finite rank completely decomposable modules if and only if for every maximal ideal P of R, there is a unique maximal ideal in the integral closure of R lying over P.

By the *dimension* of R we mean the Krull dimension. Throughout the paper, unless indicated otherwise, R is assumed to be noetherian and to have Krull dimension one (i.e., every nonzero prime ideal is maximal). These assumptions on R allow us to use Matlis' important development of divisible modules [14], which is critical in our results below. Also vital to the discussion below is the fact that the integral closure of R, denoted by $\bar{R}$, in the quotient field Q of R, is Dedekind [16].

2 Quasi-isomorphism of rank one modules

In this section we review the notion of quasi-isomorphism of torsion-free modules. Recall that two such modules A and B are called *quasi-isomorphic* if there exists $\lambda \in Q\ \mathrm{Hom}(A, B)$ and $\delta \in Q\ \mathrm{Hom}(B, A)$ such that $\delta\lambda = 1_A$ and $\lambda\delta = 1_B$, where $Q\ \mathrm{Hom}(A, B) = \{\lambda : QA \to QB \mid \exists 0 \neq r \in R$ such that $r\lambda \in \mathrm{Hom}(A, B)\}$.

Proposition 2.1 *[2] Let A be a torsion-free R-module over any integral domain R having finite rank. Then any monomorphism $f : A \to A$ has bounded cokernel.*

Proof. The ideals $QE(A) \supseteq fQE(A) \supseteq f^2QE(A) \supseteq \ldots$ must stablize. Hence there is an integer n and $\lambda \in QE(A)$ such that $f^n = f^{n+1}\lambda$; i.e., $1 = f\lambda$. For some $0 \neq r \in R$, $r\lambda \in E(A)$ and so $rA = rf\lambda(A) \subseteq f(A) \subseteq A$ as claimed. □

Since R is 1-dimensional noetherian, one can improve proposition 1 by showing that $A/f(A)$ is actually finitely generated (a similar argument appears in the proof of proposition 4.2).

Corollary 2.2 *The following are equivalent for two torsion-free modules A and B of finite rank over an integral domain R:*

(1) A is quasi-isomorphic to B.

(2) A is isomorphic to a submodule of B and B is isomorphic to a submodule of A.

(3) There is a submodule C of B that is isomorphic to A such that B/C is bounded.

Proof. (1) → (2). If there are $\lambda \in Q\ \mathrm{Hom}(A, B)$ and $\delta \in Q\ \mathrm{Hom}(B, A)$ such that $\delta\lambda = 1 \in QE(A)$ and $\lambda\delta = 1 \in QE(B)$, then there are $0 \neq r, s \in R$ such that $r\lambda : A \to B$ and $s\delta : B \to A$. Since λ and δ are monomorphisms, (2) follows.

(2) → (3). Regarding $A \leq B$, by (2) there is an embedding f of B into A. By proposition 1, $B/f(B)$ is bounded, and so is B/A.

(3) → (1). Regard $rB \subseteq A \subseteq B$ for some $0 \neq r \in R$. The embedding map $\iota : A \to B$ and the map $\mu : B \to A$ which is multiplication by r satisfy $\mu\iota = r1_A$ and $\iota\mu = r1_B$. Thus $\frac{1}{r}\iota$ and μ are maps satisying the criterion for A and B to be quasi-isomorphic. □

Of course, the preceeding results hold for any integral domain, but henceforth the assumption that R is 1-dimensional noetherian will be in force.

Lemma 2.3 *If D is any module between R and Q is such that D/R is divisible, then D is a 1-dimensional noetherian domain, and any R-submodule A of Q containing D is a D-module.*

Proof. Let A be an R-module containing D. If $a/b \in A$, then from $bD + R = D$ we obtain $D + (1/b)R = (1/b)D$, and $(a/b)D = aD + (a/b)R \subseteq A$. This shows that D is a ring (when $D = A$) and A is a D-module. That D is 1-dimensional noetherian is a consequence of the Krull-Akizuki Theorem.□

Several results from [14] will find application here. The first is that any submodule T of $K = Q/R$ has a maximal divisible submodule D', such that T/D' is finitely generated as an R-module.

Lemma 2.4 *If R is local with maximal ideal P, then for any proper submodule X of Q, $PX \neq X$.*

Proof. If D is an overring of R in Q, different from Q, then for any nonzero prime P' of D, $P' \cap R$ is a nonzero prime of R, hence must be P. Therefore, $PD = (P' \cap R)D \subseteq P' \neq D$. For general X, up to isomorphism we may assume that $R \subseteq X$. Put D/R equal to the maximal divisible submodule of X/R.

By lemma 2.3, D is a 1-dimensional noetherian domain, and as mentioned above, X/D is finitely generated over R. Then X is a finitely generated D-module and from the first paragraph of the proof, PD is contained in the Jacobson radical of D, and so $PX = PDX \neq X$, by Nakayama's Lemma. □

It follows, under the hypotheses of lemma 2.4, that $\cap_n P^n X = 0$ if $PX \neq X$, and, $PX = X$ implies $X = Q$. This is because $P(\cap_n P^n X) = \cap_n P^n X$ so $\cap_n P^n X = 0$ or Q. There exist valuation domains V whose maximal ideal P satisfies $P^2 = P$ [9], so the conclusion of lemma 2.4 does not hold even for general valuation domains.

Definition 2.5 Given a rank one module X and $0 \neq x \in X$, we define a height function $\mathbf{h}^X(x) : spec(R) \to \mathbb{N} \cup \{\infty\}$ as follows: $\mathbf{h}_P^X(x) = n$ provided $x \in P^n X \setminus P^{n+1} X$, and, $\mathbf{h}_P^X(x) = \infty$, in case $x \in \cap_n P^n X$. Two functions $\mathbf{h}, \mathbf{h}' : spec(R) \to \mathbb{N} \cup \{\infty\}$ are called equivalent, if $\Sigma_P |\mathbf{h}_P - \mathbf{h}'_P| < \infty$ (under the usual convention that $\infty - \infty = 0$, and $|k - \infty| = \infty$ for any natural number k).

A rank one module X over a Dedekind domain is characterized up to quasi-isomorphism by the equivalence class of a height function $\mathbf{h}^X(x)$ for $0 \neq x \in X$. We next show that any two height functions $\mathbf{h}^X(x)$ and $\mathbf{h}^X(y)$ for $0 \neq x, y$ belonging to a rank one module X over a 1-dimensional noetherian domain are equivalent. In theorem 2.7 below we characterize R in order for the rank one modules to be determined up to quasi-isomorphism by their height functions. It is relatively easy to see that, because R is 1-dimensional noetherian, every nonzero ideal of R is contained in only finitely many maximal ideals of R.

Lemma 2.6 *If X and Y are quasi-isomorphic rank one R-modules, then for any $0 \neq x \in X$ and $0 \neq y \in Y$, $\mathbf{h}^X(x)$ and $\mathbf{h}^Y(y)$ are equivalent.*

Proof. By corollary 2.2 we may regard $aX \leq Y \leq X$ for some $0 \neq a \in R$. There are only finitely many maximal ideals of R containing a, so $X_P = Y_P$ for almost all maximal ideals P and so $\mathbf{h}_P^X(y) = \mathbf{h}_P^Y(y)$ for almost all P. If P contains a, then by the comment following lemma 2.4, both $\mathbf{h}_P^X(y)$ and $\mathbf{h}_P^Y(y)$ are infinite, or both are finite. Therefore it is enough to assume that $X = Y$.

The element $q = x/y \in Q$ is a unit in R_P for almost all primes P. This is due to the fact that there are only finitely many maximal ideals containing either x or y. Thus, $\mathbf{h}_P^X(x) = \mathbf{h}_P^X(y)$ for almost all primes (since x is a unit times y in almost all R_P). For all other primes, by lemma 2.4, $\mathbf{h}_P^X(x)$ and $\mathbf{h}_P^X(y)$ are either both infinite, or both finite. Therefore, $\mathbf{h}^X(x)$ and $\mathbf{h}^X(y)$ are equivalent. □

Under our assumptions on R, R is called *analytically unramified*, if the integral closure of R in Q, denoted by $\bar{R}$ throught this paper, is finitely generated as a module over R. When S is a subring of Q containing R and P is a maximal ideal of R, the notation S_P means $R_P S$, which coincides with the localization of the module S at P.

Theorem 2.7 *The following are equivalent:*

(1) Any rank one R-module X is determined, up to quasi-isomorphism, by the equivalence class of $\mathbf{h}^X(x)$ *for* $0 \neq x \in X$.

(2) R is analytically unramified, and for any maximal ideal P of R, there is a unique maximal ideal of $\bar{R}$ *lying over P.*

Proof. (1) $\to$ (2). Let P be a maximal ideal of R. If $\bar{R}_P$ is not local, then the *pid* $\bar{R}_P$ has at least two irreducible elements, a_1, a_2, which are non-associates. Let X_j denote the localization of $\bar{R}_P$ at the maximal ideal $\bar{R}_P a_j$, $j = 1, 2$. Then, $P'X_j = X_j$ for any maximal ideal P' of R unequal to P, and by lemma 2.4, $PX_j \neq X_j$. Therefore, for any $0 \neq x_j \in X_j$, $\mathbf{h}^{X_1}(x_1)$ and $\mathbf{h}^{X_2}(x_2)$ are equivalent but clearly X_1 and X_2 are not quasi-isomorphic (even as $\bar{R}$-modules).

If P is any maximal ideal of R, then by lemma 2.4, $P\bar{R} \neq \bar{R}$. So, $1 \in \bar{R} \setminus P\bar{R}$, and therefore, $\mathbf{h}^R(1) = \mathbf{h}^{\bar{R}}(1)$. By (1) and corollary 2.2, there exists $0 \neq r \in R$ such that $r\bar{R} \subseteq R$. This shows that $\bar{R}$ is finitely generated over R.

(2) $\to$ (1). Let X, Y be two submodules of Q containing R with equivalent height vectors. By lemma 2.6, $\mathbf{h}^X(1)$ and $\mathbf{h}^Y(1)$ are equivalent. Also, by (2), $r_0\bar{R} \subseteq R$ for some $0 \neq r_0 \in R$, and so $\bar{R}_P = R_P$ for almost all maximal ideals P (i.e., for those P which do not contain r). Therefore, by (2), for almost all maximal ideals P, both $R_P = \bar{R}_P$ and $\mathbf{h}_P^X(1) = \mathbf{h}_P^Y(1)$ are satisfied. For all such ideals, R_P is a *dvr*, and because the rank one modules are linearly ordered, $\mathbf{h}_P^X(1) = \mathbf{h}_P^Y(1)$ implies $X_P = Y_P$. Let $P_1, \ldots, P_n$ denote the maximal ideals of R such that $X_{P_j} \neq Y_{P_j}$, and note that $X_P = Q$ exactly when $\mathbf{h}_P^X(1) = \infty$, so $X_{P_j}, Y_{P_j} \neq Q$. Let j be given.

Consider $\bar{X} = \bar{R}X$ and $\bar{Y} = \bar{R}Y$, and recall that $r_0\bar{R} \subseteq R$. Then $r_0\bar{X} \subseteq X$. Since $\bar{R}_{P_j}$ is a *dvr*, there is an $0 \neq r_j \in R$ such that $r_j\bar{X} \subseteq \bar{R}_{P_j}$. Then $r = r_0 \cdot r_1 \cdots r_n$ satisfies $rX_{P_j} \subseteq r_0\bar{R}_{P_j} \subseteq R_{P_j} \subseteq Y_{P_j}$. Consequently $rX \subseteq Y$ since the containment holds locally, so by corollary 2.2, X and Y are quasi-isomorphic. □

It is intriguing although seemingly unrelated, that the 1-dimensional noetherian domains R such that the completion H_P of R_P is a *strongly homogeneous* extension of R_P (in the sense of Arnold [2]) are precisely those satisfying condition (2) of the preceeding theorem. Condition (2) also crops up in the determination of when the class of Butler modules coincides with the class of pure submodules of finite rank completely decomposable modules (corollary 4.8).

3 When the lattice of types is distributive

We will say that R is a *distributive* ring if the lattice of submodules of Q containing R is distributive. The distributive integral domains are necessarily Pruefer [13], so the local distributive domains are precisely the *dvr*'s. A weeker notion than distributivity of the lattice of submodules of Q is that of the distributivity of the lattice of types.

A *type*, written $[X]$, is the quasi-isomorphism class of a rank one R-module X, and the set of types form a lattice with partial order $[X] \leq [Y]$ when X is isomorphic to a submodule of Y (that this relation is anti-symmetric follows from corollary 2.2). Additionally, the lattice operations are $[X] \wedge [Y] = [X \cap Y]$ and $[X] \vee [Y] = [X + Y]$.

If the lattice of types of R is distributive, we will say that R is *type distributive.* It follows from corollary 2.2 that R is type distributive if and only if for any submodules A, B and C of Q containing R, there exists $0 \neq r \in R$ for which $r(C \cap (A + B)) \subseteq C \cap A + C \cap B$. This is because $C \cap A + C \cap B \subseteq C \cap (A + B)$ for any such A, B and C.

Although some of our results hold for general domains, for simplicity we stick to the hypothesis that R is 1-dimensional noetherian. A *semi-local* ring is a ring with only finitely many maximal ideals.

Proposition 3.1 *Let R be semi-local. Then R is type distributive if and only if for each maximal ideal P, R_P is type distributive.*

Proof. Let $P_1, \ldots, P_m$ be all of the maximal ideals of R. Given submodules A, B and C of Q, if each localization of R is type distributive, there exist $0 \neq r_1, \ldots, r_m \in R$ such that $r_i(C_{P_i} \cap (A_{P_i} + B_{P_i})) \subseteq C_{P_i} \cap A_{P_i} + C_{P_i} \cap B_{P_i}$. Then for $r = r_1 \cdots r_m$, $r(C \cap (A + B)) \subseteq C \cap A + C \cap B$ (since the containment holds locally), and so R is type distributive. The converse is clear. □

An *R-order in Q* is any overring S of R in Q which is finitely generated as an R-module.

Lemma 3.2 *Let S be an R-order in Q. Then R is type distributive if and only if S is type distributive.*

Proof. There is an element $0 \neq r_0 \in R$ such that $r_0 S \subseteq R$. If S is type distributive and A, B and C are R-submodules of Q containing R, then there exists $0 \neq s \in S$ such that $s(SC \cap (SA + SB)) \subseteq (SC \cap SA) + (SC \cap SB)$. Then $r_0 s(C \cap (A + B)) \subseteq r_0((SC \cap SA) + (SC \cap SB)) \subseteq C \cap A + C \cap B$. If $s = r/r'$ with $r, r' \in R$, then $r_0 r(C \cap (A + B)) \subseteq A \cap C + B \cap C$, and R is type distributive. The reverse direction is clear since any rank one S-module is a rank one R-module. □

We will now draw upon the development of divisible modules from [14], alluded to earlier. Let $K = Q/R$. In [14] it is shown that any reduced submodule of K is finitely generated. A *simple divisible* module is a divisible module T which has no proper divisible submodules. So if T_1 and T_2 are distinct simple divisible submodules of K, then $T_1 \cap T_2$ must be reduced, and therefore finitely generated.

Two divisible modules T_1 and T_2 are called *equivalent*, if each is an epimorphic image of the other. In [14], two epimorphic images T_1, T_2 of divisible submodules of K are shown to be equivalent if and only if there is an epimorphism $f : T_1 \to T_2$ such that $ker\ f$ is finitely generated (bounded). Furthermore, it is shown that a divisible torsion module is simple divisible if and only if it is equivalent to Q/V for some valuation overring V of R in Q.

Any divisible submodule T of K has a composition series of divisible submodules:

$$T_0 = 0 \subseteq T_1 \subseteq \cdots \subseteq T_m = T,$$

where each T_i/T_{i-1} is simple divisible. The *divisible length* of T is m. More generally, if C' is a submodule of K, then the divisible length of C' is taken to be the divisible length of the maximal divisible submodule of C'. Finally, T is called *uniserial divisible* if its divisible submodules form a chain with respect to inclusion.

When R is local with maximal ideal P, one can form the P-adic completion H of R. It follows that K (and any torsion module) has a unique H-module structure extending its R-module structure [15].

Definition 3.3 When R is local, K will be called locally uniserial, if $H_{P'} \otimes K$ is uniserial divisible, for each minimal prime P' of H.

Theorem 3.4 *Let R be local. Then R is type distributive if and only if K is locally uniserial divisible.*

Proof. Let $P_1, \ldots, P_n$ be all of the rank 0 primes of H, and put $B_j/R = P_jK$. Since B_j/R is divisible, lemma 2.3 shows that B_j is an overring of R in Q. Furthermore, by Theorem 11.8 in [14], $U = \cap_j B_j$ is an R-order in Q; U is semi-local with maximal ideals $M_1, \ldots, M_n$; $U_{M_j} = B_j$; and the integral closure of B_j in Q is a *dvr*. Finally, by Theorem 11.8 again, Q/B_j is equivalent to $H_{P_j} \otimes K$.

By lemma 3.2, R is type distributive if and only if U is type ditributive. By the proposition 3.1, U is type distributive if and only if each B_j is type distributive. Therefore, in order to satisfy the theorem, it is enough to assume that $R = B_j$, and show: R is type distributive if and only if K is uniserial divisible. That is, we may assume that the integral closure of R is a *dvr* (in this case there is only one rank 0 prime P of H).

Let R be type distributive and suppose that D'_1 and D'_2 are two simple divisible submodules of K which are unequal. Because $\bar{R}$ is a *dvr*, as mentioned above, there is exactly one simple divisible torsion module, up to equivalence; namely $Q/\bar{R}$. Therefore, there is an epimorphism $f : D'_1 \to D'_2$. Then $D' = \{y \mid y = x + f(x),\ x \in D'_1\}$ is simple divisible since the kernel of the epimorphism from D'_1 onto D' is contained in the finitely generated module $D'_1 \cap D'_2$. Also, since each D'_i is simple divisible, either $D'_i \subseteq D'$ or $D' \cap D'_i$ is finitely generated for $i = 1, 2$. Since D' has divisible length 1, and $D'_1 + D'_2$ has divisible length 2, neither D'_1 nor D'_2 is contained in D'. Thus, $D' \cap D'_j$ is finitely generated for each j. With $A_i/R = D'_i$ and $C/R = D'$, $(A_1 + A_2) \cap C = C$ is not finitely generated over R, while $A_1 \cap C + A_2 \cap C$ is; a contradiction. Hence, there is a unique simple divisible module, T_1.

Now suppose D'_1 and D'_2 are two divisible submodules of K of divisible length 2. Each contains T_1 and D'_1/T_1 is equivalent to D'_2/T_1 (since there is only one simple divisible module up to equivalence). If $f : D'_1/T_1 \to D'_2/T_1$ is an epimorphism, define $D' \leq K$ by $D'/T_1 = \{x + f(x) \mid x \in D'_1/T_1\}$. Then $(D' \cap D_j)/T_1$ is finitely generated for each j, while $D' \cap (D'_1 + D'_2) = D'$ and D'/T_1 is not finitely generated. As above we can find submodules A_1, A_2 and C of Q which violate the type distributive property of R. Since there is an absolute bound on the divisible length of divisible submodules of K, repeating the above eventually establishes that the divisible submodules of K form a chain.

Now assume that the divisible submodules of K form a chain, starting with the simple divisible module $D' = D/R$. Recall that any submodule of K is finitely generated or has a nonzero divisible submodule [14]. Let A, B and C be submodules of Q containing R. In order to show there is an $0 \neq r \in R$ such that $r((A + B) \cap C) \subseteq A \cap C + B \cap C$, we may assume that $C \subseteq A + B$. Either $C = C \cap (A + B)$ is finitely generated over R, in which case we can easily find such an r, or, $C' = C/R$ contains the divisible module D'.

If $A' = A/R$ and $B' = B/R$ are reduced, then they are finitely generated (bounded) hence so is $A' + B'$. Thus one of A' or B' is not reduced and hence must contain D'. If one of A' or B' is reduced, A' say, then it is bounded, so there is an $r \neq 0$ such that $rA \subseteq B \cap C$. Then $r((A + B) \cap C) \subseteq B \cap C$ and we are finished. Therefore we may now assume that all of A, B and C contain D. By lemma 2.3, D is a 1-dimensional noetherian domain and any R-module containing D is a D-module.

Any divisible module over D is a divisible R-module so $K'' = Q/D = K/D'$ is uniserial divisible as a D-module. Therefore, we may apply induction on the divisible length of the maximal divisible submodule of C' in order to find a $0 \neq d \in D$ such that $d((A + B) \cap C) \subseteq A \cap C + B \cap C$. For $d = r/s$ with $r \in R$, r knocks $(A+B) \cap C$ into $A \cap C + B \cap C$, as needed. □

When A, B and C are submodules of Q, with $r(C \cap (A + B)) \subseteq C \cap A +$

$C \cap B$, then $C_P \cap (A_P + B_P) = C_P \cap A_P + C_P \cap B_P$, for all primes ideals P which do not contain r. In fact this holds universally for a type distributive domain in that for almost all primes P, given any A, B and C between Q and R, $C_P \cap (A_P + B_P) = C_P \cap A_P + C_P \cap B_P$.

Theorem 3.5 *The following are equivalent:*

(1) R is type distributive.

(2) For each maximal ideal P, R_P is type distributive, and, for almost all maximal ideals P, R_P is distributive.

(3) For each maximal ideal P, R_P is type distributive, and, for almost all maximal ideals P, $R_P = \bar{R}_P$.

Proof. (1) $\to$ (2). Clearly every overring of R in Q is type distributive, so we are left with showing that R_P is distributive for almost all P. Decompose $K = \oplus_{P \in spec(R)} K_P$ where $K_P = Q/R_P$ (theorem 22 in [15]). If there are infinitely many maximal ideals $P_1, P_2, \ldots$, such that R_{P_j} is not distributive for $j = 1, 2, \ldots$, then select submodules A'_j, B'_j and C'_j of K_{P_j} such that $C'_j \cap (A'_j + B'_j) \neq A'_j \cap C'_j + B'_j \cap C'_j$.

Let $A_j \subseteq Q$ such that $A'_j = A_j/R_{P_j}$, and set $A/R = \oplus_j A'_j$. We note that $A_{P_j} = A_j$ since $(K_{P_j}) \otimes R_{P_i} = 0$ for distinct i and j [15]. With B and C defined similarly, for no $0 \neq r \in R$ can $r(C \cap (A+B)) \subseteq A \cap C + B \cap C$. This is due to the fact that r is contained in only finitely many maximal ideals and thus is a unit in some R_{P_j}. However, $C_j \cap (A_j + B_j) \neq A_j \cap C_j + B_j \cap C_j$.

(2) $\to$ (3) As mentioned at the start of the section, when R is local and distributive, then R coincides with its integral closure. Therefore, (2) $\to$ (3) is concluded.

(3) $\to$ (1). For any maximal ideal P, $\bar{R}_P$ is a *pid* with finitely many primes. Since any *dvr* is distributive, it follows along the lines of proposition 3.1 that $\bar{R}_P$ is distributive. Hence, given submodules A, B and C of Q containing R, there are only finitely many maximal ideals $P = P_1, \ldots, P_n$ of R for which $C_P \cap (A_P + B_P) \neq A_P \cap C_P + B_P \cap C_P$. On the other hand, for each j, there is an $r_j \neq 0$ such that $r_j(C_{P_j} \cap (A_{P_j} + B_{P_j})) \subseteq A_{P_j} \cap C_{P_j} + B_{P_j} \cap C_{P_j}$. Therefore, with $r = r_1 \cdots r_n$, $r(C \cap (A + B)) \subseteq A \cap C + B \cap C$, since the containment holds locally. □

In [7], Butler shows that the polynomial ring over $\mathbb{Z}$ with countably many indeterminants is not type distributive. As a consequence of theorem 3.5 we can readily produce examples of type distributive and of non type distributive domains.

Example 3.6 If R is any 1-dimensional noetherian domain such that $R_P \neq \bar{R}_P$ for infinitely many prime ideals P, then R is not type distributive.

In particular, if R is the ring of example 8 in [17], then R is not type distributive.

Example 3.7 If R is analytically unramified, then R is type distributive.

Proof. While it is easy to show directly that R is type distributive, we will offer an alternate proof using theorem 3.5. Since $R_P = \bar{R}_P$ for almost all maximal ideals P of R when R is analytically unramified, we must show that each R_P is type distributive by theorem 3.5. Therefore we will assume that R is local.

In this case, K is equivalent to $\oplus_{i=1}^{n} Q/V_i$ where $V_1, \ldots, V_n$ are the valuation overrings of $\bar{R}$. Referring to the proof of the theorem 3.4, $B_j/R = P_j K$ [14], and it follows that $H_{P_j} \otimes K$ is equivalent to the simple divisible module Q/V_j which is clearly uniserial divisible. □

Rush has shown that $\bar{R}/R$ has a distributive lattice of submodules, precisely when the ideals of R are 2-generated [18]. In contrast to this, the lattice of submodules of Q containing R is distributive exactly when $R = \bar{R}$. However, when R is semi-local and every ideal is 2-generated, then R is type distributive. In addition, although R must coincide with its integral closure at almost all primes when R is type distributive, R need not be analytically unramified.

Example 3.8 If R is semi-local and every ideal of R is 2-generated, then R is type distributive. In particular, the ring of example 3.5 in [10], is an example of an analytically ramified, type distributive ring.

Proof. By proposition 3.1 we may assume that R is local. In this case either R is analytically unramified (hence type distributive by example 3.7), or, $\bar{R}$ is a *dvr* and $\bar{R}/R$ is simple divisible ([15], page 92). Moreover, it follows that $\bar{R}/R$ is the only proper divisible submodule of K, so K is uniserial divisible. □

4 Butler modules

The point of view taken in [7] is that the appropriate context in which to study Butler modules is in the quasi-category. In this section we examine this view-point. As in the previous sections, R is assumed to be 1-dimensional noetherian.

Definition 4.1 $\mathcal{P}$ is the class of pure submodules of finite rank completely decomposable R-modules; $\mathcal{R}$ is the class of torsion-free, epimorphic images of finite rank completely decomposable R-modules; $Q\mathcal{P}$ is the closure of $\mathcal{P}$ under quasi-isomorphism; and $\mathcal{QR}$ is the closure of $\mathcal{R}$ under quasi-isomorphism.

We will overlook the distinctions in the various notions of purity and call a submodule of a torsion-free module, *pure*, if the factor module is torsion-free. Butler's Theorem 3 from [7] asserts that $\mathcal{QR}$ is closed under the formation of pure submodules if and only if R is type distributive. Thus theorem 3.5 offers a characterization of when $\mathcal{QR}$ is closed under the formation of pure submodules. Since $\mathcal{P}$ and $\mathcal{R}$ both contain the finite rank completely decomposable modules, Butler's theorem implies that $\mathcal{QP} \subseteq \mathcal{QR}$ when R is type distributive.

Proposition 4.2 *(1)* $\mathcal{R} = \mathcal{QR}$.

(2) When R is Dedekind, $\mathcal{P} = \mathcal{R} = \mathcal{QP} = \mathcal{QR}$.

(3) When R is analytically unramified, $\mathcal{R} = \mathcal{QP} = \mathcal{QR}$.

Proof. (1). Let $M \in \mathcal{QR}$ be quasi-isomorphic to $N \in \mathcal{R}$. Without loss of generality we may assume that $N \subseteq M \subseteq QN$ with M/N bounded by $0 \neq r \in R$. Let F be a free submodule of N such that N/F is torsion. By Theorem 5.5 in [14], QF/F has DCC on submodules and by Corollary 5.2 in [14], $(M/F)/h(M/F)$ is finitely generated, where $h(M/F)$ is the maximal divisible submodule of M/F. It follows that M/N is finitely generated and has DCC on submodules, so consequently, there is a composition series for M/N.

There is a chain of submodules $N = M_0 \subseteq M_1 \subseteq \cdots \subseteq M_k = M$ such that M_j/M_{j-1} is simple. It suffices to show that $M_1 \in \mathcal{R}$ and so we may assume that M/N is simple. There is a finite rank, completely decomposable module C and an epimorphism $\phi : C \to N$. Pick $a \in M \setminus N$. For $C' = C \oplus R$, define $\theta : C' \to M$ by $\theta(c, r) = \phi(c) + ra$. It follows that $M \in \mathcal{R}$.

(2). By (1), theorem 3.5, and Butler's Theorem mentioned above, $\mathcal{QP} \subseteq \mathcal{R} = \mathcal{QR}$. We need to show that $\mathcal{R} \subseteq \mathcal{P}$, and the proof of this from [2] for abelian groups, goes through. We recall the proof for the reader's convenience. Let $M = A_1 + \ldots + A_n$ where $A_1, \ldots, A_n$ are rank one pure submodules of M. If one shows that the map $f : M \to M/A_1 \oplus \cdots \oplus M/A_n$ has pure image, then by induction on the rank of M we will have $M \in \mathcal{P}$. Call $A = M/A_1 \oplus \cdots \oplus M/A_n$ and assume M has rank at least 2.

If $A/f(M)$ is not torsion-free, then localizing at some prime ideal P yields the non-torsion-free R_P-module $A_P/f_P(M_P)$. Therefore, we may assume that R is a *dvr* with maximal ideal $pR_P = PR_P$ (some $p \in R$). There are $x, x_1, \ldots, x_n \in M$ such that $p(x_1 + A_1, \ldots, x_n + A_n) = (x + A_1, \ldots, x + A_n)$ and $x \in M \setminus pM$. Write $x - px_j = y_j \in A_j$, and note that $y_j \in A_j \setminus pA_j$ for each j.

Since R is a *dvr*, $A_j = Ry_j$ for each j, and so, $M = A_1 + \cdots + A_n = Ry_1 + \cdots + Ry_n = Rx + pM$. Thus, $M/pM \cong R/pR$. However, as an epimorphic image of the free module $A_1 \oplus \cdots \oplus A_n$, M is free, so M/pM must have dimension equal to rank M; a contradiction.

(3). To obtain the equality $Q\mathcal{P} = Q\mathcal{R}$ when R is analytically unramified, by Butler's theorem mentioned above and example 3.7, we need to argue that $Q\mathcal{R} \subseteq Q\mathcal{P}$. Given $M \in \mathcal{R}$, form $\bar{M} = \bar{R}M \leq QM$. Since $\bar{R}$ is finitely generated over R, there is an element $0 \neq r_0 \in R$ such that $r_0\bar{R} \subseteq R$. Therefore M is quasi-isomorphic to $\bar{M}$ by corollary 2.2. But M is an image of $X_1 \oplus \cdots \oplus X_n$ with $X_j \subseteq Q$, and so $\bar{M}$ is the image of $Y_1 \oplus \cdots \oplus Y_n$ where $Y_j = \bar{R}X_j \leq Q$. Since $\bar{R}$ is Dedekind, by (2), $\bar{M}$ embeds, as a pure submodule, into a completely decomposable $\bar{R}$-module C, and because C is a completely decomposable R-module, $M \in Q\mathcal{P}$. □

When $R = \mathbb{Z}$, Butler shows in [8] that $\mathcal{P} = \mathcal{R}$, so (3) recovers this result in the quasi-category for any analytically unramified R. In the remainder of this article, we will examine a condition on R under which $\mathcal{P} = \mathcal{R}$; the condition in mind is that of theorem 2.7 (2).

Lemma 4.3 *Let A, B and C be torsion-free modules and $f : A \to B$ and $g : A \to C$ be maps for which f is a pure embedding. Then $h = (f, g) : A \to B \oplus C$ is also a pure embedding.*

Proof. If $r(b,c) \in Im\ h$, $r(b,c) = h(a)$ say, then $rb = f(a)$. Thus, $b \in Im\ f$ implying that $b = f(a')$ where a' solves the equation $ra' = a$ in A. Since $g(a) = rg(a') = rc$, $c = g(a')$ and so $(b,c) = h(a')$. Clearly h is one-to-one. □

Recall, an ideal I of R is called a *canonical ideal* if any ideal J satisfies $J \cong_{nat}$ Hom(Hom$(J, I), I)$. Because R is noetherian, this latter condition is equivalent to $M \cong_{nat}$ Hom(Hom$(M, I), I)$ for every finitely generated torsion-free module M.

Proposition 4.4 *If R is local and type distributive, then R has a canonical ideal.*

Proof. Theorem 15.7 in [14] asserts that R has a canonical ideal if and only if Q/R contains exactly one simple divisible module from each isomorphism class of simple divisible modules. Referring to the notation of theorem 3.4, there is an epimorphism of $K = Q/R$ onto $Q/B_1 \oplus \cdots \oplus Q/B_n$ with bounded kernel (theorem 11.5 in [14]). Since Q/B_j is uniserial divisible for each j and for each $i \neq j$, the simple divisible modules Q/V_i and Q/V_j

are nonequivalent, we conclude that R has exactly one simple divisible submodule for each isomorphism type, and therefore R has a canonical ideal. □

Theorem 4.5 *Let R be type distributive, and let M be an R-module such that for any maximal ideal P, $M_P = D \oplus M'$ with D divisible and M' finitely generated. If $M \in \mathcal{R}$, then $M \in \mathcal{P}$.*

Proof. By theorem 3.5 there are finitely many maximal ideals $P_1, \ldots, P_m$ of R such that $R_P = \bar{R}_P$ for every prime ideal $P \neq P_1, \ldots, P_m$. With $R_0 = \bigcap_{P \neq P_1, \ldots, P_m} R_P$, $M_0 = R_0 M \leq QM$ is a torsion-free epimorphic image of a finite rank completely decomposable R_0-module, and since R_0 is Dedekind, M_0 embeds as a pure submodule of a finite rank completely decomposable module C_0 (proposition 4.2).

We will show below, that each $M_j = M_{P_j}$ embeds as a pure submodule in a finite rank completely decomposable R_{P_j}-module, C_j. Assuming this for now, let $\nu_j : M \to M_j$ be the natural maps, and let $\phi_j : M_j \to C_j$ denote pure embeddings, $j = 0, \ldots, n$. Define the map $\theta = (\theta_0, \ldots, \theta_n) : M \to C = C_0 \oplus \cdots \oplus C_n$ using the component maps $\theta_j = \phi_j \nu_j$. Since localizing θ at a prime P, θ_P, has a component map that is a pure embedding, θ_P must be a pure embedding by lemma 4.3. Therefore *coker* θ has zero P-primary torsion component for each P, and consequently θ must be a pure embedding.

Write $M_P = D \oplus M'$ with D dvisible and M' finitely generated. In order to obtain a pure embedding of M_P into a finite rank, completely decomposable module, it suffices to consider M'. Therefore, to finish the proof, it is enough to show that when R is local and M is torsion-free, finitely generated, then M embeds as a pure submodule of a finite rank completely decomposable R-module A.

By proposition 4.4, type distributive rings have canonical ideals; call the canonical ideal for R, I. Consider the free resolution of $Hom(M, I)$:

$$0 \to K \to F \to Hom(M, I) \to 0,$$

where F is a finite rank free module. Then $M \cong Hom(Hom(M, I), I)$ is a pure submodule of a finite direct sum, $A = Hom(F, I)$, of copies of I, and the proof is complete. □

Proposition 4.6 *Assume that R is local. The following are equivalent:*

(1) Any $M \in \mathcal{R}$ decomposes as $D \oplus M'$ with M' finitely generated and D divisible.

(2) Any proper submodule of Q is finitely generated.

(3) R is analytically unramified and $\bar{R}$ is local.

Proof. (1) → (2) is clear. For (2) → (3), if $\bar{R}$ is not local, then $\bar{R}$ is a *pid* with at least two non-associate primes. Let a be one such prime. Then the overring, $\bar{R}[a^{-1}]$, of $\bar{R}$ generated by $a^{-1} \in Q$, is neither finitely generated nor is it Q (since $\bar{R}$ has other primes) . Therefore, (3) follows from (2).

Assume (3) and let $M \in \mathcal{R}$. Since the maximal divisible submodule D of M is injective, we can factor $M = D \oplus N$ with N reduced. Now N is of bounded index in $\bar{N} = \bar{R}N \leq QN$ and is also the image of a finite rank completely decomposable module (over $\bar{R}$). Because $\bar{R}$ is a *dvr*, up to isomorphism, the rank one $\bar{R}$-modules are $\bar{R}$ and Q, and since $\bar{N}/N$ is bounded, $\bar{N}$ is reduced. Therefore, $\bar{N}$ is a finitely generated, hence free $\bar{R}$ module, from which it follows that N is a finitely generated R-module. Therefore (1) is implied by (3). □

Theorem 4.7 *If $\mathcal{P} = \mathcal{R}$, then for every maximal ideal P of R, there exists a unique maximal ideal of $\bar{R}$ lying over P.*

Proof. Suppose P is a maximal ideal of R that splits into $P_1^{e_1} \cdots P_k^{e_k}$ in $\bar{R}$ with $k \geq 2$, $e_j \geq 1$ for all j, and $P_1, \ldots, P_k$ distinct maximal ideals of $\bar{R}$; i.e., $P\bar{R} = P_1^{e_1} \cdots P_k^{e_k}$. We must find a module $M \in \mathcal{R} \setminus \mathcal{P}$, and we do this by producing a rank two module M such that every rank one image of M is an $\bar{R}$-module though M itself is not an $\bar{R}$-module (thus M cannot be a pure submodule of the direct sum of some of its rank one images). Our construction is broken down into cases as follows.

Case 1: For some j, $\bar{R}/P_j$ has dimension at least 2 as an R/P-module.

Without loss, assume $j = 1$. We will find two submodules X, Y of Q such that:

(1) $X + Y = Q$.

(2) X is not an $\bar{R}$-module.

(3) $X + R$ is an $\bar{R}$-module.

(4) $Y + R$ is an $\bar{R}$-module.

(5) $R \cap Y \subseteq X$.

Once we have found such X and Y, set $M = X \oplus Y + R(1,1)$. Every rank one image of M is an $\bar{R}$-module since by (1), any homomorphism $\phi : M \to Q$ where neither X nor Y are sent to zero, must be an epimorphism, while (3) and (4) insure that the component projection maps from M into Q have $\bar{R}$-module images. It is easy to see that condition (5) implies that $X = X \oplus 0$ is a pure submodule of M, and therefore by (2), M is not an $\bar{R}$-module.

Set m equal to the dimension of $\bar{R}/P_1$ over R/P and recall the assumption of case 1 that $m \geq 2$. Take X to be an R-module between $P_1\bar{R}_{P_1}$ and

$\bar{R}_{P_1}$ such that the R/P-subspace $X' = X/P_1\bar{R}_{P_1}$ is complementary to the subspace spanned by the element $1 + P_1\bar{R}_{P_1}$. Since $\bar{R}/P_1 = \bar{R}_{P_1}/P_1\bar{R}_{P_1}$ is a field and X' is a proper submodule of $\bar{R}_{P_1}/P_1\bar{R}_{P_1}$, X cannot be an $\bar{R}$-module. Furthermore, $X + R = \bar{R}_{P_1}$ is an $\bar{R}$-module, and so (2) and (3) are satisfied.

We define Y conditionally: if $\bar{R}/P_2$ has R/P-dimension ≥ 2, then define Y between $P_2\bar{R}_{P_2}$ and $\bar{R}_{P_2}$ in a manner analogous to the definition of X; otherwise, take $Y = P_2\bar{R}_{P_2}$. In either case, $Y + R = \bar{R}_{P_2}$ and so (4) holds. Since $Q = PQ \subseteq P_1\bar{R}_{P_1} + P_2\bar{R}_{P_2} \subseteq X + Y$, (1) holds. Finally, $R \cap Y = R \cap P_2 = P \subseteq X$ establishing (5).

Case 2: $\bar{R}/P_j$ has dimension 1 for each j.

First suppose one of the e_j's is at least 2; $e_1 \geq 2$ say. Set $X = R + P_1^2\bar{R}_{P_1}$ and $Y = P_2\bar{R}_{P_2}$. Because $\bar{R}/P_1 \cong R/P$, $\bar{R}/P_1^2$ has dimension 2 as an R/P-module. Also, while $P \subseteq P_1^2$, we have $X/P_1^2 \cong R/P$, so X is not an $\bar{R}$-module. There is an element $a \in \bar{R}$ such that $X + Ra = \bar{R}$ since $\bar{R}/P_1^2$ has dimension 2 as an R/P-module. Set $M = X \oplus Y + R(a, 1)$. Projection onto each coordinate yields an $\bar{R}$-module, and because $X + Y = Q$, any other rank one image of M is Q. Furthermore, the analogue of condition (5) holds, namely that $(Y \cap R)a \subseteq X$, implying that X is a pure submodule of M. Check: If $r[(x, y) + r'(a, 1)] = (x', 0)$, then $y = -r' \in Y \cap R = P$, and so, $x + r'a \in X$. Thus, $M \in \mathcal{R} \setminus \mathcal{P}$.

We may now assume that $e_j = 1$ for each j. We deviate slightly from above: Let $S = \bar{R}_{P_1} \cap \bar{R}_{P_2}$ so that P_1S and P_2S are the maximal ideals of S. Take $X = R + P_1P_2S$; i.e., X/P_1P_2S is the one dimensional R/P-submodule of S/P_1P_2S generated by $1 + P_1P_2S$. There is an element $a \in \bar{R}$ such that $X + Ra = S$. With $Y = P_2\bar{R}_{P_2}$, we consider $M = X \oplus Y + R(a, 1)$.

To check that X is pure in M, observe that $Y \cap R = P$ and $Pa \subseteq X$. We need to argue that any image of M is an $\bar{R}$-module, which is not readily afforded (unlike the examples above, $X + Y \neq Q$). Evidently, each coordinate projection is an $\bar{R}$-module, and any map from $M \to Q$ is of the form $\sigma((x, y) + r(a, 1)) = u(x + ra) + v(y + r)$ for some $u, v \in Q$. To show that Im σ is an $\bar{R}$-module, we may assume that $0 \neq u, v \in R$ (by clearing denominators if necessary). Note that Im $\sigma \subseteq Y$ and contains vY.

Due to the fact that $\bar{R}_{P_2}$ is a *dvr* and $P\bar{R}_{P_2} = P_2\bar{R}_{P_2}$ (i.e., $e_2 = 1$), the ideals of $\bar{R}_{P_2}$ are $P_2^m\bar{R}_{P_2} = P^m\bar{R}_{P_2}$. But $\bar{R}_{P_2}/P_2\bar{R}_{P_2} \cong R/P$ and so the only R-submodules between $P_2^m\bar{R}_{P_2}$ and $P_2^{m-1}\bar{R}_{P_2}$ are $P_2^m\bar{R}_{P_2}$ and $P_2^{m-1}\bar{R}_{P_2}$ themselves. It follows that any R-submodule of $\bar{R}_{P_2}$ containing $P_2^m\bar{R}_{P_2}$ for some m, is an $\bar{R}$ submodule, and therefore Im σ is an $\bar{R}$-module. Thus $M \in \mathcal{R} \setminus \mathcal{P}$. □

The following is a consequence of theorem 4.5, proposition 4.6 and theorem 4.7.

Corollary 4.8 *Let R be analytically unramified. Then $\mathcal{R} = \mathcal{P}$ if and only if every maximal ideal of R has a unique maximal ideal of $\bar{R}$ lying over it.*

In closing we propose that this material is germaine to the study of abelian groups. Any subring R of an algebraic number field is an analytically unramified 1-dimensional noetherian domain [2]. If G is a (standard) Butler group, then $R \otimes_{\mathbb{Z}} G$ is a Butler R-module, and is in fact a pure submodule of a finite rank completely decomposable R-module. These Butler modules offer an interesting class of torsion-free groups. In particular, when G is a so called *bracket* group, or $\mathcal{B}^{(1)}$-group, (as studied by Richman, Arnold-Vinsonhaler, Hill-Megibben, Fuchs-Metelli and Goeters-Ullery among others), the class of groups (modules) of the form $M = R \otimes_{\mathbb{Z}} G$ are shown in [1] to be characterized by the ranks of their socles $M(\tau) = \{ x \in M \mid \text{type } x \geq \tau \}$ where τ is the type of a subgroup of $\mathbb{Q}$.

In [6] Butler modules are defined using the vanishing of Bext, and the various definitions of Butler modules are equivalent when R is $\mathbb{Z}$. It would be interesting to know precisely when the various definitions for finite rank Butler modules are equivalent.

References

[1] U. Albrecht, A. Giovannitti, and H. P. Goeters, *A class of torsion-free groups characterized by the ranks of their socles*, preprint.

[2] D. M. Arnold, *Finite Rank Torsion Free Abelian Groups and Rings*, LNM **931**, Springer-Verlag (1982).

[3] D. M. Arnold, *Pure subgroups of finite rank completely decomposable groups*, in Abelian Group Theory, LNM **874**, Springer-Verlag (1981).

[4] D. M. Arnold and C. Vinsonhaler, *Pure subgroups of finite rank completely decomposable groups* II, in Abelian Group Theory, LNM **1006**, Springer-Verlag (1983).

[5] L. Bican, *Purely finitely generated abelian groups*, Comment. Mat. Univ. Carolinae **11** (1970), 1–8.

[6] L. Bican and L. Salce, *Butler modules in torsion theories*, Accad. Naz. d. Science **108** (1990), 121–137.

[7] M. C. R. Butler, *Torsion-free modules and diagrams of vector spaces*, Proc. London Math. Soc. (3) **18** (1968), 635–652.

[8] M. C. R. Butler, *A class of torsion-free abelian groups of finite rank*, Proc. London Math. Soc. (3) **15** (1965), 680–698.

[9] M. Fontana, J. A. Huckaba, and I. J. Papick, *Prüfer Domains*, Marcel-Dekker (1997).

[10] W. Heinzer and D. Lantz, *Jonsson extensions of one-dimensional semi-local domains*, Journal of Algebra **117** (1988), 179–197.

[11] J. Koehler, *The type set of a torsion-free group of finite rank*, Illinois J. Math. **9** (1965), 66–86.

[12] E. L. Lady, *Extension of scalars for torsion-free modules over Dedekind domains*, Symposia Mathematica **XXIII** (1979), 287–305.

[13] W. J. Lewis and T. S. Shores, *Serial modules and endomorphism rings*, Duke Math. J. **41** (1974), 889–909.

[14] E. Matlis, *One Dimensional Cohen-Macaulay Rings*, LNM **327**, Springer-Verlag (1973).

[15] E. Matlis, *Torsion Free Modules*, The University of Chicago Press (1972).

[16] H. Matsumura, *Commutative Ring Theory*, Cambridge University Press (1980).

[17] M. Nagata, *Local Rings*, Interscience Publishers, Number **13** (1962).

[18] D. E. Rush, *Rings with two-generated ideals*, Journal of Pure and Appl. Algebra **73** (1991), 257–275.

Trends in Mathematics, © 1999 Birkhäuser Verlag Basel/Switzerland

Completely decomposable summands of almost completely decomposable groups

Adolf Mader * and Loyiso G. Nongxa

Abstract

Almost completely decomposable groups are torsion-free finite extensions of completely decomposable groups of finite rank. We answer completely and in a constructive fashion the question when an almost completely decomposable group has non-zero completely decomposable direct summands. A new invariant, the rank-width difference of X at τ, given by

$$\operatorname{rwd}_\tau(X) = \operatorname{rk} \frac{X(\tau)}{X^\sharp(\tau)} - \operatorname{width} \frac{X^\sharp[\tau]}{X(\tau) + X[\tau]}$$

is the exact rank of a maximal τ-homogeneous (completely decomposable) direct summand of X. We show that the rank-width difference can be effectively computed for an almost completely decomposable group given in "standard description" $X = A + \mathbb{Z}N^{-1}a^\natural$ provided that A is regulating in X. We also establish an algorithmic criterion for deciding whether an almost completely decomposable group given in standard description is completely decomposable.

1 Introduction

This paper deals with a special class of torsion-free abelian groups, and "group" means abelian group in this article. ***Almost completely decomposable groups*** are finite torsion-free extensions of completely decomposable groups of finite rank. We answer completely and in a constructive fashion the question when an almost completely decomposable group has non-zero completely decomposable direct summands (Theorem 3.10, Theorem 4.5). This supplements the known criteria for the case of almost completely decomposable groups with cyclic regulating quotient ([MV95, Theorem 3.2], [DMM97, Theorem 4.4]) and the case of two critical types

*The hospitality and support of the University of the Western Cape during February 1998 is gratefully recognized.

([Lew93], [MV94]). We introduce a new invariant, the ***rank–width difference*** of X at τ, given by

$$\operatorname{rwd}_\tau(X) = \operatorname{rk}\frac{X(\tau)}{X^\sharp(\tau)} - \operatorname{width}\frac{X^\sharp[\tau]}{X(\tau)+X[\tau]},$$

and show that it is equal to the exact rank of a maximal τ–homogeneous (completely decomposable) direct summand of X (Theorem 3.17). We show that the rank–width difference can be effectively computed (Theorem 4.5) for an almost completely decomposable group given in "standard description" $X = A + \mathbb{Z}N^{-1}a^{\iota}$ (Lemma 4.1) provided that A is regulating in X. Thus the ranks of τ–homogeneous direct summands can be effectively computed. We also establish an algorithmic criterion for deciding whether an almost completely decomposable group given in standard description is completely decomposable (Theorem 3.19, Theorem 4.6).

2 Background

For a detailed survey on almost completely decomposable groups we refer the reader to [Mad95].

E.L. Lady [Lad74] started the systematic and general study of almost completely decomposable groups and in particular introduced the so–called regulating subgroups as follows. Let X be an almost completely decomposable group. For each type τ there is a ***τ–Butler decomposition***

$$X(\tau) = A_\tau \oplus X^\sharp(\tau) \tag{1}$$

with A_τ, a ***τ–Butler complement***, being τ–homogeneous completely decomposable. A type τ is a ***critical type*** of X if $A_\tau \neq 0$. The set of critical types is denoted by $\mathrm{T}_{\mathrm{cr}}(X)$. The sums $A = \sum_{\rho\in \mathrm{T}_{\mathrm{cr}}(A)} A_\rho$ turn out to be direct, and are the ***regulating subgroups*** of X. Lady showed that the regulating subgroups are exactly the completely decomposable subgroups of X that have least index $\operatorname{rgi}(X)$ among all the completely decomposable subgroups of finite index in X. Lady also showed that in any decomposition $X = X_{cd}\oplus X_{cl}$ such that X_{cd} is completely decomposable and X_{cl} is ***clipped***, i.e., X_{cl} has no completely decomposable direct summands, the completely decomposable summand X_{cd} is unique up to isomorphism which in turn means that X_{cl} is unique up to near–isomorphism. This result, together with the fact that indecomposable groups must above all lack completely decomposable summands, motivate our interest in such summands. It is also noteworthy that cyclic extensions of completely decomposable groups with critical typesets that are anti–chains, ***block–rigid crq–groups*** for short, have decompositions into indecomposable summands that are unique if they are clipped. In contrast, block–rigid crq–groups that are not clipped may have a large number of essentially different direct decompositions with indecomposable summands (see [BM94]).

In addition to the τ–socles $X(\tau)$ and $X^\sharp(\tau)$, there are also the τ–radicals $X[\tau]$ and $X^\sharp[\tau]$ to work with. For an almost completely decomposable group X containing the completely decomposable group $A = \bigoplus_{\rho\in \mathrm{T_{cr}}(A)} A_\rho$ as a subgroup of finite index, we have explicitly (H_*^G denotes the purification of H in G)

$$\begin{aligned}
A(\tau) &= \bigoplus\{A_\rho : \rho \geq \tau\}, \quad X(\tau) = (A(\tau))_*^X, \\
A^\sharp(\tau) &= \bigoplus\{A_\rho : \rho > \tau\}, \quad X^\sharp(\tau) = (A^\sharp(\tau))_*^X, \\
A[\tau] &= \bigoplus\{A_\rho : \rho \not\leq \tau\}, \quad X[\tau] = (A[\tau])_*^X, \\
A^\sharp[\tau] &= \bigoplus\{A_\rho : \rho \not< \tau\}, \quad X^\sharp[\tau] = (A^\sharp[\tau])_*^X.
\end{aligned}$$

Note that consequently, $X(\tau)/X^\sharp(\tau) \cong X^\sharp[\tau]/X[\tau]$, and $X^\sharp(\tau) \subset X[\tau]$. We will often use the abbreviation $X^\circ(\tau) = X(\tau)/X^\sharp(\tau)$ but avoid this hardly suggestive notation in the statement of results.

It turns out that the radicals play an essential role for the study of completely decomposable direct summands. It is well–known (see [AV84, Theorem 1.7(c)]) that there are also ***τ–co–Butler decompositions***

$$\frac{X}{X[\tau]} = \frac{X^\sharp[\tau]}{X[\tau]} \oplus \frac{Z_\tau}{X[\tau]} \tag{2}$$

where $X^\sharp[\tau]/X[\tau]$ is τ–homogeneous completely decomposable and Z_τ is a suitable subgroup of X.

We note some special cases. If τ is maximal in $\mathrm{T_{cr}}(X)$, then $X^\sharp(\tau) = 0$ and conversely. If τ is minimal in $\mathrm{T_{cr}}(X)$, then $X^\sharp[\tau] = X$ and conversely. If $\mathrm{T_{cr}}(X)$ happens to contain a (unique) smallest type μ, then $X(\mu) = X$ and if $\mathrm{T_{cr}}(X)$ happens to contain a (unique) largest type ν, then $X[\nu] = 0$. We have the immediate well–known consequences of these observations and the Butler decompositions.

Proposition 2.1 *Let X be a Butler group. If X has a smallest critical type μ, then X has a completely decomposable direct summand isomorphic to $X(\mu)/X^\sharp(\mu)$. If X has a largest critical type ν, then X has a completely decomposable direct summand isomorphic with $X^\sharp[\nu]/X[\nu] = X^\sharp[\nu]$.*

Proof. In the Butler decompositions $X = X(\mu)$ and $X = X/X[\nu]$. □

Our notation is mostly standard but note that $X^\sharp[\tau]$ is in some publications denoted by $X^*[\tau]$ or $X[\tau^*]$. An ***indecomposable decomposition*** is a direct decomposition with indecomposable summands. By a ***rational***

group we always mean an additive subgroup of $\mathbb{Q}$ containing $\mathbb{Z}$. We identify a rational group τ with its isomorphism class. Thus we may speak of the type τ although τ is strictly speaking a rational group. For convenience we call a direct summand isomorphic to the rational group τ a ***τ–summand*** and this can also be understood as a rank–one summand of type τ. A group is ***τ–clipped*** if it has no τ–summand, so that "clipped" is equivalent to "τ–clipped for each τ". The symbol $\mathbb{P}$ stands for the set of all prime numbers.

3 Criteria

We begin with some observations that narrow down the problem of finding rank–one direct summands.

Lemma 3.1 *Let X be an almost completely decomposable group and τ a type. Then X has a τ–homogeneous summand D if and only if there exists a summand E of a given τ–Butler complement A_τ of X such that $X = E \oplus Y$ and $E \cong D$.*

Proof. If E exists, then $D = E$ is a τ–homogeneous summand of X isomorphic with E.

Conversely assume that $X = D \oplus Y$ where D is τ–homogeneous. Then, writing $Y(\tau) = D_\tau \oplus Y^\sharp(\tau)$,

$$X(\tau) = (D \oplus D_\tau) \oplus Y^\sharp(\tau) = (D \oplus D_\tau) \oplus X^\sharp(\tau) = A_\tau \oplus X^\sharp(\tau).$$

Hence there is $\phi \in \mathrm{Hom}(D \oplus D_\tau, X^\sharp(\tau))$ such that $(1 + \phi : D \oplus D_\tau \to A_\tau$ being an isomorphism)

$$A_\tau = (D \oplus D_\tau)(1 + \phi) = D(1 + \phi) \oplus D_\tau(1 + \phi).$$

Set $E = D(1 + \phi)$. Then E is a direct summand of A_τ and since $\mathrm{Im}\,\phi \subset X^\sharp(\tau) = Y^\sharp(\tau) \subset Y$, it follows that $E = D(1 + \phi)$ is a complementary summand of Y in X. □

Lemma 3.1 says that one may take any τ–Butler complement A_τ and find a τ–summand of X (if there is one) by studying the decompositions of A_τ.

Lemma 3.1 becomes a rather strong tool in a special case. An arbitrary torsion–free abelian group G is called **slim at** τ or **τ–slim** if $\mathrm{rk}(G^\circ(\tau)) \leq 1$ where $G^\circ(\tau) = G(\tau)/G^\sharp(\tau)$. A group is **slim** if it is τ–slim for each τ.

Corollary 3.2 *Suppose that X is slim at τ. Then X has a τ–summand if and only if any τ–Butler complement is a direct summand of X.*

It is possible to reduce the search for completely decomposable summands to summands of maximal type as the following proposition shows. Note that a τ–homogeneous summand of an almost completely decomposable group is automatically completely decomposable.

Proposition 3.3 *Let X be an almost completely decomposable group and τ a type. Then the following hold.*

1. *If $X = E \oplus Y$ with E τ-homogeneous, then*

$$\frac{X}{X^\sharp(\tau)} = \frac{E + X^\sharp(\tau)}{X^\sharp(\tau)} \oplus \frac{Y}{X^\sharp(\tau)} \cong E \oplus \frac{Y}{X^\sharp(\tau)}.$$

2. *If*

$$\frac{X}{X^\sharp(\tau)} = \frac{E}{X^\sharp(\tau)} \oplus \frac{Y}{X^\sharp(\tau)} \quad \text{and} \quad \frac{E}{X^\sharp(\tau)} \subset \frac{X}{X^\sharp(\tau)}(\tau),$$

then $X = (E \cap A_\tau) \oplus Y$ where A_τ is any τ-Butler complement of X, and $E \cap A_\tau \cong \frac{E}{X^\sharp(\tau)}$ is τ–homogeneous.

Proof. Suppose first that $X = E \oplus Y$ where E is τ–homogeneous of rank r. Then $X^\sharp(\tau) = Y^\sharp(\tau) \subset Y$ and the claim is evident.

Conversely, assume that

$$\frac{X}{X^\sharp(\tau)} = \frac{E}{X^\sharp(\tau)} \oplus \frac{Y}{X^\sharp(\tau)}, \quad \text{where} \quad \frac{E}{X^\sharp(\tau)} \subset \frac{X}{X^\sharp(\tau)}(\tau).$$

Let $A = \bigoplus_{\rho \in \mathrm{T_{cr}}(A)} A_\rho$ be a regulating subgroup of X and collect homogeneous summands to get a decomposition $A = A^\flat(\tau) \oplus A^\sharp(\tau)$. Then $(A^\flat(\tau) \oplus X^\sharp(\tau))/X^\sharp(\tau) \cong A^\flat(\tau)$ is a completely decomposable subgroup of finite index in $X/X^\sharp(\tau)$. Hence τ is a maximal critical type of $X/X^\sharp(\tau)$. It follows that

$$\left(\frac{X}{X^\sharp(\tau)}\right)(\tau) = \left(\frac{A_\tau \oplus X^\sharp(\tau)}{X^\sharp(\tau)}\right)_*^{X/X^\sharp(\tau)} = \frac{X(\tau)}{X^\sharp(\tau)}.$$

Now

$$\frac{X(\tau)}{X^\sharp(\tau)} = \left(\frac{X}{X^\sharp(\tau)}\right)(\tau) \supset \frac{E}{X^\sharp(\tau)},$$

and this implies that $E \subset X(\tau)$. Hence $E = E \cap X(\tau) = (E \cap A_\tau) \oplus X^\sharp(\tau)$. We conclude that $X = E + Y = (E \cap A_\tau) + Y$, and $(E \cap A_\tau) \cap Y \subset A_\tau \cap X^\sharp(\tau) = 0$. Hence $X = (E \cap A_\tau) \oplus Y$. Furthermore, $E \cap A_\tau \cong X/Y \cong$

$E/X^\sharp(\tau)$ is τ–homogeneous. □

The next proposition shows that the existence of a critical type that is comparable with every other critical type results in the existence of rank–one summands. This is a slight extension of Proposition 2.1.

Proposition 3.4 *Let X be an almost completely decomposable group. Suppose that τ is a critical type of X that is comparable with every other critical type and let A_τ be a τ–Butler complement of X. Then A_τ is a direct summand of X.*

Proof. It is well–known and easily checked that the critical typeset of $X/X^\sharp(\tau)$ is $\{\rho \in \mathrm{T_{cr}}(X) : \rho \not> \tau\}$. The hypothesis on τ implies that $\mathrm{T_{cr}}(X/X^\sharp(\tau))$ has the unique maximal type τ. Hence (Proposition 2.1) $(X/X^\sharp(\tau))(\tau) = X(\tau)/X^\sharp(\tau)$ is a direct summand of $X/X^\sharp(\tau)$ and there is a decomposition

$$\frac{X}{X^\sharp(\tau)} = \frac{X(\tau)}{X^\sharp(\tau)} \oplus \frac{Y}{X^\sharp(\tau)}.$$

By Proposition 3.3 we have $X = (X(\tau) \cap A_\tau) \oplus Y = A_\tau \oplus Y$. □

Recall the **width** of a finite abelian group G:

$$\mathrm{width}(G) = \max\{\dim G[p] : p \in \mathbb{P}\}.$$

If G is written in principal divisor form $G = G_1 \oplus \cdots \oplus G_w$ where G_i is cyclic of order d_i and d_i divides d_{i+1} for each i, then $w = \mathrm{width}(G)$. In particular, if X is a finite essential extension of a torsion–free group A, then $\mathrm{width}(X/A) = w$ if and only if there are elements $x_i \in X$ such that

$$\frac{X}{A} = \bigoplus_{i=1}^{w} \langle x_i + A \rangle\,, \quad 1 < \mathrm{ord}(x_i + A) = d_i, \quad d_i \mid d_{i+1}. \tag{3}$$

Lemma 3.5 *Let X be a torsion-free finite extension of a group A, and suppose that B is a subgroup of A that is pure in X. Then*

$$\mathrm{width}(X/A) \le \mathrm{rk}\, X - \mathrm{rk}\, B.$$

Proof. Note that $(X/B)/(A/B) \cong X/A$ is finite. Hence $X/B \le \mathbb{Q}(A/B)$. By standard arguments it follows that for any prime p

$$\dim \frac{X}{A}[p] = \dim \frac{X/B}{A/B}[p] \le \mathrm{rk}\, \frac{A}{B} = \mathrm{rk}\, A - \mathrm{rk}\, B = \mathrm{rk}\, X - \mathrm{rk}\, B. \qquad \square$$

Corollary 3.6 *Let* $A \in \operatorname{Regg}(X)$. *Then*

$$\operatorname{width}\left(\frac{X}{A}\right) \leq \min_{\tau \in \mathrm{T}_{\mathrm{cr}}(X)} \sum_{\rho \neq \tau} \operatorname{rk} \frac{X(\rho)}{X^\sharp(\rho)} < \operatorname{rk} X.$$

Proof. Let $A = \bigoplus_{\rho \in \mathrm{T}_{\mathrm{cr}}(A)} A_\rho$ be a homogeneous decomposition of A. Then A_τ is pure in X and $\operatorname{rk} A_\tau = \operatorname{rk} X^\circ(\tau)$. □

A first indicator for the existence of τ–summands is obtained by comparing width and critical ranks.

Lemma 3.7 *Let X be almost completely decomposable and $A \in \operatorname{Regg}(X)$. Let τ be a minimal critical type of X. Then X has a τ-homogeneous summand of rank*

$$\geq \operatorname{rk} \frac{X(\tau)}{X^\sharp(\tau)} - \operatorname{width}\left(\frac{X}{X(\tau) + X[\tau]}\right) \geq 0.$$

Proof. Choose a homogeneous decomposition $A = \bigoplus_{\rho \in \mathrm{T}_{\mathrm{cr}}(A)} A_\rho$ of A. Note that $X(\tau) + X[\tau] = A_\tau \oplus X[\tau]$ and that $X(\tau) + X[\tau] \supset A$ by minimality of τ. Hence $X/(X(\tau) + X[\tau])$ is finite. Lemma 3.5 implies that

$$\operatorname{rk} \frac{X(\tau)}{X^\sharp(\tau)} \geq \operatorname{width} \frac{X}{A_\tau \oplus X[\tau]}$$

since $X[\tau]$ is pure in X. Thus the difference in the inequality is ≥ 0.

According to (3) we write

$$X = (A_\tau \oplus X[\tau]) + \sum_{i=1}^{w} \mathbb{Z} x_i \text{ where } w = \operatorname{width}\left(\frac{X}{A_\tau \oplus X[\tau]}\right).$$

Let $e = \exp(X/(A_\tau \oplus X[\tau]))$. Project $eX \subset A_\tau \oplus X[\tau]$ into A_τ. The purification of the projection is a direct summand of A_τ of rank $\leq w$. Any complementary summand is a τ–homogeneous summand of X of rank $\geq \operatorname{rk} X^\circ(\tau) - \operatorname{width}(X/A)$. □

A satisfactory criterion exists for the existence of a rank–one summand of minimal critical type.

Lemma 3.8 *Let X be an almost completely decomposable group and τ a minimal critical type of X. Then X is τ-clipped if and only if*

$$\operatorname{rk} \frac{X(\tau)}{X^\sharp(\tau)} = \operatorname{width}\left(\frac{X}{X(\tau) + X[\tau]}\right). \tag{4}$$

Proof. Let A_τ be any τ–Butler complement of X, and let A be a regulating subgroup with A_τ as τ–homogeneous component. Since τ is a minimal critical type we have that $A = A_\tau \oplus A[\tau]$. Furthermore, $X(\tau) + X[\tau] = A_\tau + X[\tau] \supset A$ has finite index in X.

By Lemma 3.7.2 the absence of a τ–summand implies the identity (4).

Conversely, set

$$w := \operatorname{rk}(X(\tau)/X^\sharp(\tau)) = \operatorname{width}(X/(X(\tau) + X[\tau])).$$

Assume, by way of contradiction, that $X = B \oplus Y$ where B is τ–homogeneous. By Lemma 3.1 we assume without loss of generality that $B \leq A_\tau$. Then $X[\tau] = Y[\tau] \leq Y$ and $Y \cap (A_\tau \oplus X[\tau]) = (Y \cap A_\tau) \oplus X[\tau]$. Hence

$$\frac{Y}{(Y \cap A_\tau) \oplus X[\tau]} \cong \frac{(Y + A_\tau) + X[\tau]}{A_\tau \oplus X[\tau]} = \frac{X}{A_\tau \oplus X[\tau]} = \frac{X}{X(\tau) + X[\tau]}.$$

Since $w = \operatorname{width}(X/(X(\tau) + X[\tau])$,

$$\operatorname{width} \frac{Y}{(Y \cap A_\tau) \oplus X[\tau]} = w.$$

By Lemma 3.5

$$w \leq \operatorname{rk}(Y \cap A_\tau) < \operatorname{rk}(B \oplus (Y \cap A_\tau)) \leq \operatorname{rk} A_\tau = w,$$

a contradiction. □

The following essential lemma reduces the question of the existence of a τ–summand to the case when τ is a minimal critical type.

Lemma 3.9 *Let X be an almost completely decomposable group, τ a critical type of X and B a τ-homogeneous subgroup of X. Then B is a summand of X if and only if B is a summand of $X^\sharp[\tau]$.*

Proof. Suppose first that B is a direct summand of X. Then $B \subset X(\tau) \subset X^\sharp[\tau]$ and hence B is a direct summand of $X^\sharp[\tau]$.

Now assume that

$$X^\sharp[\tau] = B \oplus Y. \tag{5}$$

We can conclude immediately that $X^\sharp[\tau][\tau] = Y[\tau]$. Furthermore, $X^\sharp[\tau][\tau] = X[\tau]$ since both groups are the purification of the group $\bigoplus_{\rho \not\leq \tau} A_\rho$ where $A = \bigoplus_{\rho \in \mathrm{T}_{cr}(A)} A_\rho$ is any regulating subgroup of X. We thus have

$$X[\tau] = Y[\tau] \subset Y. \tag{6}$$

By the co–Butler decomposition (2) we also have

$$\frac{X}{X[\tau]} = \frac{X^\sharp[\tau]}{X[\tau]} \oplus \frac{Z_\tau}{X[\tau]}. \tag{7}$$

Substituting (5) in (7) we obtain

$$\frac{X}{X[\tau]} = \frac{B \oplus Y}{X[\tau]} \oplus \frac{Z_\tau}{X[\tau]}.$$

It follows at once that $X = B + (Y + Z_\tau)$. We claim that the sum is direct. In fact, suppose that $x = y + z \in B \cap (Y + Z_\tau)$, where $y \in Y$ and $z \in Z_\tau$. Then $x - y = z \in (B \oplus Y) \cap Z_\tau = X[\tau] = Y[\tau] \subset Y$. Hence $x = y + z \in B \cap Y = 0$. □

Combining Lemma 3.9 with Lemma 3.8 we obtain the following necessary and sufficient condition for the existence of τ–summands in all generality.

Theorem 3.10 *Let X be an almost completely decomposable group. Then X is τ–clipped if and only if*

$$\operatorname{rk}\frac{X(\tau)}{X^\sharp(\tau)} = \operatorname{width}\left(\frac{X^\sharp[\tau]}{X(\tau) + X[\tau]}\right).$$

Proof. By Lemma 3.9 X is τ–clipped if and only if $X^\sharp[\tau]$ is τ–clipped. By Lemma 3.8 $X^\sharp[\tau]$ is τ-clipped if and only if

$$\operatorname{rk}\frac{X^\sharp[\tau](\tau)}{X^\sharp[\tau]^\sharp(\tau)} = \operatorname{width}\left(\frac{X^\sharp[\tau]}{X^\sharp[\tau](\tau) + X^\sharp[\tau][\tau]}\right).$$

The claim follows by making the substitutions $X^\sharp[\tau](\tau) = X(\tau)$, $X^\sharp[\tau]^\sharp(\tau) = X^\sharp(\tau)$, $X^\sharp[\tau](\tau) = X(\tau)$, and $X^\sharp[\tau][\tau] = X[\tau]$. □

Corollary 3.11 *An almost completely decomposable group X is clipped if and only if*

$$\forall \tau \in \mathrm{T}_{\mathrm{cr}}(X),\ \operatorname{rk}\frac{X(\tau)}{X^\sharp(\tau)} = \operatorname{width}\left(\frac{X^\sharp[\tau]}{X(\tau) + X[\tau]}\right).$$

For τ–slim groups there is a simplified necessary and sufficient condition for the existence of a τ–summand.

Corollary 3.12 *Let X be a τ–slim almost completely decomposable group. Then X has a τ–summand if and only if $X(\tau) + X[\tau]$ is pure in X.*

Proof. By assumption $\operatorname{rk}\frac{X(\tau)}{X^\sharp(\tau)} = 1$. By Theorem 3.10 X has a τ–summand if and only if $\operatorname{width}(X^\sharp[\tau]/(X(\tau) + X[\tau]) = 0$. This is the case if and only if $X^\sharp[\tau] = X(\tau) + X[\tau]$. Since $X^\sharp[\tau]/(X(\tau) + X[\tau])$ is always a torsion group and $X^\sharp[\tau]$ is pure, the quotient $X^\sharp[\tau]/(X(\tau) + X[\tau])$ is trivial if and only if $X(\tau) + X[\tau]$ is pure in X. □

The preceding results suggest the following definition.

Definition 3.13 Let X be an almost completely decomposable group and τ any type. The ***rank–width difference of*** X ***at*** τ is denoted and defined by

$$\operatorname{rwd}_\tau(X) = \operatorname{rk}\frac{X(\tau)}{X^\sharp(\tau)} - \operatorname{width}\frac{X^\sharp[\tau]}{X(\tau) + X[\tau]}.$$

Recall that $\operatorname{rwd}_\tau(X) \geq 0$ by Lemma 3.5. While socles and radicals distribute over direct sums, the width does not do so in general. As a result there is only a limited "sum formula" for rwd_τ.

Lemma 3.14 *Let X and Y be almost completely decomposable groups and τ any type. The formula*

$$\operatorname{rwd}_\tau(X \oplus Y) = \operatorname{rwd}_\tau(X) + \operatorname{rwd}_\tau(Y).$$

is valid in the following cases.

1. *X or Y is completely decomposable.*
2. *The regulating indices of both X and Y are powers of the same prime p.*
3. *$X \cong Y$.*

Proof. Set $Z = X \oplus Y$. It is straightforward to derive that

$$\frac{Z(\tau)}{Z^\sharp(\tau)} \cong \frac{X(\tau)}{X^\sharp(\tau)} \oplus \frac{Y(\tau)}{Y^\sharp(\tau)}$$

and

$$\frac{Z^\sharp[\tau]}{Z(\tau) + Z[\tau]} \cong \frac{X^\sharp[\tau]}{X(\tau) + X[\tau]} \oplus \frac{Y^\sharp[\tau]}{Y(\tau) + Y[\tau]}.$$

Therefore $\operatorname{rk}(Z^\circ(\tau)) = \operatorname{rk}(X^\circ(\tau)) + \operatorname{rk}(Y^\circ(\tau))$ in general, and

$$\operatorname{width}\left(\frac{Z^\sharp[\tau]}{Z(\tau) + Z[\tau]}\right) = \operatorname{width}\left(\frac{X^\sharp[\tau]}{X(\tau) + X[\tau]}\right) + \operatorname{width}\left(\frac{Y^\sharp[\tau]}{Y(\tau) + Y[\tau]}\right)$$

in the three cases stated. □

The sum formula together with Theorem 3.10 has several nice consequences. For one thing $\mathrm{rwd}_\tau(X^n) = n\,\mathrm{rwd}_\tau(X)$, and consequently $\mathrm{rwd}_\tau(X)$ *is a near-isomorphism invariant.*

Corollary 3.15 *The direct sum of a τ–clipped completely decomposable group and a τ–clipped almost completely decomposable group is again τ–clipped.*

Proof. Theorem 3.10 says that a group X is τ–clipped if and only if $\mathrm{rwd}_\tau(X) = 0$. The sum formula does the rest. □

Similarly, a theorem of Lady ([Lad74, Theorem 9]) is an immediate consequence. An almost completely decomposable group is p*–**local*** if its regulating index is a power of p.

Corollary 3.16 (Lady) *The direct sum of p–local τ–clipped almost completely decomposable groups is again τ–clipped.*

Corollary 3.17 *Let X be an almost completely decomposable group and τ any type. Write $X = E \oplus Y$ such that E is τ–homogeneous and Y is τ–clipped. Then*

$$\mathrm{rk}(E) = \mathrm{rwd}_\tau(X).$$

Proof. Since E is τ–homogeneous it follows that $\mathrm{rwd}_\tau(E) = \mathrm{rk}(E)$ and since Y is τ–clipped we have $\mathrm{rwd}_\tau(Y) = 0$. By Lemma 3.14 $\mathrm{rwd}_\tau(X) = \mathrm{rk}(E)$. □

We include a proof of Lady's result ([Lad74, Corollary 7]) thereby removing a small gap in Lady's proof that tacitly assumes Corollary 3.15.

Corollary 3.18 (Lady) *Let X be an almost completely decomposable group and $X = X_{cd} \oplus X_{cl}$ be any decomposition of X as a direct sum of a completely decomposable group X_{cd} and a clipped group X_{cl}. Then X_{cd} is unique up to isomorphism. More specifically, the invariants of X_{cd} are*

$$\mathrm{rk}\,\frac{X_{cd}(\tau)}{X^\sharp_{cd}(\tau)} = \mathrm{rwd}_\tau(X).$$

Proof. Let $X_{cd} = \bigoplus_\rho A_\rho$ be a homogeneous decomposition of X_{cd} and let τ be a critical type of X. Then $X_{cl} \oplus \bigoplus_{\rho \neq \tau} A_\rho$ is τ–clipped and hence $\operatorname{rk} A_\tau = \operatorname{rwd}_\tau(X)$. □

In a similar vein we will prove a criterion ([AV84, Corollary 2.1(a)]) determining when the almost completely decomposable group X is actually completely decomposable.

Theorem 3.19 (Arnold–Vinsonhaler) *An almost completely decomposable group X is completely decomposable if and only if $X(\tau) + X[\tau] = X^\sharp[\tau]$ for each critical type τ of X.*

Proof. Note first that $X(\tau) + X[\tau] = X^\sharp[\tau]$ is equivalent to $\operatorname{rwd}_\tau(X) = \operatorname{rk}(X^\circ(\tau))$.

Suppose that X is completely decomposable. Then every τ–Butler complement, whose rank is known to be $\operatorname{rk}(X^\circ(\tau))$, is a maximal τ–homogeneous direct summand and hence has rank $\operatorname{rwd}_\tau(X)$.

Conversely, let τ be a critical type of X and $\operatorname{rwd}_\tau(X) = \operatorname{rk}(X^\circ(\tau))$. By Corollary 3.17 $X = A_\tau \oplus Y$ where A_τ is τ–homogeneous completely decomposable, $\operatorname{rk}(A_\tau) = \operatorname{rk}(X^\circ(\tau))$ and Y is τ–clipped. By Lemma 3.14 and induction on rank, the group Y is completely decomposable. □

4 Algorithms

In this section we will demonstrate that the criteria that have been found are practical. Indeed, the rank–width difference can be computed according to a straight–forward algorithm. One basic ingredient of the algorithm is the computation of greatest common divisors of integral matrices which in turn is achieved by the column reduction of certain integral matrices (see [BM98]). The other basic ingredient is the computation of the Smith Normal Form of a square integral matrix. Both of these can be accomplished by using the Maple routines `ihermit` and `ismith`.

The starting point of the algorithm is the so–called **standard description** of an almost completely decomposable group which is explained and established in [BM98]. The symbol $\vec{\mathbb{Z}}$ stands for the set of integral row vectors $\vec{\alpha} = [\alpha_1, \ldots, \alpha_k]$, N denotes a $k \times k$ integral matrix with non–zero determinant, and $a^\downarrow = [a_1, \ldots, a_k]^{\mathrm{tr}}$ is a column matrix of elements of A. Juxtaposition of matrices means that the matrices are multiplied according to the usual rules. The matrix N^{-1} is a matrix with rational coefficients, but the product $N^{-1}\, a^\downarrow$ still makes sense in a divisible hull $\mathbb{Q}A$ of A and must be so understood. It is perhaps helpful to note that $N^{-1}\, a^\downarrow = \frac{1}{\det N} \operatorname{adj}(N)\, a^\downarrow$, where the adjoint adj$(N)$ is an integral matrix and therefore adj$(N)\, a^{\downharpoonleft}$ is a column vector of integral linear combinations of the entries of $a^{\downharpoonleft}$.

Lemma 4.1 The Standard Description. *Every almost completely decomposable group X has a description*

$$X = A + \overrightarrow{\mathbb{Z}} N^{-1} a^{\downharpoonleft} \leq \mathbb{Q}A, \quad \gcd^A(N, a^{\downharpoonleft}) = I,$$

$$A = \bigoplus_{\rho \in \mathrm{T_{cr}}(X)} A_\rho, \quad a^{\downharpoonleft} = \sum_{\rho \in \mathrm{T_{cr}}(X)} a^{\downharpoonleft}_\rho, \quad a^{\downharpoonleft}_\rho = [a_{\rho 1}, \dots, a_{\rho k}]^{tr}, \quad a_{\rho i} \in A_\rho.$$

Here A need not be a regulating subgroup of X. Examples of almost completely decomposable groups are exclusively given in standard description. While there are effective tests determining whether A is a regulating subgroup or not, it is as yet unclear how to construct a regulating subgroup if A turns out not to be regulating.

Example. Let $\sigma_1, \dots, \sigma_{15}$ be rational groups (additive subgroups of $\mathbb{Q}$ containing $\mathbb{Z}$). Let $\tau_1 = \sigma_1$, $\tau_2 = \sigma_{15}$, $\tau_3 = \sigma_8$, $\tau_4 = \sigma_2 = \sigma_3 = \sigma_4$, $\tau_5 = \sigma_5 = \sigma_6 = \sigma_7$, $\tau_6 = \sigma_9 = \sigma_{10} = \sigma_{11}$, $\tau_7 = \sigma_{12} = \sigma_{13} = \sigma_{14}$, and let the order relationships of the τ_i as types be given by the Hasse diagram

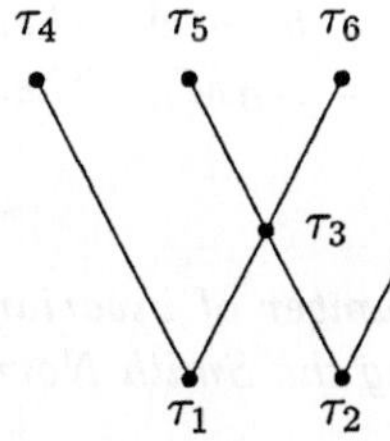

Let

$$A = \sigma_1 v_1 \oplus \cdots \oplus \sigma_{15} v_{15},$$

a completely decomposable group of rank 15. We assume for obvious reasons that $\gcd^A(2, v_i) = 1$, which is the same as $\frac{1}{2} \notin \sigma_i$. The example is

$$X = A + \overrightarrow{\mathbb{Z}} N^{-1} M\, v^{\downharpoonleft}$$

where

$$N = \begin{bmatrix} 8 & 0 & 0 & 0 & 0 & 0 & 0 \\ 0 & 32 & 0 & 0 & 0 & 0 & 0 \\ 0 & 0 & 64 & 0 & 0 & 0 & 0 \\ 0 & 0 & 0 & 128 & 0 & 0 & 0 \\ 0 & 0 & 0 & 0 & 256 & 0 & 0 \\ 0 & 0 & 0 & 0 & 0 & 512 & 0 \\ 0 & 0 & 0 & 0 & 0 & 0 & 1024 \end{bmatrix}$$

and

$$M =$$

$$\begin{bmatrix} 5 & 6 & 9 & 4 & 1 & 1 & 4 & 9 & 4 & 1 & 1 & 4 & 9 & 6 & 5 \\ 0 & -1 & -1 & -1 & 0 & 0 & -1 & -1 & -1 & 0 & 0 & -1 & -1 & -1 & 0 \\ 1 & 1 & 1 & 1 & 0 & 0 & 1 & 1 & 1 & 0 & 0 & 1 & 1 & 1 & 1 \\ 3 & 3 & 3 & 2 & 1 & 1 & 2 & 3 & 2 & 1 & 1 & 2 & 3 & 3 & 3 \\ 0 & 0 & 1 & 0 & 0 & 0 & 0 & 1 & 0 & 0 & 0 & 0 & 1 & 0 & 0 \\ 13 & 12 & 13 & 9 & 5 & 5 & 8 & 13 & 8 & 5 & 5 & 9 & 13 & 12 & 13 \\ 26 & 25 & 27 & 19 & 11 & 10 & 17 & 27 & 17 & 10 & 11 & 19 & 27 & 25 & 26 \end{bmatrix}$$

Note first that $\operatorname{rk} X^\circ(\tau_1) = \operatorname{rk} X^\circ(\tau_2) = \operatorname{rk} X^\circ(\tau_3) = 1$, and $\operatorname{rk} X^\circ(\tau_4) = \operatorname{rk} X^\circ(\tau_5) = \operatorname{rk} X^\circ(\tau_6) = \operatorname{rk} X^\circ(\tau_7) = 3$. It is easily checked by computer that $\gcd^A(N, M\, v^{\rfloor}) = \operatorname{gcld}(N, M) = I$, hence $X/A \cong \mathbb{Z}(2^3) \oplus \mathbb{Z}(2^5) \oplus \mathbb{Z}(2^6) \oplus \mathbb{Z}(2^7) \oplus \mathbb{Z}(2^8) \oplus \mathbb{Z}(2^9) \oplus \mathbb{Z}(2^{10})$, a group of order 2^{48} and width 7. □

The following results will enter the considerations.

Lemma 4.2 (Purification Lemma) *Let X be an almost completely decomposable group given in standard description as in 4.1. Also suppose that $A = B \oplus C$ and write $a^{\rfloor} = b^{\rfloor} + c^{\rfloor}$, where $b^{\rfloor} \in B^{\rfloor}$, $c^{\rfloor} \in C^{\rfloor}$. Let $N_B = \gcd^A(N, c^{\rfloor})$ and write $N = N_B M_B$. Then the following hold.*

1. $|X/A| = |\det N|$.

2. $\operatorname{width}(X/A)$ *equals the number of invariant factors $\neq 1$ of N. These can be found by computing the Smith Normal Form of N.*

3. $B_*^X = B + \vec{\mathbb{Z}} N_B^{-1} b^{\rfloor}$. *Furthermore,* $\gcd^A(N_B, b^{\rfloor}) = I$, *and consequently* $|B_*^X/B| = |\det N_B|$.

4. $X = (B_*^X \oplus C) + \vec{\mathbb{Z}} M_B^{-1}(N_B^{-1} b^{\rfloor} + N_B^{-1} c^{\rfloor})$, *where* $N_B^{-1} b^{\rfloor} \in (B_*^X)^{\rfloor}$, $N_B^{-1} c^{\rfloor} \in C^{\rfloor}$, *and* $\gcd^{B_*^X \oplus C}(M_B, N_B^{-1} b^{\rfloor} + N_B^{-1} c^{\rfloor}) = I$.

Proof. Items 1. through 3. can be found in [BM98].

4. We have

$$X = (B_*^X \oplus C) + \vec{\mathbb{Z}} N^{-1}(b^{\rfloor} + c^{\rfloor}) = (B_*^X \oplus C) + \vec{\mathbb{Z}} M_B^{-1}(N_B^{-1} b^{\rfloor} + N_B^{-1} c^{\rfloor}),$$

where $N_B^{-1} b^{\rfloor} \in (B_*^X)^{\rfloor}$, $N_B^{-1} c^{\rfloor} \in C^{\rfloor}$ which is the first claim. To verify the remaining claim that $\gcd^{B_*^X \oplus C}(M_B, N_B^{-1} b^{\rfloor} + N_B^{-1} c^{\rfloor}) = I$, it suffices to show that $[X : (B_*^X \oplus C)] = |\det M_B|$. We find, as desired, that

$$\left|\frac{X}{B_*^X \oplus C}\right| = \frac{|X/A|}{|B_*^X/B|} = \frac{|\det N|}{|\det N_B|} = |\det M_B|. \qquad \square$$

The Purification Lemma provides the means for testing whether a completely decomposable subgroup $A = \bigoplus_{\rho \in \mathrm{T_{cr}}(A)} A_\rho$ of an almost completely decomposable group X is regulating or not. One computes the indices $[X(\tau) : (A_\tau \oplus X^\sharp(\tau))]$ that turn out to be the quotients of two determinants. We are in the possession of a MAPLE procedure that easily handles our example. It is also easy to compute the Burkhardt invariants that demonstrate that the example entails some complexity.

Example continued. The group A is regulating in X. The Burkhardt invariants of X are $\beta_{\tau_1}^X = \beta_{\tau_2}^X = \beta_{\tau_3}^X = 128$, $\beta_{\tau_4}^X = \beta_{\tau_5}^X = \beta_{\tau_6}^X = \beta_{\tau_7}^X = 1$. By a formula of Mutzbauer X contains 2^{45} different regulating subgroups. □

Various criteria and the rank–width difference involved the width of the quotient $X^\sharp[\tau]/(X(\tau) + X[\tau])$ for a minimal critical type τ. We show next how to compute this width.

Lemma 4.3 *Let X be an almost completely decomposable group given in standard description (4.1). Assume further that A is regulating in X and that τ is minimal in $\mathrm{T_{cr}}(X)$. If $M = \gcd^A(N, a_\tau^\downarrow)$ and $N = ML$, then the width of $X/(X(\tau) + X[\tau])$ equals the number of non-1 invariant factors of L.*

Proof. Since A is regulating in X, we have $X(\tau) + X[\tau] = A_\tau \oplus X[\tau]$ and $X[\tau] = A[\tau]_*^X$. Since τ is minimal in $\mathrm{T_{cr}}(X)$, it follows that $A = A_\tau \oplus A[\tau]$. By Lemma 4.2.4,

$$X = (A_\tau \oplus X[\tau]) + \vec{\mathbb{Z}}L^{-1}\left(M^{-1}a_\tau + M^{-1}\sum_{\rho \neq \tau} a_\rho\right),$$

with $\gcd^A(L, M^{-1}a_\tau + M^{-1}\sum_{\rho\neq\tau} a_\rho) = I$, and the claim now follows easily from Lemma 4.2.2. □

Example continued. Lemma 4.3 suffices to compute the rank–width difference at the minimal critical types τ_1 and τ_2.
Set

$$M_{\tau_1} = \gcd{}^A(N, \begin{bmatrix} 5 \\ 0 \\ 1 \\ 3 \\ 0 \\ 13 \\ 26 \end{bmatrix} v_1) = \mathrm{gcld}(N, \begin{bmatrix} 5 \\ 0 \\ 1 \\ 3 \\ 0 \\ 13 \\ 26 \end{bmatrix}).$$

Then

$$M_{\tau_1} = \begin{bmatrix} 1 & 0 & 0 & 0 & 0 & 0 & 0 \\ 0 & 32 & 0 & 0 & 0 & 0 & 0 \\ 5 & 0 & 8 & 0 & 0 & 0 & 0 \\ 15 & 0 & 24 & 64 & 0 & 0 & 0 \\ 0 & 0 & 0 & 0 & 256 & 0 & 0 \\ 65 & 0 & 104 & 64 & 0 & 128 & 0 \\ 130 & 0 & 208 & 128 & 0 & 256 & 1024 \end{bmatrix},$$

$$L_{\tau_1} = M_{\tau_1}^{-1} N = \begin{bmatrix} 8 & 0 & 0 & 0 & 0 & 0 & 0 \\ 0 & 1 & 0 & 0 & 0 & 0 & 0 \\ -5 & 0 & 8 & 0 & 0 & 0 & 0 \\ 0 & 0 & -3 & 2 & 0 & 0 & 0 \\ 0 & 0 & 0 & 0 & 1 & 0 & 0 \\ 0 & 0 & -5 & -1 & 0 & 4 & 0 \\ 0 & 0 & 0 & 0 & 0 & -1 & 1 \end{bmatrix}$$

and the Smith Normal Form of L_{τ_1} is $\begin{bmatrix} 1 & 0 & 0 & 0 & 0 & 0 & 0 \\ 0 & 1 & 0 & 0 & 0 & 0 & 0 \\ 0 & 0 & 1 & 0 & 0 & 0 & 0 \\ 0 & 0 & 0 & 1 & 0 & 0 & 0 \\ 0 & 0 & 0 & 0 & 1 & 0 & 0 \\ 0 & 0 & 0 & 0 & 0 & 1 & 0 \\ 0 & 0 & 0 & 0 & 0 & 0 & 512 \end{bmatrix}$. This means that $\operatorname{width}(X/(X(\tau_1)+X[\tau_1])) = 1$ and $\operatorname{rwd}_{\tau_1}(X) = 0$. Hence X has no rank–one summand of type τ_1.

Set

$$M_{\tau_2} = \operatorname{gcd}^A(N, \begin{bmatrix} 5 \\ 0 \\ 1 \\ 3 \\ 0 \\ 13 \\ 26 \end{bmatrix} v_{15}) = \operatorname{gcld}(N, \begin{bmatrix} 5 \\ 0 \\ 1 \\ 3 \\ 0 \\ 13 \\ 26 \end{bmatrix}) = M_{\tau_1}.$$

Then, as before, X has no rank–one summand of type τ_2. □

Lemma 4.3 will be applied with $X^\sharp[\tau]$ in place of X. The computational version of Corollary 3.12 is as follows.

Proposition 4.4 *Let X be a τ–slim almost completely decomposable group given in standard description (4.1). Assume that A is regulating in X. Let $N_{[\tau]} = \operatorname{gcd}^A(N, \sum_{\rho \le \tau} a^{\downarrow}_{\rho})$ and $N^\sharp_{[\tau]} = \operatorname{gcd}^A(N, \sum_{\rho<\tau} a^{\downarrow}_{\rho})$. Then X has a τ–summand if and only if $|\det N_{[\tau]}| = |\det N^\sharp_{[\tau]}|$.*

Proof. By Corollary 3.12 it is a fact that X has a τ–summand if and only if $X^\sharp[\tau] = X(\tau) + X[\tau]$. This in turn is equivalent to

$$\frac{X^\sharp[\tau]}{A^\sharp[\tau]} = \frac{X(\tau) + X[\tau]}{A^\sharp[\tau]} = \frac{A_\tau \oplus X[\tau]}{A_\tau \oplus A[\tau]} \cong \frac{X[\tau]}{A[\tau]},$$

which is true if and only if $|\det N^\sharp_{[\tau]}| = |\det N_{[\tau]}|$. □

The next results explains how to compute rank–width differences and thereby provides the computational version of Theorem 3.17.

Theorem 4.5 *Let X be an almost completely decomposable group given in standard description (4.1). Assume that A is regulating in X. Let*

$$N^\sharp_{[\tau]} = \gcd{}^A\left(N, \sum_{\rho<\tau} a^\downarrow_\rho\right), \quad N_{[\tau]} = \gcd{}^A\left(N^\sharp_{[\tau]}, a^\downarrow_\tau\right) = \gcd{}^A(N, \sum_{\rho\le\tau} a^\downarrow_\rho),$$

and let w be the number of non–1 invariant factors of $N^{-1}_{[\tau]} N^\sharp_{[\tau]}$. Then

$$\operatorname{rwd}_\tau(X) = \operatorname{rk} A_\tau - w.$$

Proof. It is easily verified from the definition of greatest common divisor that $\gcd^A\left(N^\sharp_{[\tau]}, a^\downarrow_\tau\right) = \gcd^A(N, \sum_{\rho\le\tau} a^\downarrow_\rho)$. Set $L_\tau = N^{-1}_{[\tau]} N^\sharp_{[\tau]}$. This matrix is integral. By Lemma 4.2.3,

$$X^\sharp[\tau] = (A^\sharp[\tau])^X_* = A^\sharp[\tau] + \vec{\mathbb{Z}}\left(N^\sharp_{[\tau]}\right)^{-1}\left(\sum_{\rho\not<\tau} a^\downarrow_\rho\right),$$

and by Lemma 4.2.4,

$$\begin{aligned} X^\sharp[\tau] &= (A[\tau] \oplus A_\tau)^{X^\sharp[\tau]}_* \\ &= (X[\tau] \oplus A_\tau) + \vec{\mathbb{Z}} L^{-1}_\tau\left(N^{-1}_{[\tau]} a^\downarrow_\tau + N^{-1}_{[\tau]} \sum_{\rho\not\le\tau} a^\downarrow_\rho\right). \end{aligned}$$

The claim now follows from Lemma 4.3. □

Example continued. We now compute the rank–width difference at the remaining critical types $\tau_3, \tau_4, \tau_5, \tau_6$ and τ_7.

Let

$$N^{\sharp}_{[\tau_3]} = \gcd^A(N, \begin{bmatrix} 5 & 5 \\ 0 & 0 \\ 1 & 1 \\ 3 & 3 \\ 0 & 0 \\ 13 & 13 \\ 26 & 26 \end{bmatrix} \begin{bmatrix} v_1 \\ v_{15} \end{bmatrix}) = \operatorname{gcld}(N, \begin{bmatrix} 5 & 5 \\ 0 & 0 \\ 1 & 1 \\ 3 & 3 \\ 0 & 0 \\ 13 & 13 \\ 26 & 26 \end{bmatrix}).$$

Then

$$N^{\sharp}_{[\tau_3]} = \begin{bmatrix} 1 & 0 & 0 & 0 & 0 & 0 & 0 \\ 0 & 32 & 0 & 0 & 0 & 0 & 0 \\ 5 & 0 & 8 & 0 & 0 & 0 & 0 \\ 15 & 0 & 24 & 64 & 0 & 0 & 0 \\ 0 & 0 & 0 & 0 & 256 & 0 & 0 \\ 65 & 0 & 104 & 64 & 0 & 128 & 0 \\ 130 & 0 & 208 & 128 & 0 & 256 & 1024 \end{bmatrix}$$

and

$$\begin{aligned} N_{[\tau_3]} &= \gcd^A(N^{\sharp}_{[\tau_3]}, \begin{bmatrix} 6 \\ -1 \\ 1 \\ 3 \\ 0 \\ 12 \\ 25 \end{bmatrix} v_8) = \operatorname{gcld}(N^{\sharp}_{[\tau_3]}, \begin{bmatrix} 6 \\ -1 \\ 1 \\ 3 \\ 0 \\ 12 \\ 25 \end{bmatrix}) \\ &= \begin{bmatrix} 1 & 0 & 0 & 0 & 0 & 0 & 0 \\ 0 & 1 & 0 & 0 & 0 & 0 & 0 \\ 5 & 4 & 8 & 0 & 0 & 0 & 0 \\ 15 & 12 & 24 & 64 & 0 & 0 & 0 \\ 0 & 31 & 0 & 0 & 32 & 0 & 0 \\ 65 & 52 & 104 & 64 & 0 & 128 & 0 \\ 130 & 135 & 208 & 128 & 32 & 0 & 256 \end{bmatrix}. \end{aligned}$$

Next

$$L_{\tau_3} = N^{-1}_{[\tau_3]} N^{\sharp}_{[\tau_3]} = \begin{bmatrix} 1 & 0 & 0 & 0 & 0 & 0 & 0 \\ 0 & 32 & 0 & 0 & 0 & 0 & 0 \\ 0 & -16 & 1 & 0 & 0 & 0 & 0 \\ 0 & 0 & 0 & 1 & 0 & 0 & 0 \\ 0 & -31 & 0 & 0 & 8 & 0 & 0 \\ 0 & 0 & 0 & 0 & 0 & 1 & 0 \\ 0 & 0 & 0 & 0 & -1 & 1 & 4 \end{bmatrix}$$

and the Smith Normal Form of L_{τ_3} is $\begin{bmatrix} 1 & 0 & 0 & 0 & 0 & 0 & 0 \\ 0 & 1 & 0 & 0 & 0 & 0 & 0 \\ 0 & 0 & 1 & 0 & 0 & 0 & 0 \\ 0 & 0 & 0 & 1 & 0 & 0 & 0 \\ 0 & 0 & 0 & 0 & 1 & 0 & 0 \\ 0 & 0 & 0 & 0 & 0 & 1 & 0 \\ 0 & 0 & 0 & 0 & 0 & 0 & 1024 \end{bmatrix}$. This means that $\text{width}(X^\sharp[\tau_3]/(X(\tau_3) + X[\tau_3])) = 1$ and $\text{rwd}_{\tau_3}(X) = 0$. Hence X has no rank-one summand of type τ_3.

Let

$$N^\sharp_{[\tau_4]} = \gcd{}^A(N, \begin{bmatrix} 5 \\ 0 \\ 1 \\ 3 \\ 0 \\ 13 \\ 26 \end{bmatrix} v_1) = \text{gcld}(N, \begin{bmatrix} 5 \\ 0 \\ 1 \\ 3 \\ 0 \\ 13 \\ 26 \end{bmatrix}).$$

Then

$$N^\sharp_{[\tau_4]} = \begin{bmatrix} 1 & 0 & 0 & 0 & 0 & 0 & 0 \\ 0 & 32 & 0 & 0 & 0 & 0 & 0 \\ 5 & 0 & 8 & 0 & 0 & 0 & 0 \\ 15 & 0 & 24 & 64 & 0 & 0 & 0 \\ 0 & 0 & 0 & 0 & 256 & 0 & 0 \\ 65 & 0 & 104 & 64 & 0 & 128 & 0 \\ 130 & 0 & 208 & 128 & 0 & 256 & 1024 \end{bmatrix}$$

and

$$\begin{aligned} N_{[\tau_4]} &= \gcd{}^A(N^\sharp_{[\tau_4]}, \begin{bmatrix} 6 & 9 & 4 \\ -1 & -1 & -1 \\ 1 & 1 & 1 \\ 3 & 3 & 2 \\ 0 & 1 & 0 \\ 12 & 13 & 9 \\ 25 & 27 & 19 \end{bmatrix} \begin{bmatrix} v_2 \\ v_3 \\ v_4 \end{bmatrix}) \\ &= \text{gcld}(N^\sharp_{[\tau_4]}, \begin{bmatrix} 6 & 9 & 4 \\ -1 & -1 & -1 \\ 1 & 1 & 1 \\ 3 & 3 & 2 \\ 0 & 1 & 0 \\ 12 & 13 & 9 \\ 25 & 27 & 19 \end{bmatrix}) \end{aligned}$$

$$= \begin{bmatrix} 1 & 0 & 0 & 0 & 0 & 0 & 0 \\ 0 & 1 & 0 & 0 & 0 & 0 & 0 \\ 0 & 0 & 1 & 0 & 0 & 0 & 0 \\ 0 & 0 & 0 & 1 & 0 & 0 & 0 \\ 3 & 3 & 3 & 2 & 8 & 0 & 0 \\ 3 & 4 & 7 & 5 & 8 & 32 & 0 \\ 6 & 7 & 14 & 10 & 16 & 32 & 128 \end{bmatrix}.$$

Next

$$L_{\tau_4} = N_{[\tau_4]}^{-1} N_{[\tau_4]}^{\sharp} = \begin{bmatrix} 1 & 0 & 0 & 0 & 0 & 0 & 0 \\ 0 & 32 & 0 & 0 & 0 & 0 & 0 \\ 5 & 0 & 8 & 0 & 0 & 0 & 0 \\ 15 & 0 & 24 & 64 & 0 & 0 & 0 \\ -6 & -12 & -9 & -16 & 32 & 0 & 0 \\ 0 & -1 & 0 & -4 & -8 & 4 & 0 \\ 0 & 0 & 0 & -1 & -2 & 1 & 8 \end{bmatrix}$$

and the Smith Normal Form of L_{τ_4} is $\begin{bmatrix} 1 & 0 & 0 & 0 & 0 & 0 & 0 \\ 0 & 1 & 0 & 0 & 0 & 0 & 0 \\ 0 & 0 & 1 & 0 & 0 & 0 & 0 \\ 0 & 0 & 0 & 1 & 0 & 0 & 0 \\ 0 & 0 & 0 & 0 & 64 & 0 & 0 \\ 0 & 0 & 0 & 0 & 0 & 256 & 0 \\ 0 & 0 & 0 & 0 & 0 & 0 & 1024 \end{bmatrix}$.

This means that $\operatorname{width}(X^{\sharp}[\tau_4]/(X(\tau_4) + X[\tau_4])) = 3$ and $\operatorname{rwd}_{\tau_4}(X) = 0$. Hence X has no rank–one summand of type τ_4.
Let

$$\begin{aligned} N_{[\tau_5]}^{\sharp} &= \operatorname{gcd}^A(N, \begin{bmatrix} 5 & 9 & 5 \\ 0 & -1 & 0 \\ 1 & 1 & 1 \\ 3 & 3 & 3 \\ 0 & 1 & 0 \\ 13 & 13 & 13 \\ 26 & 27 & 26 \end{bmatrix} \begin{bmatrix} v_1 \\ v_8 \\ v_{15} \end{bmatrix}) \\ &= \operatorname{gcld}(N, \begin{bmatrix} 5 & 9 & 5 \\ 0 & -1 & 0 \\ 1 & 1 & 1 \\ 3 & 3 & 3 \\ 0 & 1 & 0 \\ 13 & 13 & 13 \\ 26 & 27 & 26 \end{bmatrix}). \end{aligned}$$

Then

$$N^\sharp_{[\tau_5]} = \begin{bmatrix} 1 & 0 & 0 & 0 & 0 & 0 & 0 \\ 0 & 1 & 0 & 0 & 0 & 0 & 0 \\ 5 & 4 & 8 & 0 & 0 & 0 & 0 \\ 15 & 12 & 24 & 64 & 0 & 0 & 0 \\ 0 & 31 & 0 & 0 & 32 & 0 & 0 \\ 65 & 52 & 104 & 64 & 0 & 128 & 0 \\ 130 & 135 & 208 & 128 & 32 & 0 & 256 \end{bmatrix}$$

and

$$\begin{aligned} N_{[\tau_5]} &= \operatorname{gcd}^A(N^\sharp_{[\tau_5]}, \begin{bmatrix} 1 & 1 & 4 \\ 0 & 0 & -1 \\ 0 & 0 & 1 \\ 1 & 1 & 2 \\ 0 & 0 & 0 \\ 5 & 5 & 8 \\ 11 & 10 & 17 \end{bmatrix} \begin{bmatrix} v_5 \\ v_6 \\ v_7 \end{bmatrix}) \\ &= \operatorname{gcld}(N^\sharp_{[\tau_5]}, \begin{bmatrix} 1 & 1 & 4 \\ 0 & 0 & -1 \\ 0 & 0 & 1 \\ 1 & 1 & 2 \\ 0 & 0 & 0 \\ 5 & 5 & 8 \\ 11 & 10 & 17 \end{bmatrix}) \\ &= \begin{bmatrix} 1 & 0 & 0 & 0 & 0 & 0 & 0 \\ 0 & 1 & 0 & 0 & 0 & 0 & 0 \\ 0 & 0 & 1 & 0 & 0 & 0 & 0 \\ 1 & 0 & 2 & 4 & 0 & 0 & 0 \\ 0 & 0 & 1 & 1 & 2 & 0 & 0 \\ 5 & 0 & 8 & 20 & 0 & 128 & 0 \\ 0 & 0 & 0 & 0 & 0 & 0 & 1 \end{bmatrix}. \end{aligned}$$

Next $L_{\tau_5} = N^{-1}_{[\tau_5]} N^\sharp_{[\tau_5]} = \begin{bmatrix} 1 & 0 & 0 & 0 & 0 & 0 & 0 \\ 0 & 1 & 0 & 0 & 0 & 0 & 0 \\ 5 & 4 & 8 & 0 & 0 & 0 & 0 \\ 1 & 1 & 2 & 16 & 0 & 0 & 0 \\ -3 & 13 & -5 & -8 & 16 & 0 & 0 \\ 0 & 0 & 0 & -2 & 0 & 1 & 0 \\ 130 & 135 & 208 & 128 & 32 & 0 & 256 \end{bmatrix}$ and the

Smith Normal Form of L_{τ_5} is $\begin{bmatrix} 1 & 0 & 0 & 0 & 0 & 0 & 0 \\ 0 & 1 & 0 & 0 & 0 & 0 & 0 \\ 0 & 0 & 1 & 0 & 0 & 0 & 0 \\ 0 & 0 & 0 & 1 & 0 & 0 & 0 \\ 0 & 0 & 0 & 0 & 32 & 0 & 0 \\ 0 & 0 & 0 & 0 & 0 & 64 & 0 \\ 0 & 0 & 0 & 0 & 0 & 0 & 256 \end{bmatrix}$. This means that $\mathrm{width}(X^\sharp[\tau_5]/(X(\tau_5) + X[\tau_5])) = 3$ and $\mathrm{rwd}_{\tau_5}(X) = 0$. Hence X has no rank–one summand of type τ_5.

Similarly, we find that $\mathrm{rwd}_{\tau_6}(X) = \mathrm{rwd}_{\tau_7}(X) = 0$ and conclude that X is clipped. □

Finally we state the constructive version of Theorem 3.19 if X is given in standard description. In this case A must be allowed to be an arbitrary completely decomposable subgroup of finite index since the problem is trivial if A is known to be regulating.

Theorem 4.6 *Let X be an almost completely decomposable group given in standard description* (4.1). *Then the following are equivalent.*

1. *X is completely decomposable.*
2. *For every $\tau \in \mathrm{T}_{\mathrm{cr}}(X)$,*
$$[X^\sharp[\tau] : A^\sharp[\tau]] \cdot [X^\sharp(\tau) : A^\sharp(\tau)] = [X[\tau] : A[\tau]] \cdot [X(\tau) : A(\tau)].$$
3. *For every $\tau \in \mathrm{T}_{\mathrm{cr}}(X)$,*
$$\left|\det N^\sharp_{[\tau]}\right| \cdot \left|\det N^\sharp_\tau\right| = \left|\det N_{[\tau]}\right| \cdot \left|\det N_\tau\right|,$$
where
$$N^\sharp_{[\tau]} = \mathrm{gcd}^A\left(N, \textstyle\sum_{\rho<\tau} a^\natural_\rho\right), \qquad N^\sharp_\tau = \mathrm{gcd}^A\left(N, \textstyle\sum_{\rho\not>\tau} a^\natural_\rho\right),$$
$$N_{[\tau]} = \mathrm{gcd}^A\left(N, \textstyle\sum_{\rho\leq\tau} a^\natural_\rho\right), \qquad N_\tau = \mathrm{gcd}^A\left(N, \textstyle\sum_{\rho\not\geq\tau} a^\natural_\rho\right).$$

Proof. The equivalence of 2. and 3. is immediate from Lemma 4.2 since $X^\sharp[\tau] = (A^\sharp[\tau])^X_*$, $X[\tau] = (A[\tau])^X_*$, $X^\sharp(\tau) = (A^\sharp(\tau))^X_*$, and $X(\tau) = (A(\tau))^X_*$. For example,

$$X^\sharp[\tau] = A^\sharp[\tau] + \vec{\mathbb{Z}}\left(N^\sharp_{[\tau]}\right)^{-1}\left(\sum_{\rho\not<\tau} a^\natural_\rho\right), \quad \text{and}$$

$$X[\tau] = A[\tau] + \vec{\mathbb{Z}} N^{-1}_{[\tau]}\left(\sum_{\rho\not\leq\tau} a^\natural_\rho\right).$$

Hence $[X^\sharp[\tau] : A^\sharp[\tau]] = |\det N^\sharp_{[\tau]}|$ and so forth.

By Theorem 3.19 X is completely decomposable if and only if $X^\sharp[\tau] = X(\tau) + X[\tau]$ for each τ. This equivalent with

$$\frac{X^\sharp[\tau]}{A^\sharp[\tau]} = \frac{X(\tau) + X[\tau]}{A^\sharp[\tau]} = \frac{X(\tau) + X[\tau]}{A(\tau) + A[\tau]}, \tag{8}$$

where Theorem 3.19 is applied to A. Consider the exact sequence

$$\frac{X^\sharp(\tau) + A(\tau) + A[\tau]}{A(\tau) + A[\tau]} \rightarrowtail \frac{X(\tau) + X[\tau]}{A(\tau) + A[\tau]} \twoheadrightarrow \frac{X(\tau) + X[\tau]}{X^\sharp(\tau) + A(\tau) + A[\tau]}. \tag{9}$$

The cardinality of the left end of (9) can be computed as follows. Write $A[\tau] = A^\flat(\tau) \oplus A^\sharp(\tau)$ for some group $A^\flat(\tau)$. Then

$$\frac{X^\sharp(\tau) + A(\tau) + A[\tau]}{A(\tau) + A[\tau]} = \frac{A_\tau \oplus A^\flat(\tau) \oplus X^\sharp(\tau)}{A_\tau \oplus A^\flat(\tau) \oplus A^\sharp(\tau)} \cong \frac{X^\sharp(\tau)}{A^\sharp(\tau)}.$$

The right hand end of (9) is dealt with by using that $X(\tau) \cap X[\tau] = X^\sharp(\tau)$. We have

$$\begin{aligned}
\frac{X(\tau) + X[\tau]}{X^\sharp(\tau) + A(\tau) + A[\tau]} &\cong \frac{(X(\tau) + X[\tau])/X^\sharp(\tau)}{(X^\sharp(\tau) + A(\tau) + A[\tau])/X^\sharp(\tau)} \\
&= \frac{\frac{X(\tau)}{X^\sharp(\tau)} \oplus \frac{X[\tau]}{X^\sharp(\tau)}}{\frac{A(\tau)+X^\sharp(\tau)}{X^\sharp(\tau)} \oplus \frac{A[\tau]+X^\sharp(\tau)}{X^\sharp(\tau)}} \\
&\cong \frac{X(\tau)}{A(\tau) + X^\sharp(\tau)} \oplus \frac{X[\tau]}{A[\tau] + X^\sharp(\tau)} \\
&\cong \frac{X(\tau)/A(\tau)}{(A(\tau) + X^\sharp(\tau))/A(\tau)} \oplus \frac{X[\tau]/A[\tau]}{(A[\tau] + X^\sharp(\tau))/A[\tau]}.
\end{aligned}$$

The cardinalities of the two summands in the last line can now be expressed in terms of computable indices as follows.

$$\frac{A(\tau) + X^\sharp(\tau)}{A(\tau)} = \frac{A_\tau \oplus X^\sharp(\tau)}{A_\tau \oplus A^\sharp(\tau)} \cong \frac{X^\sharp(\tau)}{A^\sharp(\tau)},$$

$$\frac{A[\tau] + X^\sharp(\tau)}{A[\tau]} = \frac{A^\flat(\tau) \oplus X^\sharp(\tau)}{A^\flat(\tau) \oplus A^\sharp(\tau)} \cong \frac{X^\sharp(\tau)}{A^\sharp(\tau)}.$$

Combining all these yields the index identity that is equivalent to (8). □

References

[AV84] D. M. Arnold and C. Vinsonhaler, *Pure subgroups of finite rank completely decomposable groups II*, in Proceedings of the Honolulu Abelian Groups Conference, Lecture Notes in Mathematics **1006** (Springer-Verlag 1984), 97–143.

[BM94] E. A. Blagoveshchenskaya and A. Mader, *Decompositions of almost completely decomposable groups*, Contemporary Mathematics **171** (1994), 21–36.

[BM98] K. Benabdallah and A. Mader. *Almost completely decomposable groups and integral linear algebra*, J. Algebra **204** (1998), 440–482.

[DMM97] U. Dittmann, A. Mader, and O. Mutzbauer, *Almost completely decomposable groups with a cyclic regulating quotient* Communications in Algebra **25** (1997), 769–784.

[Lad74] E. L. Lady, *Almost completely decomposable torsion-free abelian groups* Proc. Amer. Math. Soc. **45** (1974), 41–47.

[Lew93] W. S. Lewis, *Almost completely decomposable groups with two critical types* Communications in Algebra **21** (1993), 607–614.

[Mad95] A. Mader, *Almost completely decomposable torsion-free abelian groups* in Abelian Groups and Modules, Proceedings of the 1994 Padova Conference, Kluwer Academic Publishers (1995), 343–366.

[MV94] A. Mader and C. Vinsonhaler, *Classifying almost completely decomposable groups*, J. Algebra **170** (1994), 754–780.

[MV95] A. Mader and C. Vinsonhaler, *Almost completely decomposable groups with a cyclic regulating quotient*, J. Algebra **177** (1995), 463–492.

Trends in Mathematics, © 1999 Birkhäuser Verlag Basel/Switzerland

Some matrix rings associated with ACD groups

James D. Reid *

Abstract

In this paper we describe a way of parameterizing the complete systems of orthogonal idempotents in certain matrix rings. As applications of the results we show how the classes of inequivalent decompositions of certain torsion free abelian groups into indecomposable summands can be completely classified.

1 Introduction

Let R be a principal ideal domain and let q be a product of distinct primes in R, say $q = \Pi_{\nu=1}^{m} q_\nu$. Denote the ring of $k \times k$ matrices over R by $M_k(R)$ and write $M_k^*(R)$ for the group of non-singular matrices. We consider the ring $\Lambda = \Lambda(R, q)$ of all $k \times k$ matrices $A = [a_{ij}]$ over R satisfying $a_{i1} \equiv 0$ (Mod q) for each $i > 1$. Note that the standard idempotents $E_1, E_2, \ldots, E_k$ lie in Λ, where by E_l we mean $E_l = [e_{ij}]$ with $e_{ij} = 0$ unless $i = j = l$, and $e_{ll} = 1$;

$$E_l = \begin{bmatrix} 00\cdots0\cdots0 \\ \vdots \\ 00\cdots1\cdots0 \\ \vdots \\ 00\cdots0\cdots0 \end{bmatrix} l. \quad (\text{column } l)$$

By a *complete orthogonal system* of matrices in $M_k(R)$ we mean a k-tuple $(F_1, F_2, \ldots, F_k)$ of non-zero idempotent matrices such that $F_iF_j = 0$ if $i \neq j$, and $\sum_i F_i = 1$ in $M_k(R)$. The Krull-Schmidt Theorem for the free module of rank k over R is reflected in the matrix ring $M_k(R)$ by the fact that each complete orthogonal system of matrices, in our sense, has the form $(BE_1B^{-1}, BE_2B^{-1}, \ldots, BE_kB^{-1})$ for a suitable $B \in M_k^*(R)$. We say that the system (BE_lB^{-1}) is *conjugate* to the standard system, or that the two systems are *conjugate relative to* $M_k^*(R)$. Our aim here is to parameterize

*The author thanks A. Paras, C.I. Vinsonhaler and W.J. Wickless for their interest in earlier drafts of this paper.

the conjugacy classes of complete orthogonal systems in Λ with respect to the action of the group of units of Λ. This is done in Theorem 4.3. In the last section we indicate briefly how our results may be applied to parameterize the classes of inequivalent decompositions of certain abelian groups into indecomposable summands.

These results admit generalization in several directions. For example, the q used to define Λ need not be square-free, and more general forms of matrices can be handled; but to save space we limit ourselves here to the indicated case. We hope to provide more details, and more applications in this direction, later. We remark that these results have been a long time in preparation. Earlier drafts, for example, have been referred to in the literature (e.g. [1], p. 38).

2 The set Ω

We are interested in complete orthogonal systems of idempotents in Λ and the question of their conjugacy relative to the group Λ^* of units in Λ. We want to classify the equivalence classes of complete orthogonal systems in Λ relative to Λ^*. To this end, we make the following

Definition 2.1 Denote by Ω the set of matrices $B \in M_k^*(R)$ such that $BE_lB^{-1} \in \Lambda$ for each $l = 1, \dots, k$.

Here is a useful observation. For $B = [b_{ij}] \in M_k^*(R)$, write $B^{-1} = [b_{ij}^*]$, say, and note that $BE_lB^{-1} = [b_{il}b_{lj}^*]$ for $l = 1, \dots, k$. Then $BE_lB^{-1} \in \Lambda$ if and only if $b_{il}b_{l1}^* \equiv 0$ (Mod q) for all $i > 1$ so that $B \in \Omega$ if and only if $b_{il}b_{l1}^* \equiv 0$ (Mod q) for all $i > 1$ and all l. The following proposition clears up a detail.

Proposition 2.2 *Let $A = [a_{ij}] \in \Lambda$ and suppose that $A \in M_k^*(R)$. Then $A^{-1} \in \Lambda$, i.e. $\Lambda \cap M_k^*(R) = \Lambda^*$, the group of units of Λ.*

Proof. Since $A \in \Lambda$, we have $a_{i1} \equiv 0$ (Mod q) for all $i > 1$. As above, write $A^{-1} = [a_{ij}^*] \in M_k^*(R)$. Then

$$1 = \textstyle\sum_j a_{mj}^* a_{jm} \text{ for each } m \text{ and } 0 = \sum_j a_{mj}^* a_{jn} \text{ for } m \neq n.$$

From the first of these, with $m = 1$, we get $1 = \sum_j a_{1j}^* a_{j1} \equiv a_{11}^* a_{11}$(Mod q) since $a_{j1} \equiv 0$ (Mod q) for $j > 1$. Similarly the second equation gives $0 \equiv a_{m1}^* a_{11}$ (Mod q) for $m > 1$.

Thus we observe on the one hand that a_{11} is a unit Mod q and on the other that $a_{m1}^* a_{11} \equiv 0$ (Mod q) for $m > 1$ so that $a_{m1}^* \equiv 0$ (Mod q) for $m > 1$. This shows that $A^{-1} \in \Lambda$ as required. □

Corollary 2.3 *If $B \in \Omega$ and $A \in \Lambda^*$, then $AB \in \Omega$.*

Proof. By hypothesis $BE_lB^{-1} \in \Lambda$ for each l. Since both A and A^{-1} lie in Λ, we also have, for each l, $ABE_lB^{-1}A^{-1} \in \Lambda$, whence $AB \in \Omega$. □

Note that, since each E_l lies in Λ, Proposition 2.2 implies that $AE_lA^{-1} \in \Lambda$ as well and this implies that $\Lambda^* \subseteq \Omega$. These results give

Proposition 2.4 *The set Ω is a union of right cosets of Λ^*: $\Omega = \cup \Lambda^* B$, the union taken over certain elements $B \in \Omega$.* □

3 Partitions

Recall again the notation $q = \prod_\nu q_\nu$ with the q_ν distinct primes in R. We have the following basic definition.

Definition 3.1 For $B \in M_k^*(R)$, put $X_B(l) = \{q_\nu | b_{l1}^* \not\equiv 0 \text{ (Mod } q_\nu)\}$.

Proposition 3.2 *Let $B \in M_k^*(R)$. Then $B \in \Omega$ if and only if the family $\{X_B(l)\}$ is a partition of the set $\{q_\nu\}$of prime divisors of q.*

Proof. First suppose that $B \in \Omega$. Should it happen that $q_\mu \notin X_B(l)$ for all l then $b_{1l}^* \equiv 0 \text{ (Mod } q_\mu)$ for each l. But then we would have $1 = \sum_l b_{1l} b_{l1}^* \equiv 0 \text{ (Mod } q_\mu)$, a contradiction. Hence the union of the sets $X_B(l)$ is the full set of prime factors of q. On the other hand, if $q_\mu \in X_B(l) \bigcap X_B(l')$ for some pair $l \neq l'$ then neither b_{l1}^* nor $b_{l'1}^*$ is congruent to zero modulo q_μ. The fact that $B \in \Omega$ then requires that both b_{il} and $b_{il'}$ be congruent to zero for all $i > 1$. Expanding $\det B$ by minors of its first row would now give $\det B \equiv 0 \text{ (Mod } q_\mu)$, contradicting the fact that $\det B$ is a unit in R. Thus the family $\{X_B(l)\}$ is indeed a partition.

Conversely if $\{X_B(l) | l = 1, \ldots, k\}$ is a partition then a given q_μ lies in one of the sets $X_B(l)$ for a unique value $l = l_0$. This means that $b_{l1}^* \equiv 0$ (Mod q_μ) for all $l \neq l_0$ but that $b_{l_01}^*$ is not congruent to zero modulo q_μ. Then for any $i > 1$, we have $0 = \sum_l b_{il} b_{l1}^* \equiv b_{il_0} b_{l_01}^* \text{ (Mod } q_\mu)$. It follows that each of the products $b_{il} b_{l1}^*$ is congruent to zero modulo q_μ so that $B \in \Omega$ as required. □

Proposition 3.3 *Let $\{X(l) | 1 \leq l \leq k\}$ be a partition of the set of primes q_ν. Then there exists $B \in \Omega$ such that $X(l) = X_B(l)$ for each $l = 1, \ldots, k$.*

Proof. Define

$$v_i = \prod_{\nu \in X(i)} q_\nu \text{ and } u_i = \prod_{\mu \notin X(i)} q_\mu.$$

Now define a map $\zeta : R^k \longrightarrow R$ by $\zeta(a_1, \dots, a_k) = \sum_i a_i u_i$. This is an epimorphism since the elements u_i are relatively prime. Then the kernel of ζ has rank $k-1$ with basis $\mathfrak{y}_i = (b_{i1}, \dots, b_{ik}), 2 \le i \le k$, say, and there are elements $b_{11}, b_{12}, \dots, b_{1k}$ of R such that $\sum_i b_{1i}u_i = 1$ since $\zeta(R^k) = R$. Taking $\mathfrak{y}_1 = (b_{11}, b_{12}, \dots, b_{1k})$ we have altogether a basis of R^k.

For $s > 1$ we have $\sum_i b_{si}u_i = 0$ so for each i, $-b_{si}u_i = \sum_{i \ne i'} b_{si'}u_{i'}$. Since $v_i | u_{i'}$ for $i \ne i'$, we get $b_{si}u_i \equiv 0(\mathrm{Mod}\, v_i)$ and since $\gcd(u_i, v_i) = 1$, we conclude that $b_{si} \equiv 0(\mathrm{Mod}\, v_i)$. This holds for all $s > 1$ and all i. Let $\zeta_1, \dots, \zeta_k$ be the dual basis so $\zeta_i(\mathfrak{y}_{i'}) = \delta_{ii'}$.Write

$$\zeta_s(a_1, \dots, a_k) = \sum_t a_t b^*_{ts} \text{ with } b^*_{ts} \in R.$$

Then $\zeta_1(\mathfrak{y}_1) = 1$ and $\zeta_1(\mathfrak{y}_i) = 0, i > 1$ so $\zeta_1 = \zeta$. Since therefore $\zeta_1(a_1, \dots, a_k) = \sum_t b^*_{t1}a_t = \sum_t u_t a_t$, we have $b^*_{t1} = u_t$ for all t. Hence the matrix $B^* = [b^*_{ij}]$ has first column consisting of the u_i. Evidently now $b_{it}b^*_{t1} = b_{it}u_t \equiv 0(\mathrm{Mod}\, v_t u_t)$ for all t and $i > 1$, and since $v_t u_t = \prod_\nu q_\nu = q$, we have indeed $B \in \Omega$ as required. □

4 The classification

We do not wish to distinguish between complete orthogonal systems that differ only by a permutation of the members so that, in our basic relation, we must take permutations into account. Therefore for a permutation π of the integers $\{1, 2, \dots, k\}$ put $P_\pi = [\delta_{\pi^{-1}(i)j}] \in M^*_k(R)$ (Kronecker δ). Then the map $\pi \to P_\pi$ is an embedding of the symmetric group of degree k into $M^*_k(R)$ and for each l we have $P_\pi E_l P_\pi^{-1} = E_{\pi(l)}$. We denote by S_k the subgroup of $M^*_k(R)$ consisting of the matrices P_π. Clearly S_k acts on the right of Ω in the sense that $\Omega P \subseteq \Omega$ for each $P \in S_k$ and we can define an action of the direct product $\Delta = \Lambda^* \times S_k$ on Ω by $B \to ABP^{-1}$ for $B \in \Omega$ and $(A, P) \in \Delta$. This is a group action, as can be easily checked.

Definition 4.1 Define the relation $\sim_\Lambda$on the set Ω as follows: $B \sim_\Lambda \overline{B}$ if there exists $A \in \Lambda^*$ and a permutation π of $\{1, 2, \dots, k\}$ such that $\overline{B}E_l\overline{B}^{-1} = A(BE_{\pi(l)}B^{-1})A^{-1}$ for each $l = 1, \dots, k$.

Otherwise put, $B \sim_\Lambda \overline{B}$ if and only if $\overline{B}E_l\overline{B}^{-1} = A(BPE_lP^{-1}B^{-1})A^{-1}$ for the permutation matrix $P = P_\pi$. It will be important for us that the equivalence classes here are just the classes defined in Ω by the action of the group $\Delta = \Lambda^* \times S$, i.e.

Theorem 4.2 *Let $B, \overline{B} \in \Omega$. Then $B \sim_\Lambda \overline{B}$ if and only if $\overline{B} = ABP$ for some $A \in \Lambda^*$ and $P \in S_k$.*

Proof. Suppose first that $B \sim_\Lambda \overline{B}$ so there is $A \in \Lambda^*$ and $P \in S_k$ such that $\overline{B}E_l\overline{B}^{-1} = A(BPE_lP^{-1}B^{-1})A^{-1}$ (as above) for all l. Put $C = BP$ so we have $\overline{B}E_l\overline{B}^{-1} = A(CE_lC^{-1})A^{-1}$. This is equivalent to $\overline{B}^{-1}ACE_l = E_l\overline{B}^{-1}AC$ for all l which implies in turn that $D = \overline{B}^{-1}AC$ is a diagonal matrix, say $D = [d_1, \ldots , d_k]$. Then we have $\overline{B}^{-1}A = DC^{-1}$, or

$$\textstyle\sum_s \overline{b}^*_{is}a_{sj} = d_ic^*_{ij} \text{ for all } i, j.$$

Taking $j = 1$, and using the fact that $a_{s1} \equiv 0$ (Mod q) for $s > 1$, we get $\overline{b}^*_{i1}a_{11} \equiv d_ic^*_{i1}$ (Mod q) for all i.

Now d_i is a unit of R (since D is nonsingular) and $A \in \Lambda^*$ so a_{11} is congruent to a unit Mod q (hence modulo each prime factor q_ν of q) and $\overline{b}^*_{i1}a_{11} \equiv d_ic^*_{i1}$ (Mod q_ν), so we may conclude that $\overline{b}^*_{i1} \equiv 0$ (Mod q_ν) if and only if $c^*_{i1} \equiv 0$ (Mod q_ν). This is true for all i and all ν. Now consider the $(i, 1)$ entry in the matrix $\overline{B}C^{-1}$. This is $\sum_s \overline{b}_{is}c^*_{s1}$. If $c^*_{s1} \not\equiv 0$ (Mod q_ν) for some ν and s then $\overline{b}^*_{s1} \not\equiv 0$ (Mod q_ν) also. But $\overline{B} \in \Omega$ so $\overline{b}_{is}\overline{b}^*_{s1} \equiv 0$ (Mod q), hence $\overline{b}_{is}\overline{b}^*_{s1} \equiv 0$ (Mod q_ν) for each ν, for all $i > 1$ and we conclude that $\overline{b}_{is} \equiv 0$ (Mod q_ν). Thus in any case each term in the sum $\sum_s \overline{b}_{is}c^*_{s1}$ is congruent to zero modulo q_ν so the sum is as well. This holds for each ν, hence for the product, q, of the q_ν. This shows that $\overline{B}C^{-1} \in \Lambda$, i.e. that $\overline{B}C^{-1} = \overline{A}$, say, $\overline{A} \in \Lambda$, so that $\overline{B} = \overline{A}C = \overline{A}BP$ as required.

The converse is trivial. □

We incorporate these results into our main result as follows:

Theorem 4.3 *For $B, \overline{B} \in \Omega$, the following are equivalent:*

1. *$B \sim_\Lambda \overline{B}$;*
2. *$\overline{B} = ABP$ for some $A \in \Lambda^*$ and $P = P_\pi \in S_k$;*
3. *$X_B(\pi(i)) = X_{\overline{B}}(i)$ for each $i = 1, \ldots , k$.*

Proof. The equivalence of 1.) and 2.) has just been established. We will show that 2.) and 3.) are equivalent. Thus assume that $\overline{B} = ABP$ as in 2.) so that $\overline{B}^{-1} = P^{-1}B^{-1}A^{-1}$. Since $P^{-1}B^{-1}A^{-1} = [\sum_l \sum_s \delta_{\pi(i)s}b^*_{sl}a^*_{lj}]$ and $\delta_{\pi(i)s} = 0$ unless $\pi(i) = s$, we must have $\overline{b}^*_{ij} = \sum_l b^*_{\pi(i)l}a^*_{lj}$ for each i and j. Thus $b^*_{i1} = \sum_l b^*_{\pi(i)l}a^*_{l1} \equiv b^*_{\pi(i)1}a^*_{11}$ (Mod q) since $a^*_{l1} \equiv 0$ (Mod q) for $l > 1$ (i.e. $A^{-1} \in \Lambda$, cf. Proposition 2.2). Finally, because $a^*_{l1} \equiv 0$ (Mod

q) for $l > 1$ and A is non-singular, it is clear that a_{11}^* and q are relatively prime so we now conclude that b_{i1}^* and $\bar{b}_{i1}^*$ are either both congruent to zero modulo q_ν are neither is congruent to zero modulo q_ν for each value of ν. Thus $X_{\overline{B}}(i) = X_B(\pi(i))$ for every i as required.

An argument along similar lines shows that conversely if the partitions corresponding to matrices B and $\overline{B}$ in Ω are the same, then B and $\overline{B}$ are related as in 1.) and 2.). We leave the details to the reader. □

5 Applications

By a famous theorem of A.L.S. Corner (Cf. [3]), as refined by Butler and Zassenhaus, the ring Λ that we have been discussing, say with the base ring R taken to be a proper subring of the field Q of rational numbers, can be realized as the endomorphism ring of a torsion free abelian group G of rank k^2. It is not difficult to show that in such a case every decomposition of G into a direct sum of indecomposable subgroups involves just k summands each of which has rank k. Our results then show that there is such a decomposition of G for each partition of the set $\{q_\nu\}$ of prime divisors of q and that two decompositions are equivalent in the usual sense if and only if the corresponding partitions are the same. Thus the equivalence classes of decompositions of G into indecomposable summands are parameterized by the partitions.

However interesting this might be, our original motivation for this study was to understand better another famous result of Corner (Cf. [2]), a spectacular example of pathology in the theory of decompositions of finite rank torsion free groups into indecomposable summands. In what follows we provide a sketch indicating how our results apply to Corner's example. Many of the details are omitted in order to save space, but we hope to discuss this example, among others, elsewhere. Here is Corner's theorem:

Theorem 5.1 *(A.L.S. Corner) Given positive integers k and n, with $k \leq n$, there is a torsion free group G of rank n such that, for any partition $n = r_1 + r_2 + \cdots + r_k$ of n into positive integers r_i, there is a decomposition $G = G_1 \oplus G_2 \oplus \cdots \oplus G_k$ of G into k indecomposable summands G_i with rank(G_i)=r_i for $i = 1, 2, \ldots, k$.*

Corner's construction is the following: Choose two disjoint sets of distinct primes $\{p_0, p_1, p_2, \ldots, p_{n-k}\}$ and $\{q_1, q_2, \ldots, q_{n-k}\}$ and for each $\nu = 0, 1, \ldots, n-k$ let $R_\nu = Z[1/p_\nu]$ be the ring of rational numbers whose denominators are powers of the prime p_ν. Let $\{u_1, u_2, \ldots u_k, x_1, x_2, \ldots, x_{n-k}\}$ be a basis of an n-dimensional rational vector space. Let q be the product of the q_ν. Finally define the group G by

$$G = \oplus_{i=1}^{k} R_0 u_i \oplus \oplus_{\nu=1}^{n-k} R_\nu x_\nu + \sum_{\nu=1}^{n-k} \left\langle \frac{u_1 + x_\nu}{q_\nu} \right\rangle .$$

After a slight change of basis the endomorphism ring End(G) of this group G can be represented by the ring of all $n \times n$ matrices

$$\begin{bmatrix} A & & & & \\ & c_1 & & & \\ & & c_2 & & \\ & & & \ddots & \\ & & & & c_{n-k} \end{bmatrix}$$

where $A = [a_{ij}] \in M_k(R_0)$, $c_\nu \in R_\nu$ for $\nu = 1, \ldots, n-k$, $a_{i1} \equiv 0$ (Mod q) for all $i > 1$ and finally $a_{11} \equiv c_\nu$ (Mod q_ν) for $\nu = 1, \ldots, n-k$. Thus the matrix A here is a member of the ring $\Lambda = \Lambda(R_0, q)$ based on R_0, and, we note for the record, the 1×1 matrices c_ν are members of the analogous rings of 1×1 matrices over the rings R_ν. In any event, a matrix of this form is idempotent if and only if the upper left A is idempotent and the numbers c_ν are either 0 or 1. Furthermore, if A is idempotent then, as is easily checked, a_{11} is idempotent modulo q and hence is congruent to either 0 or 1 modulo each q_ν. Therefore the entries c_ν in an idempotent matrix are completely determined by the corresponding matrix A.

Now let $n = r_1 + r_2 + \cdots + r_k$ be a partition of n with each $r_i > 0$. Then $n - k = (r_1 - 1) + (r_2 - 1) + \cdots + (r_k - 1)$ is a partition of $n - k$ into non-negative integers, and we may interpret this as a partition of the prime factors of $q = q_1 q_2 \cdots q_{n-k}$ say by listing the q_ν in ascending order of magnitude and taking the first $r_1 - 1$ factors as $X(1)$, the second $r_2 - 1$ factors as $X(2)$, etc. Notice that some of these $X(\nu)$ may be empty (i.e. $r_\nu - 1$ may be 0), but that is allowed in our development. At any rate, this partition of the factors of q gives a system of idempotents in Λ of the form $\{BE_l B^{-1}\}$ where $B \in \Omega$ is the matrix corresponding to the partition $\{X(\nu)\}$. From this, as indicated, we obtain idempotents in End(G), hence a decomposition of G. It is easily seen that the summand corresponding to $X(\nu)$ is indecomposable and of rank r_ν.

Different partitions give rise to inequivalent decompositions. However, there may be several inequivalent decompositions corresponding to the same partition of the set of prime factors of q. The reason for this is that not every unit in Λ extends to an automorphism of G.

To state the result we need a little bit of notation. Recall that $R_\nu = Z[\frac{1}{p_\nu}]$ so that its group of units is $R_\nu^* = \{\pm p_\nu^t | t \in Z\}$. We also have, for each prime q_ν, $R_\nu / q_\nu R_\nu \cong Z/q_\nu Z$. Identifying these rings, we have a map $\phi : \Lambda \to Z/qZ = \prod_\nu Z/q_\nu Z$ given by $A \to (a_{11} + q_1 R_1, a_{11} +$

$q_2R_2, \ldots, a_{11} + q_{n-k}R_{n-k})$. This induces an epimorphism of groups $\Lambda^* \to (Z/qZ)^* = \prod_\nu (Z/q_\nu Z)^*$. Inside $(Z/q_\nu Z)^*$ lies the subgroup $\overline{\Gamma}_\nu$ generated by the cosets $\{\pm p_\nu + q_\nu Z\}$. Then $\Gamma =$Aut(G) is the inverse image of $\prod_\nu \overline{\Gamma}_\nu$ under this map ϕ and we have:

Theorem 5.2 *For each partition X of* $\{1, 2, \ldots, n-k\}$ *there are exactly* $[\Lambda^* : \Gamma]$ *inequivalent decompositions of the group G.*

We close with two examples indicating the scope of this theorem. First choose the primes p_ν and q_ν so that p_ν is a primitive element modulo q_ν for each $\nu = 1, 2, \ldots, n-k$. Then of course $\overline{\Gamma}_\nu = (Z/q_\nu Z)^*$ for each ν so $\Gamma = \Lambda^*$ and we have exactly one decomposition of G for each partition.

At the other extreme, if $p_\nu \equiv 1$ (Mod q_ν) for each ν then $\overline{\Gamma}_\nu = \{\pm 1\}$ and the index $[\Lambda^* : \Gamma] = \prod_\nu (\varphi(q_\nu)/2) = \varphi(q)/2^{n-k}$. (Here, φ is the Euler function.) This last result is related to a result of Burkhardt/Mutzbauer, 1980.

It is clear that these choices of the primes p_ν and q_ν can be made. Finally we remark that the question of determining the number of non-isomorphic decompositions of almost completely decomposable groups has been raised explicitly by Kraft/Mutzbauer (see [4]).

References

[1] D. Arnold and C. Vinsonhaler, *Finite rank Butler groups*, Abelian Groups, Lecture Notes in Pure and Applied Mathematics, **146** (Marcel Dekker 1991), 17–41.

[2] A. L. S. Corner, *A note on rank and direct decompositions of torsion-free abelian groups*, Proc. Cambridge Philos. Soc. **57** (1961), 230–233.

[3] A. L. S. Corner, *Every countable reduced ring is an endomorphism ring*, Proc. London Math. Soc. **13** (1963), 687–710.

[4] K.-J. Kraft and O. Mutzbauer, *Classification of almost completely decomposable groups*, Abelian Groups and Modules, Proceedings Udine, Springer-Verlag (1984), 151–161.

Trends in Mathematics, © 1999 Birkhäuser Verlag Basel/Switzerland

Stacked bases for a pair of homogeneous completely decomposable groups with bounded quotient

Mohamed A. Ould-Beddi * and Lutz Strüngmann †

Abstract

Let A be a homogeneous completely decomposable torsion free group of infinite rank κ and let X be a torsion free abelian group containing A such that the quotient X/A is bounded. We show that there exist stacked bases for X and A, i.e. there exist $b_i \in X$ $(i \in \kappa)$ and $d_i \in \mathbb{Z}$ $(i \in \kappa)$ such that $X = \bigoplus_{i\in\kappa} \langle b_i \rangle_*^X$ and $A = \bigoplus_{i\in\kappa} d_i \langle b_i \rangle_*^A$. This proves a stacked bases theorem for pairs of homogeneous completely decomposable torsion free abelian groups of infinite rank with bounded quotient.

1 Introduction

It is the purpose of this paper to prove a stacked bases theorem for a certain class of torsion free abelian groups. Recall that free modules $A \subseteq X$ over a principal ideal domain (PID) R have *stacked bases* if there exists a basis for X such that multiples of these basis elements constitute a basis of A. It is well known that free modules $A \subseteq X$ over a PID have stacked bases iff the quotient X/A is a direct sum of cyclics as, for example, proven by J.M. Cohen, H. Gluck [3] and by P. Hill and C. Megibben [6]. Here we study the case when X is a bounded extension of a homogeneous completely decomposable group (h.c.d.) A of infinite rank. Note that in this case X is also homogeneous of the same type as A, hence homogeneous completely decomposable of the same type and rank as A by the Baer-Kolettis Theorem (see [5, Vol. II, Theorem 86.6]). Moreover the quotient X/A is a direct sum of cyclic groups by a theorem of Baer (see [5, Vol. I, Theorem 17.2]). In

*Supported by a project No. G-0294-081.06/93 of the *German-Israeli Foundation for Scientific Research & Development* and the Deutsche Akademische Austauschdienst (DAAD).

†Supported by the Graduiertenkolleg *Theoretische und Experimentelle Methoden der Reinen Mathematik* of Essen University.

a recent paper [2], K. Benabdallah and M. A. Ould-Beddi proved a strong stacked bases theorem for the class of pairs of homogeneous completely decomposable groups $A \subseteq X$ of finite rank such that the quotient is finite. Using an approach closely related to linear algebra (see [1]) they proved that such a pair of h.c.d. groups $A \subseteq X$ have decompositions $X = \bigoplus_{i=1}^{n} \langle x_i \rangle_*^X$ and $A = \bigoplus_{i=1}^{n} d_i \langle x_i \rangle_*^A$ such that d_i divides d_{i+1} for all $i = 1, ..., n-1$. In this paper we show that if A and X are h.c.d. of infinite rank κ and the quotient is not finite but still bounded then X and A admit direct decompositions $X = \bigoplus_{i \in \kappa} \langle x_i \rangle_*^X$ and $A = \bigoplus_{i \in \kappa} d_i \langle x_i \rangle_*^A$. The divisibility condition on the d_i's can not be obtained in general.

We would like to thank Rüdiger Göbel for many helpful discussions and for first suggesting that the result of K. Benadballah and M. A. Ould-Beddi [2] can be generalized to the case of bounded quotient.

2 Simultaneous decompositions

First of all we consider the case when the quotient is bounded and countable. Let A be a τ-homogeneous completely decomposable group of infinite rank κ and n be a positive integer. If X is an n-bounded extension of A, i.e. $nX \subseteq A$, then as mentioned in the introduction X is homogeneous completely decomposable of the same type τ and the same rank as A. Furthermore the quotient X/A is a direct sum of cyclic groups, say $X/A = \bigoplus_{i \in \omega} \langle x_i + A \rangle$ where $x_i \in X$ $(i < \omega)$ and $1 < \text{ord}(x_i + A) \leq n$. The following Lemma shows that we can choose the x_i such that $d_1, ..., d_k$ form a reversed divisor chain for some $k \in \mathbb{N}$ and all d_i have the same prime divisors for $i > k$. Here we use an idea from [4, Lemma 4.1].

Lemma 2.1 *Let X be an n-bounded extension of a homogeneous completely decomposable group A of infinite rank κ such that the quotient X/A is countable. Let S be the set of all primes p for which X/A has infinite p-rank and put $m = \prod_{p \in S} p$. Then there is a decomposition $X/A = \bigoplus_{i \in \omega} \langle x_i + A \rangle$ with $d_i = ord(x_i + A)$ and an integer $k \in \mathbb{N}$ such that*

(i) $d_{i+1} \mid d_i$ for all $i = 1, ..., k-1$;

(ii) $m \mid d_i$ for all $i \in \omega$;

(iii) $p \mid d_i$ iff $p \mid m$ for all $i > k$.

We call this decomposition the canonical decomposition of X/A.

Proof. Let $X/A = \bigoplus_{p|n} (X/A)_p$ be a decomposition of the quotient X/A into its p-components. Since this is a finite direct sum we can split off those p-components for which the p-rank of X/A is finite. Hence X/A decomposes into a finite direct sum $\bigoplus_{i=1}^{k} \langle \bar{x}_i + A\rangle(\bar{x}_i \in X)$ and $\bigoplus_{p\in S} (X/A)_p$ where S is the set of all primes $p \mid n$ for which the p-rank of X/A is infinite. We can choose $\bar{x}_i$ $(i \leq k)$ such that $d_{i+1} \mid d_i$ for $(i < k)$ where $d_i = \operatorname{ord}(\bar{x}_i + A)$. Now let $C = \bigoplus_{p\in S} (X/A)_p$. Then

$$C \cong \bigoplus_{p\in S} \left(\bigoplus_{i<\omega} \mathbb{Z}\left(p^{e_{pi}}\right) \right) \cong \bigoplus_{i<\omega} \left(\bigoplus_{p\in S} \mathbb{Z}\left(p^{e_{pi}}\right) \right) \cong \bigoplus_{i<\omega} \mathbb{Z}\left(m_i\right),$$

where $e_{pi} \geq 1$ and $m_i = \prod_{p\in S} p^{e_{pi}}$. Therefore we can write

$$C = \bigoplus_{i\in\omega} \langle y_i + A\rangle$$

where $\operatorname{ord}(y_i + A) = m_i$ for some $y_i \in X$ $(i \in \omega)$. Clearly $p \mid m_i$ iff $p \mid m$ for all $i < \omega$. Again we can rearrange generators such that $m_{i+1} \mid m_i$ for all $i \leq k-1$. Chosing $x_i + A := (\bar{x}_i + y_i) + A$ for $i = 1, \ldots, k$ and $x_i := y_i$ for $i > k$ as new generators of X/A, we obtain i) to iii). □

Every homogeneous completely decomposable group A of type τ which is of infinite rank κ can be written in the form $A = \bigoplus_{i\in\kappa} a_i R$, where $R \supseteq \mathbb{Z}$ is a subgroup of the rationals of type τ. It is known that every pure rank one subgroup of a homogeneous completely decomposable group X is a direct summand, see [5, Vol. II, Theorem 86.8]. Therefore, if $X/A = \bigoplus_{i\in\omega} \langle x_i + A\rangle$ one might suggest that each $x_i R$ is pure in X, which means $\langle x_i\rangle_*^X = x_i R$, independent of i. But this is not true in general as easy examples show. However, we are able to rearrange generators such that $x_1 R$ is pure in X. This will serve as an inductive step for splitting off any finitely generated subgroup of X and decomposing A correspondingly.

Whenever X is an n-bounded extension of a homogeneous completely decomposable group $A = \bigoplus_{i\in\kappa} a_i R$, where $R \supseteq \mathbb{Z}$ is a subgroup of the rationals of type τ, we know that X is also homogeneous completely decomposable of the same type. Therefore X can be written as $X = \bigoplus_{i\in\kappa} f_i R$ with the same subgroup R coming from A. If $n = exp(X/A)$ is the exponent of X/A and A is τ-homogeneous, τ cannot be p-divisible for $p \mid n$. We choose a rational group $R \supseteq \mathbb{Z}$ such that $\frac{1}{p} \notin R$ for all $p \mid n$. By hypothesis there exists an isomorphism $\alpha : R^{\kappa} \longrightarrow A$, hence we find $a_i \in A$ such that $A = \bigoplus_{i\in\kappa} a_i R$ and by the above arguments there exist also $f_i \in X$ with $X = \bigoplus_{i\in\kappa} f_i R$.

Now we show that the generator $x_1 + A$ can be replaced by a new generator $z_1 + A$ such that $z_1 R$ is a pure subgroup of X.

Lemma 2.2 *Let X be an n-bounded extension of a homogeneous completely decomposable group $A = \bigoplus_{i\in\kappa} a_i R$, where $R \supseteq \mathbb{Z}$ is a fixed subgroup of the rationals such that $1/p \notin R$ for all $p \mid n$ and let $X/A = \bigoplus_{i\in\omega} \langle x_i + A\rangle$ be the canonical decomposition of X/A. Then there exists $z_1 \in X$ such that $z_1 R \subseteq X$ is a pure subgroup of X and $z_1 - x_1 \in A$, hence $X/A = \langle z_1 + A\rangle \oplus \bigoplus_{i\geq 2} \langle x_i + A\rangle$.*

Proof. Let $X/A = \bigoplus_{i\in\omega} \langle x_i + A\rangle$ be the canonical decomposition of X/A, where $\mathrm{ord}(x_i + A) = d_i$ and $k \in \mathbb{N}$, S, $m = \prod_{p\in S} p$ are as stated in Lemma 2.1. Then $d_{i+1} \mid d_i$ for $i = 1, \dots, k-1$ and m divides every d_i. Moreover $p \mid d_i$ iff $p \mid m$ for all $i > k$. Since $\langle x_1\rangle_*^X \subseteq X$ is a pure subgroup of X we can find $z_1 \in X$ such that $z_1 R = \langle x_1\rangle_*^X$. We want to show that $\langle x_1 + A\rangle = \langle z_1 + A\rangle$. Write

$$z_1 + A = \sum_{i\in\omega} \alpha_i x_i + A \tag{1}$$

for some integers $\alpha_i \in \mathbb{Z}$ $(i \in \omega)$, where almost all α_i are zero and those α_i which are nonzero are not divisible by d_i. Then there exists a reduced fraction $t_1/t_2 \in R$ such that $t_2 x_1 + A = t_1 z_1 + A = t_1 \left(\sum_{i\in\omega} \alpha_i x_i\right) + A$ which implies

$$d_i \mid t_1\alpha_i \text{ for } i \geq 2 \text{ and } d_1 \mid (t_2 - t_1\alpha_1). \tag{$*$}$$

We claim that $\alpha_i = 0$ for every $i \geq 2$. Suppose, by way of contradiction, that $\alpha_i \neq 0$ for some $i \geq 2$. If $i > k$ then by $(*)$ there exists a prime p dividing t_1 and m, since in the case $i \geq 2$ we have $p \mid m$ iff $p \mid d_i$. But $m \mid d_1$ and thus p divides $(t_2 - t_1\alpha_1)$ which contradicts $\gcd(t_1, t_2) = 1$. If $1 < i \leq k$ then there exists a prime p which divides t_1 and d_i, but $d_i \mid d_1$ which leads again to a contradiction. Therefore $\alpha_i = 0$ for $i \geq 2$, hence $\langle z_1 + A\rangle \subseteq \langle x_1 + A\rangle$. To show the converse inclusion $\langle x_1 + A\rangle \subseteq \langle z_1 + A\rangle$ we start with the equation $t_2 x_1 = t_1 z_1$. Assume that $\gcd(t_2, d_1) \neq 1$ and choose a prime p dividing t_2 and d_1. Then p divides n and $1/p \in R$. This contradicts our general assumption that $1/p \notin R$ for all $p \mid n$. Hence there exist integers l and k

such that $x_1 = lt_1 z_1 + kd_1 x_1$ and therefore $\langle x_1 + A\rangle \subseteq \langle z_1 + A\rangle$. Replacing x_1 by z_1 in the decomposition of X/A completes the proof. □

Next we show that a given decomposition of X/A where $x_1 R$ is pure in X can be lifted to a simultaneous decomposition of X and A.

Splitting Lemma 2.3 *Let X be an n-bounded extension of a homogeneous completely decomposable group $A = \bigoplus_{i\in\kappa} a_i R$, where $R \supseteq \mathbb{Z}$ is a fixed subgroup of the rationals such that $1/p \notin R$ for all $p \mid n$. Furthermore let $X/A = \bigoplus_{i\in\omega} \langle x_i + A\rangle$ be a decomposition into cyclic groups such that $d_i = \mathrm{ord}(x_i + A)$ and $x_1 R$ is a pure subgroup of X. Then there exists a decomposition $X = x_1 R \oplus X''$ such that*

(i) $A = d_1 x_1 R \oplus A''$ where $A'' = X'' \cap A$;

(ii) $X''/A'' = \bigoplus_{i\geq 2} \langle x_i'' + A''\rangle$, $d_i = \mathrm{ord}(x_i'' + A'')$;

(iii) $x_i - x_i'' \in A$ for all $i \geq 2$;

(iv) $X/A = \langle x_1 + A\rangle \oplus \bigoplus_{i\geq 2} \langle x_i'' + A\rangle$.

Proof. By assumption $x_1 R$ is a pure subgroup of finite rank of X, hence a direct summand (see [5, Vol. II, Lemma 86.8]). Let $X = x_1 R \oplus X'$. Then X' is again a homogeneous completely decomposable group of the same type as A by the Baer-Kolettis Theorem. We choose a basis $B := \{x_1', x_2', \dots\}$ of X', such that $X' = \bigoplus_{i\in\kappa} x_i' R$, and define

$$\phi : X' \longrightarrow x_1 R \text{ by } x_i'\phi = -t_{i,1} x_1$$

where $x_i' + A = \sum_{j\in\omega} t_{i,j} x_j + A$ is the representation of $x_i' + A$ inside the quotient X/A, so that $t_{i,j} \in \mathbb{Z}$ ($j \in \omega$) and almost all $t_{i,j}$ equal zero. Then ϕ is a well-defined homomorphism and

$$X'' = X'(1+\phi) = \bigoplus_{i\in\kappa} (x_i' - t_{i,1} x_1)\, R$$

is a new complement of $x_1 R$ in X. We leave the verification to the reader. If $\pi : X \longrightarrow X/A$ is the canonical epimorphism, then we show that $X''\pi$ is contained in $\bigoplus_{i\geq 2} \langle x_i + A\rangle$. Let $f = \sum_{i\in\kappa} r_i x_i' - \sum_{i\in\kappa} r_i t_{i,1} x_1 \in X''$ where $r_i \in R$ ($i \in \kappa$) and almost all $r_i = 0$. Then we write each r_i as a fraction $r_i = s_i/w_i$

with $s_i, w_i \in \mathbb{Z}$ relatively prime and multiply f by $v := lcm(w_1, w_2, \cdots)$ which is defined since the sums over κ are actually finite. It follows that

$$(vf)\pi = \sum_{i\in\kappa} s_i v_i \left(\left(\sum_{j\in\kappa} t_{i,j} x_j \right) - t_{i,1} x_1 \right) + A = \sum_{i\in\kappa} s_i v_i \left(\sum_{j\geq 2} t_{i,j} x_j \right) + A,$$

where $v_i = v/w_i \in \mathbb{Z}$. But this is an element of $\bigoplus_{i\geq 2} \langle x_i + A \rangle$ which is pure in X/A, hence $f\pi \in \bigoplus_{i\geq 2} \langle x_i + A \rangle$ because $\gcd(v, n) = 1$. Therefore $X''\pi \subseteq \bigoplus_{i\geq 2} \langle x_i + A \rangle$. Next we show that $A = d_1 x_1 R \oplus (A \cap X'')$. Obviously $(A \cap X'') \subseteq A$ and $d_1 x_1 R \cap (A \cap X'') \subseteq x_1 R \cap X'' = 0$. It remains to show that $d_1 x_1 R \subseteq A$ and $A \subseteq d_1 x_1 R \oplus (A \cap X'')$. Let $x = d_1 x_1 (r_1/r_2) \in d_1 x_1 R$. Then $r_2 x \in A$, hence $x \in A$ since $\gcd(n, r_2) = 1$. We conclude $d_1 x_1 R \subseteq A$. To show the inclusion $A \subseteq d_1 x_1 R \oplus (A \cap X'')$, let $a \in A \subseteq X$ and write $a = x_1(r_1/r_2) + e$ for some $r_1/r_2 \in R$ with $\gcd(r_1, r_2) = 1$ and $e \in X''$. Therefore $r_2 a = r_1 x_1 + r_2 e \in A$ and passing to the quotient shows that $d_1 \mid r_1$ since $X''\pi$ is contained in $\bigoplus_{i\geq 2} \langle x_i + A \rangle$. It follows that $a \in d_1 x_1 R \oplus (A \cap X'')$. Letting $A'' = A \cap X''$ we derive i).
Now let $i \geq 2$ and write $x_i = \alpha_i x_1 + x_i''$ where $x_i'' \in X''$ and $\alpha_i \in R$. Passing to the quotient we conclude that $\alpha_i = d_i \beta_i$ for some $\beta_i \in R$, which shows $x_i'' - x_i \in A$ and iii) holds. Now ii) and iv) follow easily by the choice of A'' and x_i'' $(i \geq 2)$ and the proof is complete. □

An immediate consequence of the Splitting Lemma 2.3 is the following

Corollary 2.4 *Let X be an n-bounded extension of a homogeneous completely decomposable group A of infinite rank κ with $n = exp(X/A)$ such that the quotient X/A is countable. Let $C \subseteq X/A$ be finitely generated. Then there exists a decomposition $X = X_1 \oplus X_2$ such that $X/A = (X_1 + A)/A \oplus (X_2 + A)/A$, $C \subseteq (X_1 + A)/A$ and $(X_1 + A)/A$ is finitely generated.*

Proof. Let $X/A = \bigoplus_{i\in\omega} \langle x_i + A \rangle$ be the canonical decomposition of X/A given by Lemma 2.1. Since n is the exponent of X/A we can find a subgroup $R \supseteq \mathbb{Z}$ of the rationals such that $A = \bigoplus_{i\in\kappa} a_i R$ where $1/p \notin R$ for all $p \mid n$. By Lemma 2.2 we may assume that $x_1 R$ is a pure subgroup of X. Now, if $C \subseteq X/A$ is finitely generated, then there exists an integer $k \in \mathbb{N}$ such that $C \subseteq \bigoplus_{i=1}^{k} \langle x_i + A \rangle$. By the Splitting Lemma 2.3 we find a decomposition $X = x_1 R \oplus X'$ such that $X'/(A \cap X') = \bigoplus_{i\geq 2} \langle x_i' + (A \cap X') \rangle$ and $x_i - x_i' \in A$ for all $i \geq 2$, i.e. $\langle x_i + A \rangle = \langle x_i' + A \rangle$. By the same arguments as above we find a second decomposition $X' = x'_2R \oplus X''$ such that $X''/(A \cap X'') = \bigoplus_{i\geq 3} \langle x''_i + (A \cap X'')\rangle$ and $x'_i - x''_i \in A$ for all $i \geq 3$, i.e. $\langle x'_i + A\rangle = \langle x''_i + A\rangle$. Hence

$$X/A = \langle x_1 + A\rangle \oplus \langle x'_2 + A\rangle \oplus \bigoplus_{i\geq 3} \langle x''_i + A\rangle$$

and $X = x_1R \oplus x'_2R \oplus X''$, where $\langle x_1 + A\rangle \oplus \langle x'_2 + A\rangle = \langle x_1 + A\rangle \oplus \langle x_2 + A\rangle$. Inductively we get the desired decomposition. □

Since we are looking for a stacked bases theorem for X and A we need the following general and easy fact which says that a simultaneous decomposition of the group X and the quotient X/A gives a decomposition of A as well. The verification is left to the reader.

Proposition 2.5 *Let $A \subseteq X$ be torsion free abelian groups and let $\pi : X \longrightarrow X/A$ be the canonical epimorphism. If $X = X_1 \oplus X_2$ such that $X/A = X_1\pi \oplus X_2\pi$ then $A = (X_1 \cap A) \oplus (X_2 \cap A)$.*

Using Lemma 2.4 we have

Lemma 2.6 *Let X be an n-bounded extension of a homogeneous completely decomposable group A of infinite rank κ with $n = exp(X/A)$ such that the quotient X/A is countable. Let $X' \subseteq_* X$ be a pure subgroup of X of finite rank. Then there exists a decomposition $X = X_1 \oplus X_2$ such that $X' \subseteq X_1$ is of finite rank and $X/A = (X_1 + A)/A \oplus (X_2 + A)/A$.*

Proof. Let $X' \subseteq_* X$ be of finite rank, then X' is a direct summand of X, say $X = X' \oplus Y$. If $\pi : X \longrightarrow X/A$ is the canonical epimorphism, then $X\pi = X'\pi + Y\pi$. Note that Y and $Y \cap A \subseteq_* A$ are again homogeneous completely decomposable groups and the quotient $Y/(Y \cap A)$ is still a direct sum of cyclics since $Y/(Y \cap A) \cong Y\pi \subseteq X/A$. Let $X' = \bigoplus_{i=1}^{l} x'_iR$ be a decomposition of X' where $R \supseteq \mathbb{Z}$ is a subgroup of $\mathbb{Q}$ such that $1/p \notin R$ for all $p|n$. It can be easily shown that $X'\pi$ is generated by $x'_1\pi, \cdots, x'_l\pi$, hence $X'\pi \cap Y\pi \subseteq (Y + A)/A \cong Y/(Y \cap A)$ is finitely generated. By Lemma 2.4 there exists a decomposition $Y = Y_1 \oplus Y_2$ such that $X'\pi \cap Y\pi \subseteq (Y_1 + A)/A$ is finitely generated and $Y\pi = Y_1\pi \oplus Y_2\pi$. Therefore

$$X = X' \oplus Y = X' \oplus (Y_1 \oplus Y_2) = (X' \oplus Y_1) \oplus Y_2.$$

Let $X_1 = X' \oplus Y_1$ and $X_2 = Y_2$. Then X_1 has finite rank and

$$X/A = X_1\pi + X_2\pi.$$

It remains to show that $X_1\pi$ and $X_2\pi$ form a direct sum. Let $w \in X_1\pi \cap X_2\pi$. Then $w \in X_1\pi = X'\pi + Y_1\pi$, hence we can write $w = w_1 + w_2$ with $w_1 \in X'\pi$ and $w_2 \in Y_1\pi$. But $X_2\pi = Y_2\pi$, hence $w \in X_2\pi = Y_2\pi \subseteq Y\pi$ and $w_2 \in Y_1\pi \subseteq Y\pi$. Therefore $w_1 = w - w_2 \in X'\pi \cap Y\pi \subseteq Y_1\pi$. This shows that $w = w_1 + w_2 \in Y_1\pi \cap X_2\pi$, hence $w = 0$ since $X_2 = Y_2$ and $Y_1\pi$ and $Y_2\pi$ form a direct sum. Thus $X_1\pi \cap X_2\pi = 0$, hence $X/A = X_1\pi \oplus X_2\pi$. □

3 Stacked bases theorem

Before we state our main result, we show that it is possible to reduce the search for a stacked bases for X and A to the case where X, A and the quotient X/A are countable. We are mainly following the ideas from [3]. First a technical lemma.

Lemma 3.1 *Let $X = B \oplus D$ be a homogeneous completely decomposable group and C a reduced torsion group. Let $0 \neq C'$ be a direct summand of C with $\beta : C \longrightarrow C'$ the canonical projection such that almost all p-components of C' are trivial. If $\alpha : X \longrightarrow C$ is any homomorphism such that $D\alpha = C'$, then there exists a homomorphism $\varphi : B \longrightarrow D$ such that*

(i) $B' := B(1_B - \varphi) \cong B$;

(ii) $B'\alpha \subseteq \ker(\beta)$.

Note that B' is a new complement of D.

Proof. Since it is easy to check that for any homomorphism $\varphi : B \longrightarrow D$ we have $B' = B(1_B - \varphi) \cong B$ it is enough to define φ such that ii) holds. By assumption almost all p-components C'_p of C' are trivial, hence the set $P := \{p : p \text{ prime}, C'_p \neq 0\}$ is finite. Since C is reduced we can choose a rational subgroup $\mathbb{Z} \subseteq R \subseteq \mathbb{Q}$ such that $\frac{1}{p} \notin R$ for all $p \in P$ and $B = \bigoplus_{i\in\kappa} Rb_i$ for some $b_i \in B$. Let $c_i := b_i\alpha\beta \in C'$ and choose $d_i \in D$ such that $d_i\alpha = c_i$ for $i \in \kappa$. Since B and D have the same type there exist $m_i, n_i \in \mathbb{Z}$ such that

$$\chi^B(m_ib_i) = \chi^D(n_id_i).$$

Moreover by the choice of R we may assume $\gcd(n_i, p) = 1$ for all $p \in P$. Let $l_i \in \mathbb{Z}$ such that

$$l_in_i \equiv 1(\mathrm{ord}(c_i)) \qquad (**)$$

Now sending b_i to $l_in_id_i \in D$ can be extended unambigeously to a homomorphism $\varphi : B \longrightarrow D$. It remains to show that $B'\alpha \subseteq \ker(\beta)$. We compute

$(b_i - b_i\varphi)\alpha\beta$.

$$(b_i - b_i\varphi)\alpha\beta = b_i\alpha\beta - (l_i n_i d_i)\alpha\beta = c_i - l_i n_i c_i = (1 - l_i n_i)c_i = 0$$

by equation (**). Hence $(b_i - b_i\varphi)\alpha \in \ker(\beta)$ for all $i \in \kappa$. Now choose an arbitrary element $b \in B$ and write

$$b = \sum_{i\in\kappa} \frac{r'_i}{s_i} b_i$$

with $r'_i/s_i \in R$, hence $\gcd(s_i, p) = 1$ for all $p \in P$. Note that all sums are finite sums, hence $s := \prod_{i\in\kappa} s_i$ exists and $\gcd(s, p) = 1$ for all $p \in P$. Let $r_i := sr'_i/s_i \in \mathbb{Z}$. We get

$$s(b - \varphi(b))\alpha\beta = (\sum_{i\in\kappa}(r_i b_i - r_i b_i\varphi)\alpha\beta = 0,$$

hence $(b - \varphi(b))\alpha\beta = 0$. Thus $B'\alpha \subseteq \ker(\beta)$. □

Lemma 3.2 *Let $X = \bigoplus_{i\in\kappa} X'_i$ be an n-bounded extension of a h.c.d. A of infinite rank κ such that $X/A = \bigoplus_{i\in\kappa} C_i$ is a direct sum of cyclic groups and $\left(\bigoplus_{i\leq j} X'_i + A\right) = \bigoplus_{i\leq j} C_i$ for all $i \in \kappa$. Then the subgroups X'_i can be replaced by isomorphic subgroups X_i of X such that*

(1) $X = \bigoplus_{i\in\kappa} X_i$;

(2) $(X_i + A)/A = C_i$.

Proof. The existence of the subgroups X_i is guaranteed by Lemma 3.1 and the rest of the proof can be found in [3, Lemma 2.4, c)-e)] and carried over verbatim to our situation. □

Lemma 3.3 *Let X be an n-bounded extension of a homogeneous completely decomposable group A of infinite rank κ. Then there exist decompositions $X = \bigoplus_{i\in\kappa} X_i$ and $A = \bigoplus_{i\in\kappa} A_i$ such that*

(i) $A_i \subseteq X_i$ for all $i \in \kappa$;

(ii) X_i, A_i are of countable rank;

(iii) X_i/A_i is countable and bounded.

Proof. Let $\{x_i : i \in \kappa\}$ be a basis of X, i.e. $X = \bigoplus_{i\in\kappa} \langle x_i\rangle_*^X$. Write $X/A = \bigoplus_{i\in\lambda} \langle c_i + A\rangle$, where $\lambda \leq \kappa$. Following the proof of Lemma Z in [3] we can define by a standard seesawing back and forth argument subgroups $X_i' \subset X$ and $C_i \subset X/A$ such that

(i) X_i' and C_i are countable for all $i \in \kappa$;

(ii) $\left(\bigoplus_{j\leq i} X_j' + A\right)/A \subseteq \bigoplus_{j\leq i} C_j$ for all $i \in \kappa$;

(iii) $X = \bigoplus_{i\in\kappa} X_i'$ and $X/A = \bigoplus_{i\in\kappa} C_i$.

Now, applying Lemma 3.2, shows that the subgroups X_i' can be replaced by isomorphic subgroups X_i for all $i \in \kappa$ such that still $X = \bigoplus_{i\in\kappa} X_i$ and $X/A = \bigoplus_{i\in\kappa} C_i$ and additionally $(X_i + A)/A = C_i$ is countable. □

Stacked Bases Theorem 3.4 *Let X be a bounded extension of a homogeneous completely decomposable group A of infinite rank κ. Then X and A have stacked bases, i.e. there exist decompositions $X = \bigoplus_{i\in\kappa} \langle x_i\rangle_*^X$ and $A = \bigoplus_{i\in\kappa} d_i \langle x_i\rangle_*^A$ where $d_i \in \mathbb{Z}$.*

Proof. By the previous Lemma we may assume that X, A and the quotient X/A are countable. We choose a basis $B = \{b_1, b_2 \cdots\}$ of X, i.e. $X = \bigoplus_{i\in\omega} \langle b_i\rangle_*^X$. By Lemma 2.6 we get a splitting $X = X_1 \oplus X'$ such that $\langle b_1\rangle_*^X \subseteq X_1$ is of finite rank and $X\pi = X_1\pi \oplus X'\pi$, where $\pi : X \longrightarrow X/A$ is the canonical epimorphism. By Proposition 2.5 we derive a corresponding decomposition $A = A_1 \oplus A'$ such that $A_1 \subseteq X_1$ and $A' \subseteq X'$. Then X' is a bounded extension of A' which is homogeneous completely decomposable. Let b_2' be the projection of b_2 into X', then we decompose X' by the same arguments as above into $X_2 \oplus X''$, where X_2 is a finite rank subgroup of X' with $\langle b_2'\rangle_*^X \subseteq X_2$. Again we get a corresponding decomposition $A' = A_2 \oplus A''$ of A' such that X_2 is a finite extension of A_2 and X'' is a bounded extension of A''. Moreover $\langle b_1\rangle_*^X, \langle b_2\rangle_*^X \subseteq X_1 \oplus X_2$. Continuing this procedure we derive for every i a splitting

$$X = \bigoplus_{j\leq i} X_j \oplus X' \text{ and } A = \bigoplus_{j\leq i} A_j \oplus A'$$

such that $\langle b_1\rangle_*^X, \langle b_2\rangle_*^X, \ldots, \langle b_i\rangle_*^X \subseteq \bigoplus_{j\leq i} X_j$. Furthermore each X_i is a finite extension of A_i and X' is a bounded extension of A'. Now $X = \bigoplus_{i\in\omega} X_i$ follows and hence we get decompositions

$$X = \bigoplus_{i\in\omega} X_i \text{ and } A = \bigoplus_{i\in\omega} A_i$$

such that each X_i is a finite extension of the homogeneous completely decomposable group A_i and thus they have stacked bases by the strong stacked bases theorem for h.c.d. groups of finite rank. This shows that also X and A have stacked bases, which completes the proof. □

Remark: We would like to mention that the techniques, used in this paper, are based on the fact, that there are only finitely many primes p involved in the quotient. Thus all the results are still valid if we replace the boundeness condition on the quotient by the assumption that X and A have the same type and the quotient is a direct sum of cyclic groups with only finitely many non-zero p-components. In fact one can generalize the results to the case when X and A have the same type, the quotient X/A is a direct sum of cylic groups and there exists a subgroup $R \supseteq \mathbb{Z}$ of $\mathbb{Q}$ of the same type as X such that $1/p \notin R$ whenever X/A has a non-zero p-component (p prime).

References

[1] K. Benabdallah and M. A. Ould-Beddi, *On finite essential extensions of torsion-free abelian groups*, Canad. J. Math. **48** (1996), 918–929.

[2] K. Benabdallah and M. A. Ould-Beddi, *A strong stacked bases theorem for hcd groups*, Proc. Int. Conf. Abelian Groups and Modules, Colorado Springs (1995), Lecture Notes in Pure and Applied Math. **182**, 143–147.

[3] J. M. Cohen and H. Gluck, *Stacked bases for modules over principal ideal domains*, J. Algebra **14** (1970), 493–505.

[4] S. Files and R. Göbel, *Gauß' theorem for two submodules*, Math. Z. **228** (1998), 511–536.

[5] L. Fuchs, *Infinite Abelian Groups*, Vols. **I** and **II**, Academic Press (1970, 1973).

[6] P. Hill and C. Megibben, *Generalizations of the stacked bases theorem*, Trans. Amer. Math. Soc. **312** (1989), 377–402.

Trends in Mathematics, © 1999 Birkhäuser Verlag Basel/Switzerland

Separability conditions for vector R-modules

Ulrich Albrecht, Tony Giovannitti, and Pat Goeters

1 Introduction

One of the recurring themes in module-theory is the problem of describing classes of modules which are obtained as the closure with respect to direct sums or products of a well-understood class $\mathcal{A}$ of modules . For instance, the classes of completely decomposable groups and vector groups arise as closures of the class $\mathcal{A}_{\mathbb{Z}}$ of subgroups of the rationals with respect to direct sums and direct products, respectively. These two classes of groups are also closed with respect to direct summands by the Baer-Kulikov-Kaplansky Theorem [8, Theorem 86.7] and O'Neill's theorem [12]. In contrast, the closure of an arbitrary class $\mathcal{A}$ under direct sums or direct products may not have this closure property as many examples demonstrate (see [8] for details). This and similar difficulties can, in many cases, be avoided by restricting the discussion to families $\mathcal{A}$ which are *semi-rigid*, i.e. which have the property that any two modules $A, B \in \mathcal{A}$ with $\mathrm{Hom}_R(A, B) \neq 0$ and $\mathrm{Hom}_R(B, A) \neq 0$ are isomorphic. Arnold, Hunter, and Richman showed in [6] that the class of finitely $\mathcal{A}$-decomposable modules is closed with respect to direct summands if $\mathcal{A}$ is semi-rigid. Here, an R-module M is *finitely $\mathcal{A}$-decomposable* if there are $A_1, \ldots, A_n \in \mathcal{A}$ with the property that $M \cong \oplus_{i=1}^n P_i$ for suitable A_i-projective modules P_i of finite A_i-rank whenever $i = 1, \ldots, n$. Similarly, an R-module V is an *$\mathcal{A}$-vector-module* if it is isomorphic to $\Pi_{i \in I} A_i$ where each $A_i \in \mathcal{A}$.

The goal of this paper is to investigate separability conditions for $\mathcal{A}$-vector-modules. To remind the reader, an R-module M is *$\mathcal{A}$-separable* if every finite subset of M is contained in a finitely $\mathcal{A}$-decomposable direct summand of M. A detailed discussion of $\mathcal{A}$-separable groups can be found in [1] and, with a particular emphasis on discrete families $\mathcal{A}$ in [2]. Here, $\mathcal{A}$ is discrete if every $A \in \mathcal{A}$ contains a finite subset X such that $\mathrm{Hom}(A/\langle X \rangle, B) = 0$ for all $B \in \mathcal{A}$. The *$\mathcal{A}$-socle of a module M* is $S_{\mathcal{A}}(M) = \Sigma_{A \in \mathcal{A}} S_A(M)$, and every $\mathcal{A}$-separable submodule of M is contained in $S_{\mathcal{A}}(M)$. The results of Section 2 characterize the $\mathcal{A}$-vector-groups which have an $\mathcal{A}$-separable $\mathcal{A}$-socle (Theorem 2.1), and give several applications of this characterization. For instance, Corollary 2.2 focuses on rings R such that the right R-module R^I is locally free for all index-sets I, i.e. every finite subset of M is contained in a free direct summand of M. Specker showed in [13] that $\mathbb{Z}$ is one of these rings, and his proof can be adapted to show that every left Noetherian ring belongs to this class. However, it has been an open question whether there exist non-Noetherian rings with this property. Corollary 2.2 gives a negative answer provided that the ring R has a slender additive group of non-measurable cardinality.

Section 3 relates the results of this paper to previous work on separable vector-groups by Mishina in [11] and by Hill and one of the authors in [4]: Given a ring R with torsion-free additive group, let $\mathcal{A}_R$ be the class of all R-modules of the form $X \otimes_{\mathbb{Z}} R$ for some $X \subseteq \mathbb{Q}$. The class $\mathcal{A}_R$ which is semi-rigid if R is a prime ring (Proposition 3.1) was introduced in [3]. It was shown that every $\mathcal{A}_R$-Butler module M is of the form $B \otimes_{\mathbb{Z}} R$ for some classical Butler group B if R is an indecomposable strongly homogeneous ring. Section 4 demonstrates that a similar result does not hold for $\mathcal{A}_R$-vector-modules in general since there are only very few rings R such that every $\mathcal{A}_R$-vector-module V is of the form $V' \otimes_{\mathbb{Z}} R$ for some vector-group V'. For a prime ring R, the $\mathcal{A}_R$-separability of an $\mathcal{A}_R$-vector-group of the form $\Pi_I A_i$ such that $A_i \in \mathcal{A}_R$ for all $i \in I$ can, nevertheless, be described in terms of the types of the A_i's just as in the classical case (Corollary 3.3).

2 Separability

Throughout this section, $\mathcal{A}$ denotes a semi-rigid class of R-modules, and $V = \Pi_{i\in I} A_i$ an $\mathcal{A}$-vector-group. The pseudo-rigidity of $\mathcal{A}$ induces a quasi-order on the index-set I by setting $i \leq j$ if $\mathrm{Hom}_R(A_i, A_j) \neq 0$. Given $i \in I$, *the cone above* i is $C(i) = \{j \in I \mid j \geq i\}$. For $i_0 \in I$, every map $\phi \in \mathrm{Hom}(A_{i_0}, V)$ can be viewed as a sequence $(\phi_i)_{i\in I}$ of maps $\phi_i \in \mathrm{Hom}(A_{i_0}, A_i)$. The *support of* ϕ is $supp(\phi) = \{i \in I \mid \phi_i \neq 0\}$. Clearly, $supp(\phi) \subseteq C(i_0)$. The collection $\{A_i \mid i \in I\}$ satisfies *the generalized cone condition with respect to* $\leq$ if, for all $i_0 \in I$ and any collection of morphisms $\{\phi_j : A_{i_0} \to A_j \mid j \in C(i_0)\}$, there are indices $i_1 \leq \ldots \leq i_m$ such that, for almost all $j \in C(i_m)$, one has $\phi_j \in \Sigma_{k=1}^m \mathrm{Hom}_R(A_{i_k}, A_j)\phi_{i_k}$.

Theorem 2.1 *Let $\mathcal{A}$ be a semi-rigid family of R-modules. Consider the following two conditions on an $\mathcal{A}$-vector-module V of the form $V = \Pi_{i\in I} A_i$.*

a) *i) Whenever $i_0 \in I$ and $\phi \in \mathrm{Hom}(A_{i_0}, V)$, then $supp(\phi)$ does not contain an infinite subset whose elements form either a strictly descending chain or are incomparable in the order $\leq$.*

ii) $\{A_i \mid i \in I\}$ satisfies the generalized cone condition with respect to $\leq$.

b) Every element of $S_{\mathcal{A}}(V)$ is contained in a finitely $\mathcal{A}$-decomposable summand of V.

Then, a) always implies b), and the converse holds if $\mathcal{A}$ is a discrete family of slender modules.

Proof. *a)* $\Rightarrow$ *b)*: Every $x \in S_{\mathcal{A}}(V)$ is of the form $x = \Sigma_{j=0}^{n}\phi_j(a_j)$ for some $A_{i_0}, \ldots, A_{i_n} \in \mathcal{A}$ and suitable maps $\phi_j : A_{i_j} \to V$ and elements a_j of A_{i_j}. Once it has been shown that each of the $\phi_j(a_j)$'s is contained in a finitely $\mathcal{A}$-decomposable summand U_j of V such that there is a cofinite subset I_j of I with $V = U_j \oplus [\Pi_{i\in I_j} A_i]$, then a straightforward induction shows that x is contained in a finitely $\mathcal{A}$-decomposable direct summand of V. Therefore, assume $n = 0$.

Write $\phi_0 = (\psi_i)_{i\in I}$ with $\psi_i \in \mathrm{Hom}_R(A_{i_0}, A_i)$ for all $i \in I$ where $\psi_i = 0$ if $i \notin C(i_0)$. Since V satisfies the generalized cone condition by ii), there are indices $i_1 \leq \ldots \leq i_m$ such that $\psi_i \in \Sigma_{k=1}^{m} \mathrm{Hom}_R(A_{i_k}, A_i)\psi_{i_k}$ for almost all $i \in C(i_m)$. Thus, for almost all $i \in C(i_m)$, one has maps $\beta_{i,k} \in \mathrm{Hom}_R(A_{i_k}, A_i)$ such that

$(*)$ $\psi_i = \Sigma_{k=1}^{m}\beta_{i,k}\psi_{i_k}$.

Setting

$$\beta_{i_j,k} = \begin{cases} 0 \text{ if } j \neq k \\ 1_{A_{i_k}} \text{ if } j = k. \end{cases}$$

extends $(*)$ to $\{i_1, \ldots, i_m\}$. Although this may require a modification of the original choice of $\beta_{i_j,k}$ if $i_j \in C(i_m)$, it guarantees that the set I_0' of exceptions to $(*)$ satisfies $I_0' \cap \{i_1, \ldots, i_m\} = \emptyset$. Write $x_0 = \overline{x}_0 + \tilde{x}_0$ where $\overline{x}_0 \in \Pi_{i\in I\setminus I_0'} A_i$ and $\tilde{x}_0 \in \oplus_{i\in I_0'} A_i$. Since I_0' is finite, $\tilde{x}$ is contained in a finitely $\mathcal{A}$-decomposable direct summand of V. Replacing x_0 by $\overline{x}_0$ if necessary allows to assume $I_0' = \emptyset$.

Let $I_0 = C(i_m) \cup \{i_1, \ldots, i_m\}$, and write $V = V_0 \oplus V^0$ where $V_0 = \Pi_{i\in I_0} A_i$ and $V^0 = \Pi_{i\in I\setminus I_0} A_i$. Then, $x = x_0 + x'$ with $x_0 \in V_0$ and $x' \in V^0$. Observe $x_0 = (\psi_i(a_{i_0}))_{i\in I_0} = \Sigma_{k=1}^{m}(\beta_{i,k}\psi_{i_k}(a_{i_0}))_{i\in I_0}$ since $I_0' = \emptyset$. Define maps $\beta_k : A_{i_0} \to V_0$ by $\beta_k = (\beta_{i,k})_{i\in I_0}$. Since $\beta_{i_k,k} = 1_{A_{i_k}}$, one obtains

$$V_0 = \beta_0(A_{i_1}) \oplus \Pi_{i\in I_0\setminus\{i_1\}} A_i = \beta_0(A_{i_1}) \oplus \ldots \oplus \beta_k(A_{i_k}) \oplus \Pi_{i\in I_0\setminus\{i_1,\ldots,i_m\}} A_i,$$

and $x_0 \in \beta_0(A_{i_1}) \oplus \ldots \oplus \beta_k(A_{i_k})$. Hence, there is a finitely $\mathcal{A}$-decomposable summand U_0 of $\Pi_{i\in I_0} A_i$ which contains x_0 and whose complement is of the form $\Pi_{I_0\setminus J_0} A_i$ for some finite subset J_0 of I_0.

If $x' \neq 0$, then the previous construction can be repeated for x' and V^0 obtaining an ascending chain $i_1' \leq \ldots \leq i_{m'}'$ of indices in $I \setminus I_0$ satisfying a condition similar to $(*)$ where one uses the family of maps $\{\psi' \mid i \in I\}$ for x' which are defined by

$$\psi_i' = \begin{cases} 0 \text{ if } i \in I_0 \\ \psi_i \text{ otherwise.} \end{cases}$$

Since $i'_{m'} \notin C(i_m)$, either $i'_{m'} < i_m$ or $\{i_m, i'_{m'}\}$ is incomparable. If x' is not contained in a finitely $\mathcal{A}$-decomposable summand of $V_1 = \Pi_{i\in I_1} A_i$ where $I_1 = \{i'_1, \dots, i'_{m'}\} \cup C(i_{m'})$, then write $x' = x'_0 + x''$ with $x'_0 \in V_1$ and $x'' \in \Pi_{I\setminus(I_0\cup I_1)} A_i$, and repeat the construction again. This process has to terminate, since otherwise $supp(\phi)$ would contain either a strictly descending chain of infinite length or an infinite subset of elements which are pairwise incomparable by $\leq$. Neither of these cases can occur because of i).

b) $\Rightarrow$ *a)*: Assume that $\mathcal{A}$ is a discrete family of slender R-modules. Consider an index $i_0 \in I$ and a map $\alpha : A_{i_0} \to V$ which induces a collection of maps $\{\alpha_i : A_{i_0} \to A_j \mid j \in C(i_0)\}$ such that $\alpha = (\alpha_i)_{i\in I}$. If X is a finite subset of A_{i_0} such that $\mathrm{Hom}_R(A_{i_0}/\langle X\rangle, A) = 0$ for all $A \in \mathcal{A}$, then there is a finitely $\mathcal{A}$-decomposable direct summand W of V which contains $\alpha(X)$. Without loss of generality, there is a cofinite subset I' of I such that $V' = \Pi_{i\in I'} A_i$ satisfies $V = W \oplus V'$ since the elements of $\mathcal{A}$ are slender. Then, $\alpha(A_{i_0}) \subseteq W$ because otherwise one could find a non-zero map $\beta_j : A_{i_0} \to A_j$ for some $j \in I \setminus I'$ with $\beta_j(X) = 0$. Suppose $I\setminus I' = \{j_1, \dots, j_m\}$, and consider the direct summand $U = A_{j_1} \oplus \ldots \oplus A_{j_m}$ of V. Let $\pi : V \to V'$ and $\delta : V \to U$ as well as $\pi' : V \to V'$ and $\delta' : V \to W$ be the projections associated with the decompositions $V = U \oplus V' = W \oplus V'$. By [8, Lemma 9.5], there is an endomorphism ϕ of V such that $\delta' = \delta + \pi\phi\delta$ and $\pi' = \pi - \pi\phi\delta$. Write $\phi = (\phi_i)_{i\in I}$ where $\phi_i : V \to A_i$ for all $i \in I$. Moreover, let π_i and η_i be the projections and embeddings associated with the given decomposition $V = \Pi_{i\in I} A_i$. Then, $\delta\alpha = \Sigma_{k=1}^m \eta_{j_k}\alpha_{j_k}$, $\pi_i\alpha = \alpha_i$ and $\pi_i\phi = \phi_i$ for all $i \in I$. Since $im\ \alpha \subseteq W$, one obtains $\delta'\alpha = \alpha$. This yields $\alpha_i = \pi_i\delta'\alpha = \pi_i\delta\alpha + \pi_i\pi\phi\delta\alpha$ for all $i \in I'$. Since $\pi_i\delta = 0$ and $\pi_i\pi = \pi_i$, one has $\alpha_i = \phi_i\delta\alpha = \Sigma_{k=1}^m \phi_i\eta_{j_k}\alpha_{j_k}$. Observe that $\phi_i\eta_{j_k} \in \mathrm{Hom}(A_{j_k}, A_i)$, and hence $\alpha_i \in \Sigma_{k=1}^m \mathrm{Hom}(A_{j_k}, A_i)\alpha_{j_k}$ for all $i \in I$.

Suppose there is a map $\phi = (\phi_i)_{i\in I} : A_{i_0} \to V$ such that $supp(\phi)$ contains either a strictly $\leq$-descending chain $j_0 > j_1 > \ldots$ of infinite length or an infinite $\leq$-incomparable subset. By the results of the last paragraph, there is $m < \omega$ such that $\phi_{j_n} \in \Sigma_{k=1}^m \mathrm{Hom}_R(A_{j_k}, A_{j_n})\phi_{j_k}$ for all $n \geq m$. In either of the two cases, $\mathrm{Hom}_R(A_{j_k}, A_{j_n}) = 0$ if $n > m$, and hence $\phi_{j_n} = 0$ for all $n > m$, which is not possible. Therefore, *i)* holds.

If the generalized cone condition is not satisfied, then there is an index i_0 and a collection $\{\phi_j \mid j \in C(i_0)\}$ of maps $\phi_j : A_{i_0} \to A_j$ with the property that, for every ascending chain of indices $i_0 \leq \ldots \leq i_m$, there are infinitely many $j \in C(i_m)$ such that $\phi_j \notin \Sigma_{k=0}^m \mathrm{Hom}(A_{i_k}, A_j)\phi_{i_k}$. Let $j_0 = i_0$ and assume that one has already found $j_0 \leq \ldots \leq j_n$ such that $\phi_{j_{k+1}} \notin \Sigma_{\ell=1}^k \mathrm{Hom}(A_{j_\ell}, A_{j_{k+1}})\phi_{j_\ell}$. The hypotheses guarantee the existence of $j_{n+1} \in C(j_n) \setminus \{j_0, \dots, j_n\}$ such that $\phi_{j_{n+1}} \notin \Sigma_{\ell=1}^n \mathrm{Hom}(A_{j_\ell}, A_{j_{n+1}})\phi_{j_\ell}$.

Define a map $\alpha = (\alpha_i)_{i\in I} : A_{i_0} \to V$ coordinatewise by

$$\alpha_i = \begin{cases} \phi_{j_n} \text{ if } i = j_n \\ 0 \text{ otherwise.} \end{cases}$$

By the results of the first paragraph, there are indices $s_1, \ldots, s_t$ in I such that $\alpha_i \in \Sigma_{\ell=1}^t \operatorname{Hom}(A_{s_\ell}, A_i)\alpha_{s_\ell}$. Let m be chosen maximal with $j_m \in \{s_1, \ldots s_t\}$. Then, $j_{m+1} \notin \{s_1, \ldots, s_t\}$, and therefore

$$\phi_{j_{m+1}} = \alpha_{j_{m+1}} \in \Sigma_{\ell=1}^t \operatorname{Hom}(A_{s_\ell}, A_{j_{m+1}})\alpha_{s_\ell} =$$

$$\Sigma_{j_i\in\{s_1,\ldots,s_t\}} \operatorname{Hom}(A_{j_i}, A_{j_{m+1}})\phi_{j_i} \subseteq \Sigma_{\ell=1}^m \operatorname{Hom}(A_{j_\ell}, A_{j_{m+1}})\phi_{j_\ell},$$

which contradicts the choice of the j_n's. Consequently, the generalized cone condition is satisfied. □

As a consequence of Theorem 2.1, one obtains the following generalization of the Baer-Specker-Theorem [13]:

Corollary 2.2 *Let R be a left Noetherian ring, then R^I is a locally free right R-module for all index-sets I. Moreover, if R^+ is slender of non-measurable cardinality, and R^I is locally free for all index-sets I of non-measurable cardinality, then R is left Noetherian.*

Proof. Apply Theorem 2.1 to the set $\mathcal{A} = \{R\}$. Then, $S_{\mathcal{A}}(R^I) = R^I$, and since $i \leq j$ for all $i, j \in I$, condition a.i) in Theorem 2.1 is satisfied. Consider a family $\{\phi_i : R \to R\}_{i\in I}$ of maps, and observe that each ϕ_i is left multiplication by a suitable $r_i \in R$. Since R is left Noetherian, there are $i_0, \ldots, i_m \in I$ such that $r_{i_0}, \ldots, r_{i_m}$ generate the left ideal generated by $\{r_i \mid i \in I\}$. Write $r_i = \Sigma_{j=0}^m t_{i,j} r_{i_j}$, and define the maps $\beta_{i,j} : R \to R$ to be left multiplication by $t_{i,j}$. Then, $\phi_i = \Sigma_{j=1}^m \beta_{i,j}\phi_{i_j}$. Thus, condition a.ii) is satisfied too.

Conversely, let K be a non-zero left ideal of R, and consider R^K. Whenever $r \in K$, define maps $\phi_r : R \to R$ by $\phi_r(s) = rs$ for all $s \in R$. Since $C(r) = K$ for all $r \in K$, Theorem 2.1 yields that there are $r_1, \ldots r_n \in K$ such that $\phi_r = \Sigma_{j=1}^n \alpha_{r,j}\phi_{r_k}$ for all $r \in K$ where $\alpha_{r,j} : R \to R$. Therefore, $\alpha_{r,j}$ is left multiplication by a suitable $t_{r,j} \in R$. Hence, $r = \phi_r(1_R) = \Sigma_{j=1}^n \alpha_{r,j}\phi_{r_k}(1_R) = \Sigma_{k=1}^n t_{r,k} r_k$, and K is finitely generated. □

Corollary 2.3 *Let A be an R-module with left Noetherian endomorphism ring. Then, $S_A(A^I)$ is locally A-projective for all index-sets I.*

Proof. Combine Theorem 2.1 and Corollary 2.2. □

Observe that $S_A(A^I) = A^I$ for all index-sets I if and only if A is finitely generated as a left $E(A)$-module. A family $\mathcal{A}$ of R-modules is *rigid* if $\mathrm{Hom}_R(A, B) = 0$ whenever A and B in $\mathcal{A}$ are non-isomorphic.

Corollary 2.4 *Let $\mathcal{A}$ be a rigid family such that $E(A)$ is left Noetherian for each $A \in \mathcal{A}$. Then, $S_{\mathcal{A}}(V)$ is $\mathcal{A}$-separable for every $\mathcal{A}$-vector group V.*

Proof. Write $V = \Pi_{i \in I} A_i$ where each $A_i \in \mathcal{A}$. Since $C(i_0) = \{i \in I \mid A_i \cong A_{i_0}\}$, condition a.i) of Theorem 2.1 is trivially satisfied, while condition ii) can be obtained as in the proof of Corollaries 2.2 and 2.3 since $S_{\mathcal{A}}(V) = \oplus_{A \in \mathcal{A}} S_A(V_A)$ where $V_A = \Pi\{A_i \mid A_i \cong A\}$. □

3 The class $\mathcal{A}_R$

For a ring R whose additive group is torsion-free, consider the family $\mathcal{A}_R = \{X \otimes_{\mathbb{Z}} R \mid X \subseteq \mathbb{Q}\}$ of submodules of $\mathbb{Q}R$. Naturally, one can restrict the discussion to those subgroups X of $\mathbb{Q}$ for which $X = pX$ for all primes in $\pi(R) = \{p \mid pR = R\}$. In [3], it was shown that $\mathcal{A}_R$ is the family of all rank 1 R-modules of a subring R of an algebraic number field exactly if R is a PID which is an IP-ring. Here a subring R of an algebraic number field is called an *IP-ring* if, for each prime p of $\mathbb{Z}$ with $pR \neq R$, the ideal $p\overline{R}$ of the integral closure $\overline{R}$ of R in $Q(R)$ is a primary ideal of $\overline{R}$. However, $\mathcal{A}_R$ may be semi-rigid even if R is neither an IP-ring nor a PID:

Proposition 3.1 *Let R be a torsion-free ring of finite rank such that $\mathbb{Q}R$ is simple. Then, $\mathcal{A}_R$ is a semi-rigid family.*

Proof. In the first step, it is shown that $X \otimes_{\mathbb{Z}} R$ is a homogeneous group whose type is the type of X whenever X is a subgroup of $\mathbb{Q}$ with $pX = X$ for all $p \in \pi(R)$.

The pure rank 1 subgroup $1_R\mathbb{Q} \cap R$ of R has idempotent type τ and $R(\tau) = R$. If R is not homogeneous, then there is a non-zero element $r \in R$ such that $\sigma = type(r) > \tau$. In particular, there are infinitely many primes $\{p_n\}_{n<\omega}$ such that $h_{p_n}(r) > 0$ but $h_{p_n}(1) = 0$ or at least one prime p with $h_p(r) = \infty$ but $h_p(1) = 0$. For each prime q arising in this way, select $0 < k \leq h_q(r)$, and write $r = q^k s_{q,k}$ for some $s_{q,k} \in R$. Then, $rt = q^k s_{q,k}t$ and $tr = q^k t s_{q,k}$ for all $t \in R$ shows that $Rr, rR \subseteq R(\sigma)$. Therefore, $R(\sigma)$ contains the non-zero two-sided ideal RrR. Since $\mathbb{Q}R$ is simple, R/RrR is finite, and the same holds for $R/R(\sigma)$ which is not possible since $R(\sigma)$ is a proper pure subgroup of R^+. Hence, R^+ is homogeneous of type τ.

Turning to $X \otimes_{\mathbb{Z}} R$, observe that every of its elements a has the form $a = \Sigma_{j=1}^m x_j \otimes r_j$ for some $x_j \in X$ and $r_j \in R$. If x_j is written as $x_j = \frac{s_j}{t_j}$ for some relatively prime integers s_j and t_j, then $\frac{1}{t_j} \in X$. Let $n = g.c.d.(t_1, \dots, t_m)$ and write $t_j = nt'_j$. Setting $\ell_i = \Pi_{j \neq i} t'_j$ for $i = 1, \dots, n$ yields a family of relatively prime integers. Thus, there are integers $k_1, \dots, k_m$ such that $1 = \Sigma_{j=1}^m k_j \ell_j$. Observe that $t_i \ell_i = t_j \ell_j$ for all $i, j \in \{1, \dots, m\}$ and denote this common value by n_0. Then, X contains $\Sigma_{j=1}^m k_j \frac{1}{t_j} = \frac{1}{n_0} \Sigma_{j=1}^m k_j \ell_j = \frac{1}{n_0}$. Thus, $a = \frac{1}{n_0} \otimes r$ where $r = \Sigma_{j=1}^m s_j \ell_j r_j$. Therefore, $U = \mathbb{Q}r \cap R$ is a pure rank 1 subgroup of R such that $a \in X \otimes_{\mathbb{Z}} U$. Since $X \otimes_{\mathbb{Z}} U$ is pure in $X \otimes_{\mathbb{Z}} R$, one has $type(a) = type(X \otimes_{\mathbb{Z}} U) = type(X)$ since $type(X) \geq \tau = type(U)$ and τ is idempotent.

Suppose that X_1 and X_2 are subgroups of $\mathbb{Q}$ such that $pX_i = A_i$ whenever $p \in \pi(R)$ and $\mathrm{Hom}_R(X_1 \otimes_{\mathbb{Z}} R, X_2 \otimes_{\mathbb{Z}} R) \neq 0$. Since $X_1 \otimes_{\mathbb{Z}} R$ and $X_2 \otimes_{\mathbb{Z}} R$ are homogeneous by the results of the last paragraph, one obtains $type(X_1) \leq type(X_2)$. Therefore, $\mathrm{Hom}(X_i \otimes_{\mathbb{Z}} R, X_j \otimes_{\mathbb{Z}} R) \neq 0$ for $i, j \in \{1, 2\}$ yields $X_1 \cong X_2$. □

The next results relate the generalized cone condition to the classical results on abelian groups in [4] and [11]. Paraphrasing the definition in [4], a family $\{A_i\}_{i \in I}$ of R-modules in $\mathcal{A}_R$ satisfies the *cone condition* if, for every $i \in I$, there is a finite subset $I_{0,i}$ of I and a finite subset $\Pi_{0,i}$ of $\pi(R)$ such that, for all $j \in C(i) \setminus I_{0,i}$ and all $p \in \pi(R) \setminus \Pi_{0,i}$, one has $A_j = pA_j$ whenever A_i contains a non-zero element of non-zero p-height.

Proposition 3.2 *Let R be a prime ring whose additive group is torsion-free reduced and has finite rank. If $\{A_i \mid i \in I\}$ is a family of elements of $\mathcal{A}_R$ which satisfies the cone condition, then it satisfies the generalized cone condition in Theorem 2.1.*

Proof. Since each $A_i \in \mathcal{A}_R$ is isomorphic to a group of the form $X_i \otimes_{\mathbb{Z}} R$ for some subgroup X_i of $\mathbb{Q}$, one may assume that each $A_i = X_i \otimes_{\mathbb{Z}} R$ such that $type(X_i) \geq type(R)$. To simplify the notation, write τ_i for the type of A_i which, of course, is the type of X_i. Observe that no generality is lost if one assumes that $pA_i \neq A_i$ for all $p \in \Pi_{0,i}$. Hence, every element of X_i has finite p-height for all $p \in \Pi_{0,i}$. In particular, $1 = m_i x_i$ where $h_p(x_i) = 0$ for all $p \in \Pi_{0,i}$.

Let $i_0 \in I$, and consider a collection of maps $\phi_i : A_{i_0} \to A_i$ for all $i \in I$. Choose finite subsets I_{0,i_0} of I and Π_{0,i_0} of Π as in the cone condition where $\Pi = \{p | R \neq pR\}$. For every $i \in I$, one has $\phi_i(x_{i_0} \otimes 1) = \frac{1}{\ell_i} \otimes r_i$ for some $r_i \in R$ and non-zero integer ℓ_i such that $\frac{1}{\ell_i} \in X_i$. Let $n_i = g.c.d.(m_i, \ell_i)$, and write $m_i = n_i m_i'$ and $\ell_i = n_i \ell_i'$. Thus, $\frac{1}{\ell_i} = \frac{m_i}{\ell_i} x_i = \frac{m_i'}{\ell_i'} x_i \in X_i$. Since there are integers b_i and c_i such that $1 = c_i m_i' + b_i \ell_i'$, one obtains $\frac{1}{\ell_i'} x_i = c_i(\frac{m_i'}{\ell_i'} x_i) + b_i x_i \in X_i$. Therefore, $\phi_i(x_{i_0} \otimes 1) = \frac{1}{\ell_i'} x_i \otimes m_i' r_i$, i.e. one may assume that $\phi_i(x_{i_0} \otimes 1) = \frac{1}{m_i} x_i \otimes r_i$ for some $r_i \in R$ and some non-zero integer m_i such that $\frac{1}{m_i} x_i \in X_i$.

There exists a finite sequence $i_1 \leq \dots \leq i_m$ of elements of I such that $r_{i_1}, \dots, r_{i_m}$ generate the left ideal of R which is generated by $\{r_i \mid i \in C(i_m)\} \cup \{r_{i_1}, \dots, r_{i_m}\}$. If this were not the case, then there would exist a strictly ascending infinite chain of left ideals of R, which is not possible since R is left Noetherian by [5, Theorem 9.4]. Write $r_i = \Sigma_{j=1}^m t_{i,j} r_{i_j}$ where $t_{i,j} \in R$. Choose the $t_{i,j}$ in such a way that

$$t_{i_j,k} = \begin{cases} 0 \text{ if } j \neq k \\ 1 \text{ otherwise,} \end{cases}$$

For each $j \in \{0, \dots, m\}$ and $i \in C(i_m) \setminus I_{0,i_0}$, the cone condition allows to define a map $\psi_{i_j}^i : A_{i_j} \to A_i$ by $\psi_{i_j}^i(x_{i_j} \otimes 1) = \frac{m_{i_j}}{m_i} x_i \otimes t_{i,j}$. Then,

$$\Sigma_{j=0}^m \psi_{i_j}^i \phi_{i_j}(x_{i_0} \otimes r) = \Sigma_{j=0}^m \psi(\frac{1}{m_{i_j}} x_{i_j} \otimes r_{i_j} r) =$$

$$\frac{1}{m_i} x_i \otimes \Sigma_{j=0}^m t_{i,j} r_{i_j} r = \phi_i(x_{i_0} \otimes r).$$

Since $A_{i_0}/(x_{i_0} \otimes 1)R$ is torsion as an abelian group, $\phi_i = \Sigma_{j=1}^m \psi_{i_j}^m \phi_{i_j}$ as desired. □

Corollary 3.3 *Let R be a prime ring whose additive group is torsion-free and has finite rank. An $\mathcal{A}_R$-vector-module $V = \Pi_I A_i$ is $\mathcal{A}_R$-separable if*

a) There are no infinite descending chains in I.

b) Every set of $\leq$-incomparable elements of I is finite.

c) $\{A_i \mid i \in I\}$ satisfies the cone condition.

Proof. Observe that no generality is lost if one assumes that $pA_i \neq A_i$ for all $p \in \Pi_{0,i}$. Hence, every element of X_i has finite p-height for all $p \in \Pi_{0,i}$. By Theorem 2.1 and Proposition 3.2, it remains to show $S_{\mathcal{A}}(V) = V$.

Consider an element $g_i = (a_i)_{i \in I}$ of V. For each $i \in I$, choose finite subsets $I_{0,i}$ of I and $\Pi_{0,i}$ of Π as in part c) where $\Pi = \{p | A \neq pA\}$, and write $A_i = X_i \otimes_{\mathbb{Z}} R$, and observe that $pX_i \neq X_i$ for all $p \in \Pi_{0,i}$. Hence, every $y_i \in X_i$ has finite p-height for all $p \in \Pi_{0,i}$. Therefore, one can write $1 = m_i x_i$ for some $x_i \in X_i$ where $h_p(x_i) = 0$ for all $p \in \Pi_{0,i}$, and assume that $a_i = \frac{1}{m_i}x_i \otimes r_i$ for some non-zero integer m_i and $r_i \in R$ such that $\frac{1}{m_i}x_i \in X_i$.

Since R is left Noetherian by [5, Theorem 9.4], there exists a finite sequence $i_1 \leq \ldots \leq i_m$ of elements of I such that $r_{i_1}, \ldots, r_{i_m}$ generate the left ideal of R which is generated by $\{r_i \mid i \in C(i_m)\} \cup \{r_{i_1}, \ldots, r_{i_m}\}$, i.e. $r_i = \Sigma_{j=1}^m t_{i,j} r_{i_j}$ where $t_{i,j} \in R$. Choose the $t_{i,j}$ in such a way that

$$t_{i_j,k} = \begin{cases} 0 \text{ if } j \neq k \\ 1 \text{ otherwise,} \end{cases}$$

and define

$$(\bar{g})_i = \begin{cases} a_i \text{ if } i \in I_{0,i_1} \cup \ldots \cup I_{0,i_m} \\ 0 \text{ otherwise.} \end{cases}$$

Set $I_j = I_{0,i_1} \cup \ldots \cup I_{0,i_m} \cup \{i_1, \ldots i_{j-1}, i_{j+1} \ldots i_m\}$, and define elements $h_1, \ldots h_m$ of $\Pi_{J_1} A_i$ as follows:

$$(h_j)_i = \begin{cases} 0 \text{ if } i \in I_j \setminus \{i_j\} \\ \frac{1}{m_{i_j}} x_{i_j} \otimes r_{i_j} \text{ if } i = i_j \\ \frac{1}{m_i} x_i \otimes t_{i,j} r_{i_j} \text{ otherwise.} \end{cases}$$

Observe that $g_1 = \bar{g} + h_1 + \ldots + h_m$. Moreover, $h_j \in \Pi_{K_j} A_i$ where $K_j = \{i_j\} \cup [J_1 \setminus I_j]$.

Since $i_j \leq i_m$, one has $C(i_m) \subseteq C(i_j)$. Moreover, $J_1 \setminus I_j \cap I_{0,i_j} = \emptyset$ yields $ht(x_{i_j}) \leq ht(\frac{1}{m_i}x_i)$ for all $i \neq i_j$ in $J_1 \setminus I_j$. Hence, there is a map $\phi_i : X_{i_j} \to X_i$ such that $\phi(x_{i_j}) = \frac{m_{i_j}}{m_i} x_i$. Define a map $\psi_i : A_{i_j} \to A_i$ by $\psi_i(x \otimes r) = \phi_i(x) \otimes t_{i,j} r$ to obtain a map $\psi^j : A_{i_j} \to \Pi_{K_j} A_i$ as follows:

$$(\psi^j(x))_i = \begin{cases} x \text{ if } i = i_j \\ \psi(x) \text{ otherwise.} \end{cases}$$

Since $h_j \in im\, \psi^j$, one obtains $S_{\mathcal{A}}(V) = V$. □

This section concludes with an example that Theorem 2.1 applies also to families of modules which do not arise as tensor-products of a ring R and rank 1 abelian groups.

Example 3.4 Let S be the endomorphism ring of $\oplus_\omega \mathbb{Z}$, and R be the subring of S generated by all lower triangular matrices in R and $1_{\oplus_\omega \mathbb{Z}}$. Since R is a countable reduced ring, there is a countable torsion-free group A with $E(A) = R$. The canonical primitive idempotents $e_n = (e_{ij})$ of R with

$$e_{ij} = \begin{cases} 1 \text{ if } i = j = n \\ 0 \text{ otherwise} \end{cases}$$

yield a family $\mathcal{A} = \{A_n = e_n(A) \mid n < \omega\}$ of direct summands of A such that $E(A_n) = \mathbb{Z}$ for all $n < \omega$ and

$$\operatorname{Hom}(A_i, A_j) = \begin{cases} \mathbb{Z} \text{ if } i \geq j \\ 0 \text{ otherwise.} \end{cases}$$

Consider the $\mathcal{A}$-vector-group $\Pi_{n<\omega} A_n$, and observe that the order induced on ω by the A_n's coincides with its natural order. Thus, condition *a.i)* of Theorem 2.1 is satisfied. The arguments used in the proofs of Proposition 3.2 and Corollary 3.3 can be adapted to establish condition *a.ii)*. By Theorem 2.1, $S_{\mathcal{A}}(V)$ is $\mathcal{A}$-separable.

4 Vector-modules and tensor products

Let R be a ring. Since the functor $- \otimes_{\mathbb{Z}} R$ commutes with direct sums, there is a natural isomorphism mapping $G \otimes_{\mathbb{Z}} R$ onto the $\mathcal{A}_R$-decomposable module $\oplus_{i\in I}[A_i \otimes_{\mathbb{Z}} R]$ whenever $G = \oplus_{i\in I} A_i$ is a completely decomposable group. Similarly, if $V = \Pi_{i\in I} A_i$ is a vector-group, then there is a natural map ψ_V from $V \otimes_{\mathbb{Z}} R$ into the $\mathcal{A}_R$-vector-module $\Pi_{i\in I}[A_i \otimes_{\mathbb{Z}} R]$. However, ψ_V need not be an isomorphism in contrast to the case of completely decomposable groups, as the results of this section show. In the following, R is an integral domain whose additive group is torsion-free. Finally, given a torsion-free group G, let

- $\pi(G)$ denote the set of integral primes p such that $pG \neq G$; and let
- $\pi_0(G)$ represent the set of primes $p \in \pi(G)$ such that $r_p(G) <$ rank G.

Lemma 4.1 *Let $\{C_i \mid i \in I\}$ be a family of subgroups of $\mathbb{Q}$ which contain $\mathbb{Z}$, and $V = \Pi_{i\in I} C_i$. If the natural map $\psi_V : V \otimes R \to \Pi_j[C_j \otimes R]$ is an isomorphism, then R has finite rank.*

Proof. By the naturality of the isomorphism, $(\Pi_{j\in J} C_j) \otimes R \cong \Pi_{j\in J}(C_j \otimes R)$ for any subset J of I. Therefore, for this argument it may be assumed that $I = \omega$. If $r_0, r_1, \ldots$ is any subset of R, then there is a finite set of elements $t_1, \ldots, t_k$ of elements of R and I-tuples, $v_i \in \Pi_j C_j$ such that $(1 \otimes r_j \mid j < \omega) = \Sigma_{i=1}^k v_i \otimes t_i$. This implies that each r_j is a rational linear combination of the t_i's, and consequently R must have finite rank. □

Lemma 4.2 *Let $\mathbb{Z} \subseteq C_j \subseteq \mathbb{Q}$ for $j \in I$. Then $(\Pi_j C_j) \otimes R \cong (\Pi_j C_j) \otimes R_S$ and $\Pi_j(C_j \otimes R) \cong \Pi_j(C_j \otimes R_S)$ where $S = \cup_j \pi(C_j)$.*

Proof. The first isomorphism follows from the fact that $\Pi_j C_j$ is a $\mathbb{Z}_S$-module, and the second is due to the fact that each $C_j \otimes R$ is a $\mathbb{Z}_S$-module. □

Theorem 4.3 *Let $V = \Pi_{j\in I} C_j$ be a vector-group where $\{C_j \mid j \in I\}$ is a collection of rank one groups. The natural map $\psi_V : V \otimes_{\mathbb{Z}} R \to \Pi_j(C_j \otimes R)$ is an isomorphism if and only if either $\pi_0(R) = \emptyset$, or, there are only finitely many indices $j \in I$ for which $pC_j \neq C_j$ for some $p \in \pi_0(R)$.*

Proof. Suppose that the natural map $\psi_V : V \otimes_{\mathbb{Z}} R \to \Pi_j(C_j \otimes R)$ is an isomorphism. Assume, to the contrary, that there exists $\{C_j\}_{j<\omega}$ such that, for each $j = 1, 2 \ldots$, one can find a prime $p = p(j) \in \pi_0(R)$ with $pC_j \neq C_j$. By the naturality of the isomorphism $(\Pi_j C_j) \otimes R \cong \Pi_j(C_j \otimes R)$, one may assume that $I = \omega$.

Case 1: There is some $p \in \pi_0(R)$ for which infinitely many C_j's are not p-divisible. In this case one may assume, by disregarding unneeded indices, that each of $C_1, C_2, \ldots$ satisfies $pC_j \neq C_j$. Furthermore, by the naturality of ψ_V, one may assume that $1/p \notin C_j$ for each of $j = 1, 2, \ldots$. By Lemma 4.1, R has finite rank. Since finite rank integral domains are quotient divisible [7], there is a maximal rationally independent subset $r_1, \ldots, r_n$ of R, such that the free subgroup $F = \mathbb{Z}r_1 \oplus \cdots \oplus \mathbb{Z}r_n$ of R has the property that R/F is divisible.

Now, from the definition of $\pi_0(R)$, the p-component of R/F is a non-zero divisible group. Therefore, for every $m = 1, 2, \ldots$, there is an element $s_m \in R$ that has order p^m modulo F such that the cosets $s_m + F$ for $m \geq 1$ generate a subgroup isomorphic to $\mathbb{Z}(p^\infty)$. If $\beta \neq (1 \otimes s_m | m = 1, 2, \ldots) \notin (\Pi_j C_j) \otimes R$, then there is a finite choice of elements $t_1, \ldots, t_k$ from R and $\alpha_1, \ldots, \alpha_k$ from $\Pi_j C_j$ such that $\beta = \Sigma_{i=1}^k \alpha_i \otimes t_i$. This implies that each s_m can be expressed as $\Sigma_{i=1}^k c_m^i t_i$ where $c_m^i \in C_m$ is the mth-component of α_i. By our assumptions, each $C_j \subseteq \mathbb{Z}_p$.

Setting $F' = \mathbb{Z}t_1 + \cdots + \mathbb{Z}t_k$, because R/F is torsion, there is an integer $0 \neq \ell$ for which $\ell F' \subseteq F$. In particular, $\ell s_m \in F_p$ for each m, since the rationals c_m^i belong to $\mathbb{Z}_p$. But this implies that $\mathbb{Z}_{(p^\infty)} \cong \Sigma_m \langle s_m + F \rangle$ is bounded, a contradiction.

Case 2: There are infinitely many distinct primes $p_1, p_2, \ldots$ from $\pi_0(R)$, and distinct indices $j_1, j_2, \ldots$ such that $p_i C_{j_i} \neq C_{j_i}$.

One may renumber and discard unneeded indices in order to assume that $j_i = i$. Additionally, by Lemma 4.2, one may assume that $1/p_j \notin C_j$. As above, there is a full free subgroup F of R for which R/F is divisible. For each p_i, R/F has a non-zero p_i-component, so one may select $s_i \in R$ which generates an element of order p_i modulo F. Then, $(1 \otimes s_i | i = 1, 2, \ldots) \notin (\Pi_i C_i) \otimes R$.

Otherwise, the arguments used above can be applied to show that there exists $t_1, \ldots, t_k$ in R and elements $c_j^1, \ldots, c_j^k \in C_j$ such that $s_j = \Sigma_i c_j^i t_i$

with $\ell \neq 0$ such that $\ell t_i \in F$ for all i, $\ell s_j \in F_{p_j}$. However, if p_j does not divide ℓ, then $0 \neq \ell s_j + F_{p_j} \in R_{p_j}/F_{p_j}$, a contradiction.

For the converse, write $I = I_0 \cup I_1$ where $I_0 = \{j \in I \mid pC_j \neq C_j$ for some $p \in \pi_0(R)\}$, and $I_1 = I \setminus I_0$. By hypothesis, I_0 is finite, so $(\Pi_{I_0} C_j) \otimes R \cong \Pi_{I_0}(C_j \otimes R)$. Also, $(\Pi_{I_1} C_j) \otimes R \cong (\Pi_{I_1} C_j) \otimes R_{\pi_1}$ and $\Pi_{I_1}(C_j \otimes R) \cong \Pi_{I_1}(C_j \otimes R_{\pi_1})$, where $\pi_1 = \cup_{j \in I_1} \pi(C_j)$ by Lemma 4.2.

Since R_p is a free $\mathbb{Z}_p$ module for every $p \in \pi_1$, and R is quotient divisible, R_{π_1} is a free $\mathbb{Z}_{\pi_1}$-module. It is easy to show then that $(\Pi_j C_j) \otimes R_{\pi_1} \cong \Pi_{I_1}(C_j \otimes R_{\pi_1})$. Thus, $(\Pi_I C_j) \otimes R \cong (\Pi_{I_0} C_j) \otimes R \oplus (\Pi_{I_1} C_j) \otimes R \cong \Pi_{I_0}(C_j \otimes R) \oplus \Pi_{I_1}(C_j \otimes R)$ as desired. □

Corollary 4.4 *Let R be an integral domain and $S = \{p \mid pR = R\}$. Then, the natural map $\psi_V : V \otimes_{\mathbb{Z}} R \to \Pi_{i \in I}[C_i \otimes_{\mathbb{Z}} R]$ is an isomorphism for all vector-groups V with $pV = V$ for all $p \in S$ if and only if R is a finitely generated $\mathbb{Z}_S$-module.*

Proof. Since the natural map $[\Pi_I \mathbb{Z}_S] \otimes_{\mathbb{Z}} R \to \Pi_I[\mathbb{Z}_S \otimes_{\mathbb{Z}} R]$ is an isomorphism for all index-sets I, R is a finitely generated $\mathbb{Z}_S$-module by [9, Proposition 5.11]. Conversely, observe that R^+ is a finitely generated free $\mathbb{Z}_S$-module, and hence $\pi_0(R) = \pi \setminus S$ where π is the set of all primes. Now apply Theorem 4.3. □

References

[1] U. Albrecht, *Products of slender modules*, Abelian Group Theory, Oberwolfach 1985, Gordon and Breach (1987), 259–274.

[2] U. Albrecht, *On direct summands of A-separable R-modules*, Forum Mathematicum **2** (1990), 103–117.

[3] U. Albrecht and H. P. Goeters, *Butler theory over Murley groups*, Journal of Algebra **200** (1998), 118–133.

[4] U. Albrecht and P. Hill, *Separable vector groups*, Contemporary Mathematics **87** (1989), 155–160.

[5] D. M. Arnold, *Finite Rank Torsion-Free Abelian Groups and Rings*, Springer LNM **931** (1982).

[6] D. M. Arnold, R. Hunter, and F. Richman, *A global Azumaya theorem in additive categories*, J. Pure and Applied Algebra **16** (1980), 232–242.

[7] R. A. Beaumont and R. S. Pierce, *Torsion-free rings*, Illinois J. Math **5** (1961), 61–98.

[8] L. Fuchs, *Infinite Abelian Groups*, Academic Press (1970, 1973).

[9] K. R. Goodearl, *Ring Theory*, Pure and Applied Mathematics **33**, Marcel Dekker (1976).

[10] M. Huber and R. Warfield, *Homomorphisms between cartesian powers of an abelian group*, Abelian Group Theory, Oberwolfach 1981, Springer LNM **874** (1981), 202–227.

[11] A. P. Mishina, *Separability of complete direct sums of torsion-free abelian groups of rank* 1, Mat. Sb **57** (1962), 375–383.

[12] J. O'Neill, *Direct summands of vector groups*, Acta Math. Hung. **55** (1990), 207–209.

[13] E. Specker, *Additive Gruppen von Folgen ganzer Zahlen*, Portugaliae Math. **9** (1950), 131–140.

Trends in Mathematics, © 1999 Birkhäuser Verlag Basel/Switzerland

Almost disjoint pure subgroups of the Baer-Specker group

Oren Kolman * and Saharon Shelah †

Abstract

We prove in ZFC that the Baer-Specker group $\mathbf{Z}^\omega$ has $2^{\aleph_1}$ non-free pure subgroups of cardinality $\aleph_1$ which are almost disjoint: there is no non-free subgroup embeddable in any pair.

In this short paper we prove the following result.

Theorem 1 *There exists a family* $\mathbf{G} = \{G_\alpha : \alpha < 2^{\aleph_1}\}$ *of non-isomorphic non-free pure subgroups of the Baer-Specker group* $\mathbf{Z}^\omega$ *such that:*
(1.1) each G_α *has cardinality* $\aleph_1$*;*
(1.2) if $\alpha < \beta$*, then* G_α *and* G_β *are almost disjoint: if* H *is isomorphic to subgroups of* G_α *and* G_β*, then* H *is free. In particular,* $G_\alpha \cap G_\beta$ *is free.*

Recall that the Baer-Specker group $\mathbf{Z}^\omega$ is the abelian group of functions from the natural numbers into the integers (see [1] and [18]). It contains the canonical pure free subgroup $\mathbf{Z}_\omega = \oplus_{n<\omega}\mathbf{Z}$. The group $\mathbf{Z}^\omega$ is not κ-free for any cardinal $\kappa > \aleph_1$, but it is $\aleph_1$-free, so the groups G_α in Theorem 1 are almost free.

Theorem 1 answers a question of the second author, and has its place in the line of recent research dealing with the lattice structure of the pure subgroups of $\mathbf{Z}^\omega$ (see [2], [3], and [5]–[8]). For example, Irwin asked whether there is a subgroup of $\mathbf{Z}^\omega$ with uncountable dual but no free summands of infinite rank. This problem was resolved recently by Corner and Göbel [5] who proved the following stronger fact.

Theorem 2 *[5] The Baer-Specker group* $\mathbf{Z}^\omega$ *contains a pure subgroup* G *whose endomorphism ring splits as* $End(G) = \mathbf{Z}\oplus Fin(G)$*, with* $|G^*| = 2^{\aleph_0}$*, where* $\mathbf{Z}$ *is the scalar multiplication by integers and* $Fin(G)$ *is the ideal of all endomorphisms of* G *of finite rank.*

*We thank the referee for constructive comments.

†This research was partially supported by the German-Israel Foundation project No. G-545-173.06/97; publication number 683.

Quotient-equivalent and almost disjoint abelian groups have been studied by Eklof, Mekler and Shelah in [9]–[11], who showed that under various set-theoretic hypotheses, there exist families of maximal possible size of almost free abelian groups which are pairwise almost disjoint. Following [11], we say that two groups A and B are almost disjoint if whenever H is embeddable as a subgroup in both A and B, then H is free. Clearly if A and B are non-free and almost disjoint, then they are non-isomorphic in a very strong way. On the other hand, the intersection of two almost disjoint groups of size $\aleph_1$ need not necessarily be countable, so group-theoretic almost disjointness differs from its set-theoretic homonym. Theorem 1 establishes in ZFC that the Baer-Specker group contains large families of almost disjoint almost free non-free pure uncountable subgroups.

Our group and set-theoretic notation is standard and can be found in [10] and [14]. For example, ${}^{\omega_1 >}2$ is the set of partial functions from ω_1 into $\{0,1\}$ whose domains are at most countable; ${}^{\omega_1}2$ is the set of all functions from ω_1 into $\{0,1\}$; for a regular cardinal χ, $H(\chi)$ is the family of all sets of hereditary cardinality less than χ.

For a set $A \subseteq H(\chi)$ for χ large enough, we write $\mathrm{dcl}_{(H(\chi),\in,<)}[A]$ for the Skolem closure (Skolem hull) of A in the structure $(H(\chi), \in, <)$, where $<$ is a well-ordering of $H(\chi)$ (for details, see [16], 400-402, or [15], 165-170).

In proving Theorem 1, we shall appeal to the well-known Engelking-Karłowicz theorem from set-theoretic topology:

Theorem 3 *[13] If $|Y| = \mu = \mu^{<\sigma} < \lambda = |X| \leq 2^\mu$, then there are functions $h_\alpha : X \to Y$ for $\alpha < \mu$ such that for every partial function f from X to Y of cardinality less than σ, for some $\alpha < \mu$, $f \subseteq h_\alpha$.*

A self-contained short proof can be found in [17], 422-423. We shall need just the case when $\mu = \sigma = \aleph_0$, and $\lambda = 2^\mu$. Since it may be less familiar to algebraists, for convenience we deduce the fact to which we appeal later on (although it also appears as Corollary 3.17 in [4]).

Lemma 4 *There exists a family $\{f_\eta : \eta \in {}^{\omega_1 >}2\}$ such that $f_\eta : \omega \to \mathbf{Z}$, and whenever $\eta_1, \dots, \eta_k$ are distinct and $a_1, \dots, a_k \in \mathbf{Z}$, then $\{i < \omega : (\forall\, l \leq k)(f_{\eta_l}(i) = a_l)\}$ is infinite.*

Proof. Take $\mu = \sigma = \aleph_0$, $\lambda = 2^\mu$, $X = {}^{\omega_1 >}2$ and $Y = \mathbf{Z}$ in the Engelking-Karłowicz theorem. Since $|{}^{\omega_1 >}2| = 2^{\aleph_0}$ and $|\mathbf{Z}| = \aleph_0$, we know that there exist functions $h_n : {}^{\omega_1 >}2 \to \mathbf{Z}$ for $n < \omega$ such that for every partial function f from ${}^{\omega_1 >}2$ to $\mathbf{Z}$ whose domain is finite, there is some $m < \omega$ such that $f \subseteq h_m$. Let $\{g_i : i < \omega\}$ be an enumeration with infinitely many repetitions of each h_n for $n < \omega$.

For each $\eta \in {}^{\omega_1 >}2$, define $f_\eta : \omega \to \mathbf{Z}$ by $f_\eta(i) = g_i(\eta)$. The family $\{f_\eta : \eta \in {}^{\omega_1 >}2\}$ is as required: for if $\eta_1, \dots, \eta_k$ are distinct and $a_1, \dots, a_k \in$

$\mathbf{Z}$ are given, then the set $f = \langle(\eta_1, a_1), \ldots, (\eta_k, a_k)\rangle$ is a finite function, so there is some m such that $f \subseteq h_m$ and it is now easy to see that $\{i < \omega : (\forall\, l \leq k)(f_{\eta_l}(i) = a_l)\}$ is infinite. □

A well-known algebraic fact will also be useful:

Lemma 5 *Let C be a closed unbounded subset of the regular uncountable cardinal κ. Suppose that H is an abelian group of cardinality κ, and $\langle H_\alpha : \alpha < \kappa\rangle$ is a κ-filtration of H (a continuous increasing chain of subgroups H_α, $|H_\alpha| < \kappa$, whose union is H). Let $S = \{\alpha \in C : H/H_\alpha$ is not κ-free$\}$. Then H is free if and only if S is non-stationary in κ.*

Proof. Well known: see Proposition IV.1.7 in [10]. □

We refer the reader to [14] for the definitions of the characteristic $\chi(g)$ and the type $\tau(g)$ of an element g in a group.

Now we prove Theorem 1.

Proof. Let $\mathbf{P}$ be the set of prime numbers, and let $\{P_\eta : \eta \in {}^{\omega_1>}2\}$ be a family of almost disjoint (infinite) subsets of $\mathbf{P}$: $\eta \neq \nu \in {}^{\omega_1>}2 \Rightarrow |P_\eta \cap P_\nu| < \aleph_0$. By Lemma 4, there exists $\{f_\eta : \eta \in {}^{\omega_1>}2\}$ such that $f_\eta : \omega \to \mathbf{Z}$, and if $\eta_1, \ldots, \eta_k$ are distinct and $a_1, \ldots, a_k \in \mathbf{Z}$, then $\{i < \omega : (\forall\, l \leq k)(f_{\eta_l}(i) = a_l)\}$ is infinite.

Define functions x_η and $x_{\eta,j}$ in $\mathbf{Z}^\omega$ as follows. Let $x_\eta = \langle \pi_{\eta,i} \cdot f_\eta(i) : i < \omega\rangle$ where $\pi_{\eta,i} = \Pi\{p \in P_\eta : p < i\}$, and let $x_{\eta,j} = \langle \pi^j_{\eta,i} \cdot f_\eta(i) : i < \omega\rangle$ where $\pi^j_{\eta,i} = \Pi\{p \in P_\eta : j \leq p < i\}$ (=0 if $i \leq j$). Note that $x_\eta = x_{\eta,0}$.

For $\eta \in {}^{\omega_1}2$, let G_η be the subgroup of $\mathbf{Z}^\omega$ generated by $\mathbf{Z}_\omega \cup \{x_{\eta|\alpha,j} : \alpha < \omega_1, 0 \leq j < \omega\}$.

We show that the family $\mathbf{G} = \{G_\eta : \eta \in {}^{\omega_1}2\}$ satisfies the conclusions of Theorem 1.

Claim 1: G_η is pure in $\mathbf{Z}^\omega$.

Proof of Claim 1: Suppose that $rx = g$ for some $x \in \mathbf{Z}^\omega$, $r \in \mathbf{N}$, and $g \in G_\eta$. Say $g = y + n_1 x_{\eta|\alpha_1,j_1} + \cdots + n_m x_{\eta|\alpha_m,j_m}$, $n_l \neq 0$, with $y \in \mathbf{Z}_\omega$. Without loss of generality (adding more elements from $\mathbf{Z}_\omega$ to the RHS if necessary), $(\forall\, l \leq m)(j_l = j)$ for some $j < \omega$, $j > r$, $y(i) = 0$ $(\forall\, i > j)$, and $x(i) = 0$ $(\forall i \leq j)$. Relabelling (if necessary), we may assume that $\alpha_1 < \cdots < \alpha_m < \omega_1$, and because $x_{\eta|\alpha_l,j}(i) = 0$ if $i \leq j$, we may write

$$rx = ry^* + n_1 x_{\eta|\alpha_1,j} + \cdots + n_m x_{\eta|\alpha_m,j}, \quad \text{for some } y^* \in \mathbf{Z}_\omega.$$

Fix $k \in \{1, \ldots, m\}$. Since $\eta|\alpha_1, \ldots, \eta|\alpha_m$ are distinct $(\alpha_1 < \cdots < \alpha_m)$, letting $a_l = \delta_{kl}$ (Kronecker delta), we know that the set $N_k = \{i < \omega :$

$(\forall\, l \neq k)(f_{\eta_l}(i) = 0,\ f_{\eta_k}(i) = 1)\}$ is infinite. For large enough i in this set (e.g. $i > \max_{1\leq l\leq m}\ [\min(P_{\eta|\alpha_l}\backslash\{0,\ldots,j\})]$), $x_{\eta|\alpha_l,j}(i)$ is zero if and only if $l \neq k$. So for infinitely many $i < \omega$, for $l \neq k$, $x_{\eta|\alpha_l,j}(i) = 0$, and $x_{\eta|\alpha_k,j}(i) \neq 0$.

Unfix k. For each $k \leq m$, for infinitely many $i \in (j,\omega) \cap N_k$, $rx(i) = n_k x_{\eta|\alpha_k,j}(i) = n_k \Pi\{p \in P_{\eta|\alpha_k} : j \leq p < i\}$. Since $r < j$, we must have $rs_k = n_k$ for some s_k in $\mathbf{Z}$, and therefore $x = y^* + s_1 x_{\eta|\alpha_1,j} + \cdots + s_m x_{\eta|\alpha_m,j} \in G_\eta$ (G_η is torsion-free). Hence G_η is pure in $\mathbf{Z}^\omega$, which establishes Claim 1.

Claim 2: G_η has cardinality $\aleph_1$, so (1.1) holds.

Proof of Claim 2: If $\xi \neq \zeta \in {}^{\omega_1>}2$, then for some $j < \omega$, $P_\xi \cap P_\zeta \subseteq j$. Pick $p, q > j$ with $p \in P_\xi$ and $q \in P_\zeta$; so the set $B = \{i < \omega : f_\xi(i) = p$ and $f_\zeta(i) = q\}$ is infinite, and if $i \in B$ is bigger than $\max\{j, p, q\}$, then $x_{\xi,j}(i) \neq x_{\zeta,j}(i)$, since $x_{\xi,j}(i)$ is non-zero and divisible by p^2 but by no prime in P_ζ, and $x_{\zeta,j}(i)$ is non-zero and divisible by q^2 but by no prime in P_ξ. It follows that G_η has cardinality $\aleph_1$. After this observation, a second's reflection on the element types of G_{η_1} and G_{η_2} (for $\eta \neq \nu$) should convince the reader that the groups are neither isomorphic nor free.

Claim 3: (1.2) holds: if $\eta_1 \neq \eta_2 \in {}^{\omega_1}2$, then G_{η_1} and G_{η_2} are almost disjoint.

Proof of Claim 3: Suppose (towards a contradiction) that for some $\eta_1 \neq \eta_2 \in {}^{\omega_1}2$, for some non-free abelian group H, there exist isomorphisms $\varphi_l : H \to \mathrm{range}(\varphi_l) \leq G_{\eta_l}$, $l = 1, 2$. Since G_{η_l} is $\aleph_1$-free, H must have cardinality $\aleph_1$. Let $\langle H_i : i < \omega_1\rangle$ be an ω_1-filtration of H. Without loss of generality, we may assume that each H_i is pure in H, so that H/H_i is torsion-free.

Let $G_{\eta,i} = \langle \mathbf{Z}_\omega \cup \{x_{\eta|\beta,j} : j < \omega,\ \beta < i\}\rangle$ for $i < \omega_1$ and $\eta \in \{\eta_1, \eta_2\}$.

Note that $\langle G_{\eta,i} : i < \omega_1\rangle$ is a ω_1-filtration of G_η, since it is increasing and continuous with union G_η, and each $G_{\eta,i}$ is countable. For large enough χ, the set C defined by $\{\delta < \omega_1 : \mathrm{dcl}_{(H(\chi),\in,<)}[\delta \cup \{G_{\eta_1}, G_{\eta_2}, \{x_\nu, f_\nu : \nu \in {}^{\omega_1>}2\}, \eta_1, \eta_2, \varphi_1, \varphi_2, \{H_i : i < \omega_1\}\}] \cap \omega_1 = \delta\}$ is a club of ω_1 (well known, or see [16], 401). Note that if $\delta \in C$, then φ_l maps H_δ into $G_{\eta_l,\delta}$. Since H is not free, it follows by Lemma 5 that $S = \{\delta \in C : H/H_\delta$ is not $\aleph_1$-free$\}$ is stationary. By Pontryagin's Criterion, for each $\delta \in S$, H/H_δ has a non-free (torsion-free) subgroup K_δ/H_δ of finite rank $n_\delta + 1$ such that every subgroup of K_δ/H_δ of rank less than $n_\delta + 1$ is free. Let $H_\delta^{\;+}/H_\delta$ be a pure subgroup of K_δ/H_δ of rank n_δ. Then $H_\delta^{\;+}/H_\delta$ is free with basis $y_0 + H_\delta, \ldots, y_{n_\delta - 1} + H_\delta$ say. So $K_\delta/H_\delta^{\;+} \simeq (K_\delta/H_\delta)/(H_\delta^{\;+}/H_\delta)$ is a torsion-free rank-1 group which is not free, and hence there is a non-zero element $y_{n_\delta} + H_\delta^{\;+}$ which is divisible in $K_\delta/H_\delta^{\;+}$ by infinitely many natural numbers. Call this set of natural numbers A.

For $l = 1,2$, for large enough $j_l(*) < \omega$, and ${\beta^l}_0 < \cdots < {\beta^l}_{k_l} < \omega_1$, $\varphi_l(y_m)$ is an element of the subgroup of G_{η_l} generated by $G_{\eta_l,\delta} \cup \{x_{\eta_l|{\beta^l}_0,j_l(*)}, \ldots, x_{\eta_l|{\beta^l}_{k_l},j_l(*)}\}$ for all $m \leq n_\delta$.

Taking large enough $\delta \in S$, we may assume that $\min\{\alpha : \eta_1|\alpha \neq \eta_2|\alpha\} < {\beta^l}_0$, $l = 1,2$. Since $\delta \in C$, we can show the following claims:
$(*)_1$: The set A does not contain infinitely many powers of one prime.
$(*)_2$: The set $Q = (\mathbf{P} \cap A) \subseteq P_{\eta_l|{\beta^l}_0} \cup \cdots \cup P_{\eta_l|{\beta^l}_{k_l}}$.

Now $(*)_1$ is true because non-zero sums of elements in $G_{\eta_l,\delta} \cup \{x_{\eta_l|{\beta^l}_0,j_l(*)}, \ldots, x_{\eta_l|{\beta^l}_{k_l},j_l(*)}\}$ are divisible by at most finitely many powers of any given prime (by the definition of the elements $x_{\eta_l|\beta,j}$). Note that $\chi(y_{n_\delta} + {H_\delta}^+) = \bigcup_{\{y \in y_{n_\delta} + {H_\delta}^+\}} \chi(y) \leq \bigcup_{\{y \in y_{n_\delta} + {H_\delta}^+\}} \chi(\varphi_l(y))$, where the characteristics are taken relative to $K_\delta / {H_\delta}^+$, K_δ and $G_{\eta_l,\delta} \cup \{x_{\eta_l|{\beta^l}_0,j_l(*)}, \ldots, x_{\eta_l|{\beta^l}_{k_l},j_l(*)}\}$ respectively. Hence $(*)_1$ holds. By $(*)_1$, since A is infinite, the set $Q = \mathbf{P} \cap A$ is infinite.

Also, the same characteristic inequality implies that $Q \subseteq P_{\eta_l|{\beta^l}_0} \cup \cdots \cup P_{\eta_l|{\beta^l}_{k_l}}$. So $(*)_2$ is true. Hence, $Q \subseteq \bigcap_{l=1,2}(\bigcup_{k \leq k_l} P_{\eta_l|{\beta^l}_k})$ which is finite (since the family $\{P_\eta : \eta \in {}^{\omega_1>}2\}$ is almost disjoint). This is a contradiction, and so Claim 3 follows, completing the proof of Theorem 1. □

Corollary 6 *Every non-slender $\aleph_1$-free abelian group G has a family $\{G_\alpha : \alpha < 2^{\aleph_1}\}$ of non-free subgroups such that:*
(6.1) each G_α is almost free of cardinality $\aleph_1$;
(6.2) if $\alpha < \beta$, then G_α and G_β are almost disjoint.

Proof. By Nunke's characterisation of slender groups (see Corollary IX.2.5 in [10] for example), G must contain a copy of the Baer-Specker group. □

Remark: For the same reason, the corollary is true for any non-slender cotorsion-free abelian group.

References

[1] R. Baer, *Abelian groups without elements of finite order*, Duke Math. J. **3** (1937), 68–122.

[2] A. Blass, *Cardinal characteristics and the product of countably many infinite cyclic groups*, J. Algebra **169** (1994), 512–540.

[3] A. Blass and R. Göbel, *Subgroups of the Baer-Specker group with few endomorphisms but large dual*, Fund. Math. **149** (1996), 19–29.

[4] W. W. Comfort and S. Negrepontis, *The Theory of Ultrafilters*, Berlin: Springer-Verlag (1974).

[5] A. L. S. Corner and R. Göbel, *Essentially rigid subgroups of the Baer-Specker group*, Manuscripta math. **94** (1997), 319–326.

[6] A. L. S. Corner and B. Goldsmith, *On endomorphisms and automorphisms of some pure subgroups of the Baer-Specker group*, in Abelian Group Theory and Related Topics (R. Göbel, P. Hill and W. Liebert, eds.), Contemp. Math. **171** (1994), 69–78.

[7] M. Dugas and J. Irwin, *On pure subgroups of cartesian products of integers*, Results in Math. **15** (1989), 35–52.

[8] K. Eda, *A note on subgroups of* $\mathbf{Z}^N$, in Abelian Group Theory (R. Göbel et al, eds.), Springer Lecture Notes in Mathematics **1006** (1983), 371–374.

[9] P. C. Eklof and A. H. Mekler, *Infinitary stationary logic and abelian groups*, Fund. Math. **112** (1981), 1–15.

[10] P. C. Eklof and A. H. Mekler, *Almost Free Modules. Set-theoretic Methods*, Amsterdam: North-Holland (1990).

[11] P. C. Eklof, A. H. Mekler, and S. Shelah, *Almost disjoint abelian groups*, Israel J. Math. **49** (1984), 34–54.

[12] P. C. Eklof and S. Shelah, *A combinatorial principle equivalent to the existence of non-free Whitehead groups*, in Abelian Group Theory and Related Topics (R. Göbel, P. Hill and W. Liebert, eds.), Contemp. Math. **171** (1994), 79–98.

[13] R. Engelking and M. Karłowicz, *Some theorems of set theory and their topological consequences*, Fund. Math. **57** (1965), 275–285.

[14] L. Fuchs, *Infinite Abelian Groups*, Vols. **I** and **II**, New York: Academic Press (1970, 1973).

[15] L. Heindorf and L. B. Shapiro, *Nearly Projective Boolean Algebras*, Springer Lecture Notes in Mathematics **1596** (1994).

[16] S. Shelah, *Classification Theory and the Number of Non-Isomorphic Models*, Revised edition, Studies in Logic and the Foundations of Mathematics, Vol. **92**, Amsterdam: North-Holland (1990).

[17] S. Shelah, *Cardinal Arithmetic*, Oxford: OUP (1994).

[18] E. Specker, *Additive Gruppen von Folgen ganzer Zahlen*, Portugaliae Math. **9** (1950), 131–140.

Abelian groups mapping onto their endomorphism rings

Shalom Feigelstock, Jutta Hausen,* and Robert Raphael †

Abstract

An abelian group G is said to be an E–group if it is the additive group of an E–ring. It is known that G is an E–group if and only if there exists a left $E(G)$–module isomorphism from G to its endomorphism ring $E(G)$. Groups which are isomorphic to the additive group of their endomorphism rings are called weak E–groups. The purpose of this article is to consider the apparently yet weaker condition that there be a homorphism from G onto the additive group of $E(G)$. Groups satisfying this condition are called EE–groups. The properties of EE–groups are studied and it is shown that they are very similar to E–groups. In fact, it is shown that every EE–group of finite torsion–free rank is a weak E–group, and that for various prominent classes of groups the concepts of EE–group and E–group coincide.

1 Introduction

Problem 45 of Laszlo Fuchs' monograph [7] asks one to characterize the rings R for which $R \simeq E(R^+)$, i.e. the rings R which are isomorphic to the endomorphism ring of their additive group R^+. It will be convenient to call such a ring a *weak E–ring*, and to call the additive group of a weak E–ring a *weak E–group*. Thus, G is a weak E–group if and only if $G \simeq E(G)^+$, and Fuchs' problem is equivalent to the characterization of the weak E–groups (which entails the determination of their endomorphism rings).

In his fundamental paper [15], Phillip Schultz gave a partial solution to Fuchs' problem. He introduced the concept of an E–ring: a ring R is said to be an *E–ring* if the left regular representation $a \mapsto a_\ell$ (the left multiplication by $a \in R$) is an isomorphism from R onto $E(R^+)$. The additive group of an E–ring is called an *E–group*. Both concepts have

*The second author wishes to thank Bar-Ilan University and Hebrew University for their hospitality and support.

†The third author acknowledges the hospitality of Bar-Ilan University and of the Hebrew University. He also acknowledges support from the NSERC (Canada).

gained prominence in abelian group theory [2, 3, 4, 5, 11, 12, 13, 14] (cf. [6]).

The object of this paper is to study abelian groups which satisfy a weaker condition. We call the abelian group G an *EE–group* if there exists an epimorphism $\varphi : G \to E(G)^+$ from G onto the additive group of its endomorphism ring $E(G)$.

We have the following results. If G is an EE–group, then the cardinalities of G, $E(G)$ as well as of G/G_t – provided G is not a torsion group – are all equal (we denote by G_t the maximal torsion subgroup of G); furthermore, the torsion–free rank of G equals that of $E(G)$, and the maximal torsion subgroups of G and of $E(G)$ are isomorphic. In addition, as is the case with the distinction between E–groups and weak E–groups, if the epimorphism $\varphi : G \to E(G)^+$ is $E(G)$–linear, then G is an E–group.

An abelian group A which is an EE–group but not a weak E–group necessarily must have the following properties: A is reduced, of infinite torsion–free rank, its p–primary components A_p are cyclic, and $A_p \neq 0$ implies $A = A_p \oplus H$ with H a p–divisible EE–group which is not a weak E–group. If A splits, then $A = T \oplus F$ where T is a cyclic group of finite order and F is a torsion–free EE–group that is not a weak E–group. And if A is torsion–free, it cannot be completely decomposable.

Recently, Rüdiger Göbel and Saharon Shelah have informed us that there exist EE–groups which are not E–groups: given any cardinal κ such that $\kappa = \kappa^{\aleph_0}$, they are able to construct an $\aleph_0$–free EE–group G of cardinality κ which is not an E–group [10].

Thus, there exists an EE–group G which is not an E–group whose torion–free rank equals the power of the continuum. In contrast, we show that every EE–group of finite torsion–free rank is a weak E–group. The following questions remain open.

Problem 1 *Does there exist an EE-group of countably infinite torsion-free rank which is not an E-group?*

Problem 2 *Does there exist a weak E-group which is not an E-group?*

Any unexplained notation can be found in the standard references [8, 9] for abelian groups. Let $r(G)$ and $r_0(G)$ denote the rank and the torsion–free rank of G, respectively.

The authors are indebted to S. Shelah for helpful discussions on this topic.

2 Properties of EE–groups

Throughout this section, we assume G is an EE–group with endomorphism ring $E = E(G)$, and we fix a group epimorphism $\varphi : G \to E^+$. Define, for all $a, b \in G$, the product $ab = \varphi(a)(b)$. One verifies that this endows G with a (not necessarily associative) ring structure. This ring shall be denoted by R_φ.

Given a ring R and $a \in R$, define a_r and a_ℓ to be the right and left multiplication by a, respectively, i.e. for all $x \in R$, $a_r(x) = xa$ and $a_\ell(x) = ax$. One easily verifies the following facts.

Lemma 2.1 *1) R_φ is a (not necessarily associative) ring with left identity e where $e \in G$ has the property that $\varphi(e) = 1$.*

2) The maps $r : a \mapsto a_r$ and $\ell : a \mapsto a_\ell$ are group homomorphisms from R_φ^+ into $E(G)^+$, and $\ell = \varphi$; if R_φ is associative then ℓ is a ring homomorphism.

3) For all $a \in R_\varphi$ there exists $b \in R_\varphi$ such that $xa = bx$ for all $x \in G$.

4) The map $r : R_\varphi^+ \to E(G)^+$ is a monomorphism.

Corollary 2.2 *Let G be an EE-group. Then G has the following properties:*

1) $|G| = |E(G)|$, and

2) $r_0(G) = r_0[E(G)^+]$.

Proof. By 2.1, G is embedded into $E(G)$, and there exists a surjection from G onto $E(G)$. Thus $|G| = |E(G)|$, and $r_0(G) \leq r_0[E(G)^+]$. Exercise 3 of [8] (p. 86) completes the proof. □

Proposition 2.3 *Let $G = H \oplus K$ such that both H and K are fully invariant. Then G is an EE-group if and only if both H and K are EE-groups.*

Proof. Let G be as stated. Then $E \simeq E(H) \oplus E(K)$ so that G is an EE–group if both H and K are. Conversely, suppose G is an EE–group. Let $\varphi : G \to E$ be an epimorphism. It suffices to show $\varphi(H) \subseteq E(H)$. Assume this were not so. Then there exists $h \in H$ such that $\varphi(h) = (\sigma, \tau)$ with $\sigma \in E(H)$ and nonzero $\tau \in E(K)$. Let $k \in K$ such that $\tau(k) \neq 0$. The evaluation map $\varepsilon_k : E(K) \to K$ defined by $\varepsilon_k(\eta) = \eta(k)$ for all $\eta \in E(K)$ is a group homomorphism. If $\pi : E \to E(K)$ denotes the projection from E onto $E(K)$ along $E(H)$, then $0 \neq \tau(k) = \varepsilon_k(\tau) = \varepsilon_k(\pi(\varphi(h)))$, and we have exhibited a nonzero homomorphism from H to K. □

Proposition 2.4 *For every EE-group G and for every prime p, G does not contain $Z(p^\infty)$.*

Proof. Assume, if possible, that $G = P \oplus H$ is an EE–group with $P = Z(p^\infty)$. Let $r = r_0(G)$. Since $E(P) \simeq J_p$ is embedded in E, the torsion–free rank of E is infinite. By 2.2, r is infinite. There exists a free subgroup of G having $\bigoplus_r Z(p^\infty)$ as an epimorphic image. Injectivity gives an exact sequence $0 \to C \to G \to Z(p^\infty)^{(r)} \to 0$ which induces the exact sequence

$$0 \to Hom(Z(p^\infty)^{(r)}, P) \to Hom(G, P).$$

It follows that $Hom(Z(p^\infty)^{(r)}, P) \simeq \prod_r J_p$ is embedded in E which is impossible since it has torsion–free rank greater than r. □

Proposition 2.5 *Let G be a group and let p be a prime such that $G_p \neq 0$. Then G is an EE–group if and only if G_p is cyclic and $G = G_p \oplus H$ with H a p-divisible EE–group which is fully invariant in G.*

Proof. Suppose, firstly, that G is an EE–group. By 2.4, G_p is reduced. The hypothesis implies that G has a decomposition $G = P \oplus H$ with P a finite p–group of rank $m \geq 1$. Suppose $p^n P = 0$, and let κ denote the rank of $H/p^n H$. Note that $E \simeq E(P) \oplus Hom(P, H) \oplus Hom(H, P) \oplus E(H)$ and $p^n E \simeq p^n E(H)$. By hypothesis, E is an epimorphic image of G and, hence, so is $E/p^n E$. It follows that $E/p^n E$ is an epimorphic image of $G/p^n G \simeq P \oplus (H/p^n H)$ which has rank $m + \kappa$. Thus, $E/p^n E$ has rank at most $m + \kappa$ (see [8], p. 138, Exercise 5). Let $\mu = r[Hom(H/p^n H, P)]$ and let $\nu = r[E(H)/p^n E(H)]$. Since $Hom(H, P) \simeq Hom(H/p^n H, P)$, we have

$$E/p^n E \simeq E(P) \oplus Hom(P, H) \oplus Hom(H/p^n H, P) \oplus [E(H)/p^n E(H)].$$

It follows that $m + \kappa \geq m^2 + \mu + \nu$. Assume κ is infinite. Then $\mu = 2^\kappa$ and the previous inequality would imply $\kappa \geq 2^\kappa$ which is impossible. We have shown κ is finite, hence $\mu = m\kappa$ which implies $m = 1$ and $\nu = 0$. It follows that $E(H) = p^n E(H)$ so that H is p–divisible. From 2.4 we deduce $H_p = 0$ which implies $G_p = P$ is cyclic and $H = p^n H = p^n G$ is fully invariant. By 2.3, H is an EE–group. Invoking 2.3 again (and recalling that cyclic groups are E-groups) completes the proof. □

Proposition 2.6 *Let $G = \langle a \rangle \oplus H$ be an EE–group with $G_p = \langle a \rangle \simeq Z(p^n) \neq 0$ and $H = pH$, and let $\varphi : G \to E^+$ be an epimorphism. Then $E^+ = \langle \varphi(a) \rangle \oplus \varphi(H)$ and $E_p = \langle \varphi(a) \rangle \simeq \langle a \rangle$. In particular, the restriction $\varphi \restriction \langle a \rangle$ is monic.*

Proof. Since both $\langle a \rangle$ and H are fully invariant, $E \simeq Z(p^n) \oplus E(H)$, and $E(H)_p = 0$ since $H_p = 0$. Thus, E_p is cyclic of order p^n. Clearly, $E^+ = \langle \varphi(a) \rangle + \varphi(H)$. It suffices to show $\varphi(H)$ contains no element of order

p. Assume the contrary. Then, being p–divisible, $\varphi(H)$ would contain a copy of $Z(p^\infty)$ contradicting the fact that E_p is cyclic. Thus, the sum is direct. □

Corollary 2.7 *Let G be an EE–group with corresponding epimorphism $\varphi : G \to E(G)^+$. Then $\varphi \restriction G_t$ is an isomorphism from G_t to $E(G)_t$. In particular, $G_t \simeq E(G)_t^+$.*

Corollary 2.8 *Let G be an EE–group such that $G \neq G_t$. Then $|G| = |G/G_t|$.*

Proof. It follows from 2.5 that $G_t \leq \mathbf{Q}/\mathbf{Z}$ is countable. By hypothesis, $|G/G_t|$ is infinite. Thus, $|G| = |G_t| \cdot |G/G_t| = |G/G_t|$. □

3 EE-groups that are E–groups

Every EE–group of finite torsion–free rank is a weak E–group as can be seen from the following result.

Theorem 3.1 *Let G be an EE–group of finite torsion-free rank. Then every epimorphism $\varphi : G \to E(G)^+$ is an isomorphism.*

Proof. Let G be an EE–group of finite torsion–free rank and let $\varphi : G \to E(G)$ be an epimorphism. Let K be the kernel of φ. Using Exercise 3, p. 86, of [8] again, it follows that $r_0(G) = r_0(K) + r_0(E(G))$. By 2.2, the torsion–free ranks of G and $E(G)$ coincide. Thus, finiteness implies $r_0(K) = 0$ proving $K \leq G_t$. Since the restriction of φ to G_t is monic (2.7), $K = 0$. Hence, φ is an isomorphism. □

Corollary 3.2 *For an abelian torsion group T, the following conditions are equivalent:*

1) T is an EE-group.

2) T is an E-group.

3) $T = Z(n)$ for some nonnegative integer n.

Proof. By 3.1, a torsion EE–group is a weak E–group. Schultz proved that torsion weak E–groups are cyclic [15] (Theorem 1). All cyclic groups are well known to be E–groups. □

Proposition 3.3 *For a nonreduced abelian group G, the following conditions are equivalent:*
1) G is an EE-group.
2) G is an E-group.
3) $G = \mathbf{Q} \oplus Z(n)$ for some nonnegative integer n.

Proof. Suppose G is a nonreduced EE–group. By 2.4, there exists a decomposition $G = \mathbf{Q} \oplus H$. Let $r = r_0(H)$. Assume, if possible, that r is infinite. Choose a free subgroup F of H such that H/F is a torsion group. Using injectivity, every homomorphism from F to $\mathbf{Q}$ extends uniquely to H. Thus

$$\bigoplus_{2^r} \mathbf{Q} \simeq \prod_r \mathbf{Q} \simeq Hom(F, \mathbf{Q}) \simeq Hom(H, \mathbf{Q}) \hookrightarrow E(G)$$

which contradicts $r_0[E(G)] = r_0(G)$ (2.2). Thus r is finite, and 3.1 implies G is a weak E–group. By Theorem 3 of [15], G satisfies 3), and groups of this form are E–groups. □

Lemma 3.4 *For a splitting mixed abelian group G, the following conditions are equivalent:*
1) G is an EE-group.
2) $G = Z(n) \oplus nG$ where n is a nonnegative integer and nG is a torsion-free EE-group.

Proof. If G satisfies 2), it follows from 2.3 that G is an EE–group. Suppose G is an EE–group which splits. Then $G = G_t \oplus H$ with H torsion–free and p–divisible for all relevant primes of G (2.5). Hence, H is fully invariant in G and 2.3 implies that G_t and H both are EE–groups. By 3.2, $G_t = Z(n)$ is cyclic, and $H = nH = nG$. □

The following proposition includes the case when G is finitely generated.

Proposition 3.5 *For a completely decomposable abelian group G, the following conditions are equivalent:*
1) G is an EE-group.
2) G is an E-group.
3) $G = Z(n) \oplus \bigoplus_{i=1}^{m} G_i$ for some nonnegative integers n and m, and the $G_i = nG_i$ are torsion-free rank-one groups of pairwise incomparable idempotent types.

Proof. Suppose G is a completely decomposable EE–group. Then G splits and the preceding lemma implies $G = Z(n) \oplus H$ with $H = nH = nG$ a torsion–free EE–group which is completely decomposable. Let $H = \bigoplus_r G_i$ where each G_i is a rank-one group. If r were infinite then $\prod_r E(G_i)$ would be embedded in $E(G)$ which is impossible since $r_0[E(G)] = r_0(G) = r$. Hence r is finite and 3.1 applies. Theorem 2 of [15] implies 3). Since each of the finitely many direct summands in the decomposition 3) is a fully invariant E–group, any such G must be an E–group. □

4 The nature of the epimorphism

Phillip Schultz singled out the E–rings among the weak E–rings by the nature of the isomorphism $R \simeq E(R^+)$ [15]. In particular, G is an E–group if and only if there exists a left $E(G)$–module isomorphism from G to $E(G)$ (cf. 4.1). Similarly, we show that the existence of an $E(G)$–module epimorphism from G to $E(G)$ forces G to be an E–group:

Theorem 4.1 *For an abelian group G, the following conditions are equivalent:*

1) There exists a left $E(G)$–module epimorphism $\varphi : G \to E(G)$.

2) G is an EE–group and $Hom_{\mathbf{Z}}(G, E(G)) = Hom_{E(G)}(G, E(G))$.

3) G is an E–group.

If one of these conditions is satisfied, then every epimorphism φ from G onto $E(G)^+$ is both a left $E(G)$–module isomorphism and a ring isomorphism from R_φ to $E(G)$.

Proof. Note that, if $\varphi : G \to E(G)^+$ is an epimorphism and $a, b \in G$, then $\varphi(a \cdot b) = \varphi(\varphi(a)(b))$. Hence, φ is a left $E(G)$–module homomorphism if and only if φ is a ring homomorphism from R_φ to $E(G)$.

Suppose 3) holds. Then $G = R^+$ for some E–ring R. The left regular representation $a \mapsto a_\ell$ is both a ring isomorphism and a left $E(G)$–module isomorphism from R to $E(G)$ ([15], Lemma 6). Also, $Hom_{\mathbf{Z}}(R, R) = Hom_R(R, R)$ ([3], 1.2). An easy change–of–ring argument (cf. [1], 4.2) shows 2) holds. Trivially, 2) implies 1). Assume the validity of 1) and let $\varphi : G \to E(G)$ be a left $E(G)$–module epimorphism. By the remark above, φ is a ring homomorphism from R_φ onto $E(G)$. Let $e \in G$ with $\varphi(e) = 1$. Since ${}_{E(G)}E(G)$ is free, $G = E(G)e \oplus K$ where K is the kernel of φ. Let $\pi : G \to E(G)e$ be the projection along K. Then $\pi = \pi \cdot 1 = \varphi(\pi(e)) = \varphi(e) = 1$ proving $K = 0$ and φ is a left $E(G)$–module isomorphism and an isomorphism of rings. By 2.1, φ is the left regular representation, and Lemma 6 of [15] implies R_φ is an E–ring. □

If R is a unital (not necessarily associative) weak E–ring, then R is an E–ring if and only if R is commutative [15]. Other conditions which are equivalent to R being an E–ring can be found in [15, 3, 14]. We conclude with a related result which illustrates that in the presence of various natural ring theoretical properties the rings R_φ are, in fact, E–rings.

Theorem 4.2 *Let G be an EE–group and let $\varphi : G \to E(G)^+$ be an epimorphism. The following conditions are equivalent:*

1) R_φ is associative.

2) R_φ is commutative.

3) R_φ is unital.

4) R_φ is an E–ring.

5) G is an E–group.

Proof. By definition, 4) implies 5). Choose $e \in R_\varphi$ such that $\varphi(e) = 1$. By 2.1, e is a left identity of R_φ.

1) $\Longrightarrow$ 2) Suppose R_φ is associative and let $a, b \in R_\varphi$. By 2.1, there exists $c \in R_\varphi$ such that $xa = cx$ for all $x \in R_\varphi$. Hence $a = ea = ce$ and associativity implies $ab = ceb = cb = ba$.

2) $\Longrightarrow$ 3) Suppose R_φ is commutative. Then e is a two–sided identity of R_φ.

3) $\Longrightarrow$ 4) Suppose R_φ is unital. By 2.1, $\ell(R_\varphi) = E(R_\varphi^+)$. Lemma 6 of [15] implies R_φ is an E–ring.

5) $\Longrightarrow$ 1) Suppose G is an E–group. By 4.1, φ is an isomorphism of rings, hence $R_\varphi \simeq E(G)$ is associative. □

References

[1] F. W. Anderson and K. R. Fuller, *Rings and Categories of Modules*, Graduate Texts in Mathematics **13**, Springer–Verlag, New York (1973).

[2] R. A. Beaumont and R. S. Pierce, *Torsion-free rings*, Illinois J. Math. 5 (1961), 61–98.

[3] R.A. Bowshell and P. Schultz, *Unital rings whose additive endomorphisms commute*, Math. Ann. **288** (1977), 197–214.

[4] M. Dugas, A. Mader, and C. Vinsonhaler, *Large E–rings exist*, J. Algebra **108** (1987), 88–101.

[5] T. G. Faticoni, *Each countable reduced torsion-free commutative ring is a pure subring of an E–ring*, Comm. in Algebra **15** (1987), 2545–2564.

[6] S. Feigelstock, *Additive Groups of Rings*, Research Notes in Mathematics, Pitman, London (1983).

[7] L. Fuchs, *Abelian Groups*, Pergamon Press, London (1967).

[8] L. Fuchs, *Infinite Abelian Groups*, Vol. **I**, Academic Press, New York (1970).

[9] L. Fuchs, *Infinite Abelian Groups*, Vol. **II**, Academic Press, New York (1973).

[10] R. Göbel and S. Shelah, *Generalized E–rings*, to appear.

[11] J. Hausen, *E-transitive torsion-free abelian groups*, J. Algebra **107** (1987), 17–27.

[12] P. A. Krylov, *Strongly homogeneous torsion-free abelian groups*, Sibirsk. Mat. Zh. **24** (1983), 77–84; English translation: Siberian Math. J. **24** (1983), 215–221.

[13] G. P. Niedzwecki and J. D. Reid, *Abelian groups projective over their endomorphism rings*, J. Algebra **159** (1993), 139–149.

[14] R. S. Pierce, *E-modules*, in **Contemporary Mathematics 87** (1989), 221–240.

[15] P. Schultz, *The endomorphism ring of the additive group of a ring*, J. Austral. Math. Soc. **15** (1973), 60–69.

Trends in Mathematics, © 1999 Birkhäuser Verlag Basel/Switzerland

Purity and Reid's theorem

Andreas Blass* and John Irwin

Abstract

We give conditions under which an abelian group is the sum of two free subgroups, one or both of which are pure.

1 Introduction

Reid [1] proved that every torsion-free abelian group of infinite rank is the sum of two free subgroups. In this paper, we consider the problem of arranging for one or both of those free subgroups to be pure.

Theorem 1.1 *For any torsion-free abelian group G of infinite rank κ, the following two statements are equivalent.*

1. *G is the sum of two free subgroups, at least one of which is pure in G.*
2. *G has a pure free subgroup of rank κ.*

Moreover, any subgroup as in 2 can serve as one of the subgroups in 1.

With a stronger condition on the rank, we get the same result with both summands pure.

Theorem 1.2 *For any torsion-free abelian group G of uncountable rank κ, the following two statements are equivalent.*

1. *G is the sum of two pure free subgroups.*
2. *G has a pure free subgroup of rank κ.*

Moreover, any subgroup as in 2 can serve as one of the subgroups in 1.

The uncountability of the rank is really needed for the implication from 2 to 1 in Theorem 1.2. We shall construct a group of countable rank where 2 holds but 1 fails.

*Partially supported by NSF grant DMS–9505118

2 Proof of Theorem 1.1

Throughout this section and the next, G is a torsion-free abelian group of rank κ.

To prove the implication from 1 to 2 in Theorem 1.1, suppose $G = E+F$ with both E and F free and with E pure in G. If E has rank κ then we have the desired conclusion 2. So assume E has rank $\mu < \kappa$. Of course then F must have rank κ. Being a subgroup of E, the intersection $E \cap F$ is freely generated by a set X of cardinality at most μ.

Since F is free, fix a basis for it, and express all elements of X in terms of this basis. Since $\mu < \kappa$, fewer than κ basis elements occur in these expressions. Let F_1 be the subgroup of F generated by these basis elements and F_2 the subgroup generated by the rest of our basis. So $F = F_1 \oplus F_2$, $E \cap F \subseteq F_1$, and F_2 has rank κ.

To complete the verification of 2, it suffices to show that F_2 is pure in G. In fact, we show that it is a direct summand of G. Since $G = E + F = E + F_1 + F_2$, it suffices to show that $(E + F_1) \cap F_2 = (0)$. Suppose, therefore, that $e \in E$, $f_1 \in F_1$, $f_2 \in F_2$, and $e + f_1 = f_2$. Then $e = f_2 - f_1 \in E \cap F \subseteq F_1$. So $e + f_1 = f_2 \in F_1 \cap F_2 = (0)$. This completes the proof that 1 implies 2.

The converse implication and the "Moreover" statement in the theorem are proved just like Reid's theorem [1], but we include the proof here for the sake of completeness.

Since G has infinite rank κ and is torsion-free, it has cardinality κ, so we can enumerate it as $G = \{g_\alpha : \alpha < \kappa\}$, where we use the customary identification of a cardinal κ with the initial ordinal of that cardinality. Let E be a pure free subgroup of G of rank κ. We shall define a certain function $e : \kappa \to E$, set $f_\alpha = g_\alpha + e(\alpha)$, and let F be the subgroup of G generated by $\{f_\alpha : \alpha < \kappa\}$. No matter how we choose e, we shall have $G = E + F$, because each element g_α of G is $-e(\alpha) + f_\alpha$ with $-e(\alpha) \in E$ and $f_\alpha \in F$. So our task is to choose e so that F is free. In fact, we shall arrange that the f_α's are independent.

For this purpose, we work in the divisible hull $\bar{G}$ of G, and we define $e(\alpha)$ by transfinite recursion on α. So assume that $e(\beta)$ and therefore f_β are already defined for all $\beta < \alpha$ and that the f_β's are linearly independent (over $\mathbb{Q}$). We must choose $e(\alpha)$ so as to preserve this independence when f_α is adjoined. That is, we must choose $e(\alpha)$ so that $f_\alpha = g_\alpha + e(\alpha)$ is not in the $\mathbb{Q}$-span of $\{f_\beta : \beta < \alpha\}$. Equivalently, $e(\alpha)$ should not be in the affine subspace of $\bar{G}$ spanned by $\{f_\beta - g_\alpha : \beta < \alpha\} \cup \{-g_\alpha\}$. That affine subspace has dimension strictly smaller than κ because $\alpha < \kappa$. So it cannot include all of the rank κ group E. Therefore appropriate choices for $e(\alpha)$ exist (in great profusion). This completes the recursive construction and thus the proof of Theorem 1.1. □

3 Proof of Theorem 1.2

In view of Theorem 1.1, we need only prove the implication from 2 to 1 and the "Moreover" assertion, on the assumption that the rank κ of G is uncountable. (As before, G is a torsion-free abelian group.) As in the corresponding part of the proof of Theorem 1.1, we begin with a pure free subgroup E of rank κ and an enumeration $\{g_\alpha : \alpha < \kappa\}$ of G, and we let F be generated by $\{f_\alpha : \alpha < \kappa\}$ where $f_\alpha = g_\alpha + e(\alpha)$ for a certain inductively defined $e : \kappa \to E$. As before, we trivially have $G = E + F$. As before, we must ensure that F is free, but in addition we must ensure that F is pure in G.

For this purpose, we shall arrange that the f_α are linearly independent modulo every prime p. In more detail, this means the following. For each prime number p, each finitely many (say k) ordinals $\alpha_1 < \alpha_2 < \cdots < \alpha_k < \kappa$, and each k integers $c_1, \ldots, c_k$, if $c_1 f_{\alpha_1} + \cdots + c_k f_{\alpha_k}$ is divisible by p in G then all the c_i are divisible by p in $\mathbb{Z}$.

If we achieve this, then it will follow that F is free and in fact the f_α are linearly independent. Indeed, any non-trivial dependence relation between them, over $\mathbb{Z}$, would also be a non-trivial dependence relation modulo any sufficiently large prime.

Furthermore, it will follow that F is pure in G. Indeed, linear independence modulo p clearly implies that, if $c_1 f_{\alpha_1} + \cdots + c_k f_{\alpha_k}$ is divisible by p in G, then it is divisible by p in F also.

So the proof will be complete if we define $e : \kappa \to E$ so that the f_α are linearly independent modulo every prime. We do this by recursion on α. Suppose we have already defined $e(\beta)$ for all $\beta < \alpha$ and that the resulting f_β's are independent modulo every prime. We wish to define $e(\alpha)$ so that this independence persists when f_α is included. This means that, for each prime p, the images $\bar{f}_\beta$ of f_β in G/pG should be linearly independent over $\mathbb{Z}/p\mathbb{Z}$ for $\beta \leq \alpha$. Since they're already independent for $\beta < \alpha$, we just have to make sure that $\bar{f}_\alpha$ is not a $\mathbb{Z}/p\mathbb{Z}$-linear combination of the earlier $\bar{f}_\beta$'s. Since $\alpha < \kappa$, there are (strictly) fewer than κ such linear combinations. Each is a coset of pG in G, and we must ensure that f_α is in none of these cosets. We use uncountability of κ to observe that, even for all primes p together, there are still fewer than κ cosets to be avoided.

Fix a basis for the free group E, and notice that the difference $b - b'$ of two basis elements is divisible by no primes p, in E and therefore in G as E is pure. Thus, none of the cosets that we must avoid can contain $g_\alpha + b$ for two different basis elements b. Thus, the fewer than κ cosets that we must avoid prohibit fewer than κ basis elements from serving as $e(\alpha)$. Since E has rank κ, the basis has κ elements, so some (in fact most) of them can serve as $e(\alpha)$. □

4 A counterexample

In this section, we show that the hypothesis of uncountable rank is needed in Theorem 1.2. We construct a group G and a pure, free, proper subgroup E, both of rank $\aleph_0$, such that every element of $G \setminus E$ is divisible by infinitely many primes. This clearly implies that every pure free subgroup of G is included in E, so G cannot be a sum of such subgroups.

Let $\bar{G}$ be a rational vector space with basis $\{e_0, e_1, e_2, \dots\} \cup \{g\}$, and let $\bar{E}$ be the subspace generated by $\{e_0, e_1, e_2, \dots\}$. (These will be the divisible hulls of our G and E.) Let E be the $\mathbb{Z}$-span of $\{e_0, e_1, e_2, \dots\}$, a free abelian group of rank $\aleph_0$. We shall obtain G as the union $\bigcup_{n=0}^{\infty} G_n$ of an increasing sequence of subgroups G_n, each generated by E together with some finite subset of $\bar{G}$. Notice that, if we ensure that E is pure in each G_n, then it will also be pure in G.

List the countable set $\bar{G} \setminus \bar{E}$ in a sequence $(v_k)_{k \in \mathbb{N}}$ in such a way that each element of $\bar{G} \setminus \bar{E}$ occurs infinitely often in the list. We define G_n by induction on n, starting with G_0 generated by $E \cup \{g\}$.

Suppose G_n has been defined, that it is a subgroup of $\bar{G}$ generated by E plus a finite set, and that E is pure in it. (This is clearly true for $n = 0$.) We shall define G_{n+1} having the same properties. If $v_n \notin G_n$ then do nothing, i.e., set $G_{n+1} = G_n$. Now suppose $v_n \in G_n$. Since $v_n \notin \bar{E}$, we have $\bar{G} = \bar{E} \oplus \mathbb{Q}v_n$. Let $\pi : \bar{G} \to \mathbb{Q}v_n$ be the projection homomorphism with kernel $\bar{E}$. Because G_n is generated by E plus a finite set, $\pi(G_n)$ is a finitely generated subgroup of $\mathbb{Q}v_n$. It contains v_n because G_n does. So

$$\pi(G_n) = \frac{1}{m}\mathbb{Z}v_n$$

for some positive integer m. Let p be a prime larger than both m and n, and let G_{n+1} be generated by $G_n \cup \{v_n/p\}$. Thus, v_n is divisible by p in G_{n+1} and therefore in the final group G.

This completes the definition of G. We check that it and E have the desired properties. It is clear that they have rank $\aleph_0$, that E is free, and that E is a proper subgroup of G (as $g \in G_0 \setminus E$). Also, if $v \in G \setminus \bar{E}$ then $v \in G_k$ for some k and there are infinitely many $n \geq k$ with $v_n = v$. For each such n, the construction of G_{n+1} ensured that v is divisible in G by a prime $p > n$. So v is divisible by infinitely many primes.

It remains to check that E is pure in G; this will also ensure that the preceding remarks about $v \in G \setminus \bar{E}$ apply to all $v \in G \setminus E$. As pointed out above, it suffices to check that E is pure in each G_n, and we shall do this by induction on n. The result is clear for $n = 0$, so assume E is pure in G_n and we have

$$k \in \mathbb{Z}, \quad k \geq 1, \quad x \in G_{n+1}, \quad \text{and} \quad kx \in E.$$

We must show $x \in E$. By definition of G_{n+1}, we have

$$x = y + \frac{r}{p}v_n, \quad r \in \mathbb{Z}, \quad y \in G_n,$$

where p is as in the definition of G_{n+1}. Using the notation of that definition, we have, since $kx \in E$,

$$y + \frac{r}{p}v_n = x \in \bar{E} = \text{kernel}(\pi),$$

so

$$\frac{r}{p}v_n = \pi(\frac{r}{p}v_n) = -\pi(y) \in \pi(G_n) = \frac{1}{m}\mathbb{Z}v_n.$$

Thus

$$\frac{r}{p}v_n = \frac{t}{m}v_n$$

for some integer t. Therefore $rm = pt$. But p is a prime not dividing m (recall p was chosen $> m$). So p divides r, and we have

$$\frac{r}{p}v_n \in \mathbb{Z}v_n \subseteq G_n.$$

Therefore $x \in G_n$. This and $kx \in E$ and the induction hypothesis that E is pure in G_n entail $x \in E$, as required.

References

[1] J. Reid, *A note on torsion free abelian groups of infinite rank*, Proc. Amer. Math. Soc. **13** (1962), 222–225.

Trends in Mathematics, © 1999 Birkhäuser Verlag Basel/Switzerland

Basic subgroups and a freeness criterion for torsion-free abelian groups

Andreas Blass* and John Irwin

Abstract

We prove that if an abelian group has a basic subgroup of infinite rank and if every subgroup disjoint from any basic subgroup is free then the group itself is free. We prove several corollaries and related results, including some that do not require the existence of a basic subgroup.

1 Introduction

All groups in this paper are abelian and torsion-free. A subgroup B of a group G is *basic* in G if it is a free, pure subgroup of G and the quotient G/B is divisible. A subgroup H of G is *B-high* if it is maximal among subgroups of G disjoint from B. ("Disjoint" means the intersection is $\{0\}$, not $\emptyset$ which is impossible for subgroups.) Our main result is the following theorem, which confirms a conjecture of the second author.

Theorem 1.1 *Let G be a torsion-free abelian group such that*

1. *G has a basic subgroup of infinite rank, and*
2. *for every basic subgroup B of G, all B-high subgroups of G are free.*

Then G is free.

Since subgroups of free groups are free and since Zorn's Lemma allows us to enlarge every subgroup disjoint from B to a B-high subgroup, hypothesis 2 of the theorem can equivalently be stated as "every subgroup of G that is disjoint from a basic subgroup is free."

Part of our motivation for this theorem is to "confine" the non-freeness of groups. Under hypothesis 1, the contrapositive of the theorem tells us that if a group is not free then a "reason" for this can be found disjoint from some basic subgroup.

*Partially supported by NSF grant DMS-9505118

Another part of our motivation is the desire for a torsion-free analog of a result proved for p-groups in [1]. In the context of p-groups, "direct sum of cyclic groups" plays the role of "free," for example in the definition of "basic." Theorem 1 of [1] says that if G is a p-group without elements of infinite height, if B is basic in G, and if every B-high subgroup of G is a direct sum of cyclic groups, then G is a direct sum of cyclic groups.

The situation is more complicated in the torsion-free case for several reasons. First, unlike p-groups, torsion-free groups — even very well-behaved ones, e.g., separable ones — need not have basic subgroups [3]. So the existence of a basic subgroup must be assumed. Second, we must assume that a basic subgroup has infinite rank. Although that assumption can be avoided in many interesting cases, e.g., if G is $\aleph_1$-free, it cannot be dropped altogether. An example from [4] is used in Section 3 to show what can go wrong. Third, we need to assume freeness of B-high subgroups for all (or at least for many) basic B's, not just for one. Whether this third difference can be eliminated remains an open problem.

In addition to the theorem mentioned above, we shall present several corollaries of the theorem or of its proof. One of these is similar to the theorem but does not require the existence of a basic subgroup; the price for this is that hypothesis 2 must be strengthened.

2 Proof of the main theorem

The proof of Theorem 1.1 is in two parts, the first of which is a construction carried out in an arbitrary free abelian group of infinite rank. This part has essentially nothing to do with the general group G of the theorem. The second part applies this construction inside a basic subgroup of G. Since the first part may be of some interest in its own right, we formulate as a proposition the result of this construction.

Proposition 2.1 *Let B be a free abelian group of infinite rank. There is a subset S of B and there is a partition of S into countably many pieces S_k ($k \in \mathbb{N}$) with the following properties.*

1. *S freely generates a subgroup $\langle S \rangle$ of B.*

2. *The purification $\langle S \rangle_*$ of $\langle S \rangle$ in B is all of B.*

3. *For each n, let $S(\geq n) = \bigcup_{k \geq n} S_k$. Then $B/\langle S(\geq n) \rangle$ is divisible.*

Proof. Fix a basis W of the free group B, and let κ be the cardinality of W; so κ is infinite by hypothesis. Let $\preceq$ be a well-ordering of W such that each element has strictly fewer than κ predecessors. In other words, the order type of $(W, \preceq)$ is the initial ordinal of cardinality κ.

Assign to each $w \in W$ countably many later (in $\preceq$) members of W, which we call $f(w,k)$ for $k \in \mathbb{N}$, such that all the $f(w,k)$ are distinct. Thus, we have

- $f : W \times \mathbb{N} \to W$ is one-to-one, and
- $f(w,k) \succ w$ for all $w \in W$ and all $k \in \mathbb{N}$.

One way to produce such an f is to partition W into κ subsets A_w (one for each $w \in W$), each of size κ and therefore each cofinal in $(W \preceq)$. Then let $f(w,k)$ be the kth element of A_w that is $\succ w$.

Also fix an enumeration of the prime numbers $\langle p_k : k \in \mathbb{N}\rangle$, such that each prime is listed infinitely often.

For each $k \geq 1$, we define

$$S_k = \{w - p_k \cdot f(w,k) \mid w \in W\},$$

and for $k = 0$ we define

$$S_0 = \{w - p_0 \cdot f(w,0) \mid w \in W\} \cup R$$

where $R = W - \text{range}(f)$. Finally, we set $S = \bigcup_{k \in \mathbb{N}} S_k$, and we claim that S and the S_k are as required in the proposition.

Because W is an independent set and f is one-to-one, it is clear by inspection that the various S_k are disjoint, so they form a partition of $S \subseteq B$. It remains to verify statements (1), (2), and (3) of the proposition.

Verification of (1): Suppose, toward a contradiction, that we had a non-trivial $\mathbb{Z}$-linear relation, say $\sum_i c_i s_i = 0$, where the s_i are finitely many distinct elements of S and the c_i are non-zero integers. Thinking of each s_i as a linear combination of members of W, let w^* be the last (with respect to $\preceq$) element of W involved in any of the s_i. Consider all s_i in which w^* occurs, and consider what these occurrences could look like. *A priori*, there are three possibilities.

- $s_i = w^* \in R$.
- $s_i = w^* - p_k \cdot f(w^*, k)$ for some k.
- $s_i = w - p_k \cdot w^*$ for some $w \in W$ and $k \in \mathbb{N}$ such that $f(w,k) = w^*$.

The second of these possibilities, however, contradicts our choice of w^*, because $f(w^*,k) \succ w^*$, so we need only consider the first and third. Only one of these can arise, since the first requires $w^* \in R = W - \text{range}(f)$ while the third requires $w^* = f(w,k) \in \text{range}(f)$. Furthermore, in the third case, w and k are uniquely determined, because f is one-to-one. Therefore, w^* can occur in only one s_i in our alleged linear relation. But then this linear relation cannot hold in B, because the coefficient of the basis element w^* cannot be cancelled. This contradiction completes the proof of (1).

Verification of (2): It suffices to prove that $\langle S\rangle_*$ contains every element w of W, and we do this by transfinite induction along the well-ordering $\preceq$. So let $w \in W$ and assume $w' \in \langle S\rangle_*$ for all $w' \prec w$. If $w \notin \text{range}(f)$, then $w \in R \subseteq S_0 \subseteq S \subseteq \langle S\rangle_*$, and we are done. So we may assume $w = f(w', k)$ for some w' and k. By our choice of f, we know $w' \prec w$ and so $w' \in \langle S\rangle_*$. We also have $w' - p_k \cdot w \in S_k \subseteq S \subseteq \langle S\rangle_*$. Being a subgroup, $\langle S\rangle_*$ must contain $p_k \cdot w$. Being pure, it must contain w.

Verification of (3): Fix n and let $[x]$ denote the image in $B/\langle S(\geq n)\rangle$ of $x \in B$. Since W generates B, it suffices to show that $[w]$ is divisible by p in $B/\langle S(\geq n)\rangle$, for all $w \in W$ and all primes p. Consider a specific $[w]$ and a specific p. Our enumeration of primes p_k listed every prime infinitely often, so we can fix some $k \geq n$ such that $p_k = p$. Then

$$w - p \cdot f(w,k) = w - p_k \cdot f(w,k) \in S_k \subseteq S(\geq n) \subseteq \langle S(\geq n)\rangle.$$

Thus in $B/\langle S(\geq n)\rangle$ we have $[w] = p \cdot [f(w,k)]$. This completes the proof of the proposition. □

Using the proposition, we can now prove Theorem 1.1.

Proof. Let G be as in the hypothesis of Theorem 1.1, let B be any basic subgroup of it of infinite rank, and let H be any B-high subgroup of G. Because B is free and has infinite rank, we can apply Proposition 2.1 to it; let S and S_k be as in that proposition. In addition to the notation $S(\geq n)$ used there, we shall use the analogous notation $S(< n) = \bigcup_{k<n} S_k$. As before, we shall use a subscript $*$ to indicate purification. For subgroups of B, it makes no difference whether we purify in B or in G because B is pure in G.

Temporarily fix an arbitrary $n \in \mathbb{N}$.

Claim 1: $\langle S(\geq n)\rangle_*$ is a basic subgroup of G.

Proof. $\langle S(\geq n)\rangle_*$ is certainly pure, being the purification of something, and free, being a subgroup of the free group B. We know that G/B is divisible, as B is basic in G, and that $B/\langle S(\geq n)\rangle_*$ is divisible, as it is a quotient of $B/\langle S(\geq n)\rangle$, which is divisible, according to (3) of Proposition 2.1. Therefore $G/\langle S(\geq n)\rangle_*$ is divisible.

Claim 2: $H + \langle S(< n)\rangle$ is disjoint from $\langle S(\geq n)\rangle$.

Proof. Suppose $h + s = s'$ with $h \in H$, $s \in \langle S(< n)\rangle$, and $s' \in \langle S(\geq n)\rangle$. Then $h = s' - s \in \langle S\rangle \subseteq B$. But H and B are disjoint, so $h = s' - s = 0$. Because S is linearly independent, by (1) of Proposition 2.1, we conclude $s' = 0 = s$.

Claim 3: $(H + \langle S(< n)\rangle)_*$ is disjoint from $\langle S(\geq n)\rangle_*$.

Proof. In any torsion-free group, disjoint subgroups have disjoint purifications.

Claim 4: $(H + \langle S(< n)\rangle)_*$ is free.

Proof. Being disjoint from $\langle S(\geq n)\rangle_*$ by Claim 3, $(H + \langle S(<n)\rangle)_*$ is included in some $\langle S(\geq n)\rangle_*$-high subgroup by Zorn's Lemma. That high subgroup is free, by hypothesis (2) and Claim 1. So its subgroup $(H+\langle S(<n)\rangle)_*$ is also free.

Now un-fix n.

Claim 5: $\bigcup_{n\in\mathbb{N}} (H + \langle S(< n)\rangle)_* = G$.

Proof. The left side of the claim, call it L, is the union of an increasing chain of pure subgroups of G, so it is itself a pure subgroup of G. It obviously includes H. It also includes

$$\bigcup_{n\in\mathbb{N}} \langle S(< n)\rangle_* = \langle S\rangle_* = B,$$

where we used part (2) of Proposition 2.1 and the fact that purification commutes with union of chains.

To finish the proof of the claim, we check that the only pure subgroup L of G that includes both B and H is all of G. Consider any $g \in G$; we want to show $g \in L$. If $g \in H$, we're done, so assume $g \notin H$. As H is B-high, $H + \langle g\rangle$ is not disjoint from B. Say $h + mg = b \neq 0$ with $h \in H$, $m \in \mathbb{Z}$, $b \in B$, and $m \neq 0$ (as H and B are disjoint). Then $mg = b - h \in L$ and $g \in L$ because L is pure.

Claim 5 exhibits G as the union of a countable, increasing sequence of pure subgroups, each free by Claim 4. By a well-known result of Hill [5, Theorem 2], it follows that G is free. □

Every free group of infinite rank obviously satisfies the hypotheses of Theorem 1.1. From this and the theorem, it follows that those hypotheses exactly characterize free groups of infinite rank. There are several immediate consequences of this — for example that the hypotheses of Theorem 1.1 are preserved by infinite-rank subgroups and by direct sums — which seem to have no easy direct proofs, i.e., no proofs avoiding Theorem 1.1.

3 Variations

In this section, we present several corollaries of the main theorem or of its proof. The first eliminates the hypothesis that the basic subgroup must have infinite rank, at the cost of assuming that G is $\aleph_1$-free, i.e., that all countable subgroups of G are free.

Unfortunately some cost is necessary; the main theorem becomes false if one simply drops the assumption that a basic subgroup has infinite rank. This follows from an example constructed in [4], a torsion-free abelian group of rank 2 such that all subgroups of rank 1 are cyclic and all torsion-free quotients of rank 1 are divisible. In such a group, every pure subgroup B of rank 1 is basic, and every B-high subgroup is free, yet the whole group clearly cannot be free.

Corollary 3.1 *Let G be an $\aleph_1$-free abelian group such that*

1. *G has a basic subgroup and*

2. *for every basic subgroup B of G, all B-high subgroups of G are free.*

Then G is free.

Proof. Let G satisfy the hypotheses of the corollary, and let B be any basic subgroup of G. If B has infinite rank, then we simply invoke Theorem 1.1, which is applicable since $\aleph_1$-free obviously implies torsion-free. We're also done if $B = G$, since a basic subgroup is free by definition. So we may assume that B has finite rank and that $G/B \neq \{0\}$. Since G/B is divisible and torsion-free, it includes a copy of $\mathbb{Q}$. Consider the preimage in G of this copy. It has finite rank, because both B and $\mathbb{Q}$ do, but it is not finitely generated, because $\mathbb{Q}$ isn't. So it is countable but not free, and this contradicts the hypothesis that G is $\aleph_1$-free. □

The next corollary slightly weakens the need to consider *all* basic subgroups in the main theorem's second hypothesis.

Corollary 3.2 *Let G be a torsion-free abelian group such that*

1. *G has a basic subgroup B of infinite rank and*

2. *for every basic subgroup B' of G that is a subgroup of B (and has the same rank as B), all B'-high subgroups of G are free.*

Then G is free.

In this corollary and the next one, there is a parenthesized clause to the effect that certain basic subgroups have equal rank. Our proofs establish the apparently stronger versions of the corollaries that include these clauses. Deletion of these clauses would not, however, really weaken the corollaries, because, as Dugas and Irwin pointed out in [3], all basic subgroups of the same group necessarily have the same rank.

Proof. In the proof of Theorem 1.1, hypothesis 2 was applied only to the basic subgroups $\langle S(\geq n)\rangle_*$, which are subgroups of B. They have the same rank as B because, in the proof of Proposition 2.1, each of the pieces S_k of S has cardinality equal to the cardinality of W, which is the rank of B. Thus, hypothesis 2 of the present corollary covers all the applications actually made of the apparently stronger hypothesis 2 of the main theorem. □

Next we use a result from [2] to infer that, under suitable assumptions, at least one B-high subgroup must be free, so the requirement that they all be free is satisfied if they all are isomorphic.

Corollary 3.3 *Let G be a torsion-free abelian group of uncountable rank κ such that*

1. *G has a basic subgroup of the same rank κ, and*
2. *for each basic subgroup B (of rank κ), all the B-high subgroups of G are isomorphic.*

Then G is free.

Proof. It suffices to show that, if G and a basic subgroup B have the same uncountable rank κ, then there exists a free, B-high subgroup of G. Once this is shown, we can simply invoke the preceding corollary to complete the proof.

Given such G and B, we obtain, from Theorem 2 of [2], a free, pure subgroup F of G such that $G = B + F$. (It is for the purity of F that we must go beyond what Reid proved in [6] and that we need the uncountability of κ. A counterexample in [2] shows that when $\kappa = \aleph_0$ then there may be no suitable F.)

Let H be a $(B \cap F)$-high subgroup of F. Since F is free, so is H, and it remains only to check that H is B-high in G. Obviously, $H \cap B = \{0\}$, so we need only check that, if $g \in G - H$ then $H + \langle g\rangle$ contains a non-zero element of B.

There is a trivial case, namely when $g \in F$. Then, as H is $(B \cap F)$-high in F, $H + \langle g\rangle$ contains a non-zero element of $B \cap F$.

So assume from now on that $g \notin F$. As $G = B + F$, we can write $g = b + f$ with $b \in B$, $f \in F$, and (as $g \notin F$) $b \notin F$. Now if $f \in H$ then $H + \langle g\rangle$ contains $-f + g = b$, which is a non-zero element of B, and we're done.

So assume from now on that $f \notin H$. As H is $(B \cap F)$-high in F, $H + \langle f \rangle$ contains a non-zero element of $B \cap F$. Say $h + mf \in B - \{0\}$ with $h \in H$ and $m \in \mathbb{Z}$. Notice that $m \neq 0$, for otherwise we would have h in both $B - \{0\}$ and H which is absurd since H is disjoint from B. Now $H + \langle g \rangle$ contains $h + mg = (h + mf) + mb \in B$, so we're done unless this element of B is 0. But then $mb = -h - mf \in F$ and, as F is pure, $b \in F$, a contradiction. □

Finally, we present a version of the main theorem applicable to groups that have no basic subgroup. For this we need the notion of a *pure independent set* in a torsion-free group G. This is an independent set $I \subseteq G$ that generates a pure subgroup $\langle I \rangle$ of G; in other words, it is a basis for a pure, free subgroup of G. We shall have to consider maximal pure independent sets. Notice that a pure independent set $I \subseteq G$ is maximal if and only if the quotient $G/\langle I \rangle$ has no pure subgroup isomorphic to $\mathbb{Z}$. (In particular, a basis for a basic subgroup is maximal pure independent.) Notice also that, if I is a maximal pure independent set in G, it need not follow that $\langle I \rangle$ is a maximal pure free subgroup of G; there may well be pure free subgroups of G properly extending $\langle I \rangle$, but they cannot have bases extending I. Finally notice that, by Zorn's Lemma, every torsion-free abelian group has a maximal pure independent subset, and in fact every pure independent set is included in a maximal one.

Corollary 3.4 *Let G be a torsion-free abelian group with a pure free subgroup of infinite rank. Assume that, for every maximal pure independent $I \subseteq G$, all $\langle I \rangle$-high subgroups of G are free. Then G is free.*

Proof. The hypothesis implies, via Zorn's Lemma, that there is an infinite maximal pure independent set $I \subseteq G$. Fix such an I and write B for $\langle I \rangle$. Then proceed as in the first two paragraphs of the proof of Theorem 1.1, introducing S, S_k, etc. and temporarily fixing n as there. Of course we cannot simply copy Claim 1 from the earlier proof, since B is no longer basic, but we have the following substitute.

Claim 1′: $G/\langle S(\geq n) \rangle_*$ has no pure subgroup isomorphic to $\mathbb{Z}$.

Proof. Because $\langle I \rangle$ is maximal pure independent, G/B contains no pure copy of $\mathbb{Z}$. From part 3 of Proposition 2.1 it follows that $B/\langle S(\geq n) \rangle_*$ is divisible. Thus $G/\langle S(\geq n) \rangle_*$ has a divisible subgroup such that the quotient by it contains no pure copy of $\mathbb{Z}$. It follows easily that $G/\langle S(\geq n) \rangle_*$ itself contains no pure copy of $\mathbb{Z}$. (We point out that this has a direct proof, making no use of the fact that a divisible subgroup is always a direct summand. If a group X has a divisible subgroup Y and X/Y contains no pure $\mathbb{Z}$, then for each $x \in X$, since its image $[x] \in X/Y$ doesn't generate a pure $\mathbb{Z}$, some prime p must divide $[x]$. Say $[x] = p[v]$ in X/Y. Then $x - pv \in Y$ and, as Y

is divisible, $x - pv = py$ for some $y \in Y$. Then $x = pv + py$, being divisible by p, cannot generate a pure $\mathbb{Z}$ in X.) □

Since $\langle S(\geq n)\rangle_*$ is free (being a subgroup of B) and pure in G, Claim 1′ implies that any basis for it is a maximal pure independent subset of G. By the hypothesis of the corollary, every $\langle S(\geq n)\rangle_*$-high subgroup of G is free.

Now the rest of the proof of Theorem 1.1, starting with Claim 2, goes through without further changes, showing that G is free. □

The same proof applies more generally.

Corollary 3.5 *Let $\mathcal{C}$ be a class of abelian groups such that, whenever $C \in \mathcal{C}$, D is divisible, and $A \cong C \oplus D$ then $A \in \mathcal{C}$. Let G be a torsion-free abelian group such that*

1. *G has a pure free subgroup B of infinite rank such that $G/B \in \mathcal{C}$, and*

2. *for each such B, all B-high subgroups of G are free.*

Then G is free.

This corollary includes the preceding one as the special case where $\mathcal{C}$ consists of the groups containing no pure copy of $\mathbb{Z}$. It also includes the main theorem as the special case where $\mathcal{C}$ consists of the divisible groups.

References

[1] K. Benabdallah and J. M. Irwin, *An application of B-high subgroups of abelian p-groups*, J. Algebra **34** (1975), 213–216.

[2] A. Blass and J. M. Irwin, *Purity and Reid's theorem*, this volume.

[3] M. Dugas and J. M. Irwin, *On basic subgroups of* $\prod Z$, Comm. Algebra **19** (1991), 2907–2921.

[4] L. Fuchs and F. Loonstra, *On the cancellation of modules in direct sums over Dedekind domains*, Nederl. Akad. Wetensch. Proc. Ser. A 74 (also Indag. Math. 33) (1971), 163–169.

[5] P. Hill, *On the freeness of abelian groups: A generalization of Pontryagin's theorem*, Bull. Amer. Math. Soc. **76** (1970), 1118–1120.

[6] J. Reid, *A note on torsion free abelian groups of infinite rank*, Proc. Amer. Math. Soc. **13** (1962), 222–225.

Trends in Mathematics, © 1999 Birkhäuser Verlag Basel/Switzerland

Absolutely rigid systems and absolutely indecomposable groups

Paul C. Eklof* and Saharon Shelah†

Abstract

We give a new proof that there are arbitrarily large indecomposable abelian groups; moreover, the groups constructed are absolutely indecomposable, that is, they remain indecomposable in any generic extension. However, any absolutely rigid family of groups has cardinality less than the partition cardinal $\kappa(\omega)$.

0 Introduction

Mark Nadel [11] asked whether there is a proper class of torsion-free abelian groups $\{A_\nu : \nu \in Ord\}$ with the property that for any $\nu \neq \mu$, A_ν and A_μ are not $L_{\infty\omega}$-equivalent; this is the same as requiring that A_ν and A_μ do not become isomorphic in any generic extension of the universe. In that case we say that A_ν and A_μ are absolutely non-isomorphic . This is not hard to achieve for torsion abelian groups, since groups of different p-length are absolutely non-isomorphic. (See section 1 for more information.)

Nadel's approach to the question in [11] involved looking at known constructions of rigid systems $\{A_i : i \in I\}$ to see if they had the property that for $i \neq j$, $\mathrm{Hom}(A_i, A_j)$ remains zero in any generic extension of the universe. We call these *absolutely rigid systems.* Similarly we call a group *absolutely rigid* (resp. *absolutely indecomposable*) if it is rigid (resp. indecomposable) in any generic extension. Nadel showed that the Fuchs-Corner construction in [4, §89] constructs an absolutely rigid system $\{A_\nu : \nu < 2^\lambda\}$ of groups of cardinality λ, where λ is less than the first strongly inaccessible cardinal. But he pointed out that other constructions, such as Fuchs' construction [5] of a rigid system of groups of cardinality the first measurable or Shelah's [13] for an arbitrary cardinal involve non-absolute notions like direct products or stationary sets; so the rigid systems constructed may not remain rigid when the universe of sets is expanded. The same comment applies to any construction based on a version of the Black Box.

*Partially supported by NSF Grants DMS-9501415 and DMS-9704477
†Partially supported by NSF Grant DMS-9704477. Pub. No. 678

Here we show that there do *not* exist arbitrarily large absolutely rigid systems. The cardinal $\kappa(\omega)$ in the following theorem is defined in section 2; it is an inaccessible cardinal much larger than the first inaccessible, but small enough to be consistent with the Axiom of Constructibility.

Theorem 0.1 *If κ is a cardinal $\geq \kappa(\omega)$ and $\{A_\nu : \nu < \kappa\}$ is a family of non-zero abelian groups, then there are $\mu \neq \nu$ in κ such that in some generic extension $V[G]$ of the universe, V, there is a non-zero (even one-one) homomorphism $f : A_\nu \to A_\mu$.*

This cardinal $\kappa(\omega)$ (called the "first beautiful cardinal" by the second author in [14]) is the precise dividing line:

Theorem 0.2 *If κ is a cardinal $< \kappa(\omega)$ and λ is any cardinal $\geq \kappa(\omega)$, there is a family $\{A_\mu : \mu < \kappa\}$ of torsion-free groups of cardinality λ such that in any generic extension $V[G]$, for all $\mu \in \kappa$, A_μ is indecomposable and for $\nu \neq \mu$, $\mathrm{Hom}(A_\nu, A_\mu) = 0$.*

Despite the limitation imposed by Theorem 0.1, the construction used to prove Theorem 0.2 yields the existence of a proper class of absolutely different torsion-free groups, in the following strong form. This answers the question of Nadel in the affirmative, and also provides a new proof of the existence of arbitrarily large indecomposables.

Theorem 0.3 *For each uncountable cardinal λ, there exist 2^λ torsion-free absolutely indecomposable groups $\{H_{i,\lambda} : i < 2^\lambda\}$ of cardinality λ such that whenever $\lambda \neq \rho$ or $i \neq j$, $H_{i,\lambda}$ and $H_{j,\rho}$ are absolutely non-isomorphic.*

We show that the groups A_μ in Theorem 0.2 and the groups $H_{i,\lambda}$ in Theorem 0.3 are absolutely indecomposable by showing that in any generic extension the only automorphisms they have are 1 and -1. (The proof of Theorem 0.3 does not depend on results from [14].) However, we cannot make the groups absolutely rigid:

Theorem 0.4 *If κ is a cardinal $\geq \kappa(\omega)$ and A is a torsion-free abelian group of cardinality κ, then in some generic extension $V[G]$ of the universe, there is an endomorphism of A which is not multiplication by a rational number.*

Theorems 0.1 and 0.4 are proved in section 3, Theorem 0.2 in section 4, and Theorem 0.3 in section 5.

1 Infinitary logic and generic extensions

We will confine ourselves to the language of abelian groups. Thus an *atomic formula* is one of the form $\sum_{i=0}^{n} c_i x_i = 0$ where the c_i are integers and the x_i are variables.

$L_{\omega\omega}$ consists of the closure of the atomic formula under negation ($\neg$), finite conjunctions ($\wedge$) and disjunctions ($\vee$), and existential ($\exists x$) and universal ($\forall x$) quantification (over a single variable — or equivalently over finitely many variables). $L_{\infty\omega}$ consists of the closure of the atomic formula under negation, *arbitrary* (possibly infinite) conjunction ($\bigwedge$) and disjunction ($\bigvee$), and under existential and universal quantification ($\exists x, \forall x$). Rather than give formal definitions of other model-theoretic concepts, we will illustrate them with examples. Thus the formula $\varphi(y)$:

$$\forall x \exists z (2z = x) \wedge (\neg \exists z (3z = y))$$

is a formula of $L_{\omega\omega}$ with one free variable, y, which "says" that every element is 2-divisible, but y is not divisible by 3. More formally, if A is an abelian group and $a \in A$ we write $A \models \varphi[a]$ and say "a satisfies φ in A", if and only if every element of A is divisible by 2 and there is no $b \in A$ such that $3b = a$.

Also, the formula $\psi(x)$:

$$\exists y (py = x) \wedge (\bigwedge_{n \geq 1} \exists z (p^n z = y)) \wedge (x \neq 0)$$

is a formula of $L_{\infty\omega}$ with free variable x such that $A \models \psi[a]$ if and only if $a \in p^{\omega+1}A - \{0\}$.

A *sentence* is a formula which has no free variables; if φ is a sentence of $L_{\infty\omega}$,we write $A \models \varphi$ if and only if φ is true in A. We write $A \equiv_{\infty\omega} B$ to mean that every sentence of $L_{\infty\omega}$ true in A is true in B (and conversely because $\neg\varphi$ true in A implies $\neg\varphi$ true in B.) Obviously, if there is an isomorphism $f : A \to B$, then $A \equiv_{\infty\omega} B$. A necessary and sufficient condition for $A \equiv_{\infty\omega} B$ is given by the following ([8], or see [1, pp. 13f]):

Lemma 1.1 *$A \equiv_{\infty\omega} B$ if and only if there is a set P of bijections $p : A_p \to B_p$ from a finite subset A_p of A onto a finite subset B_p of B with the following properties:*

(i)[the elements of P are partial isomorphisms] *for every atomic formula $\varphi(x_1, ..., x_m)$ and elements $a_1, ..., a_m$ of* dom(p), *$A \models \varphi[a_1, ..., a_m]$ if and only if $B \models \varphi[p(b_1), ..., p(b_m)]$;*

(ii)[the back-and-forth property] *for every $p \in P$ and every $a \in A$ (resp. $b \in B$), there is $p' \in P$ such that $p \subseteq p'$ and $a \in$* dom(p') *(resp. $b \in$* rge(p')*).*

It is an easy consequence that if A and B are countable, then $A \equiv_{\infty\omega} B$ if and only if $A \cong B$. Also, this implies that if it is true in V that $A \equiv_{\infty\omega} B$, then $A \equiv_{\infty\omega} B$ remains true in $V[G]$. The converse is easy, and direct, since a sentence of $L_{\infty\omega}$ which, in V, holds true in A but false in B has the same status in $V[G]$, since no new elements are added to the groups.

(By a generic extension we mean an extension of the universe V of sets defined by the method of forcing. In general, more sets are added to the universe; possibly, for example, a bijection between an uncountable cardinal λ and the countable set ω. So cardinals of V may not be cardinals in $V[G]$; but the ordinals of $V[G]$ are the same as the ordinals of V. Also, the elements of any set in V are the same in V or $V[G]$.)

There exist non-isomorphic uncountable groups A and B (of cardinality $\aleph_1$ for example) such that $A \equiv_{\infty\omega} B$. (See for example [3].) However, for any groups A and B in the universe, V, there is a generic extension $V[G]$ of V in which A and B are both countable (cf. [7, Lemma 19.9, p. 182]). Therefore we can conclude that $A \equiv_{\infty\omega} B$ if and only if A and B are "potentially isomorphic", that is, there is a generic extension $V[G]$ of the universe in which they become isomorphic. Barwise argues in [1, p. 32] that potential isomorphism (that is, the relation $\equiv_{\infty\omega}$) is "a very natural notion of isomorphism, one of which mathematicians should be aware. If one proves that $A \not\cong B$ but leaves open the question [of whether A and B are potentially isomorphic] then one leaves the possibility that A and B are not isomorphic for trivial reasons of cardinality. Or to put it the other way round, a proof that [A is not potentially isomorphic to B] is a proof that $A \not\cong B$ for nontrivial reasons."

As an example, consider reduced p-groups A_ν (ν any ordinal) such that the length of A_ν is ν, that is, $p^\nu A_\nu = 0$ but for all $\mu < \nu$, $p^\mu A_\nu \neq 0$. Then for any $\nu_1 \neq \nu_2$, the groups A_{ν_1} and A_{ν_2} are not even potentially isomorphic: this is because for any ν there is a formula $\theta_\nu(x)$ such that $\exists x(\theta_\nu(x) \wedge x \neq 0)$ is true in a p-group A if and only if A has length $\geq \nu$. Indeed, we define, by induction, θ_ν to be

$$\exists y(py = x \wedge \theta_\mu(y))$$

if $\nu = \mu + 1$ and if ν is a limit ordinal define θ_ν to be

$$\bigwedge_{\mu<\nu} \exists y(py = x \wedge \theta_\mu(y)).$$

Thus there is a proper class of pairwise absolutely non-isomorphic p-groups. Although there is not available a standard group-theoretic notion which will serve the same purpose for torsion-free groups, we will prove in section 5 that there is a proper class of indecomposable torsion-free abelian groups $\{H_\lambda : \lambda \text{ a cardinal}\}$ such that for any $\lambda \neq \rho$, the groups H_λ and H_ρ are not $L_{\infty\omega}$-equivalent.

2 Quasi-well-orderings and beautiful cardinals

A *quasi-order* Q is a pair $(Q, \leq_Q)$ where $\leq_Q$ is a reflexive and transitive binary relation on Q. There is an extensive theory of well-orderings of quasi-orders developed by Higman, Kruskal, Nash-Williams and Laver among others (cf. [12], [9]). A generalization to uncountable cardinals is due to the second author ([14]). The key notion that we need is the following: for an infinite cardinal κ, Q is called *κ-narrow* if there is no *antichain* in Q of size κ, i.e., for every $f : \kappa \to Q$ there exist $\nu \neq \mu$ such that $f(\nu) \leq_Q f(\mu)$. (Note that this use of the terminology "antichain" — in [10, p.32] for example — is different from its use in forcing theory.)

A *tree* is a partially-ordered set $(T, \leq)$ such that for all $t \in T$, pred(t)= $\{s \in T : s < t\}$ is a well-ordered set; moreover, there is only one element r of T, called the *root* of T, such that pred(r) is empty. The order-type of pred(t) is called the height of t, denoted ht(t); the height of T is $\sup\{\mathrm{ht}(t) + 1 : t \in T\}$.

If Q is a quasi-order, a *Q-labeled tree* is a pair (T, Φ_T) consisting of a tree T of height $\leq \omega$ and a function $\Phi_T : T \to Q$. On any set of Q-labeled trees we define a quasi-order by: $(T_1, \Phi_1) \preceq (T_2, \Phi_2)$ if and only if there is a function $\theta : T_1 \to T_2$ which preserves the tree-order (i.e. $t \leq_{T_1} t'$ implies $\theta(t) \leq_{T_2} \theta(t')$) as well as the height of elements and also is such that for all $t \in T_1$, $\Phi_1(t) \leq_Q \Phi_2(\theta(t))$.

One result from [14] that we will use implies that for sufficiently large cardinals κ and sufficiently small Q, any set of Q-labeled trees is κ-narrow. In order to state the result precisely we need to define a certain (relatively small) large cardinal.

Let $\kappa(\omega)$ be the first ω-Erdös cardinal, i.e., the least cardinal such that $\kappa \longrightarrow (\omega)^{<\omega}$; in other words, the least cardinal such that for every function F from the finite subsets of κ to 2 there is an infinite subset X of κ such that there is a function $c : \omega \to 2$ such for every finite subset Y of X, $F(Y) = c(|Y|)$. It has been shown that this cardinal is strongly inaccessible (cf. [7, p. 392]). Thus it cannot be proved in ZFC that $\kappa(\omega)$ exists (or even that its existence is consistent). If it exists, there are many weakly compact cardinals below it, and, on the other hand, it is less than the first measurable cardinal (if such exists). Moreover, if it is consistent with ZFC that there is such a cardinal, then it is consistent with ZFC + V = L that there is such a cardinal ([15]). If $\kappa(\omega)$ does not exist, then Theorem 2.1 is uninteresting. On the other hand, Theorem 2.2 then applies to every cardinal κ, and its consequences, given in section 4, are still of interest.

The following is a consequence of results proved in [14] (cf. Theorem 5.3, p. 208 and Theorem 2.10, p. 197):

Theorem 2.1 *If Q is a quasi-order of cardinality $< \kappa(\omega)$, and $\mathcal{S}$ is a set of Q-labeled trees, then $\mathcal{S}$ is $\kappa(\omega)$-narrow.*

On the other hand, it follows from results in [14] that for any cardinal smaller than $\kappa(\omega)$, there is an absolute antichain of that size:

Theorem 2.2 *If $\kappa < \kappa(\omega)$, there is a family $\mathcal{T} = \{(T_\mu, \Phi_\mu) : \mu < \kappa\}$ of ω-labeled trees, each of cardinality $< \kappa(\omega)$, such that in any generic extension of V, for all $\mu \neq \nu$, $(T_\mu, \Phi_\mu) \not\preceq (T_\nu, \Phi_\nu)$.*

Here ω is the quasi-order consisting of the set of natural numbers with the equality relation. Some commentary is needed on the absoluteness of the antichain $\mathcal{T} = \{(T_\mu, \Phi_\mu) : \mu < \kappa\}$, since this is not dealt with directly in [14]. $\mathcal{T}$ is constructed in a concrete, absolute, way from a function F which is an example witnessing the fact that $\kappa < \kappa(\omega)$. First, a κ-D-barrier B and function $q : B \to \omega$ is constructed (B is a kind of elaborate indexing for an antichain cf. [14, proof of 2.5, p. 195]). This gives rise ([14, proof of 1.12, pp. 192f]) to an example showing that $\mathcal{P}_\beta(\omega)$ is not κ-narrow for some $\beta < \kappa(\omega)$; this example is embedded into the quasi-order of ω-labeled trees, giving rise to $\mathcal{T}$ ([14, p. 221]). The proof that $\mathcal{T}$ is an antichain reduces to the key property of F, a property which is absolute by an argument of Silver [15]. Using an equivalent definition of $\kappa(\omega)$, F is taken to be a function from the finite subsets of κ to ω such that there is no one-one function $\sigma : \omega \to \omega$ such that for all $n \in \omega$, $F(\{\sigma(0), ..., \sigma(n-1)\}) = F(\{\sigma(1), ..., \sigma(n)\})$; this property of F is preserved under generic extensions because it is equivalent to the well-foundedness of a certain tree; more precisely, the tree of finite partial attempts at σ has no infinite branch.

3 A bound on the size of absolutely rigid systems

In this section we will prove Theorems 0.1 and 0.4. Suppose that $\{A_\nu : \nu < \kappa\}$ is a family of non-zero abelian groups, where we can assume that $\kappa = \kappa(\omega)$. For each $\nu < \kappa$, let T_ν be the tree of finite sequences of elements of A_ν; that is, the elements of T_ν are 1-1 functions $s : n_s \to A_\nu$ for some $n_s \in \omega$ and $s \leq t$ if and only if $n_s \leq n_t$ and $t \restriction n_s = s$.

Let Q_{ab} be the set of all quantifier-free n-types of abelian groups; that is, $Y \in Q_{ab}$ if and only if for some abelian group G, some $n \in \omega$, and some function $s : n \to G$, Y is the set $\mathrm{tp}_{qf}(s/G)$ of all quantifier-free formulas $\varphi(x_0, ..., x_{n-1})$ of $L_{\omega\omega}$ such that $G \models \varphi[s(0), ..., s(n-1)]$. Partially-order Q_{ab} by the relation of inclusion.

Define $\Phi_\nu : T_\nu \to Q_{ab}$ by letting $\Phi_\nu(s) = \mathrm{tp}_{qf}(s/A_\nu)$. Now we can apply Theorem 2.1 to the family of Q_{ab}-labeled trees $\mathcal{S} = \{(T_\nu, \Phi_\nu) : \nu < \kappa\}$. (Note that the cardinality of Q_{ab} is $2^{\aleph_0}$ which is $< \kappa(\omega)$ since $\kappa(\omega)$ is strongly-inaccessible.) Therefore there exists $\nu \neq \mu$ such that $(T_\nu, \Phi_\nu) \preceq$

(T_μ, Φ_μ), say $\theta : T_\nu \to T_\mu$ is such that $s \leq t$ implies $\theta(s) \leq \theta(t)$ and for all $s \in T_\nu$, $\Phi_\nu(s) \subseteq \Phi_\mu(\theta(s))$.

Now move to a generic extension $V[G]$ in which A_ν is countable. In $V[G]$, let $\sigma : \omega \to A_\nu$ be a surjection. We will define an embedding $f : A_\nu \to A_\mu$ by letting $f(\sigma(n)) = \theta(\sigma \restriction n+1)(n)$ for all $n < \omega$. To see that f is an embedding, note that $f(\sigma(n)) = \theta(\sigma \restriction k)(n)$ for all $k > n$ since θ preserves the tree ordering; moreover, for any $a, b, c \in A_\nu$, there is a k such that $a, b, c \in \text{rge}(\sigma \restriction k)$ so since

$$\Phi_\nu(\sigma \restriction k) \subseteq \Phi_\mu(\theta(\sigma \restriction k)) = \Phi_\mu(\langle f(0), ..., f(k-1)\rangle)$$

every quantifier-free formula satisfied by a, b, c in A_ν (e.g. $a \neq 0$, $a - b = c$, $ab = c$) is satisfied by $f(a), f(b), f(c)$ in A_μ. This completes the proof of Theorem 0.1 □

The argument is very general and could be applied to any family of structures (for example, to those in [6]). If we start with a torsion-free group A of cardinality $\kappa \geq \kappa(\omega)$, and apply the argument to the family of structures $\{\langle A, a_v\rangle : \nu < \kappa(\omega)\}$ where $\{a_\nu : \nu < \kappa(\omega)\}$ is a linearly independent subset of A, then we obtain $\nu \neq \mu$ such that in a generic extension in which A becomes countable we have an embedding $f : A \to A$ taking a_ν to a_μ. This proves Theorem 0.4. □

Ernest Schimmerling has pointed out that there is a "soft" proof of these results (not relying on Theorem 2.1) using a model of set theory (with $\{A_\nu : \nu < \kappa\}$ as additional predicate) and a set of indiscernibles given by the defining property of $\kappa = \kappa(\omega)$.

4 Existence theorem

In this section we will prove Theorem 0.2. So let $\kappa < \kappa(\omega)$ and let λ be a cardinal $\geq \kappa(\omega)$. Let $\{(T_\mu, \Phi_\mu) : \mu < \kappa\}$ be the family of ω-labeled trees as in Theorem 2.2. We can assume that each node of T_μ of height m is a sequence of length m and the tree-ordering is extension of sequences; so the root of the tree is the empty sequence $<>$.

Let $\langle p_{n,m,j} : n, m \in \omega, j \in \{0,1\}\rangle$ and $\langle q_{n,m,\ell,j} : n, m, \ell \in \omega, j \in \{0,1\}\rangle$ be two lists, with no overlap, of distinct primes.

For any ordinal α, Z_α be the tree of finite strictly decreasing non-empty sequences z of ordinals $\leq \alpha$ such that $z(0) = \alpha$. (Thus for some $n \in \omega$, $z : n \to \alpha$ such that $\alpha = z(0) > z(1) > ... > z(n-1)$.)

For $n \in \omega$ let $g_n : \lambda \to \mathcal{P}([\lambda n, \lambda(n+1)))$ such that for each $\nu < \lambda$, $g_n(\nu)$ is a subset of $[\lambda n, \lambda(n+1))$ which is cofinal in $\lambda(n+1) = \lambda n + \lambda$. (Here the operations are ordinal addition and multiplication, so, in particular, λn is less than λ^+, the cardinal successor of λ.) We also require that for $\mu \neq \nu$, $g_n(\mu) \cap g_n(\nu) = \emptyset$. For $n > 0$, let $Y_n = \bigcup \mathrm{rge}(g_n)$, and let $Y_0 = g_0(0)$.

For $\mu < \kappa$, let W_μ be the $\mathbb{Q}$-vector space with basis $\bigcup_{n \in \omega} \mathcal{A}_n \cup \mathcal{B}_{n,\mu}$ where for $n > 0$

$$\begin{aligned} \mathcal{A}_n &= \{a_z^\alpha : \alpha \in Y_{n-1}, z \in Z_\alpha\}, \\ \mathcal{B}_{n,\mu} &= \{b_{\eta,\mu}^\alpha : \alpha \in Y_{n-1}, \eta \in T_\mu - \{<>\}\} \end{aligned}$$

and $\mathcal{A}_0 = \{a_\mu^0\} = \mathcal{B}_{0,\mu}$. We are going to define A_μ to be a subgroup of W_μ. Since μ is fixed throughout the construction, we will usually omit the subscript μ from what follows (until we come to consider $\mathrm{Hom}(A_\nu, A_\mu)$).

For each $n > 0$, let h_n be a bijection from $\mathcal{A}_n \cup \mathcal{B}_n$ onto λ; let $h_0(a^0) = 0$. Then for any $w \in \mathcal{A}_n \cup \mathcal{B}_n$, and any $\alpha \in g_n(h_n(w))$, we will use $a_{<>}^\alpha$ or $b_{<>}^\alpha$ as a notation for w. (So $a_{<>}^\alpha = b_{<>}^\alpha$; moreover, $a_{<>}^\alpha = a_{<>}^\beta$ if and only if α and β belong to the same member of the range of g_n.) Now we can define A ($= A_\mu$) to be the subgroup of W generated (as abelian group) by the union of

$$\bigcup_{n \geq 0} \{\frac{1}{p_{n,m,0}^k} a_z^\alpha : m, k \in \omega, z \in Z_\alpha \cup \{<>\}, \alpha \in Y_n, \mathrm{dom}(z) = m\} \tag{1}$$

$$\bigcup_{n \geq 0} \{\frac{1}{p_{n,m,1}^k}(a_z^\alpha + a_{z\restriction m-1}^\alpha) : m, k \in \omega - \{0\}, z \in Z_\alpha, \alpha \in Y_n, \mathrm{dom}(z) = m\} \tag{2}$$

and

$$\bigcup_{n \geq 0} \{\tfrac{1}{q_{n,m,\ell}^k}(b_\eta^\alpha + b_{\eta\restriction m-1}^\alpha) : m, k \in \omega - \{0\}, \eta \in T_\mu, \alpha \in Y_n, \\ \mathrm{dom}(\eta) = m, \Phi_\mu(\eta) = \ell\}.$$

(where $b_{\eta\restriction -1}^\alpha = 0$). We will use the sets (1) and (2) to prove that **(I)** A_μ is absolutely indecomposable and the last set to prove that **(II)** $\mathrm{Hom}(A_\mu, A_\nu) = 0$ for all $\mu \neq \nu$.

If $x \in A_\mu$, we will write $p^\infty | x$ if for every $k \in \omega$, there exists $v \in A_\mu$ such that $p^k v = x$. For example, if $w \in \mathcal{A}_n \cup \mathcal{B}_n$ and $\alpha \in g_n(h_n(w))$ and $\Phi_\mu(<>) = \ell_0$, then $p_{n,0,0}^\infty | w$ and $q_{n,0,\ell_0}^\infty | w$. Assertions about divisibility in A_μ are easily checked by considering the coefficients of linear combinations over $\mathbb{Q}$ of elements of the basis $\bigcup_{n \in \omega} \mathcal{A}_n \cup \mathcal{B}_n$ of W_μ; for example, $p_{n,m,0}^\infty | x$ if and only if $x = \sum_{i=1}^r c_i a_{z_i}^{\alpha_i}$ for some $\alpha_1, \ldots, \alpha_r$ in Y_n, z_i of length m, and $c_i \in \mathbb{Q}$ (with denominator a power of $p_{n,m,0}$).

Proof of (I)

We will show, in fact, that in any generic extension $V[G]$ the only automorphisms of A ($= A_\mu$) are the trivial ones, 1 and -1. This part of the proof does not use the trees in $\mathcal{T}$; the absoluteness is a consequence of an argument using formulas of $L_{\infty\omega}$, which therefore works in any generic extension. We will use the following claim:

> (1A) there are formulas $\psi_{n,\alpha}(x)$ of $L_{\infty\omega}$ ($n \in \omega$, $\alpha \in Y_n$) such that for any $u \in A$, $A \models \psi_{n,\alpha}[u]$ if and only if there are $w_1, \ldots, w_r \in \mathcal{A}_n \cup \mathcal{B}_n$, and $c_1, \ldots, c_r \in \mathbb{Z} - \{0\}$, such that $u = \sum_{i=1}^r c_i w_i$ and $\alpha \in \bigcup_{i=1}^r g_n(h_n(w_i))$.

Assuming the claim for now, suppose that in $V[G]$ there is an automorphism F of A. For any $n \in \omega$, consider any $w \in \mathcal{A}_n \cup \mathcal{B}_n$; since $w = a^\alpha_{<>}$ for $\alpha \in g_n(h_n(w))$, $p^\infty_{n,0,0}|w$; therefore $p^\infty_{n,0,0}|F(w)$, and hence $F(w) = \sum_{i=1}^r c_i w_i$ for some distinct $w_i \in \mathcal{A}_n \cup \mathcal{B}_n$. Moreover, by (1A), $A \models \psi_{n,\alpha}[w]$ if and only if $\alpha \in g_n(h_n(w))$ if and only if $A \models \psi_{n,\alpha}[F(w)]$. Thus, since the elements of the range of g_n are disjoint, we must have that $r = 1$ and $w_1 = w$, that is, $F(w) = cw$ for some $c = c(w) \in \mathbb{Q}$.

If we can show that $c(w) = c(a^0)$ for all $w \in \bigcup_{n\in\omega} \mathcal{A}_n \cup \mathcal{B}_n$, then F is multiplication by $c(a^0)$, and then it is easy to see that $c(a^0)$ must be ± 1. It will be enough to show that if $w = a^\alpha_z$ (resp. $w = b^\alpha_\eta$) for some $\alpha \in Y_{n-1}$, then $c(w) = c(a^\alpha_{<>})$, for then $c(w) = c(w')$ for some $w' \in \mathcal{A}_{n-1} \cup \mathcal{B}_{n-1}$ (namely, the unique w' such that $\alpha \in g_{n-1}(h_{n-1}(w'))$) and by induction $c(w') = c(a^0)$.

So suppose $w = a^\alpha_z$; the proof will be by induction on the length of z that $c(a^\alpha_z) = c(a^\alpha_{<>})$. Suppose that the length of $z = m > 0$. Let $c = c(a^\alpha_z)$ and $c' = c(a^\alpha_{z\restriction m-1})$. By induction it is enough to prove that $c = c'$. Since $p^\infty_{n,m,1}|(a^\alpha_z + a^\alpha_{z\restriction m-1})$, it is also the case that $p^\infty_{n,m,1}$ divides

$$F(a^\alpha_z) + F(a^\alpha_{z\restriction m-1}) = ca^\alpha_z + c'a^\alpha_{z\restriction m-1} = c(a^\alpha_z + a^\alpha_{z\restriction m-1}) + (c' - c)a^\alpha_{z\restriction m-1}$$

so $p^\infty_{n,m,1}|(c' - c)a^\alpha_{z\restriction m-1}$, which is impossible unless $c = c'$.

The proof is similar if $w = b^\alpha_\eta$, but uses the primes $q_{n,m,\ell}$. So it remains to prove (1A). We will begin by defining some auxiliary formulas of $L_{\infty\omega}$. First, we will use $p^\infty|x$ as an abbreviation for

$$\bigwedge_{k\in\omega} \exists v_k (p^k v_k = x).$$

Define $\varphi_{n,m,0}(y)$ to be $p^\infty_{n,m,0}|y$. Then for $u \in A$, $A \models \varphi_{n,m,0}[u]$ if and only if u is in the subgroup ($\mathbb{Z}$-submodule) generated by $\{\frac{1}{p^k_{n,m,0}} a^\alpha_z : \alpha \in Y_n, k \in \omega, z \in Z_\alpha, \mathrm{dom}(z) = m\}$. Define $\varphi_{n,m,\beta}(y)$ for each $m > 0$ by recursion on β: if $\beta = \gamma + 1$, $\varphi_{n,m,\beta}(y)$ is

$$\varphi_{n,m,\gamma}(y) \wedge \exists y' (\varphi_{n,m+1,\gamma}(y') \wedge (p^\infty_{n,m+1,1}|(y + y')).$$

If β is a limit ordinal, let $\varphi_{n,m,\beta}(y)$ be

$$\bigwedge_{\gamma<\beta} \varphi_{n,m,\gamma}(y).$$

Then for $u \in A$, $A \models \varphi_{n,m,\beta}[u]$ if and only if u is in the subgroup generated by

$$\{a_z^\alpha : \alpha \in Y_n, z \in Z_\alpha, \mathrm{dom}(z) = m \text{ and } z(m-1) \geq \beta\}.$$

In particular, for $m = 1$, recalling that $z \in Z_\alpha$ satisfies $z(0) = \alpha$, we have that $A \models \varphi_{n,1,\beta}[u]$ if and only if u is in the subgroup generated by

$$\{a_{<\alpha>}^\alpha : \alpha \in Y_n, \alpha \geq \beta\}.$$

Now define $\psi_{n,\alpha}(x)$ to be

$$\varphi_{n,0,0}(x) \wedge \exists y[p_{n,1,1}^\infty | (x+y) \wedge \varphi_{n,1,\alpha}(y) \wedge \neg\varphi_{n,1,\alpha+1}(y)].$$

If $u = \sum_{i=1}^r c_i w_i$, for some $w_i \in \mathcal{A}_n \cup \mathcal{B}_n$, then $p_{n,1,1}^\infty | (u+y)$ iff $y = \sum_{i=1}^r c_i a_{<\alpha_i>}^{\alpha_i}$ for some $\alpha_i \in g_n(h_n(w_i))$; using the cofinality of members of the range of g_n, it follows easily that $\psi_{n,\alpha}(x)$ has the desired property.

Proof of (II)

Suppose that there is a non-zero homomorphism $H : A_\nu \to A_\mu$. We are going to use H to define $\theta : T_\nu \to T_\mu$ showing that $(T_\nu, \Phi_\nu) \preceq (T_\mu, \Phi_\mu)$, contrary to the choice of the family of labeled trees. Now $H(w) \neq 0$ for some $w \in \mathcal{A}_n \cup \mathcal{B}_{n,\nu}$ for some $n \in \omega$. Thus for some $\alpha \in Y_n$, $H(b_{<>,\nu}^\alpha) \neq 0$. Fix such an α (which can in fact be any member of $g_n(h_n(w))$). Let $\Phi_\nu(<>) = \ell_0$. Then $q_{n,0,\ell_0,0}^\infty | b_{<>,\nu}^\alpha$ so $q_{n,0,\ell_0,0}^\infty | H\,(b_{<>,\nu}^\alpha)$, and hence $H(b_{<>,\nu}^\alpha)$ must be of the form $\sum_{i=1}^r c_i b_{<>,\mu}^{\alpha_i}$ where $c_i \in \mathbb{Q} - \{0\}$, and $\Phi_\mu(<>) = \ell_0$. So letting $\theta(<>) = <>$ (as it must), we have confirmed that $\Phi_\mu(\theta(\eta)) = \Phi_\nu(\eta)$ for $\eta = <>$.

Now suppose that for some $m \geq 0$, $\theta(\eta)$ has been defined for all nodes η of T_ν of height $\leq m$ such that $\Phi_\mu(\theta(\eta)) = \Phi_\nu(\eta)$. Moreover, suppose that for every η of height $\leq m$, the coefficient of $b_{\theta(\eta),\mu}^\alpha$ in $H(b_{\eta,\nu}^\alpha)$ is non-zero. Now consider any node ζ of T_ν of height $m+1$; let $\eta = \zeta \restriction m$. In A_ν, for $\ell = \Phi_\nu(\zeta)$, $q_{n,m+1,\ell,1}^\infty | b_{\eta,\nu}^\alpha + b_{\zeta,\nu}^\alpha$ so $q_{n,m+1,\ell,1}^\infty | H\,(b_{\eta,\nu}^\alpha) + H\,(b_{\zeta,\nu}^\alpha)$ in A_μ. Since the coefficient — call it c — of $b_{\theta(\eta),\mu}^\alpha$ in $H(b_{\eta,\nu}^\alpha)$ is non-zero, there must be a node ζ' in T_μ of height $m+1$ such that $\zeta' \restriction m = \theta(\eta)$ and the coefficient of $b_{\zeta',\mu}^\alpha$ in $H(b_{\zeta,\nu}^\alpha)$ is c, and moreover such that $\Phi_\mu(\zeta') = \ell$. So we can let $\theta(\zeta) = \zeta'$. This completes the proof of Theorem 0.2. □

5 Absolutely non-isomorphic indecomposables

In this section we sketch how to modify the construction in the preceding section in order to prove Theorem 0.3. (Note that this construction does not require the trees T_μ of Theorem 2.2.) Let $\langle p_{n,m,j} : n, m \in \omega, j \in \{0,1\}\rangle$ be a list of distinct primes. Fix an uncountable λ; for any $\alpha < \lambda\omega$, let Z_α be defined as before. Let $\langle S_{i,\lambda} : i < 2^\lambda\rangle$ be a list of 2^λ distinct subsets of λ, each of cardinality λ (and hence cofinal in λ).

For $n \neq 1$, let $g_n : \lambda \to \mathcal{P}([\lambda n, \lambda(n+1)))$ be defined as before. For $i < 2^\lambda$, define $g_{1,i} : \lambda \to \mathcal{P}([\lambda, \lambda+\lambda)))$ as before but with the additional stipulation that for all $\nu < \lambda$, $g_{1,i}(\nu) \subseteq \{\lambda + \gamma : \gamma \in S_{i,\lambda}\}$. (Here again the operation is ordinal addition.) Let $Y_{1,i} = \bigcup \mathrm{rge}(g_{1,i})$ we will also choose $g_{1,i}$ such that $Y_{1,i} = \{\lambda + \gamma : \gamma \in S_{i,\lambda}\}$. For convenience, for $n \neq 1$ we let $Y_{n,i}$ denote Y_n (independent of i).

For each $n > 0$, let $h_{n,i}$ be a bijection from $\{a^\alpha_z : \alpha \in Y_{n-1,i},\ z \in Z_\alpha\}$ onto λ; use these bijections to make identifications as in the previous construction.

Then $H_{i,\lambda}$ is defined to be the subgroup of the $\mathbb{Q}$-vector space with basis

$$\{a^0\} \cup \{a^\alpha_z : n > 0,\ \alpha \in Y_{n-1,i},\ z \in Z_\alpha\}$$

which is generated (as abelian group) by the union of

$$\bigcup_{n\geq 0}\{\frac{1}{p^k_{n,m,0}} a^\alpha_z : m, k \in \omega, z \in Z_\alpha \cup \{<>\}, \alpha \in Y_{n,i}, \mathrm{dom}(z) = m\}$$

and

$$\bigcup_{n\geq 0}\{\frac{1}{p^k_{n,m,1}}(a^\alpha_z + a^\alpha_{z\restriction m-1}) : m, k \in \omega - \{0\}, z \in Z_\alpha, \alpha \in Y_{n,i}, \mathrm{dom}(z) = m\}.$$

As before, the groups $H_{i,\lambda}$ are absolutely indecomposable. It remains to show that for $\lambda \neq \rho$ or $i \neq j$, $H_{i,\lambda}$ and $H_{j,\rho}$ are not $L_{\infty\omega}$-equivalent (and hence not isomorphic in any generic extension). For this we use the formulas $\psi_{1,\alpha}(x)$. If $\lambda = \rho$ and $i \neq j$, without loss of generality there exists $\gamma \in S_{i,\lambda} - S_{j,\lambda}$; let $\alpha = \lambda + \gamma$. If $\lambda < \rho$, let $\alpha = \lambda + \gamma$ for any γ in any $S_{i,\lambda}$. In either case, $\exists x \psi_{1,\alpha}(x)$ is true in $H_{i,\lambda}$ but not in $H_{j,\rho}$. □

References

[1] J. Barwise, *Back and forth through infinitary logic*, in Studies in Model Theory (M. Morley, ed.), MAA (1973), 5–34.

[2] J. Barwise and P. Eklof, *Infinitary properties of Abelian torsion groups*, Ann. Math. Logic **2** (1970), 25–68.

[3] P. Eklof, *Infinitary equivalence of abelian groups*, Fund. Math. **81** (1974), 305–314.

[4] L. Fuchs, *Infinite Abelian Groups*, Vols.**I** and **II**, Academic Press (1970, 1973).

[5] L. Fuchs, *Indecomposable abelian groups of measurable cardinalities*, Symposia Math. **XIII** (1974), 233–244.

[6] R. Göbel and W. May, *Four submodules suffice for realizing algebras over commutative rings*, J. Pure Appl. Algebra **65** (1990), 29–43.

[7] T. Jech, *Set Theory*, Academic Press (1978).

[8] C. Karp, *Finite quantifier equivalence*, in The Theory of Models (J. W. Addison et al, ed.), North-Holland (1965), 407–412.

[9] R. Laver, *On Fraïssé's order type conjecture*, Ann. Math. **93** (1971), 89–111.

[10] R. Laver, *Better-quasi-orderings and a class of trees*, in Studies in Foundations and Combinatorics (G-C Rota, ed.), Academic Press (1978), 31–48.

[11] M. E. Nadel, *Scott heights of abelian groups*, J. Symb. Logic **59** (1994), 1351–1359.

[12] C. St. J. A. Nash-Williams, *On well-quasi-ordering infinite trees*, Proc. Camb. Phil. Soc. **61** (1965), 697–720.

[13] S. Shelah, *Infinite abelian groups, Whitehead problem and some constructions*, Israel J. Math **18** (1974), 243–325.

[14] S. Shelah, *Better quasi-orders for uncountable cardinals*, Israel J. Math. **42** (1982), 177–226.

[15] J. Silver, *A large cardinal in the constructible universe*, Fund. Math. **69** (1970), 93–100.

Trends in Mathematics, © 1999 Birkhäuser Verlag Basel/Switzerland

Around nonclassifiability for countable torsion free abelian groups

Greg Hjorth *

Abstract

The classification problem for countable torsion free abelian groups is as complicated as that for arbitrary countable structures. The isomorphism relation on general finite rank torsion free abelian groups is not *hyperfinite* – and thus does not admit invariants of the same structure as in rank 1.

1 Introduction

This paper came about as a result of Paul Eklof's invitation to exposit a construction from §1 of [10], which in some sense embeds into the isomorphism relation of countable torsion free abelian groups that of any class of countable structures. This is taken up in §3 below, and §4 presents a new result on the finite rank torsion free abelian groups, but before continuing with any further discussion of either of these results I want to say something about why *I* – who am not an abelian group theorist – find them interesting.

The issue is that of determining which kinds of invariants can *in principle* be assigned to certain classes of objects.

One side of this problem is when a complete invariant is assigned in an explicit manner and the experts agree that this process is some how *satisfactory.* Abelian group theorists appear to accept the Ulm invariants as providing a satisfactory system of complete invariants for the countable abelian p-groups. Similarly there are complete invariants for the rank one torsion free abelian groups, consisting more or less of infinite subsets of

$$\text{Primes} \times \mathbb{N}$$

considered up to finite differences (see §2 below); judging from the desire of [7] to produce a similar classification for rank 2, these would seem to be a reasonable system of complete invariants for the rank one TFA (torsion free abelian) groups.

*Partially supported by NSF Grant DMS 96-22977

At the risk of sounding banal, the way to prove that a class of objects admits a certain kind of complete invariant is to produce the classification theorem. We know that the countable torsion p-groups may be assigned a function from a countable ordinal to $\mathbb{N} \cup \{\infty\}$ as a complete invariant, because this is exactly the object arising as the Ulm invariant.

This present paper is concerned with the converse: Precise theorems stating that certain classes of algebraic objects cannot be given complete invariants of a prescribed form.

To have any hope of making progress mathematically with this kind of question it is necessary to provide a framework for which kinds of assignments we may be willing to consider *reasonable* or *non-pathological.* If we allow unbridled use of the axiom of choice, then the only restriction on classification schemes would be cardinality: Any two classes of abelian groups with cardinality $2^{\aleph_0}$ would immediately allow the same families of complete invariants.

In fact the various plausible ways of providing a general framework for this question make little difference. The next section briefly discusses some of the competing *definability* conditions we might ask that all reasonable schemes of classification should share; the non-classifiability results I have in mind hold for them all.

The first of these nonclassifiability results is a broad brush painting out many kinds of objects we might hope to assign as complete invariants for countable infinite rank TFA groups. §3 recalls a construction to embed the isomorphism relation of arbitrary countable structures into the isomorphism relation on countable TFA groups. From papers such as [6] there is a lore regarding which kinds of complete invariants will suffice for arbitrary countable structures, and thus one can obtain as a corollary results such as from [17] that the countable TFA groups cannot be given bounded subsets of $\aleph_1$ as complete invariants.

The second concerns isomorphism of *finite rank* TFA groups. §4 embeds into isomorphism on finite rank TFA groups a kind of equivalence relation known from [2] and [15] to be *untreeable.* This in particular implies that this relation is not *hyperfinite*, that it cannot be reduced to the set of all subsets of $\mathbb{N}$ mod finite, nor even subsets of A mod finite for any countable set A. In this sense one has that there is no classification of finite rank TFA groups parallel to Baer's analysis for rank 1 TFA groups. §5 discusses the specific case of rank 2, while some more recent results are discussed in §6.

Acknowledgments : I wish to thank Alexander Andretta, Alexander Kechris, and Simon Thomas for reading through the first draft of this paper and making many helpful comments; Simon Thomas in particular pointed out a serious oversight in the original proof of the isomorphism relation not being hyperfinite in rank 2; moreover Alexander Kechris after hearing about the method of proof pointed out the relevance of 5.1. I am also grateful to the referee for a detailed and insightful report and to the editor, Paul Eklof, for an addendum containing many further remarks.

2 Some definitions and background facts

The rest of the paper will be phrased in the Borel category; this section precisely states the relevant definitions.

Definition 2.1 Given a countable relational language $\mathcal{L}$, consisting of various relations $R_1, R_2, ...$ with respective arities $a_1, a_2, ...$ we let $\text{Mod}(\mathcal{L})$ be the space of all $\mathcal{L}$-structures with underlying set $\mathbb{N}$. In the topology generated by taking as subbasic open sets those of the form

$$\{\mathcal{M} \in \text{Mod}(\mathcal{L}) : \mathcal{M} \models R_n(k_1, ..., k_{a_n})\}$$

$$\{\mathcal{M} \in \text{Mod}(\mathcal{L}) : \mathcal{M} \models \neg R_n(k_1, ..., k_{a_n})\}$$

this is a *Polish space* – it is separable and admits a complete metric; it may in fact be naturally identified with the compact metrizable space[1]

$$\prod_{i\in\mathbb{N}} (2^{\mathbb{N}^{a_i}})$$

where we associate to each $\mathcal{M}$ the sequence of functions $f_1^{\mathcal{M}}, f_2^{\mathcal{M}}, ...$ governed by $f_i^{\mathcal{M}}(k_1, ...k_{a_i}) = 1$ if and only if $\mathcal{M} \models R_i(k_1, ...k_{a_i})$.

This can be extended to non-relational languages in a natural way, but let us do so solely for the variety of Abelian groups.

Let Ab be the space of all abelian groups with underlying set $\mathbb{N}$ for whom $1 \in \mathbb{N}$ is the identity.[2] (There is nothing special about the choice of $\mathbb{N}$ – any countable set will do; the decision that the same element of that countable set is always the group theoretic identity eases the notation in the proof of 2.6.) Analogously we can take the topology generated by subbasic open sets of the form

$$\{G \in \text{Ab} : k_1 +^G k_2 = k_3\}$$

$$\{G \in \text{Ab} : k_1 +^G k_2 \neq k_3\},$$

$$\{G \in \text{Ab} : (-k)^G = l\}$$

[1] For A a set, 2^A indicates the functions from A to $\{0, 1\}$ in the topology of pointwise convergence.

[2] Here and elsewhere, $\mathbb{N} = \{1, 2, 3, ...\}$.

$$\{G \in \mathrm{Ab} : (-k)^G \neq l\},$$

and under the assignment $G \mapsto (f_G, h_G)$ where $f_G : \mathbb{N}^2 \to \mathbb{N}$ is given by $f_G(k_1, k_2) = k_1 +^G k_2$ and $h_G : \mathbb{N} \to \mathbb{N}$ is given by $h_G(k) = (-k)^G$ we have that Ab is naturally homeomorphic to a closed subspace of $\mathbb{N}^{\mathbb{N}^2} \times \mathbb{N}^{\mathbb{N}}$ – for instance commutativity of addition corresponds to f_G being in the closed set

$$\bigcap_{l_1<l_2\in\mathbb{N}} \bigcap_{k_1<k_2\in\mathbb{N}} \mathbb{N}^{\mathbb{N}^2} \setminus (\mathcal{O}_{k_1,k_2,l_1} \cap \mathcal{O}_{k_2,k_1,l_2}),$$

where $\mathcal{O}_{j,k,l}$ is the subbasic open set $\{f \in \mathbb{N}^{\mathbb{N}^2} : f(j,k) = l\}$. Since the class of Polish spaces is closed under taking countable products, we have that $\mathbb{N}^{\mathbb{N}^2} \times \mathbb{N}^{\mathbb{N}}$ is Polish. Since any closed subspace of a Polish spaces is again Polish, Ab is Polish.

I will further use the notation TFA_n for the rank $\leq n$ torsion free abelian groups in Ab and TFA_∞ for the collection of all torsion free members of Ab. TFA_∞ is a closed subspace of Ab, and therefore again Polish.

Lemma 2.2 *TFA_n is a G_δ subset of TFA_∞.*

Proof. Unwinding the definition of the topology on Ab we have that for all $k_1, k_2, ..., k_{n+1} \in \mathbb{N}$, the set of $G \in \mathrm{Ab}$ for which there are non-zero $l_1, ..., l_{n+1}$ with

$$l_1 \cdot^G k_1 +^G .... +^G l_{n+1} \cdot^G k_{n+1} = 0^G$$

is an open set in TFA_∞. Then TFA_n is equal to

$$\bigcap_{k_1<...<k_{n+1}} \exists \vec{l} \in \mathbb{Z}^{n+1} \neq 0^{n+1} (l_1 \cdot^G k_1 +^G .... l_{n+1} \cdot^G k_{n+1} = 0^G),$$

and is therefore defined by a countable intersection of open sets. □

Thus by 3.C of [14], TFA_n is a Polish space in the subspace topology. (Another reference here is §2.1 and §3.1 of [9].)

Definition 2.3 A subset of a Polish space is *Borel* if it arises in the σ-algebra generated by the open sets. The *Borel functions* are those that pull open sets back to Borel. Given E and F equivalence relations on Polish spaces X and Y we say that E is *Borel reducible* to F, written $E \leq_B F$, if there is a Borel function $\theta : X \to Y$ such that for all $x_1, x_2 \in X$

$$x_1 E x_2 \Leftrightarrow \theta(x_1) F \theta(x_2).$$

Naturally we write $E <_B F$ if $E \leq_B F$ holds and $F \leq_B E$ fails.

The catch phrase in certain circles is that $E \leq_B F$ indicates that the *classification problem for X/E is no harder than that of Y/F*. Of course the reader may wonder about the decision to focus on *Borel* reducibility of one equivalence relation to another, and may prefer to work with stronger notions of reducibility, such as arising from Δ_1 functions, projective functions, functions definable in the least admissible, functions definable in ZFC, or ordinal definable in ZFC, or even all those that exist in $L(\mathbb{R})$. My own preference for Borel lies in the fact that these broader classes either presume logical sophistication to even define, or lead us into the realm of independence results, thereby requiring assumptions stronger than ZFC to obtain a structure theory, or both. It is moreover true that the great majority of classification theorems arising in mainstream mathematics use functions that are Borel with respect to some natural Borel structure. A reducibility theory parallel to Borel reducibility arises with so-called *absolute* Δ^1_2 reducibility (see [9]) working in just ZFC, $L(\mathbb{R})$ reducibility assuming large cardinals or $\mathrm{AD}^{L(\mathbb{R})}$, or ordinal definable reducibility working in a specific model of ZFC such as the Solovay model.

Definition 2.4 For A a countable set we define 2^A to be the Polish space of all $f : A \rightarrow \{0,1\}$ in the product topology and define $E_0(A)$ to be the equivalence relation of agreement mod finite. So

$$f \; E_0(A) \; g$$

if there is some finite $F \subset A$ such that for all $a \in A \setminus F$

$$f(a) = g(a).$$

E_0 is shorthand for $E_0(\mathbb{N})$.

It is easily checked that for any two countably infinite A and B we have $E_0(A) \leq_B E_0(B)$. In some natural sense we may identify the quotient object $2^{\mathbb{N}}/E_0$ with $\mathcal{P}(\mathbb{N})/\mathrm{Fin}$, the space of all subsets of $\mathbb{N}$ considered modulo finite difference.

Definition 2.5 For G a group equipped with an action on a space X, I will use E^X_G to indicate the orbit equivalence relation. A special case of this is when S_∞, the group of *all* permutations of $\mathbb{N}$, acts on $\mathrm{Mod}(\mathcal{L})$ as governed by the rule

$$(\pi \cdot \mathcal{M}) \models R_i(n_1, ..., n_{a_i}) \Leftrightarrow \mathcal{M} \models R_i(\pi^{-1}(n_1), ..., \pi^{-1}(n_{a_i})).$$

Here the orbit equivalence relation $E^{\mathrm{Mod}(\mathcal{L})}_{S_\infty}$ is the relation of isomorphism, and will be just denoted by $\cong|_{\mathrm{Mod}(\mathcal{L})}$. Similarly $\cong|_{\mathrm{TFA}_n}$ and $\cong|_{\mathrm{TFA}_\infty}$ indicate the isomorphism relation on rank $\leq n$ and infinite rank abelian groups with underlying set $\mathbb{N}$. Just as for a general equivalence relation E

on X we may write X/E for the space of all E-equivalence classes, in the case of a group action G on X we can write X/G for the set $\{[x]_G : x \in X\}$, where each $[x]_G = G \cdot x$ is the orbit of x.

Lemma 2.6 *(Essentially Baer)* $\cong|_{\mathrm{TFA}_1} \leq_B E_0$.

Proof. Let A be the set $\{(p,n) : p \text{ prime}, n \geq 1\}$. For each $G \in \mathrm{TFA}_1$ let $\theta(G) = \{(p,n) \in A : \exists k \in \mathbb{N}(p^n \cdot^G k = 2)\}$. (The choice of 2 in this definition is arbitrary; any fixed element of $\mathbb{N}$ not equal to the identity of the group would do.) This is clearly Borel, and it is a classical fact, tracing back to [3], that for any two $G_1, G_2 \in \mathrm{TFA}_1$

$$G_1 \cong G_2 \Leftrightarrow (\theta(G_1))\, E_0(A)\, (\theta(G_2)).$$

□

Definition 2.7 An equivalence relation on Polish X is said to be *Borel* if it is Borel as a subset of $X \times X$ in the product topological structure – that is to say, $E \subset X \times X$ appears in the smallest σ-algebra containing all subsets of the form $U \times V \subset X \times X$ where $U, V \subset X$ are both open. We say that an equivalence relation E on a space X is *countable* if for all $x \in X$ the equivalence class

$$[x]_E =_{df} \{y \in X : xEy\}$$

is countable. An equivalence relation F is said to be *essentially* countable if for some countable Borel equivalence relation E we have $F \leq_B E$.

Lemma 2.8 *(Compare [11]) For all n, $\cong|_{\mathrm{TFA}_n}$ is essentially countable.*

Proof. Given G, let $k_{1,G}, ...k_{r(G),G}$ ($r(G) \leq n$ being the rank of the group G) be the first available sequence of independent elements of $\mathbb{N}$ whose rational combinations span G. Then let $\theta(G) = \{(q_1, ..., q_{r(G)}) \in \mathbb{Q}^{r(G)} : \exists k \in \mathbb{N}(q_1 \cdot^G k_{1,G} +^G ...q_{r(G)} \cdot^G k_{r(G),G} = k)\}$. θ is a Borel function from TFA_n to the Polish space obtained as the disjoint union of $2^{\mathbb{Q}^l}$ for $l \leq n$. It is easily checked that $G_1 \cong G_2$ if and only if $r(G_1) = r(G_2)$ and there is $M \in \mathrm{GL}_{r(G_1)}(\mathbb{Q})$ with $\{M \cdot \vec{q} : \vec{q} \in \theta(G_1)\} = \theta(G_2)$. □

Definition 2.9 An equivalence relation E is said to be *smooth* if $E \leq_B \mathrm{id}(\mathbb{R})$ – that is to say there is Borel $\theta : X \to \mathbb{R}$ such that for all $x, y \in X$

$$xEy \Leftrightarrow \theta(x) = \theta(y).$$

An equivalence relation E on Polish X is *hyperfinite* if there is an increasing sequence of Borel equivalence relations – $E_1 \subset E_2 \subset ...$ – all of whose equivalence classes are finite (i.e. for all $n \in \mathbb{N}, x \in X$(cardinality of $[x]_{E_n} < \aleph_0$)) with $E = \bigcup_{n \in \mathbb{N}} E_n$. A countable equivalence relation E is said to be *treeable* if there is some Borel $\mathcal{T} \subset X \times X$ providing an acyclic graph on X such that the E-equivalence classes are exactly the connected $\mathcal{T}$-components. A countable Borel equivalence relation E is said to be *universal* if for any other countable Borel equivalence relation F we have

$$F \leq_B E.$$

Until very, very recently it was open whether there are only 5 countable Borel equivalence relations with infinitely many classes up to $\leq_B$ bi-reducibility. Namely:

(i) $\mathrm{id}(\mathbb{R})$;
(ii) E_0;
(iii) the $\leq_B$ maximal treeable equivalence relation, $E_{\infty\mathcal{T}}$ (see e.g. [11] or [13] for a more precise definition);
(iv) the product of that equivalence relation with itself, $E_{\infty\mathcal{T}} \times E_{\infty\mathcal{T}}$;
(v) the universal countable Borel equivalence relation, E_∞ (again compare [11] or [13]).

In the last few months this has been settled by [1], which established a plethora of $\leq_B$-incomparable non-treeable countable Borel equivalence relations.

Here I need to collect together various facts, due to various people, some of which are far less trivial than others. Due to the entanglement of contributions, I will make no attempt to portion precise credit here, but instead refer the reader to [13], [15], and [5], and note that apart from these authors, Adams, Slaman, Steel, and Weiss have also played an important part in founding this body of knowledge.

Lemma 2.10 *(I) Among the countable Borel equivalence relations, the classes of smooth, hyperfinite, and treeable are closed downwards under* $\leq_B$.

(II) If $E \subset F \subset X \times X$ *are two Borel equivalence relations and* F *is hyperfinite (respectively treeable), then* E *is hyperfinite (respectively treeable).*

(III) A countable Borel equivalence relation E *is hyperfinite if and only if* $E \leq_B E_0$.

(IV) The hyperfinite equivalence relations are strictly included in the treeable. The smooth equivalence relations are strictly included in the hyperfinite.

In view of (II) it seems natural to say that a general equivalence relation, possibly non-countable, is *essentially hyperfinite* or *essentially treeable* if it is Borel reducible to a countable equivalence relation which is respectively hyperfinite or treeable in the original sense.

Thus there is a natural Borel translation of the question of whether in general an equivalence relation E allows the same kinds of invariants as in the case of rank 1 TFA groups: Is E hyperfinite? In §4 we show that that $\cong|_{\mathrm{TFA}_n}$ is in general not even treeable.

Definition 2.11 In the case that X is a Polish space equipped with a Borel probability measure ν let us say that E is *ν-smooth,* or *ν-hyperfinite,* or *ν-treeable* if there is a ν-measurable function, an increasing union of finite equivalence ν-measurable relations, or a ν-measurable graph witnessing the respective definitions. Equivalently E is respectively ν-smooth, or ν-hyperfinite, or ν-treeable if it is genuinely smooth, hyperfinite, or treeable, on a ν-measure one set.

3 A construction for the infinite rank case

In what follows let $\mathcal{L}$ be the language generated by a single binary relation R; so the $\mathcal{L}$-structures are directed graphs. I restrict myself to this simplest of cases because it is least bother notationally, but in fact (see [6]) for any other countable language $\mathcal{K}$ we have

$$\cong|_{\mathrm{Mod}(\mathcal{K})} \leq_B \cong|_{\mathrm{Mod}(\mathcal{L})}.$$

That is to say, isomorphism on $\mathcal{L}$ structures is as complicated as isomorphism on arbitrary countable structures.

Definition 3.1 We define a function θ from $\mathrm{Mod}(\mathcal{L})$ (the Polish space of $\mathcal{L}$ models with underlying set $\mathbb{N}$) to TFA_∞ (the Polish space of torsion free abelian groups with underlying set $\mathbb{N}$). It will be notationally somewhat easier to describe a function θ_0 from $\mathrm{Mod}(\mathcal{L})$ with each $\theta_0(\mathcal{M})$ a subgroup of $\mathbb{Q}^{<\mathbb{N}}$, the $\aleph_0$ dimensional $\mathbb{Q}$-vector space; so I will concentrate on producing θ_0, and only later address side issues such as reorganizing $\theta(\mathcal{M})$ to be a group whose underlying set is $\mathbb{N}$.

For the purpose of defining θ_0 we choose a countable basis for $\mathbb{Q}^{<\mathbb{N}}$ consisting of two parts, each indexed by finite sequences of natural numbers: $\{\mathbf{e}_s : s \in \mathbb{N}^{<\mathbb{N}}\} \bigcup \{\mathbf{f}_s : s \in \mathbb{N}^{<\mathbb{N}}\}$. We moreover want to have the following method of encoding how a finite tuple appears inside some $\mathcal{M} \in \mathrm{Mod}(\mathcal{L})$: Given the sequence $\vec{n} = (n_1, ..., n_k) \in \mathbb{N}^{<\mathbb{N}}$ we let $a_{\mathcal{M}}(\vec{n})$ (or just $a(\vec{n})$ when context indicates $\mathcal{M}$) be the natural number

$$\sum\{10^{2^i 3^j 5^k} : i, j \leq k, \mathcal{M} \models R(n_i, n_j)\}.$$

By considering the decimal representation of $a(\vec{n})$ we have that $a_{\mathcal{M}}(n_1, ..., n_k) = a_{\mathcal{N}}(m_1, ..., m_l)$ if and only if $k = l$ and for all $i, j \leq k$

$$\mathcal{M} \models R(n_i, n_j) \Leftrightarrow \mathcal{N} \models R(m_i, m_j).$$

We let $p_1, p_2, ...,$ $q_1, q_2, ...,$ $k_1, k_2, ...,$ $r_1, r_2, ...,$ be increasing disjoint sequences of primes.

We then set $\theta_0(\mathcal{M})$ to be the subgroup of $\mathbb{Q}^{<\mathbb{N}}$ consisting of all finite sums of the form

$$g = \sum_{s \in \mathbb{N}^{<\mathbb{N}}} n_s \mathbf{e}_s (p_{l(s)})^{-m_s} + \sum_{s^\frown a \in \mathbb{N}^{<\mathbb{N}}} (\mathbf{e}_s + \mathbf{e}_{s^\frown a}) \bar{n}_{s^\frown a} (q_{l(s)})^{-\bar{m}_{s^\frown a}}$$

$$+ \sum_{s \in \mathbb{N}^{<\mathbb{N}}} (\mathbf{e}_s + \mathbf{f}_s) \hat{n}_s (k_{l(s)})^{-\hat{m}_s} + \sum_{s \in \mathbb{N}^{<\mathbb{N}}} \mathbf{e}_s n'_s (r_{a(s)})^{-m'_s}$$

such that the following conditions all hold:

(0) if $s = \vec{n} = (n_1, ..., n_k)$ then $s^\frown a$ indicates $(n_1, ..., n_k, a)$, the result of following s by a;

(i) each n_s, $\bar{n}_s$, $\hat{n}_s$, $n'_s \in \mathbb{Z}$; each m_s, $\bar{m}_s$, $\hat{m}_s$, $m'_s \in \{0, 1, 2, 3, ...\}$;

(ii) all but finitely many of the n_s, $\bar{n}_s$, $\hat{n}_s$, n'_s are zero;

(iii) if m_s (or $\bar{m}_s$, $\hat{m}_s$, m'_s) is non-zero, then n_s does not divide $p_{l(s)}$ and is non-zero (respectively, $\bar{n}_s$ does not divide $q_{l(s)}$ and is non-zero, $\hat{n}_s$ does not divide $k_{l(s)}$ and is non-zero, n'_s does not divide m'_s and is non-zero);

(iv) $\bar{n}_{s^\frown a}(q_{l(s)})^{-\bar{m}_{s^\frown a}}$, $\hat{n}_s(k_{l(s)})^{-\hat{m}_s}$ are strictly less than one and greater than or equal to 0.

Note that the only part of the definition of $\theta_0(\mathcal{M})$ which actually depends on $\mathcal{M}$ is the coefficients of the $\mathbf{f}_s$ appearing in the final summand above.

Theorem 3.2 *For all $\mathcal{L}$-structures $\mathcal{M}$ and $\mathcal{N}$ we have*

$$\mathcal{M} \cong \mathcal{N} \Leftrightarrow \theta_0(\mathcal{M}) \cong \theta_0(\mathcal{N}).$$

The proof of this theorem can be found in [10]. The argument – which *is* non-trivial – resists easy summary, but it might be helpful to say just that it relies on coding $\mathcal{M}$ into $\theta_0(\mathcal{M})$ by tracking which elements of $\theta_0(\mathcal{M})$ are infinitely divisible by a given prime. The group $\theta_0(\mathcal{M})$ is generated by something like finite sequences from $\mathcal{M}$; we use the primes $p_1, p_2, ...$ to indicate for us which level some $\mathbf{e}_s$ belongs, in that if $\mathbf{e}_s$ is divisible by every power of p_l then l is the length of s; we use the $q_1, q_2, ...$ to indicate the tree structure of these finite sequences, in that if $\mathbf{e}_s + \mathbf{e}_t$ is divisible by every power of $q_{l(s)}$ then t has the form $t = s^\frown a$; similarly $\mathbf{e}_s + \mathbf{f}_t$ is divisible by every power of $k_{l(s)}$ if $t = s$; and finally the unique $n \in \mathbb{N}$ such that $\mathbf{f}_s$ is divisible by every power of r_n tells us which of the various elements of s are $R^{\mathcal{M}}$- interrelated.

In this construction there is a kinship between the isomorphisms between $\mathcal{M}$ and $\mathcal{N}$ and those between $\theta_0(\mathcal{M})$ and $\theta_0(\mathcal{N})$. For instance if $\mathcal{M}$ is rigid (in the sense of there being no non-trivial $R^{\mathcal{M}}$ respecting bijections of $\mathbb{N}$), then $\theta_0(\mathcal{M})$ will be auto-rigid (in the sense of admitting $g \mapsto g^{-1}$ as the only non-trivial isomorphism from $\theta_0(\mathcal{M})$ to $\theta_0(\mathcal{M})$). It is not true that every non-trivial isomorphism between $\theta_0(\mathcal{M})$ and $\theta_0(\mathcal{N})$ arises from one between $\mathcal{M}$ and $\mathcal{N}$, but rather any such isomorphism will be analogous to a "linear combination" of isomorphisms between $\mathcal{M}$ and $\mathcal{N}$.

It is routine to choose in a Borel manner a bijection between $\theta_0(\mathcal{M})$ and $\mathbb{N}$ and obtain $\cong|_{\mathrm{Mod}(\mathcal{L})} \leq_B \cong|_{\mathrm{TFA}_\infty}$. Actually we may do better than Borel:

Theorem 3.3 *There is a continuous* $\theta : \mathrm{Mod}(\mathcal{L}) \to \mathrm{TFA}_\infty$ *such that for all* $\mathcal{M} \in \mathrm{Mod}(\mathcal{L})$ *we have*

$$\theta(\mathcal{M}) \cong \theta_0(\mathcal{M}),$$

and thus for all $\mathcal{M}, \mathcal{N} \in \mathrm{Mod}(\mathcal{L})$

$$\mathcal{M} \cong \mathcal{N} \Leftrightarrow \theta(\mathcal{M}) \cong \theta(\mathcal{N}).$$

And actually we may do better than simply continuous. The function θ can be chosen to be *recursive* – by which I mean for instance that we may build a computer which as it is fed an infinite tape consisting of a 1 marked at the nth position exactly when $n = 2^k 3^l$ for some l, k with $\mathcal{M} \models R(l, k)$ will slowly start producing an infinite tape with 1's marked at exactly each $n = 2^k 3^l 5^m$ where $k +^{\theta(\mathcal{M})} l = m$.

A version can be completed for arbitrary countable models. Given a countably infinite model $\mathcal{M}$ with underlying set S we may in an *effective fashion* (for instance uniformly definable over the least *admissible* set containing $\mathcal{M}$) produce a subgroup $G(\mathcal{M}) < \{f : S \times 2 \to \mathbb{Q} | f$ has finite support$\}$. Again, $\mathcal{M} \cong \mathcal{N}$ if and only if $G(\mathcal{M}) \cong G(\mathcal{N})$.

4 Finite rank TFA groups

For the sake of notation we do everything below just in the case that $n = 3$. Clearly it proceeds unchanged for all $n \geq 3$, since the arguments below require only that $\mathrm{SL}_n(\mathbb{Z})$ be Kazhdan.

Definition 4.1 For future reference let $\pi_3 : \mathbb{Q}^3 \twoheadrightarrow \mathbb{Q}^3/\mathbb{Z}^3$ be the canonical surjection obtained by setting elements of $\mathbb{Z}^3$ equal to 0. Let $\mathcal{S}_3$ be the set $\{G : G < \mathbb{Q}^3/\mathbb{Z}^3\}$ of all subgroups of the quotient group $\mathbb{Q}^3/\mathbb{Z}^3$ in its additive structure. We naturally identify the set $\mathcal{P}(\mathbb{Q}^3/\mathbb{Z}^3)$ of all subset of $\mathbb{Q}^3/\mathbb{Z}^3$ with $2^{\mathbb{Q}^3/\mathbb{Z}^3}$, the set of all functions from $\mathbb{Q}^3/\mathbb{Z}^3$ to $\{0, 1\}$ in the product topology. Here $2^{\mathbb{Q}^3/\mathbb{Z}^3}$ is Polish and under the indicated identification $\mathcal{S}_3$ is a closed subset, and hence Polish.

We then let $\mathrm{SL}_3(\mathbb{Z})$ act on $\mathcal{S}_3$ by induced pointwise multiplication, so that for $G \in \mathcal{S}_3$ and $M \in \mathrm{SL}_3(\mathbb{Z})$ we let

$$M \cdot G = \{M \cdot h : h \in G\}$$

where if $\vec{q} \in \mathbb{Q}^3$ with $\pi_3(\vec{q}) = h$ then we are using $M \cdot h$ to indicate $\pi_3(M(\vec{q}))$, where $M(\vec{q})$ is the usual result of applying the linear transformation indicated by M to $\vec{q}$. If $G_1, G_2 \in \mathcal{S}_3$ and there is some $M \in \mathrm{SL}_3(\mathbb{Z})$ with $M \cdot G_1 = G_2$ then G_1 and G_2 are isomorphic. Thus $E^{\mathcal{S}_3}_{\mathrm{SL}_3(\mathbb{Z})}$ – the orbit equivalence relation arising from the action of $\mathrm{SL}_3(\mathbb{Z})$ on $\mathcal{S}_3$ – is included in the isomorphism relation on $\mathcal{S}_3$.

Let Γ_3 denote the dual group of $\mathbb{Q}^3/\mathbb{Z}^3$. I will choose to think of this as space of all homomorphisms from $\mathbb{Q}^3/\mathbb{Z}^3$ to $\mathbb{T} = \mathbb{R}/\mathbb{Z}$ with the pointwise topology and the group operation of pointwise addition. Since $\mathbb{Q}^3/\mathbb{Z}^3$ is discrete, countable, and torsion we have that Γ_3 is compact, metrizable, and zero dimensional. (See [8] for a general discussion of Pontryagin duality.) Let μ be Haar measure on Γ_3. Let Ker: $\Gamma_3 \rightarrow \mathcal{S}_3$ be the map assigning to each $\psi \in \Gamma_3$ the subgroup

$$\mathrm{Ker}(\psi) = \{h \in \mathbb{Q}^3/\mathbb{Z}^3 : \psi(h) = 0\}.$$

We let ν be the Borel probability measure on $\mathcal{S}_3$ obtained as the pushout along Ker of the measure μ – so that for any Borel $B \subset \mathcal{S}_3$

$$\nu(B) = \mu(\{\psi \in \Gamma_3 : \mathrm{Ker}(\psi) \in B\}) = \mu(\mathrm{Ker}^{-1}[B]).$$

Lemma 4.2 *ν is invariant under the action of* $\mathrm{SL}_3(\mathbb{Z})$.

Proof. Fix $M \in \mathrm{SL}_3(\mathbb{Z})$ and define $\rho_M : \Gamma_3 \rightarrow \Gamma_3$ by

$$(\rho_M(\psi))(h) = \psi(M^{-1} \cdot h).$$

Then ρ_M is a homomorphic homeomorphism from Γ_3 to Γ_3. By uniqueness of Haar measure it therefore follows that for any measurable $C \subset \Gamma_3$ we have $\mu(\rho_M[C]) = \mu(C)$.

Hence if B is a subset of $\mathcal{S}_3$, then

$$\begin{aligned}\nu(M \cdot B) &= \mu(\{\psi \in \Gamma_3 : \mathrm{Ker}(\psi) \in M \cdot B\}) \\ &= \mu(\{\psi \in \Gamma_3 : \{h : \psi(h) = 0\} \in M \cdot B\}) \\ &= \mu(\{\psi \in \Gamma_3 : \{M^{-1}(h) : \psi(h) = 0\} \in B\}) \\ &= \mu(\{\psi \in \Gamma_3 : \{h : \psi(M(h)) = 0\} \in B\}) \\ &= \mu(\{\psi \in \Gamma_3 : \mathrm{Ker}(\rho_M^{-1}(\psi)) \in B\} \\ &= \mu([\rho_M[\mathrm{Ker}^{-1}[B]]),\end{aligned}$$

which, by the uniqueness of μ as above, must equal

$$\mu(\mathrm{Ker}^{-1}[B]) = \nu(B),$$

as required. □

Lemma 4.3 *If $h \in \mathbb{Q}^3/\mathbb{Z}^3$ has order n, then*

$$\nu(\{G \in \mathcal{S}_3 : h \in G\}) = \frac{1}{n}.$$

Proof. Let $\Gamma_3^1 = \{\psi \in \Gamma_3 : \psi(h) = 0\}$. Let ψ_1 be a character with

$$\psi_1(h) = \frac{1}{n}.$$

Then the cosets of Γ_3^1 are exactly

$$\Gamma_3^1, \psi_1 + \Gamma_3^1, ..., (n-1) \cdot \psi_1 + \Gamma_3^1.$$

Since each coset has the same Haar measure, we are done. □

Lemma 4.4 *If $h_1, h_2 \in \mathbb{Q}^3/\mathbb{Z}^3$, p a prime with $p \cdot h_1 = p \cdot h_2 = 0$, then*

$$\nu(\{G \in \mathcal{S}_3 : h_1 \in G \text{ or } h_2 \in G \text{ or } \exists k(h_1 + k \cdot h_2) \in G\}) = 1.$$

Proof. Let $G = \mathrm{Ker}(\psi)$. Assume $\psi(h_1), \psi(h_2) \neq 0$, or we are finished at once.

Then $\psi(h_1) = k_1/p$ and $\psi(h_2) = k_2/p$, some integers $k_1, k_2 \in \{1, 2, ..., p-1\}$. Now we may find $k \in \{1, 2, ..., p-1\}$ with $k_1 + kk_2 \equiv 0 \mod p$, and so $\psi(h_1) + k\psi(k_2) = \psi(h_1 + k \cdot h_2) = 0$, as required. □

Lemma 4.5 *There is an infinite subgroup $L \subset \mathrm{SL}_3(\mathbb{Z})$ such that for all non-trivial $M \in L$,*

$$\nu(\{G \in \mathcal{S}_3 : M \cdot G \neq G\}) = 1.$$

Proof. Let L consist of all linear transformations of the form M_l, some $l \in \mathbb{Z}$, where M_l maps

$$(1,0,0) \mapsto (1,l,0),$$

$$(0,1,0) \mapsto (0,1,0),$$

$$(0,0,1) \mapsto (0,0,1).$$

This is an infinite abelian subgroup of $\mathrm{SL}_3(\mathbb{Z})$, since $M_l \circ M_m = M_{l+m}$.

Fix $M_l \in L$. Let $p_1, p_2, p_3, ...$ enumerate the primes bigger than $|l|$. For each $i > |l|$ let

$$h_i = \pi_3((p_i)^{-1}, 0, 0),$$

$$g_i = \pi_3(0, (p_i)^{-1}, 0).$$

Claim: For each $i > |l|$ the probability that there exists some $k < p_i$ with

$$h_i + k \cdot g_i \in G \Leftrightarrow M_l(h_i) + M_l(k \cdot g_i) \notin G$$

is at least

$$1 - \frac{1}{p_i}.$$

Proof of claim. Let $p = p_i$, $h = h_i$, $g = g_i$. Note that $M_l(h) = h + lg$. There are exactly 4 possibilities to consider: (A) $h, g \notin G$; (B) $h \in G$, $g \notin G$; (C) $g \in G$, $h \notin G$; (D) $h, g \in G$.

The probability of (C)+(D) is $1/p$. If (B), we finish at once with the choice $k = 0$ (and note that if h, $h + lg \in G$, then since $l \neq 0 \bmod p$ we have $g \in G$, contradicting case assumption). If (A), then 4.4 implies that with relative probability 1 there is some $k \in \{1, 2, ..., p-1\}$ with

$$h + kg \in G.$$

So it suffices to show

$$M_l(h) + M_l(kg) =_{df} h + lg + kg \notin G.$$

However, if not then we obtain $lg \in G$, and since $l \neq 0 \bmod p$ we obtain $g \in G$, against the case assumption. (Claim □)

Thus letting $i \to \infty$ we obtain with probability 1 that there is some i and k for which

$$h_i + kg_i \in G \Leftrightarrow M_l(h_i) + kM_l(g_i) \notin G.$$

As required for the lemma. □

Corollary 4.6 *There is an invariant ν-measure one set on which $L \subset \mathrm{SL}_3(\mathbb{Z})$ acts freely.*

Corollary 4.7 *The orbit equivalence relation $E^{\mathcal{S}_3}_{\mathrm{SL}_3(\mathbb{Z})}$ arising from the action of $\mathrm{SL}_3(\mathbb{Z})$ on $\mathcal{S}_3$ is not ν-smooth – that is to say, there is no ν-measurable*

$$\theta : \mathcal{S}_3 \to \mathbb{R}$$

such that for all $G_1, G_2 \in \mathcal{S}_3$

$$\exists M \in \mathrm{SL}_3(\mathbb{Z})(M \cdot G_1 = G_2) \Leftrightarrow \theta(G_1) = \theta(G_2).$$

Proof. Suppose otherwise. Then the existence of θ may be reworked into the following equivalent form: There is a ν-measure one set $B \subset \mathcal{S}_3$ and Borel

$$\theta : B \to \mathbb{R}$$

such that for all $G_1, G_2 \in \mathcal{S}_3$

$$\exists M \in \mathrm{SL}_3(\mathbb{Z})(M \cdot G_1 = G_2) \Leftrightarrow \theta(G_1) = \theta(G_2).$$

By replacing B with

$$\bigcap_{M \in \mathrm{SL}_3(\mathbb{Z})} M \cdot B$$

we may assume B is $\mathrm{SL}_3(\mathbb{Z})$-invariant and by 4.6 we may assume that L acts freely on B. Since $E^B_{\mathrm{SL}_3(\mathbb{Z})}$ is a smooth orbit equivalence relation induced by the Borel action of a Polish group, [4] provides us with a Borel set $C \subset B$ meeting each orbit in exactly one point.

But then we have the usual contradiction.

$$\{M \cdot C : M \in \mathrm{SL}_3(\mathbb{Z})\}$$

covers the set B by countably many sets, each of which must have the same measure by 4.2. Thus this measure can not be zero. But then

$$\{M \cdot C : M \in L\}$$

gives $\aleph_0$ disjoint subsets of B each with the same non-zero measure, contradicting that ν is a *probability* measure. □

We need a version of a result from [2].

Theorem 4.8 *(Adams-Spatzier) Let $(X_0, \mathcal{B}_0)$ be a standard Borel space, and let ν_0 be a Borel probability measure on that space. Let H be a countable Kazhdan group acting on X_0 by ν_0-measure preserving transformations. Suppose that the resulting orbit equivalence relation $E_H^{X_0}$ is not ν_0-smooth.*

Then $E_H^{X_0}$ is not treeable.

Here I lie slightly. The actual statement of 4.8 found at [2] requires the stronger assumption that the action of H be *ergodic*. However, examining the proof, at least as it is presented in [11], one can see that the weaker assumption of non-smoothness suffices. In any case, the version of 4.8 here can be made to follow from the versions in [11] or [2] by appealing to the ergodic decomposition of measures.

In fact as remarked in [11], the assumptions of 4.8 are sufficient to establish that $E_H^{X_0}$ is not Borel reducible to $E_{\infty\mathcal{T}} \times E_{\infty\mathcal{T}}$, or any other finite product of treeable Borel equivalence relations. Thus the argument below suffices to show $\cong|_{\mathrm{TFA}_3}$ is not $\leq_B$ reducible to a finite product of treeable Borel equivalence relations.

Corollary 4.9 $E^{\mathcal{S}_3}_{\mathrm{SL}_3(\mathbb{Z})}$ *is not treeable.*

Proof. Citing say [21] chapter 11 we have that $\mathrm{SL}_3(\mathbb{Z})$ is Kazhdan (and more generally that $\mathrm{SL}_n(\mathbb{Z})$ is Kazhdan for all $n \geq 3$). By 4.7 we have that $E^{\mathcal{S}_3}_{\mathrm{SL}_3(\mathbb{Z})}$ is not ν-smooth, and by 4.2 we have that $\mathrm{SL}_3(\mathbb{Z})$ acts by measure preserving transformations on $\mathcal{S}_3$.

Thus $E^{\mathcal{S}_3}_{\mathrm{SL}_3(\mathbb{Z})}$ is non-treeable by 4.8. □

Corollary 4.10 *Isomorphism on* TFA_3*, the space of rank 3 torsion free abelian groups, is not treeable.*

Proof. Let $A_3 = \{G < \mathbb{Q}^3 : \mathbb{Z}^3 \subset G\}$. For each $G \in A_3$ let $\pi_3[G] = \{\pi_3(g) : g \in G\}$ be the pointwise image of G under the surjection $\pi_3 : \mathbb{Q}^3 \twoheadrightarrow \mathbb{Q}^3/\mathbb{Z}^3$. We then let F be pull back of $E^{\mathcal{S}_3}_{\mathrm{SL}_3(\mathbb{Z})}$ along the map $G \mapsto \pi_3[G]$. Thus $G_1 F G_2$ if and only if

$$\pi_3[G_1] E^{\mathcal{S}_3}_{\mathrm{SL}_3(\mathbb{Z})} \pi_3[G_2].$$

Note that $G \mapsto \pi_3[G]$ provides a Borel bijection between A_3 and $\mathcal{S}_3$ and thus F is not treeable.

However if $G_1 F G_2$ then we may find $M \in \mathrm{SL}_3(\mathbb{Z})$ with

$$M \cdot \pi_3[G_1] =_{df} \{M \cdot \pi_3(g) : g \in G_1\} = \pi_3[G_2].$$

In other words

$$\{k + M(g) : g \in G_1, k \in \mathbb{Z}^3\} = \{k + g : g \in G_2, k \in \mathbb{Z}^3\}.$$

Thus since $\mathbb{Z}^3 \subset G_1, G_2$ and $M[\mathbb{Z}^3] = \mathbb{Z}^3$

$$\{M(g) : g \in G_1\} = G_2$$

and so $G_1 \cong G_2$. In other words, M defines an isomorphism between G_1 and G_2.

Thus F is included in the isomorphism relation on subgroups of $\mathbb{Q}^3$. Since F is not treeable, neither is $\cong|_{\mathrm{TFA}_3}$ by 2.10. □

This method gives the same result restricted to the class of finite rank torsion free abelian groups where every element of is divisible by every prime save one. In other words, if we let $\mathbb{Z}_{(p)}$ be the rationals whose denominator is relatively prime to p, then one can obtain that the isomorphism relation on finite rank torsion free $\mathbb{Z}_{(p)}$-modules is non-treeable.

To see this, let p be a prime, and let $\hat{\Gamma}_3$ be the dual group for the quotient $\mathbb{Q}^3/\mathbb{Z}^3_{(p)}$. Let $\hat{\mathcal{S}}_3$ be the space of subgroups of $\mathbb{Q}^3/\mathbb{Z}^3_{(p)}$. Again we can let $\hat{\mu}$ be Haar measure of $\hat{\Gamma}_3$ and let $\hat{\nu}$ be it push out along the map

$$\mathrm{Ker} : \hat{\Gamma}_3 \to \hat{\mathcal{S}}_3,$$

$$\mathrm{Ker} : \psi \mapsto \{h : \psi(h) = 0\}.$$

Again the action of $\mathrm{SL}_3(\mathbb{Z})$ on the subgroups of $\mathbb{Q}^3$ pushes down to an action of $\mathrm{SL}_3(\mathbb{Z})$ on $\hat{\mathcal{S}}_3$. Again let

$$\hat{\pi}_3 : \mathbb{Q}^3 \twoheadrightarrow \mathbb{Q}^3/\mathbb{Z}^3_{(p)}$$

be the projection map.

Lemma 4.11 *If $G \in \hat{\mathcal{S}}_3$ then $\hat{\pi}_3^{-1}[G]$ is a $\mathbb{Z}_{(p)}$-module.*

Proof. $\hat{\pi}_3^{-1}[G]$ clearly includes $\mathbb{Z}^3_{(p)}$. Thus for any $h \in \hat{\pi}_3^{-1}[G]$ and l relatively prime to p we may find some $n \in \mathbb{N}$ such that

$$p^n \cdot h \in \mathbb{Z}^3_{(p)},$$

$$\therefore \frac{p^n}{l} \cdot h \in \mathbb{Z}^3_{(p)} \subset \hat{\pi}_3^{-1}[G].$$

Thus by the euclidean algorithm,

$$l^{-1} \cdot h \in \hat{\pi}_3^{-1}[G].$$

□

Again we can define $L = \{M_k : k \in \mathbb{Z}\}$ as before, and again we need only show that the action of L is free on a $\hat{\mu}$-measure one set.

So let $\psi \in \hat{\Gamma}_3$ be sufficiently "random" or "typical" with respect to the measure $\hat{\mu}$. Let $k \in \mathbb{Z}$. Then we may find some n such that p^n does not divide k. Let

$$\hat{h}_i = \hat{\pi}_3(p^{-i}, 0, 0),$$

$$\hat{g}_i = \hat{\pi}_3(0, p^{-i}, 0)$$

for $i \in \mathbb{N}$. We may find n_1, n_2 such that

$$\psi(\hat{h}_{n_1}) = \frac{l_1}{p^m} \neq 0,$$

$$\psi(\hat{g}_{n_2}) = \frac{l_2}{p^m} \neq 0,$$

where p does not divide l_1 or l_2, and

$$m > n + |n_1 - n_2|.$$

Therefore in particular

$$\psi(\hat{\pi}_3(l_2 k(0, p^{-n_1}, 0))) = \frac{k(l_2)^2}{p^m p^{n_1 - n_2}} \neq 0.$$

Then

$$\psi(l_2 \hat{h}_{n_1} - l_1 \hat{g}_{n_2}) = 0,$$

while

$$\psi(M_k(l_2 \hat{h}_{n_1} - l_1 \hat{g}_{n_2})) = \psi(\hat{\pi}_3(M_k(l_2 p^{-n_1}, - - l_1 p^{-n_2}, 0)))$$

$$= \psi(\hat{\pi}_3((l_2 p^{-n_1}, - - l_1 p^{-n_2}, 0))) + \psi(\hat{\pi}_3(0, l_2 k p^{-n_1}, 0))$$

$$= \psi(\hat{\pi}_3(0, l_2 k p^{-n_1}, 0)) = \frac{k(l_2)^2}{p^m p^{n_1 - n_2}} \neq 0.$$

Thus

$$\mathrm{Ker}^{-1}(\psi) \neq M_k \cdot \mathrm{Ker}^{-1}(\psi).$$

Then, as before, $E^{\hat{S}_3}_{\mathrm{SL}_3(\mathbb{Z})}$ is non-treeable. Since by 4.11 $E^{\hat{S}_3}_{\mathrm{SL}_3(\mathbb{Z})}$ pulls back to a subequivalence relation of isomorphism on collection of subgroups of $\mathbb{Q}^3$ that are $\mathbb{Z}_{(p)}$-modules, we have, as before, that isomorphism on rank 3 $\mathbb{Z}_{(p)}$-modules is non-treeable.

5 Rank 2 TFA groups

The arguments of the last section would not seem to extend to the case $n = 2$ since $\mathrm{GL}_2(\mathbb{Z})$ is *not* Kazhdan . Nevertheless Alexander Kechris pointed out the relevance of the following fact, which can be found at [15] 5.6.

Proposition 5.1 *Let $(X_0, \mathcal{B}_0)$ be a standard Borel space, and let ν_0 be a Borel probability measure on that space. Suppose a countable non-amenable group H acts freely by ν_0-measure preserving transformations on X_0.*

Then the resulting orbit equivalence relation $E^{X_0}_H$ is not hyperfinite.

Here the appropriate group for us to use is $\mathrm{PSL}_2(\mathbb{Z}) = \mathrm{SL}_2(\mathbb{Z})/\{+1, -1\}$.

Proposition 5.2 $\mathrm{PSL}_2(\mathbb{Z})$ *is non-amenable.*

Proof. In general given a countable group G and a normal subgroup H we have G amenable if and only if both H and G/H are amenable (4.1.6 [22]). As indicated by [21] 7.1 and chapter 10, $\mathrm{SL}_2(\mathbb{Z})$ is non-amenable; $\{+1, -1\}$ is quite trivially amenable. Thus $\mathrm{PSL}_2(\mathbb{Z})$ must be non-amenable. □

Since we have that for all $H \in \mathrm{TFA}_2$,

$$1 \cdot H = -1 \cdot H = H,$$

the action of $\mathrm{SL}_2(\mathbb{Z})$ pushes down to an action of $\mathrm{PSL}_2(\mathbb{Z})$ on TFA_2. Again we let $\pi_2 : \mathbb{Q}^2 \twoheadrightarrow \mathbb{Q}^2/\mathbb{Z}^2$ be the canonical surjection. Let Γ_2 denote the dual group of $\mathbb{Q}^2/\mathbb{Z}^2$. Let $\mathcal{S}_2$ denote the space of subgroups of $\mathbb{Q}^2/\mathbb{Z}^2$. We let μ_2 be Haar measure on Γ_2 and let ν_2 be its push forward along the map

$$\mathrm{Ker} : \Gamma_2 \to \mathcal{S}_2,$$

$$\mathrm{Ker} : \psi \mapsto \{h : \psi(h) = 0\}.$$

As before, the action of $\mathrm{PSL}_2(\mathbb{Z})$ on TFA_2 projects down to an action of $\mathrm{PSL}_2(\mathbb{Z})$ on $\mathcal{S}_2$. As before, ν_2 is $\mathrm{PSL}_2(\mathbb{Z})$-invariant.

Lemma 5.3 *There is a ν_2-measure one set on which* $\mathrm{PSL}_2(\mathbb{Z})$ *acts freely.*

Proof. Let $M \in \mathrm{SL}_2(\mathbb{Z})$, $M \notin \{1, -1\}$. We need to show that there is a measure one set on which M acts freely.

For some $k_1, k_2, l_1, l_2 \in \mathbb{Z}$ we have

$$M : (1, 0) \mapsto (k_1, l_1),$$

$$M : (0, 1) \mapsto (k_2, l_2).$$

Let ψ be a sufficiently randomly chosen character, and let

$$H = \{\pi_2(\vec{q}) : \psi(\pi_2(\vec{q})) = 0\}.$$

We want to obtain

$$M \cdot H \neq H =_{\mathrm{df}} \{\pi_2(M(\vec{q})) : \pi_2(\vec{q}) \in H\}.$$

Since M has determinant one but is unequal to ± 1, M cannot be an integer multiple of the identity – that is to say that there is no $k \in \mathbb{Z}$ with $k_1 = l_2 = k$ and $l_2 = k_1 = 0$.

Claim: Let p be a prime bigger than k_1, k_2, l_1, l_2. Then the ν_2-probability that

$$H \cap \langle \pi_2(p^{-1}, 0), \pi_2(0, p^{-1}) \rangle = M \cdot H \cap \langle \pi_2(p^{-1}, 0), \pi_2(0, p^{-1}) \rangle$$

is at most

$$\frac{3}{p}.$$

Proof of claim. There is only a one in p chance that $\pi_2(p^{-1}, 0) \in H$, so let us assume instead that

$$\pi_2(p^{-1}, 0) \notin H,$$

i.e.

$$\psi(\pi_2(p^{-1}, 0)) \neq 0.$$

Let $g = \pi_2(p^{-1}, 0)$ and $h = \pi_2(0, p^{-1})$. Then there will be a unique $n \in \{0, 1, 2, ..., p-1\}$ such that

$$\pi_2(np^{-1}, p^{-1}) \in H,$$

i.e.

$$\psi(ng + h) = n\psi(g) + \psi(h) = 0.$$

Keep in mind below that each value of n in $\{0, 1, 2, ..., p-1\}$ is equally probable.

Now in order for

$$H \cap \langle \pi_2(p^{-1}, 0), \pi_2(0, p^{-1}) \rangle = M \cdot H \cap \langle \pi_2(p^{-1}, 0), \pi_2(0, p^{-1}) \rangle$$

we must in fact have

$$np^{-1} \cdot \pi_2(k_1, l_1) + p^{-1} \cdot \pi_2(k_2, l_2) \in H,$$

i.e.

$$\psi(n(k_1 g + l_1 h) + k_2 g + l_2 h) = 0,$$

$$\therefore nk_1\psi(g) + nl_1\psi(h) + k_2\psi(g) + l_2\psi(h) = 0.$$

Subtracting off $nl_1 + l_2$ times the equation

$$n\psi(g) + \psi(h) = 0$$

we obtain

$$nk_1\psi(g) + k_2\psi(g) - n(nl_1 + l_2)\psi(g) = 0;$$

here using $\psi(g) \neq 0$ we obtain the quadratic

$$k_1 n + k_2 - l_1 n^2 - l_2 n \equiv 0 \mod p$$

in n; this equation is non-trivial by M not being a linear multiple of the identity, and thus it has at most two solutions over $\mathbb{Z}_p$. (Claim □)

As $p \to \infty$ this probability approaches 0, and we obtain with probability 1 that $H \neq M \cdot H$. □

And now we can start down the same sequence of observations as in the preceding section. By 5.1 we obtain $E^{\mathcal{S}_2}_{\mathrm{PSL}_2(\mathbb{Z})}$ is not hyperfinite. Since the inverse image map along π_2 sets up a Borel bijection between $\mathcal{S}_2$ and subgroups of $\mathbb{Q}^2$ containing $\mathbb{Z}^2$, and since this inverse image map respects the action of $\mathrm{PSL}_2(\mathbb{Z})$, we have that $\cong|_{\mathrm{TFA}_2}$ is not hyperfinite.

Here one could compare problem 66 of [7], which explicitly raised the issue of complete invariants for the rank two case and held up as a paradigm the classification of rank one TFA groups from 2.6. Of course the problem there is phrased in a loose and open ended manner, so it is impossible to say precisely what a negative or positive solution should entail. However it is not unreasonable to now take a pessimistic attitude towards Fuchs' problem, and to conclude on the basis of the above results that there is no system for classifying rank two TFA groups as satisfactory as presented by Baer for rank one.

The original draft of this part of the paper contained an error which was pointed out by Simon Thomas, who also arrived at a rather different approach to 5.3. His method has the advantage of easily lifting to higher dimensions and showing that in general the measures concentrate on the auto-rigid TFA groups. With his kind permission I will briefly sketch this argument.

The main idea is to show that for $\psi \in \Gamma_2$ a sufficiently random character, and $\pi_2(\vec{q}_1)$, $\pi_2(\vec{q}_2)$ distinct elements of $\mathrm{Ker}(\psi)$, we must have infinitely many primes p with

$$\pi_2(p^{-1} \cdot \vec{q}_1) \in \mathrm{Ker}(\psi) \Leftrightarrow \pi_2(p^{-1} \cdot \vec{q}_2) \notin \mathrm{Ker}(\psi).$$

It suffices to consider the case that $\pi_2(\vec{q}_1)$, $\pi_2(\vec{q}_2)$ are in $\mathbb{Z}^2$ and are linearly independent.

Then for each prime the events

$$\psi(\pi_2(p^{-1} \cdot \vec{q}_1)) = 0$$

and

$$\psi(\pi_2(p^{-1}\vec{q}_2)) = 0$$

have probability $\frac{1}{p}$ and can be shown to be μ_2-independent. Thus the probability

$$\pi_2(p^{-1} \cdot \vec{q}_1) \in \mathrm{Ker}(\psi) \Leftrightarrow \pi_2(p^{-1} \cdot \vec{q}_2) \notin \mathrm{Ker}(\psi)$$

equals

$$2\frac{p-1}{p^2}.$$

It can be shown that as p ranges over the primes, the values

$$\psi(\pi_2(p^{-1} \cdot \vec{q}_1)), \quad \psi(\pi_2(p^{-1} \cdot \vec{q}_2))$$

are all μ_2-independent. Thus we need only show that for each $l \in \mathbb{N}$

$$\prod_{p>l,\, p \text{ prime}} (1 - 2\frac{p-1}{p^2}) = 0,$$

or, equivalently,

$$\sum_{p>l,\, p \text{ prime}} \log(1 - \frac{2p-2}{p^2}) = -\infty,$$

which by considering the derivative of the log function around $x = 1$ amounts to

$$\sum_{p>l,\, p \text{ prime}} \frac{2p-2}{p^2} = \infty,$$

which in turn follows from Euler's theorem (see [12]) that

$$\sum_{p \text{ prime}} \frac{1}{p} = \infty.$$

This works equally well in all dimensions, and thus:

Proposition 5.4 *(Thomas) For each $n \in \mathbb{N}$ there is a Borel probability measure $\check{\nu}_n$ on the subgroups of $\mathbb{Q}^n$ such that*

(i) $\check{\nu}_n$ is $\mathrm{SL}_n(\mathbb{Z})$-invariant;

(ii) $\check{\nu}_n$ concentrates on the auto-rigid groups;

(iii) $\check{\nu}_n$ concentrates on the subgroups which include $\mathbb{Z}^n$, and hence have rank n.

In a different direction, Kechris has extended the above arguments to show:

Theorem 5.5 $\cong|_{\mathrm{TFA}_2}$ *is not treeable.*

Kechris' proof proceeds by representing $\mathrm{PSL}_2(\mathbb{Z}[1/2])$ as a lattice in a product group, and lifting a $\mathrm{PSL}_2(\mathbb{Z}[1/2])$-invariant probability measure to a measure invariant under the action of the larger group. From there he proves a general result regarding non-treeability of equivalence relations induced by measure preserving actions of certain kinds of product groups, which he is then able to transfer back to the non-treeability of the orbit equivalence relation induced by $\mathrm{PSL}_2(\mathbb{Z}[1/2])$ acting on the subgroups of $\mathbb{Q}^2$. Granting all these steps, the non-treeability of $\cong|_{\mathrm{TFA}_2}$ follows as before.

6 Open problems

The most aggravating loose end regards the universality of the isomorphism relation for finite rank TFA groups. This question has been previously posed in [11].

Question 6.1 *Do we have*

$$E_\infty \leq_B \cong|_{\mathrm{TFA}_2}?$$

In other words, is it the case that for any countable Borel equivalence relation E

$$E \leq_B \cong|_{\mathrm{TFA}_2}?$$

More generally, are there any $n \in \mathbb{N}$ *with*

$$E_\infty \leq_B \cong|_{\mathrm{TFA}_n}?$$

Most naturally occuring countable Borel equivalence relations tend to be either smooth, hyperfinite, or universal. Here the reader could compare [11] or [20] for examples along these lines. Thus we might expect $\cong|_{\mathrm{TFA}_2}$ to follow this pattern, and since the results above show it cannot be hyperfinite, perhaps it should be universal.

On the other hand one has

$$\cong|_{\mathrm{TFA}_1} <_B \cong|_{\mathrm{TFA}_2},$$

and so we might also wonder whether by analogy at every n

$$\cong|_{\mathrm{TFA}_n} <_B \cong|_{\mathrm{TFA}_{n+1}}.$$

In the case of *rigid* finite rank TFA groups Scott Adams and Alexander Kechris in [1] have shown just that:

Definition 6.2 For each n let TFA_n^* be the auto-rigid groups in TFA_n – that is to say, those groups for which

$$g \mapsto g$$

and

$$g \mapsto -g$$

are the only automorphisms. Let $\cong_n^*$ be the isomorphism relation on groups in TFA_n^*.

Theorem 6.3 *(Adams-Kechris) For each n*

$$\cong_n^* <_B \cong_{n+1}^* .$$

Finally some comments about what does *not* remain open, since there might be a question concerning the weight of the decision to proceed with Borel and measurable functions. Alternatively one might ask whether there exist *set recursive* functions assigning subsets of $\mathbb{N}$ mod finite as a complete invariant, despite the impossibility of a Borel reduction.

The above proofs extend to these settings. Most reasonable classes of functions will give rise solely to measurable functions when applied to Borel probability spaces, and this, along with modest closure properties, is all the above argument used. For instance it will succeed for the functions that are *absolutely* (the definitions continue to work through forcing extensions) Δ_1 definable in ZFC, or, under appropriate set theoretical assumptions, for the set *recursive* functions of [17]. Indeed using [18] and the above we may obtain models of ZF in which there is no injection from the finitely rank abelian torsion free groups up to isomorphism into $\mathcal{P}(\mathbb{N})/\mathrm{Fin}$ or models of ZFC in which there is no *definable* injection from $\mathrm{TFA}_2/\cong$ into $2^{\mathbb{N}}/E_0$.

References

[1] S. Adams and A. S. Kechris, *Linear algebraic groups and countable Borel equivalence relations*, in preparation, (1999).

[2] S. Adams and R. Spatzier, *Kazhdan groups, cocycles and trees*, American Journal of Mathematics **112** (1990), 271–287.

[3] R. Baer, *Abelian groups without elements of finite order*, Duke Journal of Mathematics **3** (1937), 68–122.

[4] J. P. Burgess, *A selection theorem for group actions*, Pacific Journal of Mathematics **80** (1979), 333-336.

[5] R. Dougherty, S. Jackson, and A. S. Kechris, *The structure of hyperfinite Borel equivalence relations*, Transactions of the American Mathematical Society **34** (1994), 193–225.

[6] H. Friedman and L. Stanley, *A Borel reducibility theory for classes of countable structures*, Journal of Symbolic Logic **54** (1989), 899–914.

[7] L. Fuchs, *Infinite Abelian Groups*, Vols. **I** and **II**, Academic Press (1970, 1973).

[8] E. Hewitt and K. A. Ross, *Abstract Harmonic Analysis,* Grundlehren der mathematischen Wissenschaft **115**, Springer-Verlag, New York (1979).

[9] G. Hjorth, *Classification and orbit equivalence relations*, unpublished manuscript (available at www.math.ucla.edu/~greg/research.html).

[10] G. Hjorth, *Invariants for measure preserving transformations*, typed notes (1998).

[11] G. Hjorth and A. S. Kechris, *Borel equivalence relations and classifications of countable models*, Annals of Pure and Applied Logic **82** (1996), 221–272.

[12] K. Ireland and M. Rosen, *A Classical Introduction to Modern Number Theory*, Graduate texts in mathematics **84**, Springer-Verlag, New York (1982).

[13] S. Jackson, A. S. Kechris, and A. Louveau, *Countable Borel equivalence relations*, in preparation, (1999).

[14] A. S. Kechris, *Classical Descriptive Set Theory*, Graduate Texts in Mathematics **156**, Springer-Verlag, New York (1995).

[15] A. S. Kechris, *Definable equivalence relations and Polish group actions*, unpublished manuscript.

[16] A. S. Kechris, *Isomorphism of rank 2 torsion free abelian groups is not treeable*, handwritten note, (1998).

[17] G. Melles, *One cannot show from ZFC that there is an Ulm-type classification of the countable torsion free abelian groups*, in Set theory of the continuum (H. Judah, W. Just, and W. Woodin, eds.), MSRI publications **26**, Springer, New York (1992).

[18] R. M. Solovay, *A model of set theory in which every set of reals is Lebesgue measurable*, Annals of Mathematics **71** (1970), 1–56.

[19] S. Thomas, *Notes on* TFA_2 *groups*, handwritten note, (1998).

[20] S. Thomas and B. Velickovic, *On the complexity of the isomorphism relation for finitely generated groups*, preprint, (1998).

[21] S. Wagon, *The Banach-Tarski Paradox*, Cambridge University Press (1993).

[22] R. Zimmer, *Ergodic Theory and Semi-Simple Groups*, Birkhauser, Basel (1984).

Trends in Mathematics, © 1999 Birkhäuser Verlag Basel/Switzerland

On the compact-open topology of Ext(C,A)

Christian Leopold *

Abstract

The compact-open topology of $\mathrm{Ext}(C, A)$ is investigated for discrete abelian groups A and C. We compute the closure of 0 which appears to be a well-known subgroup of Ext. We also characterize those pairs A and C for which $\mathrm{Ext}(C, A)$ is compact, respectively locally compact.

1 Introduction

The compact-open topology on the factor systems induces a group topology on $\mathrm{Ext}(C, A)$ which is investigated in this paper for discrete abelian groups A and C. Section 2 contains the precise definition of this topology, some elementary properties of it and a description of the closure of 0. From a topological point of view, Example 2.5 might be of intrinsic interest since it indicates how the compact-open topology of $\mathrm{Ext}(C, A)$ occurs in the study of the arcwise connected component of 0 of a compact abelian group. The determination of the closure of 0 depends heavily on a homological characterization of purity which suggests a possible generalization to non-discrete groups.

Section 3 answers the natural question, when $\mathrm{Ext}(C, A)$ is compact. The conditions on A and C are easy to describe and show that this phenomenon is not so rare as one might assume. For instance, $\mathrm{Ext}(C, A)$ is always compact if C is torsion and A is finite rank torsion-free. In this situation, it is also possible to describe the topological structure of $\mathrm{Ext}(C, A)$ by means of C and A. Furthermore, all compact, totally disconnected abelian groups are covered by some $\mathrm{Ext}(G, \mathbf{Z})$.

In Section 4 it turns out that the locally compact case does not exceed very much the compact or the discrete case. The proof of Theorem 4.1 contains a description of necessary and sufficient conditions on A and C which turn $\mathrm{Ext}(C, A)$ into a locally compact abelian group. Finally, $\mathrm{Ext}(C, A)$ is always complete (modulo the closure of 0).

Unexplained notation is as in [2]. The word "group" always means "abelian group". We use C_t to denote the torsion subgroup of a group

*Research fellow of the *Studienstiftung des deutschen Volkes* .

C and C_p to denote the p-torsion subgroup. We use "$\simeq$" to indicate an algebraic isomorphism, whereas "$\cong$" is used for topological isomorphisms only.

2 The compact-open topology

For topological abelian groups A and C we define $\Delta_{K,U} = \{f \in \mathrm{Fact}(C,A) \mid f(K) \subset U\}$ where K is a compact subset of $C \times C$ and U is a neighborhood of 0 in A. Then the system of all such $\Delta_{K,U}$ is a fundamental system of (not necessarily open) neighborhoods of 0 in $\mathrm{Fact}(C,A)$ for a group topology. This induces a group topology on $\mathrm{Ext}(C,A)$ which we call the *compact-open topology* . If C is discrete, the $\Delta_{K,U}$ are open. If, in addition, A is also discrete, we define $\Delta_K = \Delta_{K\times K,\{0\}}$ where K is a finite subset of C. Evidently, these Δ_K form a base of open neighborhoods of 0 for the compact-open topology. From now on we regard $\mathrm{Ext}(C,A)$ as being equipped with the compact-open topology.

If $\alpha : C \to D$ is a continuous homomorphism then it is easily seen that the induced homomorphism $\alpha^* : \mathrm{Ext}(D,A) \to \mathrm{Ext}(C,A)$ is continuous, too. The same is true for α_*. This can be rephrased as: $\mathrm{Ext}(.\,,.)$ is an additive bifunctor in the category of topological abelian groups where the morphisms are continuous homomorphisms.

At some occasions in the sequel it will be necessary to know that such a homomorphism is also topological, i.e. open on its image. Unfortunately, this is not always true for α^*, even if α is topological.

Lemma 2.1 *If* $C \cong C_1 \times C_2$ *then* $\mathrm{Ext}(C,A) \cong \mathrm{Ext}(C_1,A) \times \mathrm{Ext}(C_2,A)$. *If* $A \cong A_1 \times A_2$ *then* $\mathrm{Ext}(C,A) \cong \mathrm{Ext}(C,A_1) \times \mathrm{Ext}(C,A_2)$.

Proof. The topological splitting of C_1 from C is equivalent to the existence of continuous homomorphisms $\iota : C_1 \to C$ and $\pi : C \to C_1$ such that $\pi\iota = \mathrm{id}_{C_1}$ (see [3]). Since $\mathrm{Ext}(.\,,A)$ is an additive functor in the category of topological groups, the assertion follows. The other statement can be proved in the same manner. □

From now on we restrict our attention to discrete abelian groups A and C, so a typical open neighborhood of 0 is given by Δ_K where K is a finite subset of C.

Lemma 2.2 *The* $\{\Delta_K\}$ *form a base of open neighborhoods where* K *runs through all finite subgroups, respectively finitely generated subgroups.*

Proof. Denote by ι_K the inclusion map $K \to C$ where K is a subgroup of C. Then $\Delta_K = \ker \iota_K^*$. If S is a finite subset of C we denote by K the subgroup generated by S. Then $K = F \oplus E$ where F is free and E is finite. Since $\mathrm{Ext}(K, A) \simeq \mathrm{Ext}(E, A)$ we obtain $\Delta_E = \Delta_K \subset \Delta_S$, so we are done. □

If we want to prove that certain induced morphisms are topological, we have to do this in almost all cases on the level of factor systems. But unfortunately, this does not carry over to the extensions in general. An important exception is the case when the induced morphism is surjective.

Lemma 2.3 *The following homomorphisms are continuous and open:*

1. $\iota_{C_t}^* : \mathrm{Ext}(C, A) \to \mathrm{Ext}(C_t, A)$
2. $\mathrm{Ext}(\oplus C_p, A) \xrightarrow{\simeq} \prod_p \mathrm{Ext}(C_p, A)$

Proof. All our homomorphisms are continuous which follows immediately from preceding considerations. They are all onto. Hence it suffices to consider the maps between factor systems which we call ϕ for simplicity.

1. $\phi : f \mapsto f \restriction C_t \times C_t$. Let K be a finite subgroup of C. Then $\phi(\Delta_K) = \Delta_K$ since $K \leq C_t$. Hence ϕ is open.
2. $\phi : f \mapsto (f \restriction C_p \times C_p)_p$. Let K be a finite subgroup of C. Since $K = \oplus_p K_p$ we obtain $\phi(\Delta_K) = \prod_p \Delta_{K_p}$. For almost all p we have $\Delta_{K_p} = \Delta_0 = \mathrm{Ext}(C_p, A)$, so $\prod_p \Delta_{K_p}$ is open in the product topology. □

Now we are ready to answer our first question: When is $\mathrm{Ext}(C, A)$ Hausdorff?

Theorem 2.4 *The closure of* 0 *in* $\mathrm{Ext}(C, A)$ *is* $\mathrm{Pext}(C, A)$.

Proof. Let $E : 0 \to A \to B \to C \to 0$ be an element of $\mathrm{Ext}(C, A)$. $E \in \Delta_K$ means that $\iota_K^* E = 0$. We have the following commuting diagram, the vertical maps being inclusion maps:

$$\begin{array}{rccccccc} E: & 0 \to & A & \to & B & \to & C & \to 0 \\ & & \| & & \uparrow & & \uparrow & \\ \iota_K^* E: & 0 \to & A & \to & B_2 & \to & K & \to 0 \end{array}$$

Hence $E \in \bigcap \Delta_K$ iff for each $A \leq B_2 \leq B$ such that B_2/A is finite we have that A splits from B_2. The latter is equivalent to A being pure in B. □

This theorem can be generalized in the following way: let $\mathcal{K}$ be a class of groups which is closed under homomorphic images, let A be discrete and C be arbitrary. If the compact subsets of C fulfil the nice property that all Δ_K ($K \in \mathcal{K}$) constitute a neighborhood base at 0 then we see that in this situation the closure of 0 is $\mathrm{Pext}_{\mathcal{K}}(C, A)$ (cf. [5]).

Example 2.5 We investigate the case $A = \mathbf{Z}$: The exact sequence $E : 0 \to \mathbf{Z} \longrightarrow \mathbf{R} \xrightarrow{\exp} \mathbf{T} \to 0$ induces

$$\mathrm{Hom}(G, \mathbf{R}) \xrightarrow{\exp_*} \mathrm{Hom}(G, \mathbf{T}) = \hat{G} \xrightarrow{\delta_*} \mathrm{Ext}(G, \mathbf{Z}) \to 0$$

where $\hat{G}$ is the Pontrjagin dual of G. If $\mathbf{T}$ is furnished with its usual compact topology and $\hat{G}$ with the compact-open topology, $\hat{G}$ is compact.

Using Lemma 2.2 it can be shown that δ_* is open. Since E has a locally continuous representation function, δ_* is also continuous. Moreover, Dixmier [1] has shown that im $\exp_* = \hat{G}_a$, the arcwise connected component of $\hat{G}$. Hence we obtain $\mathrm{Ext}(G, \mathbf{Z}) \cong \hat{G}/\hat{G}_a$. If G is torsion then $\mathrm{Ext}(G, \mathbf{Z}) \cong \hat{G}$, whence it is compact. If G is torsion-free, $\mathrm{Pext}(G, \mathbf{Z}) = \mathrm{Ext}(G, \mathbf{Z})$ forces $\mathrm{Ext}(G, \mathbf{Z})$ to be indiscrete.

It is a well-known result that $\mathrm{Pext}(C, A)$ is the first Ulm-subgroup of $\mathrm{Ext}(C, A)$ which in turn is the closure of 0 in the $\mathbf{Z}$-adic topology. It is now quite natural to ask if there is a connection between the compact-open and the $\mathbf{Z}$-adic topology. Since $\mathrm{Ext}(C, A)/\Delta_K \simeq \mathrm{Ext}(K, A)$ is bounded, we infer that the $\mathbf{Z}$-adic topology is finer than the compact-open topology.

But the $\mathbf{Z}$-adic topology is in general not equal to the compact-open topology as is shown by the following example:

Example 2.6 Let λ be an infinite cardinal. The p-adic integers, equipped with its usual compact topology, are denoted by $\mathbf{J}_p$. $\mathrm{Ext}(\bigoplus_\lambda \mathbf{Z}(p^\infty), \mathbf{Z})$ is, according to Example 2.5, topologically isomorphic to $\prod_\lambda \mathbf{J}_p$, whence it is compact. But the $\mathbf{Z}$-adic topology on this group is clearly not the product topology. Hence it is strictly finer than the compact-open topology.

We end this section with an example which illuminates a previous remark on topological morphisms.

Example 2.7 The exact sequence $0 \to \mathbf{Z}_p \longrightarrow \mathbf{Q} \xrightarrow{\alpha} \mathbf{Z}(p^\infty) \to 0$ induces $\mathrm{Hom}(\mathbf{Z}_p, \mathbf{Z}) = 0 \longrightarrow \mathrm{Ext}(\mathbf{Z}(p^\infty), \mathbf{Z}) \xrightarrow{\alpha^*} \mathrm{Ext}(\mathbf{Q}, \mathbf{Z})$. The induced homomorphism is topological on the factor systems, because for every finite subset K of $\mathbf{Z}(p^\infty)$ we can find a finite subset L of $\mathbf{Q}$ such that $\alpha(L) = K$. Nevertheless, α^* is not a topological embedding between the Ext's since $\mathrm{Ext}(\mathbf{Z}(p^\infty), \mathbf{Z}) \cong \mathbf{J}_p$ is compact, whereas $\mathrm{Ext}(\mathbf{Q}, \mathbf{Z})$ is indiscrete.

3 The compact case

Every group C is the direct limit of its finitely generated subgroups. Thus we have a natural mapping from $\mathrm{Ext}(C, A)$ into the projective limit of the $\mathrm{Ext}(K, A)$ where K runs through all finitely generated subgroups of C. This map is onto (see [4]) and its kernel is the intersection of all Δ_K, so it equals the closure of 0.

This map is clearly continuous and all $\mathrm{Ext}(K, A)$ are discrete since obviously $\Delta_K = 0$. Furthermore, we obtain:

Proposition 3.1 $\mathrm{Ext}(C, A)/\mathrm{Pext}(C, A)$ *is a complete topological group.*

Proof. All we have to do is to show that the preceding map is topological. Since $\mathrm{Ext}(C, A)$ has a linear topology, it is known that we can embed $\mathrm{Ext}(C, A)/\mathrm{Pext}(C, A)$ into the projective limit of the $\mathrm{Ext}(C, A)/\Delta_K$ where Δ_K runs through a neighborhood base of 0. A simple appeal to Lemma 2.2 completes this proof. □

This leads to a topological proof of the following well-known theorem:

Corollary 3.2 $\mathrm{Ext}(C, A)$ *is algebraically compact if* $\mathrm{Pext}(C, A) = 0$.

Proof. Here $\mathrm{Ext}(C, A)$ is Hausdorff, both in the compact-open and in the $\mathbf{Z}$-adic topology. It is complete in the compact-open topology. Hence it is also complete in the finer $\mathbf{Z}$-adic topology. □

After this little detour, we return to our initial task. If $\mathrm{Ext}(C, A)$ is compact we have that $\mathrm{Pext}(C, A) = 0$, the Hausdorff property being included in the definition of compactness. The pure-exact sequence $0 \to C_t \to C \to C/C_t \to 0$ induces

$$\begin{array}{ccccc} \mathrm{Ext}(C/C_t, A) & \to & \mathrm{Ext}(C, A) & \to & \mathrm{Ext}(C_t, A) \to 0 \\ \| & & \uparrow & & \\ \mathrm{Pext}(C/C_t, A) & \to & \mathrm{Pext}(C, A) = 0 & & \end{array}$$

We infer from the exactness of the sequence that $\mathrm{Ext}(C, A) \simeq \mathrm{Ext}(C_t, A)$.

Lemma 2.3 tells us that then $\mathrm{Ext}(C, A)$ is topologically isomorphic to $\prod_p \mathrm{Ext}(C_p, A)$. So in the sequel we can restrict ourselves to p-groups C. Now we know that $\mathrm{Ext}(C, A)$ is topologically isomorphic to the projective limit above. The projections which correspond to our projective limit are exactly the ι_K^* induced by the inclusion maps $\iota_K : K \to C$. Hence they are onto. This means that $\mathrm{Ext}(C, A)$ is compact iff all $\mathrm{Ext}(K, A)$ are finite.

Let K be a finite subgroup of C. Since $\mathrm{Ext}(\mathbf{Z}(p^n), A) \simeq A/p^nA$ we conclude that $\mathrm{Ext}(K, A)$ is finite iff A/p^nA is finite for all n such that K contains an element of exponent n. Select a p-basic subgroup B of A. Since $A/p^nA \simeq B/p^nB$ we infer that A/p^nA is finite iff A/pA is finite. Now we have proved:

Theorem 3.3 $\mathrm{Ext}(C, A)$ *is compact in the compact-open topology if and only if* $\mathrm{Pext}(C, A) = 0$ *and* A/pA *is finite for all* p *such that* $C[p] \neq 0$. □

By examining the proof more closely we arrive at:

Corollary 3.4 $\mathrm{Ext}(C, A)$ *is quasi-compact iff* A/pA *is finite for all* p *such that* $C[p] \neq 0$.

We should remark that "quasi-compact" means compact without the Hausdorff property.

Corollary 3.5 $\mathrm{Ext}(C, A)$ *is linearly compact iff it is compact.*

Proof. One direction is trivial. For the other note that projective limits, direct products and epimorphic images of linearly compact groups are linearly compact. Furthermore, observe that a discrete group is linearly compact iff it is Artinian. Since $\mathrm{Ext}(K, A)$ is bounded, we are ready. □

Example 3.6 Let C be a p-group and A be torsion-free. This implies that $\mathrm{Pext}(C, A)$=0. Theorem 3.3 tells us that $\mathrm{Ext}(C, A)$ is compact iff A/pA is finite (for instance, let A be of finite rank). This is the case iff a p-basic subgroup B of A is finitely generated, say isomorphic to $\mathbf{Z}^n$. From the exact sequence $0 \to B \to A \to A/B$ we obtain $\mathrm{Hom}(C, A/B) = 0 \to \mathrm{Ext}(C, B) \to \mathrm{Ext}(C, A) \to \mathrm{Ext}(C, A/B) = 0$. Hence $\mathrm{Ext}(C, A) \cong \hat{C}^n$ by Example 2.5 and Lemma 2.1.

4 The locally compact case

Let $\mathrm{Ext}(C, A)$ be Hausdorff, i.e. $\mathrm{Pext}(C, A) = 0$. We have seen that $\mathrm{Ext}(C, A)$ is topologically isomorphic to the projective limit of all $\mathrm{Ext}(K, A)$ where K runs through all finite subgroups of C. If $\mathrm{Ext}(C, A)$ is locally compact, one of the Δ_K must be compact. Consequently, for all $L \geq K$ the kernel of $\mathrm{Ext}(L, A) \to \mathrm{Ext}(K, A)$ must be compact and discrete, whence finite. This is the main tool for the proof of

Theorem 4.1 $\mathrm{Ext}(C, A)$ *is locally compact iff it is a product of a discrete and a compact group.*

Proof. We can assume that C is a p-group. Now we can distinguish several cases:

1. If A/pA is finite then $\mathrm{Ext}(C, A)$ is compact.
2. If C is finite, $\mathrm{Ext}(C, A)$ is discrete.

3. If C has infinite rank and A/pA is infinite then for each K we can choose an element x which is independent of K, so that the kernel of $\mathrm{Ext}(K\oplus <x>, A) \to \mathrm{Ext}(K, A)$ is infinite. Hence $\mathrm{Ext}(C, A)$ is not locally compact in this situation.

4. Let C have finite rank, so $C \simeq E\oplus\mathbf{Z}(p^\infty)^k$ where E is finite. It suffices to consider the case $C = \mathbf{Z}(p^\infty)$. Let $B = \bigoplus_{m_0} \mathbf{Z} \oplus \bigoplus_{n>0} \bigoplus_{m_n} \mathbf{Z}(p^n)$ be a p-basic subgroup of A. Let $K \simeq \mathbf{Z}(p^k)$ and $L \simeq \mathbf{Z}(p^l)$ be two finite subgroups of $\mathbf{Z}(p^\infty)$ such that $l \geq k$. The map $\mathrm{Ext}(L, A) \to \mathrm{Ext}(K, A)$ coincides with the natural map $A/p^lA \to A/p^kA$ which is the same as $B/p^lB \to B/p^kB$. Since B/p^kB is isomorphic to

$$\bigoplus_{m_0} \mathbf{Z}(p^k) \oplus \bigoplus_{n=1}^{k} \bigoplus_{m_n} \mathbf{Z}(p^n) \oplus \bigoplus_{n>k} \bigoplus_{m_n} \mathbf{Z}(p^k)$$

we see that the kernel of the latter morphism is finite iff m_0 is finite, all m_n are finite for $n > k$ and only finitely many m_n are not equal to 0. The locally compactness ensures the existence of a natural number k such that for all $l \geq k$ all these kernels are finite. This means in particular that m_0 is finite and there are only finitely many m_n. Consequently, a p-basic subgroup of A_p has to be bounded which implies that A_p itself is bounded.

To sum up, $\mathrm{Ext}(\mathbf{Z}(p^\infty), A)$ is locally compact iff A_p is bounded and A/A_t has finitely generated p-basic subgroups.

In this case, $\mathrm{Ext}(\mathbf{Z}(p^\infty), A) \cong A_p \times \mathbf{J}_p^{m_0}$ where A_p is discrete. □

References

[1] J. Dixmier, *Quelques propriétés des groupes abéliens localement compacts* , Bull. Sci. Math. (2) **81** (1957), 38–48.

[2] L. Fuchs, *Infinite Abelian Groups* **I**, Academic Press (1970).

[3] E. Hewitt and K. A. Ross, *Abstract Harmonic Analysis* **I**, Springer Verlag (1963).

[4] R. J. Nunke, *Modules of extensions over Dedekind rings*, Ill. J. Math **3** (1959), 222–241.

[5] C. P. Walker, *Relative homological algebra and abelian groups*, Ill. J. Math **10** (1966), 186–209.

Trends in Mathematics, © 1999 Birkhäuser Verlag Basel/Switzerland

Direct decompositions of LCA groups

Peter Loth

Abstract

A closed subgroup H of an LCA (locally compact abelian) group G is called topologically pure if $\overline{nH} = H \cap \overline{nG}$ for all positive integers n. It is shown that a finite subgroup of an LCA group is a splitting subgroup if and only if it is topologically pure. A characterization of LCA groups with splitting subgroup of all compact elements is established, and the groups splitting in every LCA group in which they are contained as the subgroup of all compact elements are described. The abelian case of Lee's [8] Theorem on supplements for the identity component in locally compact groups is proved. An example of a compact abelian group G with nonsplitting identity component G_0 is given, and a supplement for G_0 is determined.

Introduction

In this paper, all considered groups will be abelian, unless otherwise stated. The class of all (discrete) abelian groups will be denoted by A, and L will denote the class of all LCA (locally compact abelian) groups. By $\mathbf{F}_p$ we mean the additive group of p-adic numbers, $\mathbf{R}$ and $\mathbf{Z}$ are the additive groups of reals and integers, respectively, and $\mathbf{Z}(p^\infty)$ is the quasicyclic group. Let H be a closed subgroup of $G \in \mathsf{L}$. Then we say that H *splits in* G (or H *is a topological direct summand of* G) if there exists a closed subgroup K of G such that $H \cap K = 0$ and that the map $\varphi : H \times K \to G$ defined by $\varphi(h, k) = h + k$ for every $(h, k) \in H \times K$ is a homeomorphism. In this case we write $G = H \oplus K$.

Baer [2] and Fomin [3] proved that the torsion part tA of a group $A \in \mathsf{A}$ is a splitting subgroup if tA is bounded. Khan [7] constructed an LCA group G whose torsion part tG is finite but *not* a topological direct summand of G, answering a question raised by Armacost in [1, 6.30]. A closed subgroup H of a group $G \in \mathsf{L}$ is called *topologically pure* if $\overline{nH} = H \cap \overline{nG}$ for each positive integer n (cf. [11]). It is shown that a finite subgroup of an LCA group is a topological direct summand if and only if it is topologically pure (Theorem 1.3).

Baer's [2] and Fomin's [3] description of torsion groups splitting in every group in A in which they are contained as torsion part and May's [10]

characterization of groups in A with splitting torsion part give rise to corresponding results on LCA groups where the torsion part is replaced by the subgroup of all compact elements (Theorems 2.3 and 2.7, see also [9]). It was shown in [9] that the dual counterpart of Theorem 2.7 is a characterization of LCA groups with splitting identity component. Lee [8] proved that every locally compact (not necessarily abelian) group G contains a compact supplement for the identity component G_0 if G/G_0 is compact. We show a proof of the abelian case (Theorem 3.2). Then we exhibit an example of a compact abelian group with nonsplitting identity component, and determine a corresponding supplement (Example 3.3).

1 Finite subgroups

Proposition 1.1 *Let $G \in \mathsf{L}$ and let H be a cyclic subgroup of order p^k where p is a prime and k is a positive integer. Then the following statements are equivalent:*

1. *H is topologically pure in G;*
2. *$H \cap \overline{p^k G} = 0$;*
3. *H splits in G.*

Proof. Clearly (1) implies (2). To prove that (2) implies (3), we assume $H \cap \overline{p^k G} = 0$ and let $\mathcal{C}$ be the set of all open subgroups A of G containing $p^k G$ such that $A \cap H = 0$. Note that $\mathcal{C}$ is nonempty because $G/\overline{p^k G}$ is totally disconnected and therefore contains a compact open subgroup $A/\overline{p^k G}$ such that $A \cap H = 0$ (cf. [6, Theorem 7.7]). By Zorn's lemma, $\mathcal{C}$ has a maximal element C. We show that $G = H \oplus C$ (cf. the proof of [4, Proposition 27.1]). Let $x \in G$, q a prime and suppose $qx = h + c$ $(h \in H, c \in C)$. If $q \neq p$, then there is a $y \in H$ so that $qy = h$ and if $q = p$, then $q^{k-1}h + q^{k-1}c = q^k x \in C$ and therefore $q^{k-1}h = 0$. In either case, $qy = h$ for some $y \in H$. Let $x' = x - y$ and assume $x' \notin C$. By the maximality of C, we have $H \cap (C + \langle x' \rangle) \neq 0$, hence there is a nonzero element $h' = c' + nx' \in H$ where $c' \in C$ and $(n, q) = 1$. Then there are integers r and s so that $rn + sq = 1$, thus $x' = r(h' - c') + sqx' \in H \oplus C$ and it follows that $x \in H \oplus C$. This shows that $G/(H \oplus C)$ is torsion-free. Assume that there is an element $g \in G \backslash (H \oplus C)$. Again, $H \cap (C + \langle g \rangle) \neq 0$ by the maximality of C, so there exists a nonzero element $h'' = c'' + mg \in H$ $(c'' \in C, m$ a nonzero integer), but then $mg \in H \oplus C$. Therefore $G/(H \oplus C)$ is a torsion group and we conclude that $G = H \oplus C$, as claimed.

It is clear that (3) implies (1). □

To prove that a finite subgroup of an LCA group is a splitting subgroup if it is topologically pure, the next lemma is needed.

Lemma 1.2 *Let H and K be subgroups of $G \in \mathsf{L}$ such that H is compact, K is closed in G and $H \subset K$. If K is topologically pure in G, then K/H is topologically pure in G/H.*

Proof. Let n be a natural number. If $\overline{nK} = K \cap \overline{nG}$, then $\overline{(H + nG)} \cap K = H + (\overline{nG} \cap K) = H + \overline{nK} = \overline{H + nK}$, hence $\overline{n(K/H)} = K/H \cap \overline{n(G/H)}$. □

Theorem 1.3 *Let H be a finite subgroup of $G \in \mathsf{L}$. Then H splits in G if and only if H is topologically pure.*

Proof. Suppose H is topologically pure and write it as a direct sum of cyclic groups whose orders are prime powers: $H = H_1 \oplus \ldots \oplus H_n$. If H_1 is cyclic of order p^k, then $H_1 \cap \overline{p^k G} = 0$. By Proposition 1.1, there exists a closed subgroup K_1 of G such that $G = H_1 \oplus K_1$. Then Lemma 1.2 shows that $H \cap K_1 \cong H/H_1$ is topologically pure in $K_1 \cong G/H_1$. By induction, K_1 contains a closed subgroup A such that $K_1 = (H \cap K_1) \oplus A$. It follows that $G = H_1 \oplus [(K_1 \cap H) \oplus A] = [(H_1 \oplus K_1) \cap H] \oplus A = H \oplus A$, therefore H splits in G. Since topological direct summands are topologically pure, the proof is complete. □

Remark 1.4 In [7], Khan constructed an LCA group with finite, nonsplitting torsion part T. By Theorem 1.3, T is not topologically pure.

2 The subgroup of all compact elements

The following description of a special class of torsion groups is well known:

Theorem 2.1 (Baer [2], Fomin [3]) *Let $T \in \mathsf{A}$ be a torsion group. Then T splits in every group in A in which T is contained as torsion part if and only if it is a direct sum of a divisible and a bounded group.*

If $G \in \mathsf{L}$, then an element $x \in G$ is called *compact* if $\overline{\langle x \rangle}$ is compact. The set of all compact elements of G is denoted by bG and is a closed subgroup of G (cf. [6, Theorem 9.10]). It is clear that bG coincides with the torsion part tG if G is discrete. Note that if K is a compact subgroup of $G \in \mathsf{L}$, then $b(G/K) = bG/K$. Pontrjagin duality shows that bG splits in G exactly if the dual group of G contains the identity component as a splitting subgroup. Hence [1, Corollary 6.9] can be used to prove the next lemma.

Lemma 2.2 *Let $G \in \mathsf{L}$ and let K be a compact subgroup of G. Then bG splits in G if and only if $b(G/K)$ splits in G/K.*

The groups $C \in \mathsf{L}$ consisting of compact elements and splitting in each group $G \in \mathsf{L}$ where $bG = C$ are characterized in the next theorem (for a proof of the equivalence of (1) and (3), see also [9]).

Theorem 2.3 *Let $C \in \mathsf{L}$ such that $bC = C$. Then the following conditions are equivalent:*

1. *C splits in every group $G \in \mathsf{L}$ where $bG = C$;*
2. *For every compact open subgroup K of C, the quotient C/K is a direct sum of a divisible and a bounded group;*
3. *There exists a compact open subgroup K of C such that C/K is a direct sum of a divisible and a bounded group.*

Proof. (1) implies (2). Suppose C splits in every group in L where C is the subgroup of all compact elements, and let K be any compact open subgroup of C. Then C/K is a discrete torsion group. Assume that A is a group in A such that the torsion part tA is isomorphic to C/K. By [4, Theorem 24.6], there is a group G containing C such that C/K is the torsion part of G/K. Now an open basis at 0 in C can be used as an open basis at 0 in G to define a topology on G (cf. [6, 4.18(h)]). The resulting group is in L and contains C and therefore K as an open subgroup. Since $bG/K = t(G/K) = C/K$, we have $bG = C$ and therefore bG splits in G. By Lemma 2.2, $t(G/K)$ splits in G/K which means that tA splits in A. Now C/K is a direct sum of a divisible and a bounded group by Theorem 2.1.

(2) clearly implies (3).

(3) implies (1). Suppose K is a compact open subgroup of $C \in \mathsf{L}$ as in (3) and let $G \in \mathsf{L}$ with $bG = C$. Then $C/K = t(G/K)$ splits in G/K by Theorem 2.1. By Lemma 2.2, bG splits in G. □

Corollary 2.4 *Let $G \in \mathsf{L}$. Then we have:*

1. *If bG contains a dense subgroup which is divisible, then bG splits in G.*
2. *If bG is a discrete torsion group which is a direct sum of a divisible and a bounded group, then bG splits in G.*

Remark 2.5 If H is a closed subgroup of $G \in \mathsf{L}$ such that $H \cong \mathbf{F}_p$ or $H \cong \mathbf{Z}(p^\infty)$, then H does not necessarily split in G (see [1, Examples 6.4 and 6.22]). However, if in addition $H = bG$, then H splits in G.

We recall the following characterization of groups in A with splitting torsion part:

Theorem 2.6 (May [10]) *Let $A \in$ A. Then the torsion part tA of A splits in A if and only if A contains an ascending sequence of subgroups $A_1 \subset \ldots \subset A_n \subset \ldots$ such that*

1. $\bigcup_{n=1}^{\infty} A_n = A$;

2. tA_n *is bounded for all* n;

3. $t(A/A_n) = (tA + A_n)/A_n$ *for all* n.

Theorem 2.6 can be extended to a characterization of LCA groups with splitting subgroup of all compact elements (for a different proof, see [9]):

Theorem 2.7 *Let $G \in$ L. Then bG splits in G if and only if G contains an ascending sequence of subgroups $K \subset A_1 \subset \ldots \subset A_n \subset \ldots$ (where K is compact and A_n is open in G for all n) such that*

1. $\bigcup_{n=1}^{\infty} A_n = G$;

2. bA_n/K *is a bounded torsion group for all* n;

3. $t(G/A_n) = (bG + A_n)/A_n$ *for all* n.

Proof. Suppose $G = bG \oplus H$ where H is a closed subgroup of G. The group bG contains a compact open subgroup K, and

$$A_n = (n!)^{-1} K \oplus H \quad (n = 1, 2, \ldots)$$

satisfies (1) - (3).

Conversely, suppose G has a sequence of subgroups as in the theorem. Write $G = V \oplus \tilde{G}$ where V is a maximal vector subgroup and $\tilde{G}$ has a compact open subgroup G' containing K. Let $C = \tilde{G}/G'$ and define $D_n = [(A_n \cap \tilde{G}) + G']/G'$ $(n = 1, 2, \ldots)$. Then $\bigcup_{n=1}^{\infty} D_n = C$ and the torsion part of every group D_n is bounded. Further, we have an isomorphism

$$t(C/D_n) \xrightarrow{\sim} \frac{t(G/A_n)}{(G' + A_n)/A_n}$$

which maps $(tC + D_n)/D_n$ onto $[(bG + A_n)/A_n]/[(G' + A_n)/A_n]$. Therefore $t(C/D_n)$ is equal to $(tC + D_n)/D_n$ for all n. By Theorem 2.6, tC splits in C, and Lemma 2.2 shows that bG splits in G. □

3 The identity component

Dualization of Theorem 2.7 yields the following characterization of LCA groups with splitting identity component (see [9]):

Theorem 3.1 *Let $G \in \mathsf{L}$. Then G_0 splits in G if and only if G contains a descending sequence of subgroups $H \supset B^1 \supset \ldots \supset B^n \supset \ldots$ (where H is open in G and B^n is compact for all n) such that*

1. $\bigcap_{n=1}^{\infty} B^n = 0$;
2. *$H/(G_0 + B^n)$ is a bounded torsion group for all n;*
3. *$(B^n)_0 = G_0 \cap B^n$ for all n.*

Not every LCA group contains a splitting identity component (see the example after the next theorem). Lee [8] showed that every locally compact (not necessarily abelian) group G with compact quotient G/G_0 contains a compact supplement for G_0, that is, there exists a compact totally disconnected subgroup D such that G is generated by G_0 and D. To prove the abelian case, the proof in [8] can be simplified considerably:

Theorem 3.2 (Lee [8]) *Let $G \in \mathsf{L}$ such that G/G_0 is compact. Then there exists a compact totally disconnected subgroup D of G such that $G = G_0 + D$.*

Proof. Suppose $G \in \mathsf{L}$ where G/G_0 is compact. Let $\mathcal{C}$ be the set of all compact subgroups C of G such that $G = G_0 + C$, together with the inclusion as partial order. Note that $G_0 = V \oplus E$ where V is a maximal vector subgroup and E is a compact connected group, hence $G/E = G_0/E \oplus H/E$ for some compact group H. Therefore the set $\mathcal{C}$ is nonempty. Now let $\{C_i\}_{i \in I}$ be a descending chain in $\mathcal{C}$. We show that the compact group $C = \bigcap_i C_i$ is contained in $\mathcal{C}$. For any $g \in G$ and $i \in I$ there exist elements $h_i \in G_0$ and $c_i \in C_i$ with $g = h_i + c_i$, and we may assume that the net $\{c_i\}_{i \in I}$ converges to an element $c \in C$. Then $\{h_i\}_{i \in I} = \{g - c_i\}_{i \in I}$ converges to $h = g - c \in G_0$, hence $G = G_0 + C$ and therefore $C \in \mathcal{C}$. By Zorn's lemma, there is a minimal element D in $\mathcal{C}$. Assume D is not totally disconnected. Then there is a nonzero element $x_0 \in D_0$. By [6, Theorem 24.7], the open set $D \backslash \{x_0\}$ contains a compact subgroup K of D such that D/K is isomorphic to $(\mathbf{R}/\mathbf{Z})^n \times N$ where $n < \omega$ and N is a finite group. Therefore, we have $D/K = (D/K)_0 \oplus F/K$ for some compact subgroup F of D, hence $D = D_0 + F$. But then $G = G_0 + F$ which implies $D = F$ because of the minimality of D. It follows that D/K is totally disconnected and therefore $x_0 \in D_0 \subset K$, a contradiction. Thus D is totally disconnected. □

For each prime p and each ordinal α, a closed subgroup $G[p^\alpha]$ of $G \in \mathsf{L}$ is defined by

$G[p^0] = 0,$
$G[p^{\alpha+1}] = \{x \in G : px \in G[p^\alpha]\}$, and
$G[p^\alpha] = \overline{\sum_{\beta<\alpha} G[p^\beta]}$ if α is a limit ordinal.

These groups may be useful in the following example.

Example 3.3 Let p be a prime. The torsion part of the discrete group $A = \langle a_1, \dots, a_n, \dots : pa_1 = \dots = p^n a_n = \dots \rangle$ does not split in A (cf. [3] or [4, vol. II, p. 188]). The dual of A is the compact group

$$G = \{(x_n) \in \prod_{i=1}^{\infty} T_i : px_1 = \dots = p^n x_n = \dots\}$$

where $T_i = \mathbf{R}/\mathbf{Z}$ for all i. We will show directly that the identity component G_0 does not split in G and exhibit a supplement for G_0. Since G_0 is the maximal divisible subgroup of G (cf. [6, Theorem 24.25]), we have

$$G_0 = \{(x_n) \in G : x_1 = px_2, \dots, x_n = px_{n+1}, \dots\}.$$

Note that $G[p^\omega] = \prod_{i=1}^{\infty} Z(p^i)$ where $Z(p^i) \subset T_i$ is the cyclic subgroup of order p^i. If $k < \omega$, then $G[p^{\omega+k}] = \{(x_n) \in G : p^{k+1}x_1 = 0\}$ and it follows that $G[p^{\omega\cdot 2}] = G$. If $H \neq G$ is any closed subgroup of G, then $(G/H)[p] \neq 0$ because $G[p^{\omega\cdot 2}] = G$. Since $G_0[p] = 0$, G_0 does not split in G. It is clear that $G[p^\omega]$ is a supplement for G_0.

References

[1] D. L. Armacost, *The Structure of Locally Compact Abelian Groups*, Marcel Dekker Inc. (1981).

[2] R. Baer, *The subgroup of the elements of finite order of an abelian group*, Ann. Math. **37** (1936), 766–781.

[3] S. V. Fomin, *Über periodische Untergruppen der unendlichen abelschen Gruppen*, Mat. Sb. **2** (1937), 1007–1009.

[4] L. Fuchs, *Infinite Abelian Groups*, Vols. **I** and **II**, Academic Press (1970, 1973).

[5] R. O. Fulp and P. Griffith, *Extensions of locally compact abelian groups* I *and* II, Trans. Amer. Math. Soc. **154** (1971), 341–363.

[6] E. Hewitt and K. Ross, *Abstract Harmonic Analysis*, Vol. **I**, Springer Verlag (1963).

[7] J. A. Khan, *The finite torsion subgroup of an LCA group need not split*, Period. Math. Hungar. **31** (1995), 43–44.

[8] D. H. Lee, *Supplements for the identity component in locally compact groups*, Math. Z. **104** (1968), 28–49.

[9] P. Loth, *Splitting in locally compact abelian groups*, Rend. Sem. Mat. Univ. Padova **99** (1998), 187–196.

[10] W. May, *Isomorphism of group algebras*, J. Algebra **40** (1976), 10–18.

[11] N. Vilenkin, *Direct decompositions of topological groups* I, Mat. Sb., N. S. **19**(61) (1946), 85–154. [English translation from the Russian by E. Hewitt in AMS Translations, Series 1, Volume 23 (1950).]

Trends in Mathematics, © 1999 Birkhäuser Verlag Basel/Switzerland

Realizing automorphism groups of metabelian groups

Rüdiger Göbel and Agnes T. Paras

Abstract

Suppose M is any group of cardinality at most κ, where $\aleph_0 \leq \kappa < 2^{\aleph_0}$. Then there exists a torsion-free metabelian group G with trivial center and cardinality at most κ such that $\operatorname{Out} G \cong M$, where $\operatorname{Out} G = \operatorname{Aut} G/\operatorname{Inn} G$.

1 Introduction

It has been shown in [5] and [6] that every group M is the outer automorphism group of a torsion-free metabelian group G with trivial center. However, when the given group M has cardinality less than $2^{\aleph_0}$, the two constructions yielded a metabelian group G with cardinality $2^{\aleph_0}$. In this paper a different construction is presented, wherein the group G will have cardinality at most κ, when M has cardinality at most κ and $\aleph_0 \leq \kappa < 2^{\aleph_0}$. This is carried out with the aid of the Magnus representation of free metabelian groups and an adaptation of an ingenious technique used by Corner in [2]. The proof is interesting in the sense that it combines well-known results and methods in the study of group rings, modules and countable, torsion-free abelian groups in order to obtain a non-abelian result.

For the convenience of the reader, the Magnus representation of free metabelian groups, some lemmas and a proposition found in [2] and [5] are included in Sections 2 and 3. In Section 3 a torsion-free, abelian group H is defined with

$$B_M{}' < H < \widehat{B_M{}'}$$

such that H is invariant under conjugation by elements of B_M and $M \leq \operatorname{Aut} H$ (where $B_M{}'$ is the free abelian derived subgroup of the free metabelian group B_M, and $\widehat{B_M{}'}$ is its completion with respect to the $\mathbb{Z}$-adic topology. Finally it is shown that if M is any group with $|M| \leq \kappa$ ($\aleph_0 \leq \kappa < 2^{\aleph_0}$), the group $G = H \cdot B_M$ has $\operatorname{Out} G \cong M$ and $|G| \leq \kappa$, where

$$H \cdot B_M = \{h \cdot b : h \in H \text{ , } b \in B_M\}$$

The first author acknowledges the support of the German-Israeli Foundation for Scientific Research and Development Project No. G-545-173.06/97, and the second author was partially supported by a UP Diamond Jubilee Faculty Grant.

with binary operation defined by $(h \cdot b)(h' \cdot b') = hh'^{b^{-1}} \cdot bb'$.

2 The Magnus representation

If F is a free group on the set $\{x_i : i \in I\}$, the quotient group $B = F/F''$ is called a free metabelian group with free generating set $\{a_i = x_iF'' : i \in I\}$, where F'' denotes the second derived group of F. The groups $B' = F'/F''$ and $\overline{B} = B/B'$ are both free abelian, and $\overline{B}$ acts on B' via conjugation:

$$x^{\bar{b}} = x^b \ , \ x \in B' \ , \ b \in B \text{ and } \bar{b} = bB' \in \overline{B}$$

Thus B' can be viewed as a $\mathbb{Z}[\overline{B}]$–module. The use of the Magnus representation of B gives insight into the structure and properties of B' as a $\mathbb{Z}[\overline{B}]$–module as shown below.

Suppose F is a non-cyclic free group with basis $\mathcal{A} = \{x_i : i \in I\}$. If $x_i \in \mathcal{A}$, let

$$s_i = x_iF' \ , \ a_i = x_iF'' \text{ and } t_{x_i} = t_i \, .$$

Let $\{s_i = x_iF' : i \in I\}$ be a generating set for $\overline{B}$ and $\oplus_{i\in I}\mathbb{Z}[\overline{B}]t_i$ be a free $\mathbb{Z}[\overline{B}]$–module of rank $|I|$. The set of matrices

$$\begin{bmatrix} \overline{B} & 0 \\ \oplus_{i\in I}\mathbb{Z}[\overline{B}]t_i & 1 \end{bmatrix} = \left\{ \begin{pmatrix} g & 0 \\ \sum_{i\in I} r_it_i & 1 \end{pmatrix} : g \in \overline{B}, \sum_{i\in I} r_it_i \in \oplus_{i\in I}\mathbb{Z}[\overline{B}]t_i \right\}$$

forms a group under formal matrix multiplication.

Lemma 2.1 ([7], [3]) *The map*

$$a_j \to \begin{bmatrix} s_j & 0 \\ t_j & 1 \end{bmatrix} \quad (j \in I) \tag{1}$$

extends to an injective homomorphism

$$\psi : B \ \to \begin{bmatrix} \overline{B} & 0 \\ \oplus_{i\in I}\mathbb{Z}[\overline{B}]t_i & 1 \end{bmatrix}. \tag{2}$$

This identification of B with a matrix group $\psi(B)$, shows that B' embeds in $\oplus_{i\in I}\mathbb{Z}[\overline{B}]t_i$ and if $x = \sum b_it_i \in B'$ and $s = \sum n_g g \in \mathbb{Z}[\overline{B}]$, then

$$x^s = (\sum n_g g) \cdot (\sum b_it_i) = \sum (\sum n_g g) b_it_i \ ,$$

where $b_i \in \mathbb{Z}[\overline{B}]$, $g \in \overline{B}$ and $n_g \in \mathbb{Z}$. Hence the action of $\mathbb{Z}[\overline{B}]$ by conjugation on B' is "multiplication" when viewed within the context of the Magnus representation. Since $\mathbb{Z}[\overline{B}]$ has no zero divisors, B' is a torsion-free $\mathbb{Z}[\overline{B}]$-module.

We consider the general form of the elements $b_i \in \mathbb{Z}[\overline{B}]$ for $\sum b_i t_i \in B'$. Since the commutator equation

$$[uv, w] = [u, w]^v [v, w] \tag{3}$$

holds for elements of any group and the action of $\overline{B}$ on B' is multiplication, it suffices to see that

$$[a_i, a_j] = (s_j - 1)t_i + (1 - s_i)t_j \ . \tag{4}$$

As shown in [1], the element

$$\sum_{i \in I} b_i t_i \in B' \ \text{ iff } \ \sum_{i \in I}(s_i - 1)b_i = 0 \ .$$

Moreover, by equations (3) and (4), each b_i is in the augmentation ideal of $\mathbb{Z}[\overline{B}]$, and so B' is properly embedded in $\oplus_{i \in I}\mathbb{Z}[\overline{B}]t_i$.

Let M be a group with $|M| < 2^{\aleph_0}$ and F be a free group on the set

$$\{m_i : (m, i) \in M \times \{0, 1\}\} \ .$$

From now on let $B_M = F/F''$ and $\overline{B_M} = B_M/B_M{}'$. The group M acts on B_M via

$$(m_i F'')^x = (mx)_i F'' \ , \quad \text{where } m, x \in M \text{ and } i \in \{0, 1\} \ .$$

The action of $x \in M$ on a commutator $[m_i, m'_j]$ is

$$[m_i, m'_j]^x = [(mx)_i, (m'x)_j] = (s_{(m'x)_j} - 1)t_{(mx)_i} + (1 - s_{(mx)_i})t_{(m'x)_j} \ ,$$

whereas the action of $b \in B_M$ on $[m_i, m'_j]$ is

$$[m_i, m'_j]^b = \bar{b}(s_{m'_j} - 1)t_{m_i} + \bar{b}(1 - s_{m_i})t_{m'_j} \ .$$

As automorphisms of B_M, it follows that the groups Inn (B_M) and M intersect trivially. Thus the semi-direct product $\overline{B_M} \rtimes M$ embeds in Aut $(B_M{}')$. From now on we let R denote the group ring $\mathbb{Z}[\overline{B_M} \rtimes M]$. It follows that the free abelian group $B_M{}'$ is an R-module and, by the Magnus representation, the group $B_M{}'$ embeds properly in $\oplus_{(m,i) \in M \times \{0,1\}}\mathbb{Z}[\overline{B_M}]t_{m_i}$.

When the group M has non-trivial torsion elements, then $B_M{}'$ will have R-torsion elements. Although $B_M{}'$ is not necessarily R-torsion-free, we have the following property.

Proposition 2.2 ([6]) *The ring homomorphism* $R \to \text{End}\,(B_M{}')$ *is faithful.*

Proof. The element $[e_0, e_1]$ is an R-torsion-free element of $B_M{}'$, where e is the identity element of the group M. □

3 Realizing automorphism groups

The completion of the ring $\mathbb{Z}$ under the $\mathbb{Z}$-adic topology is the complete ring direct product $\widehat{\mathbb{Z}} = \prod_p J_p$, where J_p is the ring of p-adic integers. Clearly $\widehat{\mathbb{Z}}$ will contain zero divisors. However the following shows that $\mathbb{Z}$ is contained in a subring P of $\widehat{\mathbb{Z}}$, which is an integral domain and has the same cardinality as $\widehat{\mathbb{Z}}$. More importantly P is a unique factorization domain.

Lemma 3.1 ([2]) *The ring $\widehat{\mathbb{Z}}$ has a pure subring P of cardinal $2^{\aleph_0}$ and containing $\mathbb{Z}$, with the property that every element of P is an associate in P of an element of $\mathbb{Z}$.*

The completion $\widehat{B_M{}'}$ of the free abelian group $B_M{}'$ under the $\mathbb{Z}$-adic topology is a $\widehat{\mathbb{Z}}$–module. Clearly $\widehat{B_M{}'}$ is a torsion-free P–module. Let $\{f_i : i \in I\}$ be a maximal P-independent subset of $B_M{}'$. If $a \in B_M{}'$, then there exist π_a , $\pi_a^i \in P$, not all zero, such that

$$\pi_a a = \sum_{i \in I} \pi_a^i f_i \, .$$

Since $\pi_a = \xi n$, for some unit $\xi \in P$ and some $n \in \mathbb{Z}$, we can assume that $\pi_a \in \mathbb{N}$. If π_a is chosen as small as possible, then π_a and π_a^i are uniquely determined by $a \in B_M{}'$.

Let $\Pi = \langle \pi_a^i \in P : a \in B_M{}' \, , \, i \in I \rangle_*$ be the pure subring generated by the π_a^i. If $\aleph_0 \leq \kappa < 2^{\aleph_0}$ and M has cardinality at most κ, then $B_M{}'$ has infinite cardinality that is at most κ. Since the cardinality of Π is at most κ and the cardinality of P is $2^{\aleph_0}$, the ring P contains $2^{\aleph_0}$ elements which are algebraically independent over Π. (A set $X \subset P$ is *algebraically independent* over Π, if for every finite subset of $\{\alpha_1, \dots, \alpha_n\}$ of X, $f(\alpha_1, \dots, \alpha_n) \neq 0$ for all non-zero $f(x_1, \cdots, x_n)$ in the polynomial ring $\Pi[x_1 \cdots, x_n]$.)

Observe that in the preceding construction of Π , the group $B_M{}'$ can very well be replaced by any reduced, torsion-free and abelian group A with cardinality less than $2^{\aleph_0}$ (cf. [4]). Thus we state Lemma 2.1 in [2], which remains valid when A is assumed to be an abelian group which is not necessarily a ring.

Lemma 3.2 ([2]) *Let A be a reduced, torsion-free and abelian group with $|A| < 2^{\aleph_0}$. Suppose that $\sum_{j=1}^n \gamma_j a_j = 0$, where the $\gamma_j \in P$ and are linearly independent over Π, and the $a_j \in A$. Then each $a_j = 0$.*

Proof. In the construction of Π, we see that there exist $q \in \mathbb{N}$ and elements $\pi_{ji} \in \Pi$ $(j = 1, \dots, n \, ; \, i \in I)$ such that

$$q a_j = \sum_{i \in I} \pi_{ji} f_i \quad (j = 1, \dots, n) \, .$$

Since $\sum_{j=1}^{n} \gamma_j a_j = 0$, it follows that $\sum_{j=1}^{n} \sum_{i\in I} \gamma_j \pi_{ji} f_i = 0$. By the linear independence of $\{f_i : i \in I\}$, we have

$$\sum_{j=1}^{n} \gamma_j \pi_{ji} = 0 \quad (i \in I)\,.$$

The linear independence of $\gamma_j \in P$ over Π implies that each $\pi_{ji} = 0$ and so $qa_j = 0$. Since A is torsion-free, each $a_j = 0$. □

Lemma 3.3 ([5]) *Let B be a free metabelian group of rank at least two. Suppose $G = H \cdot B$, where $B' \leq H < \widehat{B'}$ and H is closed under conjugation by elements of B.*

(i) If A is a normal, abelian subgroup of G, then $A \leq H$. Hence H is the largest, normal abelian subgroup of G, and so is characteristic in G.

(ii) If $\varphi \in \operatorname{Aut} G$, then $\varphi\restriction_H \neq -1 \cdot id_H$.

Whenever A is a subgroup of a torsion-free abelian group G, the *pure subgroup generated by A* is $A_* = \{g \in G : ng \in A \text{ for some } 0 \neq n \in \mathbb{Z}\}$. If $x \in \widehat{B_M{}'}$, let x^φ denote the image of x under the homomorphism φ and let the symbol Rx denote the R–submodule of $\widehat{B_M{}'}$ generated by x. Recall the notation $R = \mathbb{Z}[\overline{B_M} \rtimes M]$.

Theorem 3.4 *Suppose M is a group with $|M| \leq \kappa$ and $\aleph_0 \leq \kappa < 2^{\aleph_0}$. Then there exists a group H such that $B_M{}' < H < \widehat{B_M{}'}$, H is an R–module, $|H| \leq \kappa$, and if $\varphi \in \operatorname{Aut}(H \cdot B_M)$, then $\varphi\restriction_H = r \in R$. In particular r is a unit in R and*

$$B_M{}' < H < \oplus_{(m,i)\in M\times\{0,1\}} P[\overline{B_M}]t_{m_i}\,.$$

Proof. Fix $u \in B_M{}'$ which is R–torsion-free (say $u = [e_0, e_1]$ as in Proposition 3.4). Choose elements α_a, β_a $(a \in B_M{}')$ of P that are algebraically independent over Π. Define

$$h_a = \alpha_a \cdot u + \beta_a \cdot a \quad (a \in B_M{}')$$

and

$$H = \langle B_M{}', Rh_a : a \in B_M{}' \rangle_*\,.$$

Clearly H is a reduced, torsion-free, abelian group, $\aleph_0 \leq |H| \leq \kappa$ and $R \leq \operatorname{End}(H)$.

Let $G = H \cdot B_M$. Suppose $\varphi \in \operatorname{Aut} G$. We show that $\varphi\restriction_H \in R$. By Lemma 3.3, the group H is a characteristic subgroup of G. So the restriction

$\varphi\restriction_H \in \operatorname{Aut} H$. Now $h_a^\varphi = \alpha_a \cdot u^\varphi + \beta_a \cdot a^\varphi$. There exists a non-zero integer q such that

$$q \cdot h_a^\varphi \,, q \cdot u^\varphi \,, q \cdot a^\varphi \in \langle B_M{}' \,, Rh_a : a \in B_M{}' \rangle \,.$$

Hence

$$q \cdot h_a^\varphi = b + \sum_{x \in B_M{}'} h_x^{r_x} \,, \quad q \cdot u^\varphi = c + \sum_{x \in B_M{}'} h_x^{s_x} \,, \quad q \cdot a^\varphi = d + \sum_{x \in B_M{}'} h_x^{t_x}$$

for some $b, c, d \in B_M{}'$ and finitely many non-zero $r_x, s_x, t_x \in R$. We obtain from the last equation,

$$\begin{aligned} b + \sum_{x \in B_M{}'} \alpha_x \cdot u^{r_x} + \beta_x \cdot x^{r_x} &= \alpha_a (c + \sum_{x \in B_M{}'} \alpha_x \cdot u^{s_x} + \beta_x \cdot x^{s_x}) \\ &\qquad + \beta_a (d + \sum_{x \in B_M{}'} \alpha_x \cdot u^{t_x} + \beta_x \cdot x^{t_x}). \end{aligned}$$

Since the set $\{\alpha_x, \beta_x : x \in B_M{}'\}$ is algebraically independent over Π and only finitely many $r_x, s_x, t_x \in R$ are non-zero, Lemma 3.2 applies and yields the following equations:

$$\begin{array}{llll} u^{r_a} = c \,, & a^{r_a} = d \,, & b = 0 \,, & a^{s_a} + u^{t_a} = 0 \\ u^{r_x} = 0 \,, & u^{t_x} = 0 \,, & & \text{for all } x \neq a \\ u^{s_x} = 0 & & & \text{for all } x \in B_M{}' \end{array}$$

Since u is R-torsion-free, it follows that $r_x = t_x = 0$ for all $x \neq a$ and $s_x = 0$ for all $x \in B_M{}'$. Thus $t_a = 0$ and $q \cdot h_a^\varphi = h_a^{r_a}$, $q \cdot u^\varphi = u^{r_a}$, $q \cdot a^\varphi = a^{r_a}$ for all $a \in B_M{}'$. By the continuity of φ on H, the image $g^\varphi \in \langle Rg \rangle_*$ for all $g \in H$.

Another application of Lemma 3.2 shows that each h_a is R-torsion-free and the set $\{h_a : a \in B_M{}'\}$ is R-independent. Write $h_a = n_a f_a$, for some integer n_a, such that the p-height $h_p^H(f_a) = 0$ for all primes p. Clearly $Rf_a \leq \langle Rf_a \rangle_*$. Let $x \in \langle Rf_a \rangle_*$. Then $mx = f_a^r$, for some non-zero integer m, and $r \in R$, and $n_a mx = n_a f_a^r = h_a^r$. Without loss of generality, assume that if p divides m, then p does not divide r in R. Now $mx = f_a^r$ means that the p-height $h_p^R(f_a^r) > 0$ and

$$n_a f_a^r = \alpha_a \cdot u^r + \beta_a \cdot a^r = b + \sum (\alpha_g \cdot u^{r_g m n_a} + \beta_g \cdot g^{r_g m n_a}) = n_a mx \,,$$

for some $b, g \in B_M{}'$ and $r_g \in R$. By Lemma 3.2, we get $r = r_a m n_a$. Hence if p divides f_a^r in H, then p divides r in R, i.e., $h_p^H(f_a^r) \leq h_p^R(r)$. Thus $h_p^R(r) > 0$ for p dividing m. But it was assumed that if p divides m, then p does not divide r in R. It follows that $m = 1$ and $x \in Rf_a$. Hence $\langle Rh_a \rangle_* = \langle Rf_a \rangle_* = Rf_a$. Moreover, each f_a is R-torsion-free, the set $\{f_a : a \in B_M{}'\}$ is R-independent and $f_a^\varphi \in Rf_a$ for all $a \in B_M{}'$.

Let a and b be distinct elements of $B_M{}'$ and $f_a^\varphi = f_a^r$, $f_b^\varphi = f_b^s$ for some $r, s \in R$. We show that $r = s$. Now $(f_a + f_b)^\varphi \in \langle R(f_a + f_b)\rangle_*$, i.e., there exists a non-zero integer n such that $n(f_a + f_b)^\varphi = (f_a + f_b)^t$, for some $t \in R$. Since φ is a homomorphism,

$$f_a^{rn} + f_b^{sn} = n(f_a^r + f_b^s) = (f_a + f_b)^t \,.$$

Since the elements f_a and f_b are R-independent and R-torsion-free, it follows that $rn = t$ and $sn = t$. Since r, s and t are elements of the group ring R, the integer n must divide t and $r = s$. Thus there exists $r \in R$ such that $f_a^\varphi = f_a^r$ for all $a \in B_M{}'$, and so $h_a^\varphi = h_a^r$ for all $a \in B_M{}'$.

Now $\alpha_a \cdot u^r + \beta_a \cdot a^r = h_a^r = h_a^\varphi = \alpha_a \cdot u^\varphi + \beta_a \cdot a^\varphi$, for all $a \in B_M{}'$. Since $g^\varphi \in \langle Rg\rangle_*$ for all $g \in H$, we can apply the preceding argument and Lemma 3.2 to show that $u^\varphi = u^r$ and $a^\varphi = a^r$ for all $a \in B_M{}'$. By the continuity of φ on H , we have $\varphi\restriction_H = r \in R$. □

The following proposition shows that any two automorphisms of G which are identical when restricted to H must be congruent modulo $\operatorname{Inn} G$.

Proposition 3.5 *Let H be as in Theorem 3.4 and $G = H \cdot B_M$. If $\varphi \in \operatorname{Aut} G$ such that $\varphi\restriction_H = id_H$, then $\varphi \in \operatorname{Inn} G$.*

Proof. Suppose $\varphi \in \operatorname{Aut} G$ such that $\varphi\restriction_H = id_H$. Then we claim that φ induces $id_{G/H}$. Let $b \in B_M$ and $h \in H$. Now $b^{-1}hb = (b^{-1}hb)^\varphi = (b^{-1})^\varphi h b^\varphi$ implies that $b^\varphi b^{-1}$ commutes with every $h \in H$. By Lemma 3.3, $b^\varphi b^{-1} \in H$. The claim follows.

Let $b, c \in B_M$, $b^\varphi = hb$ and $c^\varphi = kc$ for some $h, k \in H$. Then $[hb, kc] = [b, c]^\varphi = [b, c]$ implies that

$$h^{\bar{b}(\bar{c}-1)} = k^{\bar{c}(\bar{b}-1)} \,.$$

In particular, if $\bar{b} = \bar{c}$, then $h = k$. Suppose $\bar{b} \neq \bar{c}$. By construction, the group H is a pure subgroup of $\oplus_{(m,i)\in M\times\{0,1\}}\mathbb{Z}[\overline{B_M}]t_{m_i}$. Let $n \cdot h = x + \sum h_a^{r_a}$ and $n \cdot k = y + \sum h_a^{s_a}$ for some $0 \neq n \in \mathbb{Z}$, $x, y \in B_M{}'$ and finitely many non-zero $r_a, s_a \in R$. Thus

$$x^{\bar{b}(\bar{c}-1)} + \sum(\alpha_a \cdot u^{r_a\bar{b}(\bar{c}-1)} + \beta_a \cdot a^{r_a\bar{b}(\bar{c}-1)}) =$$
$$y^{\bar{c}(\bar{b}-1)} + \sum(\alpha_a \cdot u^{s_a\bar{c}(\bar{b}-1)} + \beta_a \cdot a^{s_a\bar{c}(\bar{b}-1)}) \,.$$

By Lemma 3.2,

$$x^{\bar{b}(\bar{c}-1)} = y^{\bar{c}(\bar{b}-1)} \text{ and } r_a\bar{b}(\bar{c}-1) = s_a\bar{c}(\bar{b}-1) \,.$$

Since $\mathbb{Z}[\overline{B_M}]$ is a unique factorization domain and $\bar{b} \neq \bar{c}$, it follows that

$$x = x'^{(\bar{b}-1)} \,, \qquad y = y'^{(\bar{c}-1)} \,, \text{ for some } x', y' \in B_M{}' \text{ and}$$

$$r_a = r'_a(\bar{b}-1)\,, \qquad s_a = s'_a(\bar{c}-1)\,, \text{ for some } r'_a, s'_a \in R\,.$$

Let $z' = x' + \sum h_a^{r'_a}$. Clearly $z' \in H$, $n \cdot h = z'^{(\bar{b}-1)}$ and $n \cdot k = z'^{\bar{b}\bar{c}^{-1}(\bar{c}-1)}$. By the purity of H and the action of $\mathbb{Z}[\overline{B_M}]$ on H, there exists $z \in H$ such that $h = z^{(\bar{b}-1)}$ and $k = z^{\bar{b}\bar{c}^{-1}(\bar{c}-1)}$. Thus $b^\varphi = z^{(\bar{b}-1)}b = (b)^{z^{-b}}$ and $c^\varphi = z^{\bar{b}\bar{c}^{-1}(\bar{c}-1)}c = (c)^{z^{-b}}$. Hence φ is conjugation by $z^{-b} \in H$. □

Theorem 3.4 shows that if $\varphi \in \operatorname{Aut} G$, then $\varphi\restriction_H = \sum \sigma_m m \in R$, where $\sigma_m \in \mathbb{Z}[\overline{B_M}]$. In order to prove that $\varphi \in \operatorname{Inn} G \rtimes M$, it is enough to show, by Proposition 3.5, that all σ_m are zero, except for one $\sigma_{m'}$, which is an element of $\overline{B_M}$.

Theorem 3.6 *Let $G = H \cdot B_M$ be as in Theorem 3.4. Then* $\operatorname{Aut} G = \operatorname{Inn} G \rtimes M$.

Proof. Clearly $\operatorname{Inn} G \rtimes M \leq \operatorname{Aut} G$. Let $\varphi \in \operatorname{Aut} G$. By Theorem 3.5, $\varphi \restriction H$ is a unit $\mu \in R$. Let $h \in H$ and $b \in B_M$. Then

$$h^{\mu\bar{b}^\varphi} = (b^{-1}hb)^\varphi = (b^{-1}hb)^\mu = h^{\bar{b}\mu}$$

for all $h \in H$. By Proposition 2.2, $\mu\bar{b}^\varphi = \bar{b}\mu$, when viewed as elements of R. Let $\mu = \sum_m \sigma_m m \in \mathbb{Z}[\overline{B_M}][M] = R$, where $\sigma_m \in \mathbb{Z}[\overline{B_M}]$. Without loss of generality, assume $\sigma_e \neq 0$, where e is the identity element in M (otherwise, multiply μ by m^{-1} if $\sigma_m \neq 0$). Using the multiplication in a semi-direct product, we obtain

$$\sum_m \sigma_m \bar{b}^{\varphi m^{-1}} m = \sum_m \bar{b}\sigma_m m\,.$$

This means $\sigma_m \bar{b}^{\varphi m^{-1}} = \bar{b}\sigma_m$, for all $m \in M$ and $b \in B_M$. Since $\mathbb{Z}[\overline{B_M}]$ is an integral domain, $\bar{b}^{\varphi m^{-1}} = \bar{b}$, for all $b \in B_M$ and for all m such that $\sigma_m \neq 0$. In particular, $\sigma_e \neq 0$ and so $\bar{b}^\varphi = \bar{b}$ for all $b \in B_M$, i.e., φ induces the identity on $\overline{B_M}$. Suppose $\sigma_{m'} \neq 0$ for some $m' \neq e$. Then there exists $c \in B_M$ such that $\bar{c}^{m'} \neq \bar{c}$ (for example $c = e_0$). Hence $(\bar{c})^{\varphi m'^{-1}} \neq \bar{c}$ gives a contradiction. This shows that the M-support of μ as an element of the group ring $(\mathbb{Z}[\overline{B_M}])[M]$ is a singleton, i.e., μ is a unit in $\mathbb{Z}[\overline{B_M}]$. By Lemma 3.3, $\mu \in \overline{B_M}$. In general, $\varphi\restriction_H \in \overline{B_M} \rtimes M$. By Proposition 3.5, $\varphi \in \operatorname{Inn} G \rtimes M$. Therefore $\operatorname{Aut} G = \operatorname{Inn} G \rtimes M$. □

References

[1] S. Bachmuth, *Automorphisms of free metabelian groups*, Trans. Amer. Math. Soc. **118** (1965), 93–104.

[2] A. L. S. Corner, *Every countable reduced torsion-free ring is an endomorphism ring*, Proc. London Math. Soc. (3) **13** (1963), 687–710.

[3] R. H. Fox, *Free differential calculus* I, Ann. of Math. (2) **57** (1953), 547–560.

[4] R. Göbel and W. May, *Independence in completions and endomorphism algebras*, Forum Math. **1** (1989), 215–226.

[5] R. Göbel and A. T. Paras *Automorphisms of metabelian groups with trivial center*, Illinois J. Math. **42** (1998), 333–346.

[6] R. Göbel and A. T. Paras, *Outer automorphism groups of metabelian group*, to appear in Journal of Pure and Appl. Algebra.

[7] W. Magnus, *On a theorem of Marshall Hall*, Ann. of Math. (2) **40** (1939), 764–768.

Trends in Mathematics, © 1999 Birkhäuser Verlag Basel/Switzerland

On the class semigroups of Prüfer domains

Laszlo Fuchs

Abstract

S. Bazzoni has proved several theorems on the class semigroups of Prüfer domains of finite character. Here we extend her results, and at the same time give shortened proof of some of her theorems.

1 Introduction

In any domain R the isomorphy classes $[I]$ of fractional ideals I form a semigroup $\mathcal{S}(R)$ under multiplication. Bazzoni and Salce [6] have shown that if R is a valuation domain, then $\mathcal{S}(R)$ is of a very special kind: it is a Clifford semigroup, i.e., a semigroup that is the union of disjoint groups. In a recent paper [2], Bazzoni has extended this result to the class semigroup $\mathcal{S}(R)$ of Prüfer domains R of finite character by proving that they are likewise Clifford semigroups.

Our present aim is to generalize her result: we consider arbitrary Prüfer domains, but restrict ourselves to the subsemigroup $\mathcal{S}^*(R)$ of $\mathcal{S}(R)$ consisting of isomorphy classes representable by ideals that contain at least one element of finite character. We prove that Bazzoni's theorem holds for this subsemigroup: it is always of Clifford type. At the same time, we give a simple proof of this theorem.

The structure theorems in [3] and [4] on the idempotents and on the group constituents of the Clifford semigroup $\mathcal{S}(R)$ carries over to the general case considered here. In addition to the class groups of fractional overrings of R, the only other groups occurring are the so-called archimedean groups of the localizations R_P of R at maximal ideals P. These groups were first considered in Bazzoni-Fuchs-Salce [5] in connection with the classification of uniserial modules over valuation domains. They are the group constituents of the class semigroups of a valuation domain (see Bazzoni-Salce [6]). Recall that a fractional ideal I is called *archimedean* (Matlis [7]) if its endomorphism ring is isomorphic to R. If the maximal ideal P of a valuation domain is not principal, then the isomorphy classes of archimedean ideals form an abelian group under multiplication, with the isomorphy class $[P]$ of P as identity. (If P is principal, then there are no non-principal archimedean ideals.) Archimedean ideals of valuation domains have been investigated by Bazzoni [1].

The notion of archimedean ideal makes sense for every domain R. However, their isomorphy classes rarely form a group. This is certainly the case for most Prüfer domains. In spite of this, we can introduce archimedean groups for Prüfer domains of finite character, by using a different idea: the direct limit device.

In this note, R will denote throughout a Prüfer domain, unless stated otherwise.

1.1 The Clifford semigroup

If I is an ideal of R, we shall denote by $R^{[I]}$ the following overring of R:

$$R^{[I]} = \bigcap_{P \notin \Omega(I)} R_P.$$

As usual, R_P denotes the localization of R at the maximal ideal P, and $\Omega(I)$ stands for the set of maximal ideals containing I.

The term '*of finite character*' will be used for a ring to indicate that every non-zero element (equivalently, every non-zero ideal) in the ring is contained but in a finite number of maximal ideals. We will also say that an element is *of finite character* if it belongs only to a finite number of maximal ideals. Evidently, every ideal that contains a non-zero element of finite character must itself be of finite character.

The next lemma holds over arbitrary domains. It is a slightly stronger version of a lemma proved by Bazzoni [2].

Lemma 1.1 *Let I be an ideal of the domain R. If I contains a non-zero element a_0 of finite character, then it contains a finitely generated ideal J such that $a_0 \in J$ and $\Omega(J) = \Omega(I)$.*

Proof. Set $\Omega(I) = \{P_1, \dots, P_n\}$, and assume $P_1, \dots, P_n, P_{n+1}, \dots, P_{n+k}$ are the maximal ideals containing a_0. For each $j = 1, \dots, k$, pick an element $a_j \in I \setminus P_{n+j}$. Then the ideal $J = \sum_{j=0}^{k} a_j R$ is as desired. □

We continue with a technical lemma.

Lemma 1.2 *Any ideal $I \neq 0$ of a Prüfer domain R satisfies*

$$R : I \leq R^{[I]}.$$

If I contains an element of finite character, then we also have

$$R \leq IR^{[I]}.$$

Proof. By way of contradiction, let us assume that $q \in Q$ satisfies $qI \leq R$, but $q \notin R_P$ for some $P \notin \Omega(I)$. Since R_P is a valuation domain, $q^{-1} \in PR_P$, we can write $q^{-1} = a/b$ with $a \in P$, $b \in R \setminus P$. Then $bI \leq aR \leq P$, contradicting $b \notin P$, $I \not\leq P$. Hence $q \in R_P$, and $R : I \leq R^{[I]}$.

If I contains an element of finite character, then by the preceding lemma it also contains a finitely generated ideal J with $\Omega(J) = \Omega(I)$. The ideal J is invertible, thus $R : J \leq R^{[J]}$ implies $R \leq JR^{[I]}$, and so *a fortiori* $R \leq IR^{[I]}$. □

Turning our attention to the class semigroup $\mathcal{S}(R)$ of R, we start with quoting a criterion for the regularity of an element in a semigroup. Recall that an element a of a semigroup S is regular (in the sense of von Neumann) if there is an $x \in S$ such that $axa = a$.

Lemma 1.3 (Bazzoni [2]) *Let I be a (fractional) ideal of R. Then $[I]$ is a regular element in $\mathcal{S}(R)$ if and only if $(I : I^2)\, I^2 = I$.* □

This criterion is used to prove the following crucial fact.

Lemma 1.4 *Let R be a Prüfer domain. If the ideal $I \neq 0$ of R contains an element of finite character, then $[I]$ is a regular element of $\mathcal{S}(R)$.*

Proof. Owing to Lemma 1.3, it suffices to verify the inclusion $I \leq (I : I^2)\, I^2$. Evidently,

$$I : I^2 = (\bigcap_P I_P) : I^2 = \bigcap_P (I_P : I_P^2) = (\bigcap_{P \in \Omega(I)} (I_P : I_P^2)) \bigcap (\bigcap_{P \notin \Omega(I)} (I_P : I_P^2)),$$

where $\Omega(I)$ is a finite set. Hence we conclude

$$(I : I^2)\, I^2 = \bigcap_{P \in \Omega(I)} (I_P : I_P^2)\, I_P^2 \bigcap (\bigcap_{P \notin \Omega(I)} (I_P : I_P^2))\, I^2.$$

Bazzoni-Salce [6] have shown that $\mathcal{S}(R_P)$ is a Clifford semigroup for a valuation domain R_P. Thus the ideal I_P of R_P satisfies $(I_P : I_P^2)\, I_P^2 = I_P$ for all $P \in \Omega(I)$. Furthermore, note that for $P \notin \Omega(I)$ we have $I_P = R_P$, whence $I_P : I_P^2 = R_P$ follows. This means that the second intersection in the last displayed formula is equal to $R^{[I]} I^2$. This product contains, in view of Lemma 1.2, the ideal $RI = I$. Consequently, the last member in the displayed formula contains I. This suffices to establish the desired inclusion. □

As a result, we can claim that the isomorphy classes of fractional ideals that can be represented by ideals containing at least one element of finite character are regular elements of $\mathcal{S}(R)$. Moreover, we can establish the following result:

Theorem 1.5 *Let R be a Prüfer domain. The elements of the class semigroup $\mathcal{S}(R)$ that can be represented by ideals containing at least one element of finite character form a Clifford semigroup $\mathcal{S}^*(R)$.*

Proof. It is clear that all what we have to check is that if the ideal I contains an element a of finite character, then R contains an ideal which is isomorphic to $I : I^2$ and which also contains an element of finite character. By Lemma 1.4, $(I : I^2)I^2 = I$ holds, thus $(I : I^2)\ a^2 \leq I$. It follows that $a^2 I : I^2$ is an ideal in R. It is isomorphic to the fractional ideal $I : I^2$ and contains the element a^2 of finite character. □

An immediate corollary is Bazzoni's theorem:

Corollary 1.6 (Bazzoni [2]) *The class semigroup of a Prüfer domain of finite character is a Clifford semigroup.* □

2 The structure of $\mathcal{S}(R)$

In order to unveil the structure of a Clifford semigroup, one has to find its idempotents, the constituent groups and the bonding homomorphisms.

As far as the idempotents are concerned, Bazzoni [3] proves that each idempotent in the class semigroup $\mathcal{S}(R)$ of a Prüfer domain R of finite character can be represented by a unique idempotent fractional ideal L such that

$$L = P_1 P_2 \cdots P_n D \qquad (n \geq 0)$$

with uniquely determined factors satisfying the following conditions:

(i) the P_i are pairwise incomparable idempotent prime ideals of R;
(ii) each $P_i D$ is a maximal ideal of the domain D;
(iii) $D = L : L = \operatorname{End} L$ is a fractional overring of R;
(iv) D contains each endomorphism ring $\operatorname{End} P_i$.

The same conclusion holds in our case, i.e. for the idempotents in $\mathcal{S}^*(R)$ for an arbitrary Prüfer domain R. For instance, we verify (ii).

To see that for a prime ideal P of R, PD is a maximal ideal of the domain $D = \operatorname{End} P$ (and hence of larger overrings of R), we show that if P' is a prime $P' > P$ in R, then $P'D = D$. We go back to Lemma 1.2 to argue that $P'R^{[P']} = R^{[P']}$. Let $L_1, \ldots, L_m$ denote the maximal ideals of R containing P, but not P'. Evidently, $P'R_P = R_P$, whence we infer that

$$\begin{aligned} P'D &= P'R^{[P]} \cap P'R_P \geq P'(R^{[P]} \cap R_{L_1} \cap \cdots \cap R_{L_m}) \cap R_P \\ &= P'R^{[P']} \cap R_P = R^{[P']} \cap R_P \end{aligned}$$

contains the identity 1 of R, so $P'D$ can not be a proper ideal of D.

She also characterizes the partial order in the set of idempotents. The corresponding result in our case reads as follows.

Lemma 2.1 *Let R be a Prüfer domain, and*

$$L = P_1 \cdots P_n D \quad \text{and} \quad L' = P_1' \cdots P_m' D'$$

idempotent fractional ideals in $\mathcal{S}^*(R)$ where the factors are subject to conditions (i)-(iv). Then $[L'] \leq [L]$ in $\mathcal{S}(R)$ if and only if

(a) $D \leq D'$;

(b) *for each i ($i = 1, \ldots, n$) either $P_i D' = D'$ or there is an index j ($j = 1, \ldots, m$) such that $P_i D' = P_j' D'$.*

For an idempotent prime ideal L, G_L denotes the group constituent of $\mathcal{S}^*(R)$ whose identity is $[L]$. The main structural result by Bazzoni [4] carries over as well:

Theorem 2.2 *Let R be a Prüfer domain, and $L = P_1 \cdots P_n D \neq 0$ an idempotent fractional ideal in $\mathcal{S}^*(R)$, where the factors satisfy conditions* (i)-(iv) *above. The subgroup G_L of the semigroup $\mathcal{S}^*(R)$ fits into the exact sequence*

$$1 \to \mathcal{C}(\operatorname{End} L) \to G_L \xrightarrow{\pi} \operatorname{Arch} R_{P_1} \times \cdots \times \operatorname{Arch} R_{P_n} \to 1,$$

where π maps $[I] \in G_L$ to $([IR_{P_1}], \ldots, [IR_{P_n}])$. □

In order to obtain a full picture of $\mathcal{S}^*(R)$, it remains to identify the bonding homomorphisms between its group constituents. By definition, a bonding homomorphism $\phi^L_{L'} : G_L \to G_{L'}$ for idempotents $[L'] \leq [L]$ sends $[I] \in G_L$ to $[IL'] \in G_{L'}$. It might be helpful to point out right away:

(a) The embedding $\mathcal{C}(\operatorname{End} L) \to G_L$ of the class group of the overring $\operatorname{End} L$ in G_L may be given by multiplication by $[P_1] \cdots [P_n]$.

(b) We can conclude at once that if $L = P_1 \cdots P_n D$, then G_L is generated by the images $[P_1] \cdots [P_n] G_{P_i D}$ of the constituent groups $G_{P_i D}$.

(c) If $D < D'$ are fractional overrings of R, then $G_D = \mathcal{C}(D), G_{D'} = \mathcal{C}(D')$ and $\phi^D_{D'}$ is an epimorphism. In fact, a finitely generated ideal C' of D' can be written as $C' = a_1 D' + \ldots + a_m D'$ with $a_j \in R$, so it is the image of $C = a_1 D + \ldots + a_m D$ under $\phi^D_{D'}$.

(d) If $P' < P$ are idempotent prime ideals of R, and D is an overring with $\operatorname{End} P, \operatorname{End} P' \leq D$, then $\phi^{PD}_{P'D}$ maps G_{PD} onto the subgroup of $G_{P'D}$ corresponding to $\mathcal{C}(D)$. Indeed, multiplication of $[I] \in G_{PD}$ by $[P']$ yields $[IP'] = [P']$, because there is an $r \in R$ such that $P'R_P < r^{-1} IR_P < PR_P$, whence $(P')^2 = P' = P'P$.

By making use of (a)-(d), we can easily verify:

Proposition 2.3 *Let $L' = P'_1 \cdots P'_m D' \leq L = P_1 \cdots P_n D$ be idempotent fractional ideals in $S^*(R)$ of a Prüfer domain R where the factors satisfy conditions* (i)-(iv) *above. Define*

$$K = P_1 \cdots P_k D' = P'_1 \cdots P'_k D'$$

with $k \leq n, m$ where $P_i D' = P'_i D'$ $(i = 1, \ldots, k)$ and $P_i D' = D'$ $(i = k+1, \ldots, n)$. Then the bonding homomorphism $\phi^L_{L'} : G_L \to G_{L'}$ is the composition of a bonding epimorphism and a bonding monomorphism:

$$\phi^L_{L'} = \phi^K_{L'} \circ \phi^L_K.$$

Proof. The morphism ϕ^L_K acts via multiplication by $[D']$. By virtue of (b), it is enough to investigate how it works on $G_{P_i D}$. Obviously, it induces an epimorphism as it maps the subgroup corresponding to the class group $C(D)$ and the subgroups corresponding to the archimedean groups $G_{P_i D}$ $(i = k+1, \ldots, n)$ upon the subgroup correspondig to $C(D')$. The map $\phi^K_{L'}$ acts as multiplication by $[P'_{k+1}] \cdots [P'_m]$, inducing a monomorphism $G_K \to G_{L'}$. □

Let us mention two special cases of interest.

One is concerned with Bézout domains. Note that overrings of Bézout domains are likewise Bézout domains. For Bézout domains we can state:

Corollary 2.4 *If R is a Bézout domain, then $S^*(R)$ is a Clifford semigroup where the constituent groups are isomorphic to direct products of archimedean groups of localizations of R at prime ideals. A bonding homomorphism maps each factor of such a direct product either trivially or isomorphically.* □

The other special case concerns strongly discrete Prüfer domains. The absence of idempotent primes implies that the only fractional ideals are in the class groups of fractional overrings. Consequently,

Corollary 2.5 *If R is a strongly discrete Prüfer domain, then $S^*(R)$ is a Clifford semigroup where the constituent groups are isomorphic to class groups of fractional overrings. All bonding homomorphisms are surjective.*□

3 Archimedean ideals

We shall now exploit the results obtained above to get an insight into the collection of archimedean fractional ideals. The framework provided by

$S(R)$ is well suited to introduce archimedean groups more generally than for valuation domains.

For the sake of simplicity, we restrict our consideration to Prüfer domains of finite character.

From the discussion above it becomes evident that the isomorphy class $[A]$ of an archimedean fractional ideal A belongs to the class group of R or to a group constituent G_L where L is of the form $L = P_1 \cdots P_k$ with idempotent archimedean maximal ideals P_i of R.

Corollary 3.1 *If R is a Prüfer domain of finite character, then the isomorphy classes of archimedean ideals form a Clifford semigroup where*

(i) *the constituent groups are extensions of the class group $C(R)$ of R by direct products of archimedean groups* Arch R_{P_i} *at idempotent archimedean maximal ideals P_i;*

(ii) *all the bonding homomorphisms are monomorphisms.*

Proof. This follows simply from (1.5) and (2.2); there is no need for making additional comments. □

The Clifford semigroup of archimedean ideals suggests the introduction of archimedean groups for Prüfer domains, at least for those which are of finite character.

Instead of isomorphisms, it is reasonable to deal with a weaker notion: let us call the ideals A and B **similar** if $A = IB$ for an invertible fractional ideal I. It is pretty clear that the classes under this equivalence relation form again a Clifford semigroup which is somewhat simpler than the one in (2.2) inasmuch as the constituent groups $G_{[P_1 \cdots P_k]}$ are direct products of archimedean groups Arch R_{P_i} at idempotent archimedean maximal ideals P_i and the bonding homomorphisms are monomorphisms. This fortunate behavior of the bonding homomorphisms implements a highly non-trivial direct system. In view of this, we may define the **archimedean group** Arch R as the direct limit of this system.

What has been said provides a proof of the following theorem.

Theorem 3.2 *The archimedean group of a Prüfer domain R of finite character is a (restricted) direct product:*

$$\mathrm{Arch}R \cong \mathrm{Arch}R_{P_1} \times \cdots \times \mathrm{Arch}R_{P_i} \times \ldots,$$

with P_i running over the set of idempotent archimedean maximal ideals of R. □

Bazzoni-Salce [6] investigate the structure of archimedean groups for valuation domains. Their results show that archimedean groups must belong to certain classes groups in some situation, but in general they can be arbitrary abelian groups.

References

[1] S. Bazzoni, *Archimedean ideals of valuation domains*, Comm. Algebra **18** (1990), 3613–3618.

[2] S. Bazzoni, *Class semigroups of Prüfer domains*, J. Algebra **184** (1996), 613–631.

[3] S. Bazzoni, *Idempotents of the class semigroup of a Prüfer domain of finite character*, Abelian Groups, Module Theory, and Topology, Lecture Notes in Pure and Applied Math. **201** (Marcel Dekker, 1998), 79–89.

[4] S. Bazzoni, *Groups in the class semigroup of a Prüfer domain of finite character*, preprint.

[5] S. Bazzoni, L. Fuchs, and L. Salce, *The hierarchy of uniserial modules over a valuation domain*, Forum Math. **7** (1995), 247–277.

[6] S. Bazzoni and L. Salce, *Groups in the class semigroups of valuation domains*, Israel J. Math. **95** (1996), 135–155.

[7] E. Matlis, *Injective modules over Prüfer rings*, Nagoya Math. J. **15** (1959), 57–69.

Trends in Mathematics, © 1999 Birkhäuser Verlag Basel/Switzerland

Uniform modules, Γ-invariants, and Ziegler spectra of regular rings

Jan Trlifaj *

Abstract

The structure of uniform modules over arbitrary rings is investigated by means of the method of Γ-invariants. In particular, Γ- and Γ*- invariants are introduced for strongly uniform modules of uncountable dimension in order to measure the failure of their submodule lattices to be relatively complemented. Our main result is for dimension ω_1: for any field F, the Γ*-invariants of indecomposable injective modules over locally matricial F-algebras achieve all of the possible 2^{ω_1} many values. Some consequences for the structure of Ziegler spectra of von Neumann regular rings are derived.

Introduction

A non-zero module U is *uniform* provided that $V \cap W \neq 0$ for all non-zero submodules V and W of U. In other words, uniform modules coincide with non-zero submodules of indecomposable injective modules. Uniform modules appear in several ways in general module theory.

First, they form building blocks for a dimension theory of modules. A module M is said to have *Goldie dimension* κ provided that κ is the minimal cardinal such that $\text{card}(A) \leq \kappa$ for each independent set A consisting of non-zero submodules of M. So M is uniform iff M has Goldie dimension 1. Moreover, if $n < \omega$, then M has Goldie dimension n iff there exist independent uniform submodules $U_1, \dots, U_n \subseteq M$ such that $\oplus_{i=1}^{n} U_i$ is essential in M, [1, §25].

Second, the structure theory of uniform modules is closely related to properties of submodule lattices of indecomposable injective modules. This has model-theoretic consequences, since (isomorphism classes of) indecomposable pure injective modules coincide with the points of the so called *Ziegler spectrum*, $\text{Zg}(R)$, of the ring R. The latter is a topological space with the property that closed subsets of $\text{Zg}(R)$ correspond bijectively to complete theories of modules closed under products, [11]. The set of indecomposable injective modules is known to form a closed subset of $\text{Zg}(R)$ iff

*Research supported by grant GAUK 12/97.

R is a right coherent ring, [13, Corollary 4.4]. In particular, this is the case when R is regular (e.g., when R is a locally matricial F-algebra over a field F). In the regular case, pure injective modules coincide with the injective ones, so (isomorphism classes of) injective hulls of uniform modules actually form all of $\mathrm{Zg}(R)$. Moreover, there are no $\sum$-pure injective modules unless R is semisimple.

A trivial sufficient condition for uniformity of a module over an arbitrary ring is the existence of a minimal non-zero submodule. Such uniform modules are called *cocyclic*. Nevertheless, in general, uniform modules have a much more complex structure. For example, there is no upper bound for lengths of cofinal strictly decreasing continuous chains (*c.d.c.*'s) of non-zero submodules of uniform modules.

The main goal of the present paper is to provide a new basis for the structure theory of uniform modules, and of the indecomposable injective ones, by applying the notion of a Γ-invariant, [3]. Our Γ-invariants are defined by means of c.d.c.'s, and they measure the failure of the submodule lattices to be relatively complemented (see Section 1 below for more details). The invariants take values in the Boolean algebra, $B(\kappa)$, of all subsets of κ modulo the filter of subsets containing a closed unbounded set. Whether a given element of $B(\kappa)$ is *realized*, that is, appears as a Γ-invariant of a uniform module, depends substantially on the underlying ring R. Negative realization results are known for right noetherian and for commutative rings: if $\kappa > \omega$, then $\bar{\kappa}$ is the only element of $B(\kappa)$ which is realized as a Γ-invariant of a uniform module, [18, Theorems 2.10 and 2.12].

In Section 1, we provide positive realization results for modules over locally matricial F-algebras. Combining recent results of Goodearl-Wehrung [9] with [4], Růžička proved that all elements of $B(\omega_1)$ occur as Γ-invariants of suitable uniform modules, [14]. In Theorem 1.10, we prove a similar realization result, but concerning a modified Γ-invariant – the Γ^*-invariant – for indecomposable injective modules. On one hand, this implies that the points of $\mathrm{Zg}(R)$ are rather complex even if R is a locally matricial F-algebra. On the other hand, the Γ^*-invariants provide a way of classifying the complexity.

In Section 2, we investigate the topology on $\mathrm{Zg}(R)$ when R is an arbitrary regular ring. Several properties are obtained by means of natural embedding(s) of the maximal spectrum, $\mathrm{Spec}\, R$, of R into $\mathrm{Zg}(R)$. Then we deal with the particular case of regular rings with primitive factors artinian. This case is similar to the commutative one: for example, $\mathrm{Zg}(R)$ is isomorphic to $\mathrm{Spec}(R)$. If R is also semiartinian then we show that the Cantor-Bendixson rank of $\mathrm{Zg}(R)$ coincides with the Loewy length of R (Theorem 2.8(ii)).

For a ring R, let $\mathrm{simp}\, R$ denote a representative set of all simple (right R-) modules, and $\mathrm{Com}(R)$ a representative set of all indecomposable pure-injective (= algebraically compact) modules. Further, denote by $\mathrm{Bool}(R)$

the Boolean algebra of all central idempotents of R, and by $\mathrm{L}_2(R)$ the lattice of all (two-sided) ideals of R.

A ring R is (von Neumann) *regular* provided that each $r \in R$ has a pseudo-inverse $r' \in R$; that is, $rr'r = r$. Let F be a field. An F-algebra R is *matricial* provided that R is isomorphic to a finite direct product of full matrix algebras over F. An F-algebra R is *locally matricial* provided that R is a direct limit of matricial F-algebras.

For a module M, denote by $\mathrm{E}(M)$ the injective hull of M in $\mathrm{Mod}\,R$. Also, denote by $L(M)$ the submodule lattice of M. Put $\mathrm{Coc}\,R = \{N \in \mathrm{Com}(R) \mid \exists S \in \mathrm{simp}\,R : N \cong \mathrm{E}(S)\}$. So $\mathrm{Coc}\,R$ is a representative set of all injective cocyclic modules.

Denote by $\mathrm{Zg}(R)$ the *Ziegler spectrum* of R, [12]. Recall that $\mathrm{Zg}(R)$ is a topological space with the underlying set $\mathrm{Com}(R)$ and with a basis consisting of the sets $\{N \in \mathrm{Com}(R) \mid \varphi(N) \subset \psi(N)\}$ where φ, ψ are pp-formulas in one free variable such that $\varphi \leq \psi$. Here, we define $\phi(N) = \{n \in N \mid N \models \phi(n)\}$ for a pp-formula ϕ and $N \in \mathrm{Mod}\,R$, [11, §4.7]. The key fact about $\mathrm{Zg}(R)$ is that for any ring R, $\mathrm{Zg}(R)$ is a compact (but not necessarily T_0) space whose closed subsets correspond 1-1 to complete theories of modules closed under products, ([11, Theorems 4.66-7]: a closed subset, C, corresponds to the complete theory of the module $M = \oplus_{N \in C} N^{(\omega)}$).

Assume that R is a regular ring. Then $\mathrm{Com}(R)$ is just a representative set of all indecomposable injective modules. Moreover, $\mathrm{Zg}(R)$ has a basis consisting of the sets $V_e = \{N \in \mathrm{Com}(R) \mid Ne \neq 0\}$ where $e \in R$ is an idempotent, [11, Corollary 16.18].

There is another compact space related to any ring R, namely the *maximal spectrum*, $\mathrm{Spec}\,R$. Recall that the underlying set for $\mathrm{Spec}\,R$ is $\mathrm{Max}\,R$, the set of all maximal ideals of R. A basis for the topology consists of the sets $U_x = \{M \in \mathrm{Max}\,R \mid x \notin M\}$, where $x \in R$. So the closed sets are just $C_X = \{M \in \mathrm{Max}\,R \mid X \subseteq M\}$ where $X \subseteq R$ (or $X \in \mathrm{L}_2(R)$), and compactness is an easy consequence of $1 \in R$. Unlike the Ziegler spectrum, the maximal spectrum is always a T_1-space. Moreover, if $\bigcap \mathrm{Max}\,R = 0$, then the clopen sets of $\mathrm{Spec}\,R$ coincide with those U_e where $e \in \mathrm{Bool}(R)$.

For a regular ring R, there is a basis of topology for $\mathrm{Spec}\,R$ consisting of the sets U_e where $e \in R$ is an idempotent.

For further properties of the notions defined above we refer to [1], [5] and [11].

1 Realizations for uniform modules

Among uniform modules there is a particular subclass fitted for classification by the Γ-invariants in the sense of [3]. We recall the basic definitions and properties from [18, §2] and [4]:

Definition 1.1 Let R be a ring and $U \in$ Mod-R. A sequence of non-zero submodules of U, $\mathcal{U} = (U_\alpha \mid \alpha < \kappa)$, is called a *c.d.c.* in U provided that $\mathcal{U}$ is

- continuous ($U_0 = U$, and $U_\alpha = \cap_{\beta<\alpha} U_\beta$ for all limit ordinals $\alpha < \kappa$),
- strictly decreasing ($U_{\alpha+1} \subset U_\alpha$ for all $\alpha < \kappa$), and
- cofinal (for each non-zero submodule $V \subseteq U$ there is $\alpha < \kappa$ such that $U_\alpha \subseteq V$).

The module U is *strongly uniform* provided that there is a c.d.c. in U. The ordinal κ is called the *length* of $\mathcal{U}$. The least ordinal κ such that there is a c.d.c. $\mathcal{U}$ of length κ in U is called the *dimension* of U.

It is easy to see that any strongly uniform module U is uniform, and either $d = 1$ or d is a regular infinite cardinal, where d is the dimension of U. Also, $d = 1$ iff U is cocyclic. The case of $d = \omega$ has been considered in detail in [4, §2] and [18, §2]. So we will concentrate on the case of $d \geq \omega_1$, where the notion of Γ-invariant plays a crucial role.

Definition 1.2 Let U be a strongly uniform module of dimension $\kappa \geq \omega_1$.

(i) Let $0 \neq V \subset W \subseteq U$. Then W is *complemented* over V (in U) provided that there is a submodule $X \subseteq U$ such that $W \cap X = V$ and $W + X = U$. For example, U is complemented over any $0 \neq V \subset U$.

(ii) U is *constricted* provided that for each c.d.c. $\mathcal{U}$ of length κ in U there is $\alpha < \kappa$ such that for each $\alpha \leq \beta < \kappa$, the module U_β is not complemented over U_γ for some $\beta < \gamma < \kappa$. For example, U is constricted provided that it is uniserial.

(iii) For a subset $E \subseteq \kappa$, define

$$\bar{E} = \{D \subseteq \kappa \mid \exists C \subseteq \kappa : \ C \text{ closed and unbounded in } \kappa \ \ \& \ D \cap C = E \cap C\}.$$

Note that $\bar{E} \in B(\kappa)$.

Let $\mathcal{U} = (U_\alpha \mid \alpha < \kappa)$ be a c.d.c. in U. Let

$$E(\mathcal{U}) = \{\alpha < \kappa \mid \exists \beta : \alpha < \beta < \kappa \ \& \ \ U_\alpha \text{ is not complemented over } U_\beta\}.$$

Define $\Gamma(U) = \overline{E(\mathcal{U})}$. Then $\Gamma(U)$ does not depend on the choice of the c.d.c. $\mathcal{U}$. $\Gamma(U)$ is called the *Γ-invariant* of U.

(iv) U is *narrow* provided that $\Gamma(U) = \bar{\kappa}$. For example, any constricted module is narrow.

The realization problem asks for the possible values of the Γ-invariants of strongly uniform modules of dimension $d \geq \omega_1$. So the question is which elements of $B(d)$ are realized as Γ-invariants of such modules. The key tool for answering the question in dimension ω_1 is the following general result of Goodearl and Wehrung:

Lemma 1.3 *Let κ be an infinite cardinal. Let D be a 0-sublattice of the Boolean algebra, $P(\kappa)$, of all subsets of κ. Assume that* $\mathrm{card}(D) = \kappa$. *Then there exists a dimension vector space, V, over $\mathbb{Q}$ such that* $\mathrm{card}(V) = \kappa$ *and the maximal semilattice quotient of the positive cone of V is isomorphic to D. Moreover, the lattice, $S(V)$, of all directed convex subgroups of V is isomorphic to the ideal lattice, $Id(D)$, of D.*

Proof. Except for the equality $\mathrm{card}\, V = \kappa$, the assertion follows by [9, Theorem 4.4] and [9, Proposition 2.6]. As in [9, Definition 3.1], denote by $\mathbf{B}(D)$ the generalized Boolean subalgebra of $P(\kappa)$ generated by D. Then $V = \mathbb{Q} < D >$, where $\mathbb{Q} < D >$ is the $\mathbb{Q}$-subalgebra of $\mathbb{Q}^\kappa$ consisting of all functions, f, with finite range and with $f^{-1}(q) \in \mathbf{B}(D)$ for all $0 \neq q \in \mathbb{Q}$. It follows that $\mathrm{card}(V) = \mathrm{card}\, \mathbf{B}(D) = \mathrm{card}(D) = \kappa$. □

For $\kappa = \omega_1$, we have an important consequence due to Růžička [14]:

Theorem 1.4 *Let F be a field. Let D be a bounded distributive lattice of cardinality ω_1. Denote by $L = Id(D)$ the ideal lattice of D. Then there exists a locally matricial F-algebra, S, such that $\mathrm{L}_2(S) \cong L$.*

Proof. By Stone's Theorem, we may view D as a sublattice of $P(\omega_1)$. By Lemma 1.3, there is a dimension vector space V over $\mathbb{Q}$ such that $S(V) \cong L$. Since D is bounded, the constant function $C = 1$ is an order unit in V. Since $\mathrm{card}(V) = \omega_1$, [7, Theorem 1.5] gives a locally matricial F-algebra, S, such that the Grothendieck group $(K_0(S), [S])$ is isomorphic to (V, C) as a dimension group with order unit. By [5, Corollary 15.21], $\mathrm{L}_2(S) \cong S(V)$, so $\mathrm{L}_2(S) \cong L$. □

Another ingredient needed is a lattice construction from [4, §1]:

Definition 1.5 Let $\kappa \geq \omega$ and let $E \subseteq \kappa$ be such that $0 \notin E$. Define L_E to be the set of all subsets, B, of κ where $B = \emptyset$ or

$$B = \cup_{i \leq n} [\beta_i, \gamma_i)$$

for some $0 \leq n < \omega$ and $0 \leq \beta_0 < \gamma_0 < \beta_1 < \cdots < \gamma_n < \kappa$ where for all $i \leq n$, $\beta_i \notin E$, [4, §1]. Then L_E with the reverse order ($B \leq B'$ iff $B' \subseteq B$) is a distributive lattice. Denote by I_E the ideal lattice of L_E. Note that I_E is also distributive.

Theorem 1.6 (A realization theorem for ω_1) *Let $E \subseteq \omega_1 \setminus \{0\}$. Let F be a field. Then there exist a locally matricial F-algebra, R, and a strongly uniform module, U, of dimension ω_1 such that $L(U)$ is distributive and $\Gamma(U) = \bar{E}$.*

Proof. Put $D = L_E$. By Theorem 1.4, there is a locally matricial F-algebra, S, such that $\mathrm{L}_2(S) \cong I_E$. Put $R = S \otimes_F S^{op}$ and $U = S_R$. Then $\mathrm{L}_2(R) \cong L(U)$, and R is a locally matricial F-algebra by [7, Lemma 3.1]. By [4, Theorem 1.15], $\bar{E} = \Gamma(U)$. □

Theorem 1.6 shows that there exist narrow modules which are not constricted. Indeed, take $E_1 = \omega_1 \setminus \{0\}$ and let E_2 be the set of all limit ordinals $< \omega_1$. Then there exist strongly uniform modules of dimension ω_1, U_1 and U_2, such that $L(U_1) \cong I_{E_1}$ and $L(U_2) \cong I_{E_2}$. So both U_1 and U_2 are narrow, U_1 is constricted (even uniserial), but U_2 is not constricted (cf. [4, Corollary 1.13]).

Next, we will prove that all elements of $B(\omega_1)$ are realized as suitable invariants of indecomposable injective modules. This will show that the structure of points of Ziegler spectra of regular rings is rather complex in general.

Let U be a strongly uniform module of dimension κ. By modularity of $L(U)$, we have $\Gamma(V) \leq \Gamma(U)$ for any non-zero submodule V of U. In particular, the set

$$\mathcal{G}(U) = \{\Gamma(V) \mid 0 \neq V \subseteq U\}$$

is a lower directed subset of $B(\kappa)$. In some cases, $\mathcal{G}(U)$ has a least element:

Lemma 1.7 *Assume that there exists $0 \neq V \subseteq U$ such that $L(V)$ is distributive and V is not constricted. Then $\Gamma(V) = \min \mathcal{G}(U)$.*

Proof. Since V is not constricted, there is a c.d.c. $\mathcal{V} = (V_\alpha \mid \alpha < \kappa)$ in V such that $\kappa \setminus E(\mathcal{V})$ is cofinal in κ. Take $\alpha \in \kappa \setminus E(\mathcal{V})$. Put $\mathcal{V}' = \{V_\beta \mid \alpha \leq \beta < \kappa\}$ and $\mathcal{V}'' = \{V\} \cup \{V_\beta \mid \alpha < \beta < \kappa\}$. Then $\mathcal{V}'$ and $\mathcal{V}''$ is a c.d.c. in V_α and V, respectively.

Let $\alpha < \beta < \gamma < \kappa$. If V_β splits over V_γ in V, then it does also in V_α. Conversely, assume that V_β splits over V_γ in V_α, so $V_\beta \cap C = V_\gamma$ and $V_\beta + C = V_\alpha$, for a submodule C of V_α. Since $\alpha \in \kappa \setminus E(\mathcal{V})$, there is $B \in L(V)$ such that $V_\alpha \cap B = V_\gamma$ and $V_\alpha + B = V$. By distributivity, $V_\beta \cap (C + B) = V_\gamma + V_\gamma = V_\gamma$, and $V_\beta + (C + B) = V$. So V_β splits over V_γ in V.

This shows that $\Gamma(V_\alpha) = \Gamma(V)$ for all $\alpha \in \kappa \setminus E(\mathcal{V})$. Since the latter set is cofinal in κ, we conclude that $\Gamma(V) \leq \Gamma(W)$ for any non-zero submodule W of U. □

Definition 1.8 In the setting of Lemma 1.7, we put

$$\Gamma^*(U) = \min \mathcal{G}(U).$$

Clearly, $\Gamma^*(U)$ is an invariant of U. It is called the *Γ^*-invariant* of U.

If $L(U)$ is distributive and U is not constricted, then $\Gamma^*(U) = \Gamma(U)$ by Lemma 1.7. In general, $\Gamma^*(U)$ can be strictly less than $\Gamma(U)$:

Example 1.9 Let R be a ring. Let $E \subseteq \kappa \setminus \{0\}$ and $M \in \text{Mod-}R$ be strongly uniform of dimension κ and such that $L(M)$ is distributive and $\Gamma(M) = \bar{E} \neq \bar{\kappa}$.

Put $S = \{\left(\begin{smallmatrix} r & r' \\ 0 & r \end{smallmatrix}\right) \mid r, r' \in R\}$, $U = \{\left(\begin{smallmatrix} 0 & m \\ 0 & 0 \end{smallmatrix}\right) \mid m \in M\} \in \text{Mod-}S$ and $U' = \{\left(\begin{smallmatrix} m & m' \\ 0 & m \end{smallmatrix}\right) \mid m, m' \in M\} \in \text{Mod-}S$. Then U is essential in U', so U' is strongly uniform of dimension κ. Clearly, $L(U_S) = L(M_R)$, so $\Gamma(U_S) = \Gamma(M_R) = \bar{E}$. Moreover, if $X \in L(U'_S)$, then either $X \in L(U_S)$ or

$$\{\left(\begin{smallmatrix} 0 & n' \\ 0 & 0 \end{smallmatrix}\right) \mid n' \in N'\} \subseteq X \subseteq \{\left(\begin{smallmatrix} n & n'' \\ 0 & n \end{smallmatrix}\right) \mid n \in N, n'' \in N''\},$$

for some $0 \neq N \subseteq N' \subseteq N'' \in L(M_R)$. If follows that there are no $Z \subset Y \in L(U_S)$ such that Y is complemented over Z in U'. So U' is constricted, and $\Gamma(U'_S) = \bar{\kappa}$. Let $I = E(U_S) = E(U'_S)$. Then $\Gamma(I) = \bar{\kappa}$, while $\Gamma^*(I) = \bar{E}$ by Lemma 1.7.

Theorem 1.10 (A realization by elements of Zg(R))

Let $i \in B(\omega_1)$ and let F be a field. Then there exist a locally matricial F-algebra, R, and a strongly uniform indecomposable injective R-module, I, of dimension ω_1 such that $\Gamma^*(I) = i$.

Proof. By Theorem 1.6, there exist R as required and a strongly uniform module M of dimension ω_1 such that $L(M)$ is distributive and $\Gamma(M) = i$. Let $I = \text{E}(M)$ be the injective hull of M. Then I is strongly uniform of dimension ω_1, hence indecomposable. By Lemma 1.7, $\Gamma^*(I) = \Gamma(M) = i$. □

In the realization results of Theorems 1.6 and 1.10, the ring R is locally F-matricial, hence unit regular. There is no hope to have a realization with R commutative regular, or even R regular with primitive factors artinian: over such rings, all uniform modules are simple [17, Theorem 2.2], hence of dimension 1. The following lemma of Goodearl-Menal-Moncasi makes possible realizations where $R = S \otimes_F S^{op}$ and S is a prime regular subdirect product of simple artinian F-algebras:

Lemma 1.11 *Each regular ring is a homomorphic image of a regular subdirect product of simple artinian rings.*

Let F be a field. Then each regular F-algebra is a homomorphic image of a prime regular subdirect product of simple artinian F-algebras.

Proof. By [8, Theorems 2.5 and 4.4]. □

Corollary 1.12 *Let $i \in B(\omega_1)$ and let F be a field. Then there exist a prime regular subdirect product, S, of simple artinian F-algebras, and strongly uniform right $S \otimes_F S^{op}$-modules of dimension ω_1, M and I, such that I is indecomposable injective and $\Gamma(M) = \Gamma^*(I) = i$.*

The realization problem for (regular) cardinals $d > \omega_1$ remains open. In fact, though Lemma 1.3 works for arbitrary $\kappa > \omega_1$, the proof of Theorem 1.4 does not extend. Namely, [7, Theorem 1.5] does not generalize - a counterexample appears in [19, Corollary 2.12]. [1]

2 Ziegler spectra of regular rings

In Section 2, we have seen that the points of $\mathrm{Zg}(R)$ may have a rather complex submodule structure for a general regular ring R. Nevertheless, the poset of all closed subsets of $\mathrm{Zg}(R)$ is easily determined from the ideal lattice $\mathrm{L}_2(R)$:

Proposition 2.1 *Let R be a regular ring. Denote by $\mathcal{L}$ the class of all complete distributive lattices whose dual lattice is algebraic with greatest element compact.*

(i) Let $I \in \mathrm{L}_2(R)$. Then $\mathrm{Zg}(R/I)$ is (canonically homeomorphic to) the closed subset $\{N \in \mathrm{Com}(R) \mid NI = 0\}$ of $\mathrm{Zg}(R)$.

(ii) The closed subsets of $\mathrm{Zg}(R)$ ordered by inclusion form a lattice, $\mathrm{CL}(R)$, which is anti-isomorphic to $\mathrm{L}_2(R)$. In particular, $\mathrm{CL}(R) \in \mathcal{L}$.

(iii) Let F be a field. Then any countable lattice from $\mathcal{L}$ is isomorphic to $\mathrm{CL}(R)$ where R is a countable direct limit of matricial F-algebras. Nevertheless, there exist uncountable lattices in $\mathcal{L}$ that are not isomorphic to $\mathrm{CL}(R)$ for any unit regular ring R.

Proof. (i) Since R is regular, any injective right R/I-module is also injective as a right R-module. Consider $\nu : \mathrm{Com}(R/I) \to \mathrm{Com}(R)$ defined by $\nu(N) = N$. Then $\nu(V_{e+I}) = V_e \cap \{N \in \mathrm{Com}(R) \mid NI = 0\}$ for each $e = e^2 \in R$. So ν is a homeomorphism of $\mathrm{Zg}(R/I)$ onto $\mathrm{Im}\,\nu = \{N \in \mathrm{Com}(R) \mid NI = 0\}$. The latter is clearly a closed subset in $\mathrm{Zg}(R)$.

(ii) The basic open sets of $\mathrm{Zg}(R)$ are V_e, where $e^2 = e \in R$. Since each ideal is generated by idempotents, we infer that closed sets are exactly those of the form $\{N \in \mathrm{Com}(R) \mid NI = 0\}$, where $I \in \mathrm{L}_2(R)$. It follows that the map $\eta : \mathrm{L}_2(R) \to \mathrm{CL}(R)$ defined by $\eta(I) = \{N \in \mathrm{Com}(R) \mid NI = 0\}$ is surjective.

[1] *Added in proof:* Using a different approach, S. Shelah and the author have recently shown that for any regular cardinal $d > \omega_1$ and any $i \in B(d)$ there are a ring R and a strongly uniform module U such that $\Gamma(U) = i$. Details of the construction appear in their preprint "Spectra of the Γ-invariant of uniform modules".

If $I, J \in \mathrm{L}_2(R)$ and $I \nsubseteq J$, then there is an idempotent $e \in I \setminus J$. Let M' be a maximal R/J-submodule of $M = (eR + J)/J$ and put $N = M/M'$. Then N is a simple right R/J-module which is not annihilated by I, so $\mathrm{E}(N) \in \eta(J) \setminus \eta(I)$. It follows that the map η is injective. Since both η and η^{-1} are order reversing, η is a lattice anti-isomorphism. Since each element of $\mathrm{L}_2(R)$ is generated by idempotents, we infer that $\mathrm{L}_2(R)$ is a distributive lattice and the assertion follows.

(iii) This is by part (ii), [2, §7] and [19, Corollary 2.12]. □

It is well-known that the maximal spectrum and the Ziegler spectrum are homeomorphic in the case when R is a commutative regular ring, [11, §4.7 and §16.2], [10]. We shall extend this result to regular rings which have all primitive factors artinian, and relate the topology to algebraic properties of the ring R.

Let R be a regular ring. For short, we will say that R has *p.f.a.* provided that all right primitive factor-rings of R are right artinian. This notion is left-right symmetric. Moreover, if R has p.f.a., then the notions of a prime, left (right) primitive, and maximal, ideal coincide [5, Theorem 6.2].

For example, any abelian regular ring, and any regular ring of bounded index of nilpotence, has p.f.a. [5, Theorem 3.2 and Corollary 7.10]. Clearly, any regular ring with p.f.a. is a subdirect product of simple artinian rings.

Proposition 2.2 *Let R be a regular ring with p.f.a.*

(i) The map $\eta : \mathrm{Com}(R) \to \mathrm{Max}\, R$ defined by $\eta(N) = \mathrm{Ann}_R(N)$ is a homeomorphism of $\mathrm{Zg}(R)$ onto $\mathrm{Spec}\, R$. So $\mathrm{Zg}(R)$ is (compact and) T_1.

(ii) The following conditions are equivalent:

(a) $\mathrm{Zg}(R)$ is Hausdorff,

(b) $\mathrm{Zg}(R)$ is normal,

(c) $\mathrm{Zg}(R)$ is totally disconnected,

(d) R is a biregular ring (i.e., RxR is generated by a central idempotent for each $x \in R$).

Proof. By [17, Proposition 2.2], we may w.l.o.g. assume that $\mathrm{Com}(R) = \mathrm{simp}\, R$.

(i) The map η is well-defined because R has p.f.a. Since N embeds into the simple artinian ring $R/\eta(N)$, we infer that η is bijective. If $e = e^2 \in R$, then $\eta(V_e) = U_e$. It follows that η is a homeomorphism. The compactness and the T_1-property are clear from our introductory remarks.

(ii) Since $\mathrm{Zg}(R)$ is compact, (a) is equivalent to (b). By (i) and [6, Theorem 15], (a) is equivalent to (d).

Assume (c). Since $U_e \cap U_{e'} = U_{ee'}$ for all $e, e' \in \mathrm{Bool}(R)$, the space $\mathrm{Spec}\, R$ has a basis consisting of the sets U_e where $e \in \mathrm{Bool}(R)$. Take $x \in R$. Since R is a right V-ring [5, Proposition 6.18], each ideal of R is an intersection of right primitive (= maximal) ideals. It follows that for each $e \in \mathrm{Bool}(R)$ we have $U_e \subseteq U_x$ iff $e \in RxR$. Denote by I the sum of all ideals of the form eR, where $e \in \mathrm{Bool}(R)$ is such that $e \in RxR$. Since U_x is the union of all U_e, where $e \in RxR \cap \mathrm{Bool}(R)$, we infer that $RxR = I$. Note that $eR + e'R = fR$, where $f = e + e' - ee' \in \mathrm{Bool}(R)$, for all $e, e' \in \mathrm{Bool}(R)$. So $RxR = eR$ for some $e \in \mathrm{Bool}(R)$ and (d) holds.

Assume (d). Then for each (idempotent) $x \in R$ there is $e \in \mathrm{Bool}(R)$ such that $U_x = U_e$, so $\operatorname{Spec} R$ has a basis consisting of clopen sets, and (c) holds. □

It is tempting to try to extend 2.2 to the class of all regular subdirect products of simple artinian rings. Nevertheless, Lemma 1.11 shows this is not possible. Actually, Ziegler spectra of rings in the latter class already contain Ziegler spectra of all regular rings:

Corollary 2.3 *For any regular ring, R, there is a regular subdirect product of simple artinian rings, S, such that $\mathrm{Zg}(R)$ is homeomorphic to a closed subset of $\mathrm{Zg}(S)$.*

Proof. By 1.11 and 2.1(i). □

For an arbitrary regular ring R, there exist canonical homeomorphisms of $\operatorname{Spec} R$ on to subspaces of $\mathrm{Zg}(R)$. The subspaces are dense in $\mathrm{Zg}(R)$ when $\bigcap \operatorname{Max} R = 0$:

Proposition 2.4 *Let R be a regular ring. Let $\xi : \operatorname{Max} R \to \mathrm{Com}(R)$ be any map satisfying $\xi(M) = \mathrm{E}(N)$ for some $N \in \mathrm{simp}(R/M)$.*

(i) ξ is an injective homeomorphism of $\operatorname{Spec} R$ into $\mathrm{Zg}(R)$.

(ii) $\operatorname{Im} \xi \subseteq \operatorname{Coc} R \subseteq \mathrm{Com}(R)$. The inclusions are strict in general, already in the class of all regular subdirect products of simple artinian rings.

(iii) Assume that $\operatorname{card} R < 2^\omega$. Then ξ is surjective iff R has p.f.a.

(iv) Assume that $\bigcap \operatorname{Max} R = 0$. Then $\operatorname{Im} \xi$ (and hence $\operatorname{Coc} R$) is a dense subset of $\mathrm{Zg}(R)$.

Proof. (i) Clearly, ξ is injective and $\xi(U_e) \subseteq V_e$ for each $e = e^2 \in R$. Since R is regular, $\xi(N)$ coincides with the injective hull of N as a right R/M-module. It follows that $\xi(U_e) = V_e \cap \operatorname{Im} \xi$ for each $e = e^2 \in R$, so ξ is a homeomorphism.

(ii) Clearly, $\operatorname{Im} \xi \subseteq \operatorname{Coc} R \subseteq \mathrm{Com}(R)$. Clearly, $\operatorname{Coc} R = \mathrm{Com}(R)$ iff each uniform module is cocyclic. By 1.6, this is not true in general.

Moreover, there exist (unit) regular rings such that $\operatorname{Im} \xi \neq \operatorname{Coc} R$. For example, consider a simple countable (unit) regular ring R such that R is not artinian. Then $\operatorname{simp} R$ and $\mathrm{Com}(R)$ have cardinality 2^ω [16, Proposition 6.3], so there are 2^ω different possibilities for defining the map ξ. Since R

is simple, the topology on $\mathrm{Zg}(R)$ is trivial ("indiscrete", [13]; in particular, $\mathrm{Zg}(R)$ is not T_0). On the other hand, $\operatorname{Spec} R$ is just the one-point space.

Let S be a simple countable regular ring such that S is not artinian. We have $\operatorname{Im}\xi \cap \{N \in \mathrm{Com}(R) \mid NI = 0\} \subsetneq \operatorname{Coc} S = \operatorname{Coc} R \cap \{N \in \mathrm{Com}(R) \mid NI = 0\}$, so $\operatorname{Im}\xi \neq \operatorname{Coc} R$.

Similarly, using 2.1(i) and 1.6, we get $\operatorname{Coc} R \neq \mathrm{Com}(R)$.

The examples in the class of all regular subdirect products of simple artinian rings are now obtained by an application of Lemma 1.11.

(iii) Assume ξ is surjective. For each $M \in \operatorname{Max} R$, we have $M = \operatorname{Ann}_R(N)$, where N is the simple submodule of $\xi(M)$. It follows that each right primitive factor-ring of R is simple. If $\operatorname{card} R < 2^\omega$, then [16, Proposition 6.3] shows that $\operatorname{simp}(R/M)$ has cardinality $\geq 2^\omega$ for each $M \in \operatorname{Max} R$ such that R/M is not artinian. Since ξ is surjective, we infer that R/M is artinian for each $M \in \operatorname{Max} R$, and that R has p.f.a. The converse is clear from 2.2(i).

(iv) Let $N \in \mathrm{Com}(R) \setminus \operatorname{Im}\xi$ and $e = e^2 \in R$ be such that $N \in V_e$. By the premise, there is $M \in \operatorname{Max} R$ such that $e \notin M$. Then $\xi(M)e \neq 0$, and $\xi(M) \in V_e$. □

Of course, ξ coincides with the homeomorphism η^{-1} of Proposition 2.2(i) (and hence ξ is unique) in the case when R is regular with p.f.a.

If $\operatorname{card} R < 2^\omega$ and $\bigcap \operatorname{Max} R = 0$, then Proposition 2.4 shows that $\operatorname{Spec} R$ corresponds canonically to the space of indecomposables of a complete theory of modules closed under products iff R has p.f.a. At present, no examples are known of regular non-artinian rings R such that $\operatorname{card} \operatorname{simp} R = 1$. So the condition $\operatorname{card} R < 2^\omega$ in Proposition 2.4(iii), and in the previous comment, can actually be dropped for all known examples of regular rings. Together with Proposition 2.1, this provides a partial answer to the Question of [11, p.105].

In some particular cases, much more is known about $\operatorname{Spec} R$, and Proposition 2.4(i) makes it possible to transfer the results into $\mathrm{Zg}(R)$. Recall that a regular ring satisfies *general comparability* provided that for all $x, y \in R$ there exists $e \in \mathrm{Bool}(R)$ such that exR embeds into eyR, and $(1-e)yR$ into $(1-e)xR$ (as modules), [5, Chapter 8].

Corollary 2.5 *Let R be a regular ring. Let $\xi : \operatorname{Max} R \to \mathrm{Com}(R)$ be any map satisfying $\xi(M) = \mathrm{E}(N)$ for some $N \in \operatorname{simp}(R/M)$.*

(i) Assume R satisfies general comparability. Then $\operatorname{Im}\xi \cong \operatorname{Spec} \mathrm{Bool}(R)$ is a compact Hausdorff totally disconnected subspace of $\mathrm{Zg}(R)$.

(ii) Let R be directly finite and right self-injective. Then $\operatorname{Im}\xi$ is a dense subspace of $\mathrm{Zg}(R)$ which is homeomorphic to the maximal spectrum of the complete Boolean algebra $\mathrm{Bool}(R)$.

Proof. (i) By [5, Theorem 8.25], the map $\tau : \operatorname{Spec} R \to \operatorname{Spec} \operatorname{Bool}(R)$ defined by $\tau(M) = M \cap \operatorname{Bool}(R)$ is a homeomorphism, and 2.4(i) applies.

(ii) First, $\operatorname{Bool}(R)$ is a complete Boolean algebra by [5, Proposition 9.9]. By [5, Theorem 9.15], R satisfies general comparability, so part (i) applies. By [5, Corollary 9.27 and 2.4(iv)], $\operatorname{Im} \xi$ is dense in $\operatorname{Zg}(R)$. □

Example 2.6 If R is regular with p.f.a., then $\operatorname{simp} R = \operatorname{Coc} R = \operatorname{Com}(R)$ by Proposition 2.1. The first equality characterizes right V-rings. The second holds for any right semiartinian ring, for example.

Nevertheless, there exist semiartinian regular right V-rings such that the map ξ is not onto (that is, $\operatorname{Im} \xi \neq \operatorname{simp} R$). To see this, denote by Q the full endomorphism ring of an infinite dimensional right linear space over a field. Let R be the subring of Q generated by the finite rank endomorphisms and by scalar multiplications. Then R is well-known to be regular of Loewy length 2, and a right V-ring. Since $\operatorname{Soc}(R)$ is the only maximal ideal of R, $\operatorname{Spec} R$ consists of one point, while $\operatorname{Zg}(R)$ has just two points: $\xi(\operatorname{Soc}(R))$ and S, where S is the simple faithful module. The only subset of $\operatorname{simp} R$ which is not closed is $\{S\}$, and $\operatorname{Zg}(R)$ is not T_1.

If R is a semiartinian regular ring with p.f.a., then we can derive more details on $\operatorname{Zg}(R)$ using the following result from [15]:

Lemma 2.7 *Let R be a right semiartinian ring and $\mathcal{L} = (S_\alpha \mid \alpha \leq \sigma + 1)$ be the right Loewy chain of R. Then the following conditions are equivalent:*

(i) R is regular and has p.f.a.

(ii) For each $\alpha \leq \sigma$ there are a cardinal λ_α, positive integers $n_{\alpha\beta}$ ($\beta < \lambda_\alpha$) and skew-fields $K_{\alpha\beta}$ ($\beta < \lambda_\alpha$) such that $S_{\alpha+1}/S_\alpha \cong \oplus_{\beta<\lambda_\alpha} M_{n_{\alpha\beta}}(K_{\alpha\beta})$, as rings without unit. The pre-image of $M_{n_{\alpha\beta}}(K_{\alpha\beta})$ coincides with the β-th homogenous component of R/S_α and it is finitely generated as right R/S_α-module for all $\beta < \lambda_\alpha$. Moreover, λ_α is infinite iff $\alpha < \sigma$.

Proof. By [15, Theorem 2.1]. □

Theorem 2.8 *Let R be a right semiartinian ring and $\mathcal{L} = (S_\alpha \mid \alpha \leq \sigma + 1)$ be the right Loewy chain of R. Assume that R is regular and has p.f.a.*

(i) Let $0 \neq e = e^2 \in R$. Then the basic open set V_e coincides with the set of all $N \in \operatorname{simp} R$ such that N is a subfactor of eR.

Let $\alpha \leq \sigma + 1$ be the least (non-limit) ordinal such that $e \in S_\alpha$. Then $(eS_\beta \mid \beta \leq \alpha)$ is the Loewy chain of eR, and $eS_{\beta+1}/eS_\beta$ is infinitely generated for all $\beta < \alpha$.

(ii) The Cantor-Bendixson rank of $\operatorname{Zg}(R)$ is σ.

Proof. Let $\mathcal{L} = (S_\alpha \mid \alpha \leq \sigma+1)$ be the (right) Loewy chain of R. Let $\alpha \leq \sigma$ and $\beta < \lambda_\alpha$. Denote by $H_{\alpha\beta}$ the β-th homogenous component of R/S_α. Denote by $P_{\alpha\beta}$ a simple submodule of $H_{\alpha\beta}$, and let $A_{\alpha\beta} = \mathrm{Ann}_R(H_{\alpha\beta}) = \mathrm{Ann}_R(P_{\alpha\beta})$. By [17, Proposition 2.2] and [15, Proposition 2.6], we may w.l.o.g. assume that $\mathrm{Com}(R) = \mathrm{simp}\, R = \{P_{\alpha\beta} \mid \alpha \leq \sigma, \beta < \lambda_\alpha\}$.

(i) Clearly, $P_{\alpha\beta}e \neq 0$ is equivalent to $e \notin A_{\alpha\beta}$, and to $eR/(eR \cap A_{\alpha\beta})$ being isomorphic to a non-zero submodule of $H_{\alpha\beta}$ (Since $eR/(eR \cap A_{\alpha\beta}) \cong (eR + A_{\alpha\beta})/A_{\alpha\beta} \subseteq R/A_{\alpha\beta} \cong H_{\alpha\beta}$ as modules). The latter condition is equivalent to $P_{\alpha\beta}$ being isomorphic to a submodule of $eR/eA_{\alpha\beta} \cong C = eS_{\alpha+1}/e(S_{\alpha+1} \cap A_{\alpha\beta})$. The completely reducible module $D = eS_{\alpha+1}/eS_\alpha$ decomposes as $D = C' \oplus C''$, where $C' \cong C$ and $C'' = e(S_{\alpha+1} \cap A_{\alpha\beta})/eS_\alpha$. Since $\mathrm{Hom}_R(P_{\alpha\beta}, C'') = 0$ and D is the α-th slice of eR, we infer from [15, Proposition 2.8(ii)] that $P_{\alpha\beta}$ embeds into $eR/eA_{\alpha\beta}$ iff $P_{\alpha\beta}$ embeds into D iff $P_{\alpha\beta}$ embeds into eR.

By induction on $\beta \leq \alpha$, we prove that $(eS_\beta \mid \beta \leq \alpha)$ is the Loewy chain of eR. Since $eS_\beta = eR \cap S_\beta$, we infer that $E_\beta = eS_{\beta+1}/eS_\beta$ embeds into $S_{\beta+1}/S_\beta$, so E_β is completely reducible. Assume that $N \in \mathrm{simp}\, R$ is such that N embeds into eR/eS_β, but not into E_β. Then N embeds into $eR/eS_{\beta+1}$, and into $R/S_{\beta+1}$, so $N \cong P_{\beta+1,\gamma}$ for some $\gamma < \lambda_{\beta+1}$. On the other hand, since N embeds into R/S_β, we have $N \cong P_{\beta\gamma}$ for some $\gamma < \lambda_\beta$, a contradiction. This proves that $\mathrm{Soc}(eR/eS_\beta) = E_\beta$. Moreover, $E_\beta = 0$ iff $eS_\beta = eR$ iff $eR \subseteq S_\beta$, so α is the Loewy length of eR.

Also, E_β is not a summand in eR/eS_β for each $\beta < \alpha$, whence E_β is not finitely generated for $\beta < \alpha$.

(ii) By induction on $\gamma \leq \sigma + 1$ we prove that the γ-th Cantor-Bendixson derivative of $\mathrm{Zg}(R)$ equals $D_\gamma = \{P_{\alpha\beta} \mid \gamma \leq \alpha \leq \sigma, \beta < \lambda_\alpha\}$. For $\gamma = 0$, we have $D_\gamma = \mathrm{Zg}(R)$. Also, $D_\gamma = \cap_{\delta<\gamma} D_\delta$ for each limit ordinal $\gamma \leq \sigma$.

For γ non-limit, $\gamma = \delta + 1$, 2.7 gives existence of idempotents $e_{\delta\beta} \in R$ so that $H_{\delta\beta} = (e_{\delta\beta} + S_\delta)R/S_\delta$ and $e_{\delta\beta} + S_\delta$ is a central idempotent in R/S_δ for each $\beta < \lambda_\delta$. Then $\{P_{\delta\beta}\} = \{N \in D_\delta \mid Ne_{\delta\beta} \neq 0\}$, so $P_{\delta\beta}$ is an isolated point of D_δ (equipped with the induced topology), for each $\beta < \lambda_\delta$.

Take $\delta < \alpha \leq \sigma$, $\beta < \lambda_\alpha$ and $e \in R$ such that $P_{\alpha\beta}e \neq 0$. So $P_{\alpha\beta}(e + S_\delta) \neq \bar{0}$. In particular, there is $\beta' < \lambda_\delta$ such that $P_{\delta\beta'}$ embeds into $(e + S_\delta)(R/S_\delta) \neq \bar{0}$. Applying part (i) for R/S_δ, we infer that $P_{\delta\beta'}e \neq 0$, so $P_{\alpha\beta}$ is not isolated in D_δ.

This proves that $D_\gamma = \{P_{\alpha\beta} \mid \gamma \leq \alpha \leq \sigma, \beta < \lambda_\alpha\}$.

It follows that $\gamma = \sigma + 1$ is the least ordinal such that D_γ has no isolated points, whence the Cantor-Bendixson rank of $\mathrm{Zg}(R)$ equals σ, [11, p.120]. □

References

[1] F. W. Anderson and K. R. Fuller, *Rings and Categories of Modules*, Springer, New York (1992).

[2] G. M. Bergman, *Von Neumann regular rings with tailor-made ideal lattices*, unpublished notes (1986).

[3] P. C. Eklof, *Set-theoretic methods: the uses of Gamma-invariants*, in Abelian Groups (Fuchs, Göbel, eds.), M. Dekker, New York, LNPAM **146** (1993), 43–53.

[4] P. C. Eklof and J. Trlifaj, *Gamma invariants for dense lattices*, to appear in Algebra Universalis **40** (1999).

[5] K. R. Goodearl, *Von Neumann Regular Rings*, 2nd ed., Krieger, Melbourne (FL) (1991).

[6] K. R. Goodearl and D. E. Handelman, *Homogenization of regular rings of bounded index*, Pacific J. Math. **84** (1979), 63–78.

[7] K. R. Goodearl and D. E. Handelman, *Tensor products of dimension groups and K_0 of unit regular rings*, Canad J. Math. **38** (1986), 633–658.

[8] K. R. Goodearl, P. Menal and J. Moncasi, *Free and residually artinian regular rings*, J. Algebra **156** (1993), 407-432.

[9] K. R. Goodearl and F. Wehrung, *Representations of distributive lattices by dimension groups, regular rings, C^*-algebras and complemented modular lattices*, preprint (1998).

[10] R. S. Pierce, *Modules over Commutative Regular Rings*, Memoirs of AMS **70**, Providence (1967).

[11] M. Prest, *Model Theory and Modules*, Cambridge Univ. Press, Cambridge (1988).

[12] M. Prest, *Remarks on elementary duality*, Ann. Pure Appl. Logic **62** (1993), 183–205.

[13] M. Prest, P. Rothmaler and M. Ziegler, *Absolutely pure and flat modules and "indiscrete" rings*, J. Algebra **174** (1995), 349–372.

[14] P. Růžička, *Notes about Γ-invariants.*

[15] P. Růžička, J. Trlifaj and J. Žemlička, *Criteria of steadiness*, in Abelian Groups, Module Theory, and Topology (Dikranjan, Salce, eds.), M. Dekker, New York (1998), 359–371.

[16] J. Trlifaj, *Whitehead test modules*, Transactions of AMS **348** (1996), 1521–1554.

[17] J. Trlifaj, *Two problems of Ziegler and uniform modules over regular rings*, in Abelian Groups and Modules (Arnold, Rangaswamy, eds.), M. Dekker, New York (1996), 373–383.

[18] J. Trlifaj, *Modules over non-perfect rings*, in Advances in Algebra and Model Theory (Droste, Göbel, eds.), Gordon & Breach, Philadelphia (1997), ALA **9**, 471–492.

[19] F. Wehrung, *Non-measurability properties of interpolation vector spaces*, Israel J. Math. **103** (1998), 177–206.

Trends in Mathematics, © 1999 Birkhäuser Verlag Basel/Switzerland

Locally simple objects

Ivo Herzog*

Abstract

Let $\mathcal{A}$ be an abelian category. The Ziegler spectrum of $\mathcal{A}$ is used to introduce the notion of a *locally simple* object of $\mathcal{A}$, which is analyzed in terms of the transitive relation $A \sqsubset B$ on $\mathcal{A}$, defined to hold whenever the object A occurs at least twice as a subquotient of B. It is shown that a nonzero object is locally simple if and only if it is minimal with respect to $\sqsubset$. By analogy to the notion of simplicity, the Serre subcategory generated by the locally simple objects is used to define *local* Krull-Gabriel dimension. It is proved that an object has local Krull-Gabriel dimension if and only if it satisfies the $\sqsubset$-descending chain condition and that the Serre subcategory of objects with local Krull-Gabriel dimension corresponds in the Ziegler spectrum to the interior of the set of points E satisfying Prest's condition $(\wedge)$: The localization $\mathcal{A}/\mathcal{S}(E)$ is local, where $\mathcal{S}(E)$ denotes the Serre annihilator of E in $\mathcal{A}$. We also show that a locally finite length object satisfies Fitting's Lemma.

Introduction

An object A of an abelian category $\mathcal{A}$ is termed *locally simple* if for every $\cap$-irreducible Serre subcategory $\mathcal{S}$ of $\mathcal{A}$ such that $A \notin \mathcal{S}$, the localized object $A_{\mathcal{S}}$ is a simple object of the localization $\mathcal{A}/\mathcal{S}$. Such objects do occur in Nature, most recently in the analyses of Burke and Prest [3], Prest and Schröer (unpublished) and Schröer [12] of certain domestic string algebras Λ. There, the abelian category is taken to be $\mathrm{Ab}(\Lambda)$, the free abelian category over Λ (see [5]) or, what is equivalent, the category $\mathrm{coh}(\Lambda\text{-mod}, \mathrm{Ab})$ of coherent functors $A : \Lambda\text{-mod} \to \mathrm{Ab}$ (see [1]). Another example of a locally simple functor is the uniserial object $Z(p) \otimes_{\mathbf{Z}} - \in \mathrm{Ab}(\mathbf{Z})$ where $\mathbf{Z}$ denotes the ring of integers and $Z(p)$ the simple group of order the prime p.

The point of this article is to give a systematic treatment of the locally simple objects of an abelian category $\mathcal{A}$ in terms of the transitive relation $A \sqsubset B$, defined to hold whenever A occurs more than once as a subquotient of B; an object is locally simple if and only if it is minimal with respect to $\sqsubset$ (Proposition 1.2). By analogy to the classical definition [6] of Krull-Gabriel

*Supported by NSF Grant DMS96-26708.

dimension, the notion of locally simple object may be used to define the local Krull-Gabriel dimension $\mathrm{LKG}(A)$ of an object $A \in \mathcal{A}$.

Theorem. The following conditions for an object $A \in \mathcal{A}$ are equivalent:

1. $\mathrm{LKG}(A) < \infty$, that is, A has local Krull-Gabriel dimension.

2. The object A satisfies the $\sqsubset$-descending chain condition.

3. Every point $E \in \mathcal{O}(A)$ satisfies Prest's condition $(\wedge)$: The localization $\mathcal{A}/\mathcal{S}(E)$ is local.

The notation $\mathcal{O}(A)$ refers to the basic open subset of the Ziegler spectrum $\mathrm{Zg}(\mathcal{A})$ of $\mathcal{A}$ (see [15, 9, 10]) which is the support of A. For $E \in \mathrm{Zg}(\mathcal{A})$, $\mathcal{S}(E)$ is the Serre annihilator of E in $\mathcal{A}$, and an abelian category is called *local* if it has up to isomorphism one simple object. A consequence of the theorem is that the objects A that satisfy the $\sqsubset$-dcc form a Serre subcategory - it corresponds in the Ziegler spectrum to the interior $\mathcal{O}_\wedge$ of the set of points that satisfy $(\wedge)$.

A section of the article is devoted to showing that Fitting's Lemma holds for the objects $A \in \mathcal{A}$ that are locally of finite length.

Throughout the paper, $\mathcal{A}$ will denote a small abelian category. We will write $A \leq B$ to assert that the object A is a subobject of B and $\ell(A)$ will denote the length of A. A (full) subcategory $\mathcal{S} \subseteq \mathcal{A}$ is called *Serre* if given a short exact sequence

$$0 \to A \to B \to C \to 0$$

in $\mathcal{A}$, the object B belongs to $\mathcal{S}$ if and only if A and C do. It is clear that the Serre subcategories of $\mathcal{A}$ are closed under intersection. If $\mathcal{X} \subseteq \mathcal{A}$ is an arbitrary subcategory, then $\sqrt{\mathcal{X}}$ denotes the smallest Serre subcategory that contains $\mathcal{X}$. If $\mathcal{X} = \{A\}$ consists of a single object, we write $\sqrt{A}$ for $\sqrt{\mathcal{X}}$.

We write $A \prec B$ to assert that A is a *subquotient* of B, that is, A is a subobject $A \leq Q$ of some quotient object Q of B. It means that there is a filtration of B in which A occurs as a factor. Equivalently, there is a filtration of B

$$B = B_0 \geq B_1 \geq B_2 \geq 0$$

with $B_1/B_2 \cong A$. Evidently, the relation $A \prec B$ is transitive and reflexive and if $\mathcal{S}$ is a Serre subcategory, then $B \in \mathcal{S}$ and $A \prec B$ imply $A \in \mathcal{S}$. The category $\sqrt{\mathcal{X}}$ consists of those objects that have a finite filtration all of whose factors are subquotients of objects of $\mathcal{X}$. In particular, $\sqrt{A}$ consists of the objects of $\mathcal{A}$ that have a finite filtration with factors among the subquotients of A.

The Serre subcategories of $\mathcal{A}$ form a distributive lattice with respect to the supremum operation $\sqrt{\bigcup_i \mathcal{S}_i}$ and the infimum operation $\bigcap_i \mathcal{S}_i$. We say that the Serre subcategory $\mathcal{S}$ is $\cap$-irreducible if it is $\cap$-irreducible in this lattice, that is, if it is not the intersection of two Serre subcategories that properly contain $\mathcal{S}$.

1 Locally simple objects

Let A and B be objects of $\mathcal{A}$. Define the relation $A \sqsubset B$ to hold whenever there is a filtration of B in which A occurs at least twice as a factor. Equivalently, there is a filtration of B

$$B = B_0 \geq B_1 \geq B_2 \geq B_3 \geq B_4 \geq 0$$

such that $B_1/B_2 \cong A \cong B_3/B_4$. The relation $A \sqsubset B$ is transitive and for every object $A \in \mathcal{A}$, $0 \sqsubset A$ and $A \sqsubset A^2$. If $A \sqsubset B$ holds in $\mathcal{A}$, the same is true in $\mathcal{A}^{\mathrm{op}}$. It is also evident that $A \sqsubset B$ implies $\ell(B) \geq 2\ell(A)$. If $\mathcal{S} \subseteq \mathcal{A}$ is a Serre subcategory, then localization at $\mathcal{S}$ preserves the $\sqsubset$-relation: If $A \sqsubset B$, then $A_{\mathcal{S}} \sqsubset B_{\mathcal{S}}$. A nonzero object $A \in \mathcal{A}$ is termed *$\sqsubset$-minimal* if $B \sqsubset A$ only if $B = 0$.

Let us note that if $A \in \mathcal{A}$ is $\sqsubset$-minimal and $\mathcal{S} \subseteq \mathcal{A}$ is a Serre subcategory such that $A \notin \mathcal{S}$, then the localized object $A_{\mathcal{S}}$ is also $\sqsubset$-minimal in the localization $\mathcal{A}/\mathcal{S}$. If $X \sqsubset A_{\mathcal{S}}$ were a nonzero object of $\mathcal{A}/\mathcal{S}$, there would then be a filtration of A

$$A = A_0 \geq A_1 \geq A_2 \geq A_3 \geq A_4 \geq 0$$

such that $(A_1/A_2)_{\mathcal{S}} \cong X \cong (A_3/A_4)_{\mathcal{S}}$. By the definition of isomorphism in $\mathcal{A}/\mathcal{S}$, it follows that there is a refinement

$$A \geq A_1 \geq A'_1 \geq A'_2 \geq A_2 \geq A_3 \geq A'_3 \geq A'_4 \geq A_4 \geq 0$$

where $A'_1/A'_2 \cong B \cong A'_3/A'_4$ and A_1/A'_1, A'_2/A_2, A_3/A'_3 and A'_4/A_4 all belong to the Serre subcategory $\mathcal{S}$. Then $B \sqsubset A$ is nonzero.

Proposition 1.1 *An object $A \in \mathcal{A}$ is $\sqsubset$-minimal if and only if for every $\cap$-irreducible Serre subcategory $\mathcal{S} \subseteq \mathcal{A}$ for which $A \notin \mathcal{S}$, the localized object $A_{\mathcal{S}}$ is a simple object in the localization $\mathcal{A}/\mathcal{S}$.*

Proof. ($\Rightarrow$). By localizing at $\mathcal{S}$ we may, without loss of generality, assume that $\mathcal{S} = 0$ is $\cap$-irreducible and show that A is simple. Let $B \leq A$ be a proper subobject and consider the corresponding short exact sequence

$$0 \to B \to A \to C \to 0.$$

By hypothesis, B and C have no common subquotients and therefore $\sqrt{B} \cap \sqrt{C} = 0$. By assumption, $\sqrt{B} = 0$ and therefore $B = 0$.

($\Leftarrow$). Suppose that A is not $\sqsubset$-minimal and take a nonzero object $B \sqsubset A$. By Zorn's Lemma, there is a Serre subcategory $\mathcal{S}$ maximal with respect to the property that $B \notin \mathcal{S}$. It is $\cap$-irreducible and because $B_\mathcal{S}$ is nonzero and $B_\mathcal{S} \sqsubset A_\mathcal{S}$, the localized object $A_\mathcal{S}$ cannot be simple. □

Let $\mathcal{C} = \mathrm{colim}(\mathcal{A})$ be the category whose objects are the formal colimits $\varinjlim A_i$ of directed systems in $\mathcal{A}$. The morphisms of this category are defined so that the following two conditions are satisfied:

1. For $X \in \mathcal{C}$, $\mathrm{Hom}_\mathcal{C}(\varinjlim A_i, X) = \varprojlim \mathrm{Hom}_\mathcal{C}(A_i, X)$

2. For $A \in \mathcal{A}$, $\mathrm{Hom}_\mathcal{C}(A, \varinjlim A_i) = \varinjlim \mathrm{Hom}_\mathcal{C}(A, A_i)$

The first condition ensures that $\varinjlim A_i$ is a colimit in $\mathcal{C}$ - we think of $\mathcal{A}$ as a subcategory of $\mathcal{C}$ in the obvious way. The second ensures that the objects of $\mathcal{A}$ are precisely the finitely presented objects of $\mathcal{C}$. As they form an abelian category, the finitely presented objects are therefore coherent and $\mathcal{C} = \mathrm{colim}(\mathcal{A})$ is a locally coherent Grothendieck category [9, 10]. For example, if $(A_i, \eta_{ij})_{i \geq j \in I}$ is a directed system for which the colimit exists in $\mathcal{A}$, there is the natural $\mathcal{C}$-morphism

$$\varinjlim{}_\mathcal{C} A_i \to \varinjlim{}_\mathcal{A} A_i$$

which is an isomorphism if and only if the system $\{A_i\}_{i \in I}$ is eventually constant.

Because the category $\mathcal{C} = \mathrm{colim}(\mathcal{A})$ is a Grothendieck category [13], every object $X \in \mathcal{C}$ has an injective envelope $X \leq E(X)$. As $\mathcal{A}$ is small, there are up to isomorphism only set-many indecomposable injective objects. These are the points of the *Ziegler spectrum* $\mathrm{Zg}(\mathcal{A})$ of $\mathcal{A}$. The topology on $\mathrm{Zg}(\mathcal{A})$ is given by the basic open subsets

$$\mathcal{O}(A) := \{E \in \mathrm{Zg}(\mathcal{A}) : \mathrm{Hom}_\mathcal{C}(A, E) \neq 0\},$$

for $A \in \mathcal{A}$. There is a bijective correspondence [9, Theorem 3.8] between the Serre subcategories $\mathcal{S}$ of $\mathcal{A}$ and the open subsets $\mathcal{O}$ of $\mathrm{Zg}(\mathcal{A})$ via the rule

$$\mathcal{S} \mapsto \mathcal{O}(\mathcal{S}) := \bigcup_{A \in \mathcal{S}} \mathcal{O}(A).$$

The finitely generated Serre subcategories $\sqrt{A}$ correspond to the quasi-compact subsets $\mathcal{O}(A)$.

Let $\mathcal{S} \subseteq \mathcal{A}$ be a Serre subcategory. The localization functor [6] $A \mapsto A_{\mathcal{S}}$ from $\mathcal{A}$ to $\mathcal{A}/\mathcal{S}$ induces a localization functor

$$\varinjlim A_i \mapsto \varinjlim (A_i)_{\mathcal{S}}$$

from colim($\mathcal{A}$) to colim($\mathcal{A}/\mathcal{S}$). Because every localized object $A_{\mathcal{S}}$ is itself a colimit of objects from $\mathcal{A}$, there is a fully faithful embedding colim($\mathcal{A}/\mathcal{S}$) $\subseteq$ colim($\mathcal{A}$) which serves as right adjoint to the localization functor. Now an object $E \in$ colim($\mathcal{A}/\mathcal{S}$) is injective (respectively, indecomposable) if and only if it is so in colim($\mathcal{A}$) and, in fact [9, Proposition 3.6], the bijection $E \mapsto E$ induces a homeomorphism from Zg($\mathcal{A}/\mathcal{S}$) onto the closed subset Zg($\mathcal{A}$) $\setminus \mathcal{O}(\mathcal{S})$.

If $E \in \mathcal{C}$ is an injective object, then the *Serre annihilator* of E is the subcategory

$$\mathcal{S}(E) := \{C \in \mathcal{A} : \mathrm{Hom}_{\mathcal{C}}(C, E) = 0\};$$

it is Serre. If, furthermore, $E \in$ Zg($\mathcal{A}$), then $\mathcal{S}(E)$ is $\cap$-irreducible, because it corresponds to the maximal open subset of Zg($\mathcal{A}$) not containing E, the complement of the closure of $\{E\}$. We call an object $A \in \mathcal{A}$ *locally simple* if for every $E \in \mathcal{O}(A)$, the localized object $A_{\mathcal{S}(E)}$ is a simple object of $\mathcal{A}/\mathcal{S}(E)$. By Proposition 1.1, every $\sqsubset$-minimal object is locally simple. On the other hand, if A is not $\sqsubset$-minimal and B is nonzero such that $B \sqsubset A$, then $A_{\mathcal{S}(E)}$ won't be simple for any $E \in \mathcal{O}(B)$. As $\mathcal{O}(B) \subseteq \mathcal{O}(A)$, we get the following.

Proposition 1.2 *An object $A \in \mathcal{A}$ is locally simple if and only if it is $\sqsubset$-minimal.*

Examples. Let R be an associative ring and denote by Ab(R) the free abelian category over R. The category Ab(R) is equivalent to the category coh(R-mod, Ab) of coherent functors $A : R\text{-mod} \to \mathrm{Ab}$ from the category of finitely presented left R-modules to the category of abelian groups.

i. Let $R = \mathbf{Z}$ be the ring of integers and consider the representable functor $A = \mathrm{Hom}_{\mathbf{Z}}(Z(p), -) \in \mathrm{Ab}(\mathbf{Z})$. The Ziegler spectrum Zg(Ab($\mathbf{Z}$)) consists of the points $U \otimes_{\mathbf{Z}} -$ where U is an algebraically compact indecomposable abelian group. By Yoneda's Lemma,

$$\mathrm{Hom}(A, U \otimes_{\mathbf{Z}} -) \cong U \otimes_{\mathbf{Z}} Z(p) \cong U/pU.$$

The support $\mathcal{O}(A)$ of A therefore consists of the p-adic completion of $\mathbf{Z}$, $\overline{Z}_p \otimes_{\mathbf{Z}} -$ and the finite cyclic p-groups $Z(p^n) \otimes_{\mathbf{Z}} -$. Localized at any of these, A is simple.

ii. Let R be a von Neumann regular ring [7, 14]. The free abelian category Ab(R) is equivalent [9, p. 531] to the category proj-R of finitely generated projective right R-modules and colim($\mathcal{A}$) is equivalent to the category Mod-R of right R-modules. The object $R \in \mathrm{Ab}(R)$ is locally simple if and only if every injective indecomposable right R-module E_R is simple as a module over the endomorphism ring $\mathrm{End}_R E$. It follows that $E \cong R/I$ where I is a maximal two-sided ideal of R and the quotient ring R/I is a field. Equivalently, the ring R is abelian regular [7].

iii. The ring R is left coherent if and only if R-mod is an abelian category. If R is a domain that is not left Ore, there are nonzero $a, b \in R$ such that $Ra \cap Rb = 0$. Then $R \sqsubset Ra \oplus Rb \leq R$, so that $R \sqsubset R$ in R-mod.

Several Serre subcategories of $\mathcal{A}$ arise from these considerations. Define $\mathcal{A}_{\mathrm{LS}} \subseteq \mathcal{A}$ to be the Serre subcategory generated by the locally simple objects. If $A \in \mathcal{A}_{\mathrm{LS}}$, then A is locally bounded in the sense that there is a natural number n such that for every $\cap$-irreducible Serre subcategory $\mathcal{S}$, the length $\ell(A_\mathcal{S}) \leq n$ (in $\mathcal{A}/\mathcal{S}$) - it is enough that the same holds for the Serre annihilator $\mathcal{S}(E)$ of every $E \in \mathcal{O}(A)$. The locally bounded objects form a Serre subcategory $\mathcal{A}_{\mathrm{LB}}$. Finally, there is the subcategory $\mathcal{A}_{\mathrm{LF}}$ of locally finite (length) objects. These are the objects A all of whose localizations $A_\mathcal{S}$ at $\cap$-irreducible Serre subcategories $\mathcal{S}$ have finite length. We have the filtration

$$0 \subseteq \mathcal{A}_{\mathrm{LS}} \subseteq \mathcal{A}_{\mathrm{LB}} \subseteq \mathcal{A}_{\mathrm{LF}} \subseteq \mathcal{A},$$

but know of no examples where either of the middle two inclusions are proper.

Conjecture. Let Λ be an artin algebra. If $\mathrm{Ab}(\Lambda)_{\mathrm{LF}} = \mathrm{Ab}(\Lambda)$, then Λ is of finite representation type.

Auslander proved [2] that if Λ is not of finite representation type, there is a large (not finitely generated) indecomposable left Λ-module. Prest [11, Corollary 13.4] noted that in that case, there exists a large algebraically compact indecomposable. Crawley-Boevey showed [4] that the existence of a large indecomposable G which is endofinite, that is, of finite length as a module over the endomorphism ring $\mathrm{End}_\Lambda G$ is equivalent to the Second Brauer-Thrall Conjecture. The above conjecture asserts that if every large indecomposable module is endofinite, then Λ must also be of finite representation type. If $\mathcal{A}_{\mathrm{LF}}$ is replaced by $\mathcal{A}_{\mathrm{LB}}$, then the conjecture becomes an apparently open generalization of the First Brauer-Thrall Conjecture.

2 Fitting's lemma

A topological space is *sober* if every $\cap$-irreducible closed subset is the closure of a point. For the Ziegler spectrum $\mathrm{Zg}(\mathcal{A})$, sobriety is equivalent to the condition that every $\cap$-irreducible Serre subcategory $\mathcal{S} \subseteq \mathcal{A}$ is the Serre annihilator of an injective indecomposable object $E \in \mathrm{Zg}(\mathcal{A})$. It is proved in [8, Lemma 4.7] that if $\mathrm{Zg}(\mathcal{A})$ has a countable basis, then it is sober. The argument uses a paraphrase of the following lemma, which is a special case of Ziegler's [15, Lemma 4.7]. We furnish a proof, because of the variation of idiom.

Lemma 2.1 *Let $A \in \mathcal{A}$ and $\mathcal{F}, \mathcal{I} \subseteq \mathrm{Latt}(A)$ be a filter and an ideal, respectively, of subobjects of A. If $\mathcal{F} \cap \mathcal{I} = \emptyset$, there is an injective indecomposable $E \in \mathrm{Zg}(\mathcal{A})$ such that for every $B \in \mathcal{F}$ and $C \in \mathcal{I}$,*

$$E \in \mathcal{O}(B/(B \cap C)) = \mathcal{O}((B + C)/C)).$$

Proof. By Zorn's Lemma, there is an ideal $\mathcal{I}' \supseteq \mathcal{I}$ maximal with the property $\mathcal{I}' \cap \mathcal{F} = \emptyset$. Let

$$X = \sum_{C' \in \mathcal{I}'} C'$$

For no $B \in \mathcal{F}$ is it the case that $B \leq X$ (in $\mathrm{colim}(\mathcal{A})$). Furthermore, the quotient object A/X is uniform so that the injective envelope $E = E(A/X)$ is indecomposable, $E \in \mathrm{Zg}(\mathcal{A})$. Now if $B \in \mathcal{F}$ and $C \in \mathcal{I} \subseteq \mathcal{I}'$, the natural morphism

$$\frac{B+C}{C} \to \frac{B+X}{X}$$

is nonzero and lifts to E. Thus $E \in \mathcal{O}((B + C)/C)$. □

For example, the lemma implies that if

$$A_0 \succ A_1 \succ A_2 \succ \cdots \succ A_n \succ \cdots \tag{1}$$

is a $\prec$-descending chain of nonzero objects in $\mathcal{A}$, then $\bigcap_n \mathcal{O}(A_n) \neq \emptyset$. To see this, write $A_i = B_i/C_i$ where the B_i, C_i are subobjects of A_0 such that

$$0 = C_0 \leq C_1 \leq C_2 \leq \cdots \leq C_n \leq \cdots \leq B_n \leq \cdots \leq B_2 \leq B_1 \leq B_0 = A_0.$$

Now apply the lemma with $\mathcal{F} \subseteq \mathrm{Latt}(A)$ the filter generated by the B_i and $\mathcal{I} \subseteq \mathrm{Latt}(A)$ the ideal generated by the C_i.

Sobriety is verified in the following way: Let $\mathcal{S}$ be a $\cap$-irreducible Serre subcategory. Localizing at $\mathcal{S}$, we may assume that 0 is $\cap$-irreducible and, because of the countability condition, that there is a descending chain

$$\mathcal{O}(A_0) \supseteq \mathcal{O}(A_1) \supseteq \mathcal{O}(A_2) \supseteq \cdots \supseteq \mathcal{O}(A_n) \supseteq \cdots$$

of basic nonempty open subsets such that if $\mathcal{O}$ is a nonempty open subset, then there is a natural number n such that $\mathcal{O}(A_n) \subseteq \mathcal{O}$. If (1) holds, then there is a generic point $E \in \bigcap_n \mathcal{O}(A_n)$. Let us define $A'_n \prec A_n$ by recursion on n so that each A'_n is nonzero and for all n, $A'_{n+1} \prec A'_n$: One experiences immediate sobriety.

i. Let $A'_0 = A_0$.

ii. Suppose A'_n has already been defined satisfying the conditions. As the Serre subcategory 0 is $\cap$-irreducible, $\sqrt{A'_n} \cap \sqrt{A_{n+1}} \neq 0$, so that A'_n and A_{n+1} have a common subquotient A'_{n+1}.

Lemma 2.2 *If $\mathcal{O}(A_0) \supseteq \mathcal{O}(A_1) \supseteq \mathcal{O}(A_2) \supseteq \cdots \supseteq \mathcal{O}(A_n) \supseteq \cdots$ is a descending chain of nonempty basic (quasi-compact) open subsets of the Ziegler spectrum, then $\bigcap_n \mathcal{O}(A_n) \neq \emptyset$.*

Proof. Let $\mathcal{S} \subseteq \mathcal{A}$ be a Serre subcategory, maximal with respect to the property that for all n, $A_n \notin \mathcal{S}$. Such a Serre subcategory is $\cap$-irreducible and by the foregoing line of argument, $\bigcap_n \mathcal{O}(A_n) \neq \emptyset$ in the Ziegler spectrum $\mathrm{Zg}(\mathcal{A}/\mathcal{S})$ and hence in $\mathrm{Zg}(\mathcal{A}) \setminus \mathcal{O}(\mathcal{S})$. □

The Local Fitting's Lemma. Let $A \in \mathcal{A}$ be locally finite and $\eta : A \to A$ an endomorphism. There is a natural number n such that $A = \mathrm{Ker}\ \eta^n \oplus \mathrm{Im}\ \eta^n$ where $\eta : \mathrm{Im}\ \eta^n \to \mathrm{Im}\ \eta^n$ is an isomorphism.

Proof. A priori, we have an ascending chain

$$0 \leq \mathrm{Ker}\ \eta \leq \mathrm{Ker}\ \eta^2 \leq \cdots \leq \mathrm{Ker}\ \eta^n \leq \mathrm{Ker}\ \eta^{n+1} \leq \cdots .$$

We shall prove that this chain is stationary. A similar argument shows that the descending chain

$$A \geq \mathrm{Im}\ \eta \geq \mathrm{Im}\ \eta^2 \geq \cdots \geq \mathrm{Im}\ \eta^n \geq \cdots$$

also stops. If n is such that for all $k \geq n$, $\mathrm{Ker}\ \eta^{k+1} = \mathrm{Ker}\ \eta^k$ and $\mathrm{Im}\ \eta^{k+1} = \mathrm{Im}\ \eta^k$, the standard proof of the classical Fitting's Lemma gives the result.

Note that for all n, $\mathcal{O}(\mathrm{Ker}\ \eta^{n+1}/\mathrm{Ker}\ \eta^n) \supseteq \mathcal{O}(\mathrm{Ker}\ \eta^{n+2}/\mathrm{Ker}\ \eta^{n+1})$. If the ascending chain did not stop, then by Lemma 2.2, there would be an $E \in \mathrm{Zg}(\mathcal{A})$ with the inclusion $\mathrm{Ker}\ \eta^n(E) \leq \mathrm{Ker}\ \eta^{n+1}(E)$ proper for every n. That would contradict the assumption that the length of $A_{\mathcal{S}(E)}$ is finite. □

If A is locally bounded, say that $\ell(A_{\mathcal{S}}) \leq b$ for all $\cap$-irreducible $\mathcal{S}$, then we may take $n = b$ in Fitting's Lemma. In particular, if A is locally simple, that is, if $b = 1$, then the endomorphism ring $\mathrm{End}_{\mathcal{A}} A$ is von Neumann regular, in fact, abelian regular.

3 Local Krull-Gabriel dimension

If $\mathcal{S}$ is a $\cap$-irreducible Serre subcategory, there are various conditions that may be imposed on $\mathcal{S}$ to ensure proper behavior. The first, already encountered, is that $\mathcal{S}$ be the Serre annihilator of an indecomposable injective object $E \in \mathrm{Zg}(\mathcal{A})$. A stronger condition is that the intersection $\mathcal{S}^+$ of all Serre subcategories properly containing $\mathcal{S}$ itself contain $\mathcal{S}$ properly. Stronger yet, is the condition that the localization $\mathcal{A}/\mathcal{S}$ have a simple object S. In that case, $\mathcal{A}/\mathcal{S}$ is a *local* category, that is, it contains up to isomorphism just one simple object and $0^+ \subseteq \mathcal{A}/\mathcal{S}$ is the Serre subcategory $\sqrt{S}$. Prest says [11, p. 211] that $E \in \mathrm{Zg}(\mathcal{A})$ satisfies the condition $(\wedge)$ if the localization $\mathcal{A}/\mathcal{S}(E)$ is local or, equivalently, if the localization has a simple object S, which is then unique up to isomorphism. In that case, E is necessarily the injective envelope of S.

Proposition 3.1 *Let $A \in \mathcal{A}$. Every injective idecomposable object $E \in \mathcal{O}(A)$ satisfies $(\wedge)$ if and only if there is no infinite descending chain (with each A_n nonzero)*

$$A = A_0 \sqsupset A_1 \sqsupset \cdots \sqsupset A_n \sqsupset \cdots .$$

Proof. $(\Rightarrow)$. Suppose that, to the contrary, such a chain existed. Let $\mathcal{S} \subseteq \mathcal{A}$ be a Serre subcategory, maximal with respect to the property that for all n, $A_n \notin \mathcal{S}$. Such a Serre subcategory is $\cap$-irreducible. By localizing at $\mathcal{S}$, we may assume, without loss of generality, that $\mathcal{S} = 0$ is a $\cap$-irreducible Serre subcategory and that every nontrivial Serre subcategory contains one of the A_n. By Lemma 2.2, there is an $E \in \bigcap_n \mathcal{O}(A_n)$ and $\mathcal{S}(E) = \mathcal{S} = 0$. By hypothesis, the object E satisfies $(\wedge)$, so there is a simple object $S \in \mathcal{A}$. But then $A_n \in \sqrt{S}$ for some n, and so must have finite length. Because $\ell(A_k) \geq 2\ell(A_{k+1})$, this is a contradiction.

$(\Leftarrow)$. Let $E \in \mathcal{O}(A)$. By hypothesis, there is an object $B \sqsubset A$, $\sqsubset$-miminal with respect to the property that $E \in \mathcal{O}(B)$. In other words, $\mathrm{Hom}_{\mathcal{C}}(B, E) \neq 0$ and $C \sqsubset B$ implies $\mathrm{Hom}_{\mathcal{C}}(C, E) = 0$. Consequently, the object $B_{\mathcal{S}(E)}$ is $\sqsubset$-minimal in the category $\mathcal{A}/\mathcal{S}(E)$. By Proposition 1.1, it is simple. □

It follows from the proposition that the objects satisfying the $\sqsubset$-descending chain condition form a Serre subcategory. For example, if $A \in \mathcal{A}$ is locally finite and $E \in \mathcal{O}(A)$, then $A_{\mathcal{S}(E)}$ has finite length in $\mathcal{A}/\mathcal{S}(E)$ and therefore E satisfies the condition $(\wedge)$. By Proposition 3.1, A satisfies the $\sqsubset$-descending chain condition. A similar argument shows that if for every $E \in \mathrm{Zg}(\mathcal{A})$, the localization $A_{\mathcal{S}(E)}$ has Krull-Gabriel dimension, then A satisfies the $\sqsubset$-descending chain condition. The objects that satisfy the $\sqsubset$-descending chain condition, may be graded according to the filtration

$$0 \subseteq \mathcal{A}_{\mathrm{LS}} \subseteq \mathcal{A}^2_{LS} \subseteq \cdots \subseteq \mathcal{A}^{\alpha}_{\mathrm{LS}} \subseteq \mathcal{A}^{\alpha+1}_{\mathrm{LS}} \subseteq \cdots \subseteq \mathcal{A}$$

of $\mathcal{A}$, defined by recursion on the ordinal α as follows:

1. If $\alpha = 0$, then $\mathcal{A}^0_{\mathrm{LS}} = 0$.

2. If $\alpha = \beta+1$, is a successor ordinal, then $\mathcal{A}^{\alpha}_{\mathrm{LS}}$ consists of those objects A whose localization at $\mathcal{A}^{\beta}_{\mathrm{LS}}$ belongs to the Serre subcategory generated by the locally simple objects.

3. If $\alpha = \lambda$ is a limit ordinal, then $\mathcal{A}^{\alpha}_{\mathrm{LS}} = \bigcup_{\beta<\alpha} \mathcal{A}^{\beta}_{\mathrm{LS}}$.

Let $\mathcal{A}_{\mathrm{LKG}} = \bigcup_{\alpha} \mathcal{A}^{\alpha}_{LS}$. An object A has *local Krull-Gabriel dimension* if $A \in \mathcal{A}_{\mathrm{LKG}}$ and we define $\mathrm{LKG}(A)$ to be the least α such that $A \notin \mathcal{A}^{\alpha+1}_{\mathrm{LS}}$.

Proposition 3.2 *An object A has local Krull-Gabriel dimension if and only if it satisfies the $\sqsubset$-descending chain condition.*

Proof. Suppose A has local Krull-Gabriel dimension and $E \in \mathcal{O}(A)$. Consider the maximal α for which $E \notin \mathcal{O}(\mathcal{S})$ where $\mathcal{S} = \mathcal{A}^{\alpha}_{\mathrm{LS}}$. In the localized category $\mathcal{A}/\mathcal{S}$, there is a locally simple subquotient S of $A_{\mathcal{S}}$ for which $E \in \mathcal{O}(S)$ in $\mathrm{Zg}(\mathcal{A}/\mathcal{S})$. Thus E satisfies $(\wedge)$.

If A does not have local Krull-Gabriel dimension, localize at $\mathcal{S} = \mathcal{A}_{LKG}$. There are no $\sqsubset$-minimal objects in $\mathcal{A}/\mathcal{S}$ so by Proposition 2.2, there is an injective indecomposable $E \in \mathcal{O}(A) \setminus \mathcal{O}(\mathcal{S})$ that does not satisfy $(\wedge)$. □

By Propositions 3.1 and 3.2, the theorem stated in the introduction is now proved. Note that if R is an atomless commutative von Neumann regular ring, then the object $R \in \mathrm{Ab}(R) \cong \text{proj-}R$ is locally simple (and therefore of local Krull-Gabriel dimension zero), but does not have Krull-Gabriel dimension. However, we know of no example of an abelian category $\mathcal{A}$ and an object $A \in \mathcal{A}$ which has local Krull-Gabriel dimension, while some localization $A_{\mathcal{S}}$ at a $\cap$-irreducible Serre subcategory does not have Krull-Gabriel dimension.

Note also, that if $A \in \mathcal{C}$ is locally simple, but does not have Krull-Gabriel dimension, then the basic open subset $\mathcal{O}(A)$ has at least the cardinality of the continuum. To see this, we may assume - by localizing at the Serre subcategory of objects with Krull-Gabriel dimension - that A has no subquotients with Krull-Gabriel dimension. If $B \leq A$ is a proper nonzero subobject, then there is a proper separation

$$\mathcal{O}(A) = \mathcal{O}(B) \,\dot{\cup}\, \mathcal{O}(A/B).$$

Both B and A/B are again locally simple without subquotients with Krull-Gabriel dimension. The process may therefore be continued to produce, by Lemma 2.2, a Cantor set in $\mathcal{O}(A)$.

References

[1] M. Auslander, *Coherent functors*, in Proceedings of the Conference on Categorical Algebra, La Jolla 1965, (S. Eilenberg, D. K. Harrison, S. MacLane, and H. Röhrl, eds.) Springer, New York (1966), 189–231.

[2] M. Auslander, *Large modules over artin algebras*, in Algebra, Topology and Category Theory. A collection of papers in honor of Samuel Eilenberg, (A. Heller and M. Tierney, eds.), Academic Press, New York (1976), 1–17.

[3] K. Burke and M. Prest, *Ziegler spectra of some domestic string algebras*, in preparation, University of Manchester.

[4] W. W. Crawley-Boevey, *Modules of finite length over their endomorphism ring*, in Representations of Algebras and Related Topics, (H. Tachikawa and Sh. Brenner, eds.), London Mathematical Society Lecture Notes Series **168**, Cambridge University Press (1992), 127–184.

[5] P. Freyd, *Representations in abelian categories*, in Proceedings of the Conference on Categorical Algebra, La Jolla, 1965, (S. Eilenberg, D.K. Harrison, S. MacLane and H. Röhrl, eds.), Springer-Verlag, New York/Berlin (1966).

[6] P. Gabriel, *Des Catégories Abéliennes*, Bull. Soc. Math. France **90** (1962), 323–448.

[7] K. Goodearl, *Von Neumann Regular Rings,* Pitman, London (1979).

[8] I. Herzog, *Elementary duality of modules*, Transactions of the American Mathematical Society **340** (1993), 37–69.

[9] I. Herzog, *The Ziegler spectrum of a locally coherent Grothendieck category*, Proceedings of the London Mathematical Society (3) **74** (1997), 503–558.

[10] H. Krause, *The spectrum of a locally coherent Grothendieck category*, Journal of Pure and Applied Algebra **114** (1997), 259–271.

[11] M. Prest, *Model Theory and Modules*, London Mathematical Society Lecture Notes Series **130**, Cambridge University Press (1988).

[12] J. Schröer, *On the Krull-Gabriel dimension of an algebra*, preprint, Universität Bielefeld.

[13] B. Stenström, *Rings of Quotients*, Springer, New York (1975).

[14] J. Trlifaj, *Two problems of Ziegler and uniform modules over regular rings*, in Abelian Groups and Modules, Marcel Dekker (1996), 373–383.

[15] M. Ziegler, *Model theory of modules*, Annals of Pure and Applied Logic **26** (1984), 149–213.

Trends in Mathematics, © 1999 Birkhäuser Verlag Basel/Switzerland

On purely extending modules

John Clark *

Abstract

This note considers a generalization of the notion of an extending module, suggested by L. Fuchs, defined using purity. This leads to a characterization of right semihereditary rings R for which the maximal left quotient ring $Q_\ell(R)$ is flat as a left R-module.

Introduction

Throughout R will be an associative ring with identity. *R-Mod* denotes the category of all unitary left R-modules and M will always be an object in *R-Mod.* The notation $K \trianglelefteq M$ indicates that K is an essential submodule of M. A submodule C of M is said to be *closed* if it has no proper essential extensions in M.

Chapter 2 of Mohamed and Müller's monograph [12] begins by noting that any (quasi-)injective module M satisfies the following three conditions:

(C_1) every submodule of M is essential in a (direct) summand of M,

(C_2) if a submodule A of M is isomorphic to a summand of M then A *is* a summand of M,

(C_3) if A_1 and A_2 are summands of M with $A_1 \cap A_2 = 0$ then $A_1 \oplus A_2$ is a summand of M.

A module M which satisfies (C_2) also satisfies (C_3). If M satisfies both (C_1) and (C_2) then M is called *continuous* while if it satisfies both (C_1) and (C_3) then it is called *quasi-continuous.*

The module M satisfies (C_1) if and only if each of its closed submodules is a summand and for this reason is called a *CS* module. Here we prefer to use the alternative: a module is called *extending* if it satisfies (C_1). Extending modules have been the subject of much study recently, as evidenced by the monograph [4].

In his notes [7], Fuchs generalizes the conditions above by replacing "direct summand" by "pure submodule", where purity is in the sense due

*The author is grateful to László Fuchs for his notes [7] and to Robert Wisbauer for useful comments on this paper. The referees' comments are also gratefully acknowledged.

to P. M. Cohn (see, for example, Chapter 18 of [3] or Chapter 7 of [14]). Here we look specifically at modules satisfying his generalization of (C_1).

1 Purely extending modules

We say that a module M is *purely extending* if every submodule of M is essential in a pure submodule of M.

Trivially, any extending module is purely extending, as is any module in which every submodule is pure.

The following two lemmas are in Fuchs [7]. We sketch their proofs.

Lemma 1.1 *For any R-module M, the following conditions are equivalent:*

(a) *M is purely extending;*

(b) *every closed submodule of M is pure;*

(c) *if A is a direct summand of the injective hull $E(M)$ of M then $A \cap M$ is pure in M.*

Proof. The equivalence of (a) and (b) is immediate from the definition. The equivalence of (b) and (c) follows from the observation that if E is a module with $M \trianglelefteq E$ and A is a summand of E then $A \cap M$ is closed in M.

To see this, let $E = A \oplus B$ and K be a submodule of M with $A \cap M \trianglelefteq K$. Then $K \cap B = 0$ since $(A \cap M) \cap (K \cap B) = 0$. If $k \in K$ but $k \notin A$, $k = a + b$ with $a \in A, b \in B$ and $b \neq 0$. Then, since $M \trianglelefteq E$, for some $r \in R$ we have $rb \in M$ and $rb \neq 0$. But $rb = rk - ra \in K \cap B = 0$, a contradiction. Thus $K \subseteq A$ and so $A \cap M = K$, i.e., $A \cap M$ is closed in M. □

Lemma 1.2 *Let N be a closed submodule of the purely extending module M. Then both N and M/N are purely extending.*

Proof. The transitivity of the closed and pure submodule properties (see, for example, [4] and [3]) shows that N is purely extending. □

In what follows we will give further definitions which are "pure" modifications of terminology from the theory of extending modules.

We say that the ring R is *left purely extending* if the left R-module ${}_RR$ is purely extending.

Since all the submodules of ${}_RR$ are pure precisely when R is a (von Neumann) regular ring (see Fieldhouse [6]), it follows that regular rings are purely extending. Also, clearly, any left extending ring is purely extending.

However there are regular rings that are not (left or right) extending. To see this, we note that Utumi [13] shows (in effect) that a left extending regular ring is left continuous and gives an example of a commutative, continuous, regular ring R for which the ring of 2×2 matrices over R is neither left nor right extending. (See also [9], Chapter 13 and [12], Appendix 4.)

The next result shows that a left nonsingular left purely extending ring has flat tendencies. A ring is called *left PF* if each of its principal left ideals is flat. Jøndrup [11] has shown that any left PF ring is right PF.

Proposition 1.3 *A left nonsingular ring R is left purely extending if and only if each cyclic nonsingular left R-module is flat. Consequently a left nonsingular left purely extending ring is left PF.*

Proof. Let R be left purely extending and let $M = Ra$ be a cyclic nonsingular left R-module generated by a. To show that M is flat, it suffices to show that the left annihilator $\ell(a)$ of a is pure in R. By assumption there is a pure left ideal P of R for which $\ell(a) \trianglelefteq P$. Let $x \in P$. Then the left ideal $L = \{r \in R : rx \in \ell(a)\}$ is essential in R. However, $Lxa = 0$ and so xa has an essential left annihilator. Hence, since M is nonsingular, $xa = 0$, i.e., $x \in \ell(a)$. This shows that $\ell(a) = P$ and so $\ell(a)$ is pure, as required.

Conversely, suppose that each cyclic nonsingular left R-module is flat and let C be a closed left ideal of R. Then (see Proposition 2,4 of [9]) R/C is nonsingular so flat and hence C is pure. □

Let R be the K-subalgebra of the ring of 3×3 matrices over a field K generated by the matrix units $e_{11}, e_{22}, e_{33}, e_{13}$, and e_{23}. Chatters and Hajarnavis [1] note that R is left and right artinian hereditary (and so nonsingular and PF) but ${}_RR$ is not extending. In fact R is not left purely extending since the left ideal Re_{13} is closed but not pure in R.

Next we note that purely extending is a Morita invariant.

Proposition 1.4 *Let R and T be rings for which there is a Morita equivalence F : R-Mod$\rightarrow T$-Mod and let $M \in R$-Mod. Then M is purely extending if and only if $F(M)$ is purely extending.*

Proof. This follows since being an essential submodule and being a pure submodule are both Morita invariant properties. □

For each $n \in \mathbb{N}$, let $M_n(R)$ denote the ring of $n \times n$ matrices over R and $R^{(n)}$ denote the direct sum of n copies of R. Proposition 1.4 gives

Corollary 1.5 *For any ring R and each $n \in \mathbb{N}$, the left R-module $R^{(n)}$ is purely extending if and only if $M_n(R)$ is a left purely extending ring.*

We now see that, as for extending modules (see [4], Section 12), (finitely generated) free modules over purely extending rings may not be purely extending.

Proposition 1.6 *Let R be a commutative integral domain. Then the following statements are equivalent:*

(a) *R is a Prüfer domain;*

(b) *$R^{(2)}$ is an extending module;*

(c) *$R^{(2)}$ is a purely extending module;*

(d) *for each $n \in \mathbb{N}$, $R^{(n)}$ is an extending module;*

(e) *for each $n \in \mathbb{N}$, $R^{(n)}$ is a purely extending module.*

Proof. The equivalence of (a), (b) and (d) are given in Corollary 12.10 of [4]. Since (d) implies (e), it suffices to show that (c) implies (a).

However, if $R^{(2)}$ is a purely extending module then, by Corollary 1.5, $M_2(R)$ is a left extending ring and so a PF ring. This implies that every ideal of R which can be generated by two elements is flat. Thus, by Lemma 4.1 of [10], every ideal of R is flat and hence R is Prüfer, as required. □

2 Nonsingular Σ-purely extending rings

We say that a ring R is *left $\sum$-purely extending* if every direct sum of copies of ${}_RR$ is purely extending.

Proposition 2.1 *Let R be a left nonsingular ring. Then the following statements are equivalent:*

(a) *R is left $\sum$-purely extending;*

(b) *every projective left R-module is purely extending;*

(c) *each N in R-Mod has a singular submodule S such that N/S is flat;*

(d) *every nonsingular left R-module is flat;*

(e) *every nonsingular left R-module is purely extending.*

Proof. (*a*) $\Leftrightarrow$ (*b*), (*c*) $\Rightarrow$ (*d*) and (*e*) $\Rightarrow$ (*a*) are obvious.

Now assume (*b*) holds. Let $M \in R\text{-}Mod$, let F be a free R-module for which there is an epimorphism $f : F \to M$, and let $K = \ker(f)$. By (*b*), there is a pure submodule P of F for which $K \trianglelefteq P$. Then $S = K/P$ is a singular module. Also, since $(F/K)/S \simeq F/P$ and P is pure in F, it follows that $(F/K)/S$ is flat (see, for example, section 36 of [14]). Since $F/K \simeq M$, regarding S as a submodule of M establishes (c).

Next assume (*d*). Let A be a submodule of the nonsingular module M and let B be the submodule of G such that B/A is the singular submodule of M/A. Then, since M is nonsingular, we have $A \trianglelefteq B$ and, since R is left nonsingular, M/B is nonsingular (see Propositions 1.21 and 1.22 in [8]). Thus, by (*d*), both M and M/B are flat and so B is pure in M (see 36.6 in [14]). Since $A \trianglelefteq B$ we conclude that M is purely extending, proving (e). □

Corollary 2.2 *Let R be a left nonsingular ring. Then R is left $\sum$-purely extending if and only if R is* right *semi-hereditary and $Q_\ell(R)$ is flat as a left R-module, where $Q_\ell(R)$ denotes the maximal left quotient ring of R.*

Proof. A result of Goodearl (see Proposition 5.16 of [8]) says that every nonsingular left R-module is flat if and only if R is *right* semi-hereditary and $Q_\ell(R)$ is flat as a left R-module. □

Proposition 2.1 is modelled on results for $\sum$-extending modules (see [2], [4] and [5]). The following example, in Goodearl [8], shows that, in contrast to the $\sum$-extending theory, left-right symmetry fails in Corollary 2.2.

Let Q be the ring of all sequences over a field F and R be the subring of Q of sequences which are eventually constant. Let T be the matrix ring

$$T = \begin{bmatrix} R & 0 \\ Q & Q \end{bmatrix}.$$

Then T is both a left and right nonsingular ring and every nonsingular right T-module is flat (see Exercise 5.C.4 of [8]). Thus T is right $\sum$-purely extending. However, the left T-module $Q_\ell(T)$ is not flat and so T is not left $\sum$-purely extending.

References

[1] A. W. Chatters and C. R. Hajarnavis, *Rings in which every complement right ideal is a direct summand*, Quart. J. Math. **28** (1977), 61–80.

[2] J. Clark and R. Wisbauer, *Σ-extending modules*, J. Pure Appl.Algebra **104** (1995), 19–32.

[3] J. Dauns, *Modules and Rings*, Cambridge University Press (1994).

[4] N. V. Dung, D. V. Huynh, P. F. Smith and R. Wisbauer, *Extending Modules*, Pitman Research Notes in Mathematics No. **313**, Longman Scientific & Technical (1994).

[5] N. V. Dung and P. F. Smith, Σ-*CS-modules*, Comm. Algebra **22** (1994), 83–93.

[6] D. J. Fieldhouse, *Pure theories*, Math. Ann. **184** (1969), 1–18.

[7] L. Fuchs, *Notes on generalized continuous modules*, preprint, (1995).

[8] K. R. Goodearl, *Ring Theory. Nonsingular Rings and Modules*, Marcel Dekker, Inc. (1976).

[9] K. R. Goodearl, *Von Neumann Regular Rings, 2nd edition*, Krieger (1991).

[10] S. Jøndrup, *On finitely generated flat modules* II, Math. Scand. **27** (1970), 105–112.

[11] S. Jøndrup, *p.p. rings and finitely generated flat ideals*, Proc. Amer. Math. Soc. **28** (1971), 431–435.

[12] S. H. Mohamed and B. J. Müller, *Continuous and Discrete Modules*, London Mathematical Society Lecture Notes Series No. **147**, Cambridge University Press (1990).

[13] Y. Utumi, *On continuous regular rings and semisimple self-injective rings*, Canad. J. Math. **12** (1960), 597–605.

[14] R. Wisbauer, *Foundations of Module and Ring Theory*, Gordon and Breach (1991).

The number of submodules

Andreas Ecker*

Abstract

The question, how many submodules a module over a ring must have, is considered. We restrict ourselves to the case where the cardinality of the module is infinite and greater than that of the ring.

Introduction

R. BURNS asked, how many submodules an uncountable module over a countable ring must have. J. STEPRĀNS ([4]) and B. HARTLEY ([2]) gave examples of $\aleph_1$-sized modules over countable rings with only $\aleph_1$ submodules. Both examples can easily be extended to examples for κ^+-sized modules over rings of cardinality κ with only κ^+ submodules (where κ is an arbitrary infinite cardinal). Therefore the question we will be concerned with is:

> Let M be an infinite R-module, $\mu := |M|, \rho := |R|$ with $\mu > \rho^+$. Must M have 2^μ distinct submodules?

STEPRĀNS treated this question (for $\rho = \aleph_0$) in the second chapter of [4]. But some of his proofs are incomplete. We shall prove the statements that are made there using a different method and generalize the question for arbitrary cardinalities of the ring. We also give a new proof for a Theorem of BURNS, OKOH, SMITH and WIEGOLD that a module over a Noetherian ring must have an independent set of the same cardinality as the module if the cardinality of the module is greater than that of the ring. In the fourth chapter we treat the case $\mu = \aleph_{\alpha+\omega}$ (α an arbitrary ordinal), where we get a better result for $\mathrm{cf}(\mu) = \omega$ than in the general case.

Throughout the text the modules will be assumed to be unital left modules over rings which have a multiplicative identity 1. We will denote ordinals and cardinals with small Greek letters like $\alpha, \beta, \ldots, \kappa, \lambda, \ldots$. $\mathrm{cf}(\kappa)$ is the cofinality of a cardinal κ. The set of natural numbers will be denoted with ω. $\langle X \rangle$ denotes the span of a subset X of a module.

*The author was supported by the Landesgraduiertenförderung Baden-Württemberg.

1 Independent sets

To determine the number of submodules, we will look for an independent set I, so that every subset of I creates an unique submodule. That will show that the module has at least $2^{|I|}$ submodules. There are several possibilities to define *independence* and it will be important to choose a suitable definition.

Definition 1.1 Let M be an R-module.

(1) A set $I \subseteq M$ is called *independent* if $0 \notin I$ and for all $n \in \omega, x_i \in I$ and $r_i \in R$ $(i = 1, \ldots, n)$: If $\sum_{i=1}^{n} r_i x_i = 0$ and $x_i \neq x_j$ for $i \neq j$, then $r_i x_i = 0$ for all $i = 1, \ldots, n$.

(2) Let $N \leq M$. A set $I \subseteq M$ is called *independent modulo* N if $I \cap N = \emptyset$ and for all $n \in \omega, x_i \in I$ and $r_i \in R$ $(i = 1, \ldots, n)$: If $\sum_{i=1}^{n} r_i x_i \in N$ and $x_i \neq x_j$ for $i \neq j$, then $r_i x_i \in N$ for all $i = 1, \ldots, n$.

(3) A set $I \subseteq M$ is called *semi-independent* if for all $x \in I : x \notin \langle I - \{x\} \rangle$.

It is clear that the definition of "semi-independent" is more general than that of "independent" or "independent modulo N" for some N. On the other hand every semi-independet set is independent modulo some N.

Proposition 1.2 *Let M be an R-module and I a semi-independent set in M. Then there exists a submodule $N \leq M$, so that I is independent modulo N.*

Proof. Let $N := \{ y \in M;\ \forall x \in I :\ y \in \langle I - \{x\} \rangle \}$. Then N is a submodule of M. Because $x \notin \langle I - \{x\} \rangle$ for all $x \in I$, the intersection $I \cap N$ is empty. Now let $n \in \omega, x_i \in I$ and $r_i \in R$ $(i = 1, \ldots, n)$ so that $x_i \neq x_j$ for $i \neq j$ and $\sum_{i=1}^{n} r_i x_i \in N$. Let $j \in \{1, \ldots, n\}$. Then $\sum_{i=1}^{n} r_i x_i \in N \subseteq \langle I - \{x_j\} \rangle$. Therefore $r_j x_j \in \langle I - \{x_j\} \rangle \cap \langle \{x_j\} \rangle \subseteq N$. As this can be done for all $j = 1, \ldots, n$ the proof is complete. □

2 Independence series

It will be our aim to find "large" (semi-) independent sets. Using the Axiom of Choice there are always maximal (semi-) independent sets. But it is indeed possible that a module has maximal (semi-) independent sets of distinct cardinalities. Nevertheless something can be said about the relation between the cardinalities of two maximal independent sets (LAWRENCE [3], Theorem 1):

Let M be an R-module and I and J two maximal independent sets in M. Then $|I| \leq |J| \cdot |R|$.

(The proof in [3] is valid only if J or R is infinite, but the statement also holds for finite sets.) We state without proof that if R is infinite, the corresponding relation for two maximal semi-independent sets I, J would be $|I| \leq (|J| + 1)^{|R|}$. It is possible that maximal independent sets are still very small. The idea to handle this is to modify the method of the LOEWY series (or upper socular series) as used by HARTLEY in [2]. Instead of defining a series over maximal completely reducible submodules we will define a series with the help of maximal independent sets.

Definition 2.1 Let M be an R-module. An ascending chain $(A_\alpha)_{\alpha \leq \delta}$ of submodules of M is called an *independence series* of M if

(i) $A_0 = 0, A_\delta = M$;

(ii) $A_\alpha \subseteq A_\beta$ for $\alpha \leq \beta$, and $A_\alpha \neq A_\beta$ for $\alpha \neq \beta$;

(iii) for all $\alpha \in \delta$ there is a maximal independent set $I_{\alpha+1}$ in M/A_α with $A_{\alpha+1}/A_\alpha = \langle I_{\alpha+1} \rangle$;

(iv) and for limit ordinals $\beta \leq \delta : A_\beta = \bigcup_{\alpha \in \beta} A_\alpha$.

δ is called the *length of the series.*

Corresponding to Corollary 2.2 in [2] for LOEWY series there is a limit for the length of an independence series which only depends on the cardinality of the ring.

Proposition 2.2 *Let M be an R-module, $\rho := |R|$ and $(A_\alpha)_{\alpha \leq \delta}$ an independence series of M. Then $\delta \leq \rho^+$.*

Proof. Choose $\beta \in \delta$, $x \in M - A_\beta$ and $\alpha < \beta$. Let $I_{\alpha+1}$ be as in the previous definition. Choose a system of representatives $J_{\alpha+1}$ of $I_{\alpha+1}$ in M. Then $x \notin A_{\alpha+1} \subseteq A_\beta$ and therefore $J_{\alpha+1} \cup \{x\}$ cannot be independent modulo A_α.

We will now show that there has to be an $r_\alpha \in R$ so that $r_\alpha \cdot x \in A_{\alpha+1} - A_\alpha$. Suppose that this is not true. It follows: $\forall r \in R : (rx \in A_{\alpha+1} \to rx \in A_\alpha)$. Therefore $J_{\alpha+1} \cup \{x\}$ is independent modulo A_α, which is a contradiction to the above. (If $rx \in \langle J_{\alpha+1} \rangle \cup A_\alpha \subseteq A_{\alpha+1}$ for some $r \in R$, then $rx \in A_\alpha$.)

Fixing x we see that for all $\alpha < \beta$ an element r_α can be chosen, so that $r_\alpha x \in A_{\alpha+1} - A_\alpha$. Clearly $r_\alpha \neq r_\gamma$ for $\alpha \neq \gamma$ and $\alpha, \gamma < \beta$. It follows that $|\beta| \leq \rho$ and therefore $\delta \leq \rho^+$. □

Remark 2.3 The estimate is sharp: In the examples of HARTLEY and STEPRĀNS every independence series has length ω_1 while the ring is countable.

The preceding proposition helps us to obtain a result for certain cardinalities of the module (corresponding to the Lemmas 2.1 and 2.3 in [4]).

Theorem 2.4 *Let M be an R-module, $\mu := |M|, \rho := |R|$ and $cf(\mu) > \rho^+$. Then M has 2^μ submodules.*

Proof. Choose an independence series $(A_\alpha)_{\alpha\leq\delta}$ of M. By Proposition 2.2 we have $\delta \leq \rho^+$. Since $\mathrm{cf}(\mu) > \rho^+$ there is an ordinal $\alpha \in \delta$, so that $|A_{\alpha+1}| = \mu$. Choose α minimal with this property. Then $|A_{\alpha+1}/A_\alpha| = \mu$. $A_{\alpha+1}/A_\alpha$ is the direct sum of μ cyclic submodules. Thus there are 2^μ submodules of $A_{\alpha+1}/A_\alpha$ and M has the same number of submodules. □

Remark 2.5 Let α be the minimal ordinal in the above proof. For any $\beta < \delta$, $I_{\beta+1}$ is independent modulo A_β. For the minimal α also $|I_{\alpha+1}| = \mu$ and $|A_\alpha| < \mu$. We will use this observation in chapter 4.

3 The increase of the series

To obtain more results with the method of independence series one ought to know more about the structure of these series. In this chapter we will investigate the relations between the sizes of A_α and $I_{\alpha+1}$.

We first want to mention an example of LAWRENCE [3] (Example 3):

> There is a module of cardinality $2^{\aleph_0}$ over a countable ring, in which every maximal independent set has cardinality 1.

An examination of this example shows that every independence series of the module has length 2 with $|I_1| = 1$ and $|I_2| = 2^{\aleph_0}$. We observe that maximal independent sets may be very small, although the module may have many submodules. The first question we want to ask is how large a maximal independent set has to be.

Proposition 3.1 *Let M be an R-module, $\rho := |R|$ and $(A_\alpha)_{\alpha\leq\delta}$ an independence series of M. Let $(I_{\alpha+1})_{\alpha\in\delta}$ be the family of maximal independent sets in Definition 2.1. Then for every $1 < \alpha \in \delta$: $|I_{\alpha+1}| \leq |A_\alpha|^\rho$.*

Proof. Suppose that the statement is wrong: $|I_{\alpha+1}| > |A_\alpha|^\rho =: \lambda$. Let J be a system of representatives of $I_{\alpha+1}$ in M. Define for every $x \in J$ a function

$$f_x: \begin{array}{rcl} R & \longrightarrow & A_\alpha \\ r & \mapsto & \begin{cases} rx, & \text{if } rx \in A_\alpha, \\ 0, & \text{else.} \end{cases} \end{array}$$

As $\lambda < |J|$, at least two of the functions have to be identical. Let x and y be two elements of J, so that $f_x = f_y$.

We show now that $I_1 \cup \{x - y\}$ is independent. To prove this, we show $\langle I_1 \cup \{x-y\}\rangle = \langle I_1\rangle \oplus \langle\{x-y\}\rangle = A_1 \oplus \langle\{x-y\}\rangle$. Suppose that there exists an $r \in R$, so that $r(x-y) = rx - ry \in A_1 \subseteq A_\alpha$. Then rx and ry both lie in A_α, because J is independent modulo A_α. Since $f_x = f_y$ it follows that $rx = f_x(r) = f_y(r) = ry$, and therefore $r(x-y){=}0$. That means $I_1 \cup \{x-y\}$ is independent, which is a contradiction to the maximality of I_1. □

The following proposition is a slight improvement for Theorem 5 in [3] (we do not need the assumption, that $2^{|R|} < |M|$).

Proposition 3.2 *Let R be an infinite ring and I a maximal independent set in an R-module M. Then the following holds.*
If $|I| = 0$, then $M = \{0\}$,
if $|I| = 1$, then $|M| \leq 2^{|R|}$,
if $|I| > 1$, then $|M| \leq |I|^{|R|}$.

Proof. If $I = \emptyset$ there is nothing to show.

Otherwise set $\kappa := |I \cup \{I\}|, \mu := |M|, \rho := |R|, \lambda := \kappa^\rho$. We have to show that $\mu \leq \lambda$. Choose an independence series $(A_\alpha)_{\alpha\leq\delta}$ of M, so that $A_1 = \langle I\rangle$. As $\delta \leq \rho^+ \leq \lambda$, one can show with Propositon 3.1 via transfinite induction that $|A_\alpha| \leq \lambda$ for all $\alpha \leq \delta$. (If $\alpha = \beta + 1$ is a successor ordinal, then $|I_\alpha| \leq |A_\beta|^\rho \leq \lambda^\rho = \lambda$ and $|A_\alpha| \leq |I_\alpha| \cdot |R| \leq \lambda$. If α is a limit ordinal, then $|A_\alpha| \leq \sum_{\beta<\alpha} |A_\beta| \leq \rho^+ \cdot \lambda \leq \lambda$.) Hence we have $\mu = |M| = |A_\delta| \leq \lambda$. □

With the help of Proposition 3.2 we can find more cases when a module must have the maximal number of submodules. Under assumption of the Generalized Continuum Hypothesis (GCH) the answer to the question from the introduction is "yes". The following theorem contains the cases of Lemma 2.4 and Theorem 2.5 and partially the case of Lemma 2.2 from [4].

Theorem 3.3 *Let M be a module over the infinite ring R, $\mu := |M|, \rho := |R|$, and assume that $\mu > \rho^+$.*

(i) If for all $\kappa < \mu : \kappa^\rho < \mu$, then every maximal independent set in M has cardinality μ. Therefore M has 2^μ submodules.

(ii) Under GCH: M has 2^μ submodules.

Proof. (i): Let I be a maximal independent set in M, $\kappa := |I \cup \{I\}|$. Because of Proposition 3.2 $\kappa^\rho \geq \mu$, and hence $\kappa = \mu$. Every subset of I creates an unique submodule of M, hence we have 2^μ submodules.

(ii): There are two cases to observe. If μ is a successor cardinal, then we know from Theorem 2.4 that M has 2^μ submodules. If μ is a limit cardinal, then because of GCH for all $\kappa < \mu : \kappa^\rho < \mu$. Therefore we are in case (i) and M has 2^μ submodules. □

In [1] BURNS, OKOH, SMITH and WIEGOLD proved that a module over a Noetherian ring, that has greater cardinality than the ring, must have the maximal number of submodules. We want to present a more general result that contains the cases of the Theorems 2.4 and 3.3(i).

Theorem 3.4 *Let R be a ring so that every well-ordered strictly increasing chain of ideals in R has less than ν elements. Let M be an infinite R-module so that $\mu := |M| > |R|$.*

(i) If cf(μ) $> \nu$, then M has 2^μ submodules.

(ii) If for all $\kappa < \mu : \kappa^{<\nu} < \mu$, then every maximal independent set in M has cardinality μ. Therefore M has 2^μ submodules.

Proof. We first show that an independence series of M has length $\leq \nu$. Let $(A_\alpha)_{\alpha\leq\delta}$ be an independence series of M, $(I_{\alpha+1})_{\alpha\in\delta}$ the family of independent sets from the definition. Let $\alpha \in \delta$ and $x \in M - A_\alpha$. Then for every $\beta < \alpha$ the set $L_\beta := \{r \in R;\ rx \in A_\beta\}$ is a left ideal in R. ¿From the proof of Proposition 2.2 we know that $L_\beta \subseteq L_{\beta'}$ for $\beta \leq \beta'$ and $L_\beta \neq L_{\beta'}$ for $\beta \neq \beta'$. Hence $\alpha < \nu$ and therefore $\delta \leq \nu$.

(i) follows now exactly as the statement of Theorem 2.4.

To prove (ii) we show that $|I_{\alpha+1}| \leq (|A_\alpha| \cdot \rho)^{<\nu}$ for all $1 < \alpha \in \delta$. Suppose that the assertion is false. Let J be a system of representatives for $I_{\alpha+1}$ in M. For all $x \in J$ the set $L_x := \{r \in R;\ rx \in A_\alpha\}$ is a left ideal in R. Every left ideal in R is generated by less than ν elements (as a left R-module). Therefore there can be at most $\rho^{<\nu} < |J|$ distinct left ideals. Hence there is a $x_0 \in J$ and a subset $K \subseteq J$, so that $|K| > (|A_\alpha| \cdot \rho)^{<\nu}$ and for all $x \in K : L_x = L_{x_0}$. Let $\{r_\alpha;\ \alpha \in \lambda\}$ be a set of generators for L_{x_0}. We can now proceed analogously as in Proposition 3.1. It suffices to define the function f_x (for $x \in L$) on the set of generators. Therefore there can only be $|A_\alpha|^\lambda < |K|$ distinct functions of this kind. If $f_x = f_y$ then one can show as in the proof of Proposition 3.1 that $I_1 \cup \{x - y\}$ is independent, which is a contradiction to the maximality of I_1.

Now, if $|I_1| < \mu$, then it follows inductively that $|A_\alpha| \leq (|I_1| \cdot \rho)^{<\nu} < \mu$ for all $\alpha \leq \delta$, which means $|M| = |A_\delta| < \mu$, a contradiction. □

Remark 3.5 (i) If ν is regular, then the condition that every ascending well-ordered chain of ideals in R has less than ν elements is equivalent to the condition that every ideal in R is generated by less than ν elements.

(ii) If R is finite and M infinite, then every maximal independent set in M has the same cardinality as M, hence M has $2^{|M|}$ submodules.

Corollary 3.6 *Let R be an infinite left-Noetherian ring, M an R-module, $\rho := |R|, \mu := |M|$, so that $\rho < \mu$. Then every maximal independent set in M has cardinality μ. Therefore M has 2^μ submodules.*

Proof. Set $\nu := \omega$. Because $\kappa^{<\omega} = \kappa < \mu$ for $\aleph_0 \leq \kappa < \mu$ we can use 3.4(ii). □

Remark 3.7 LAWRENCE showed in [3] a converse to Corollary 3.6. If R is a countable ring, so that every (left) R-module with cardinality $2^{\aleph_0}$ has an uncountable independent set, then R is left Noetherian.

4 A special case

We turn now to the case $\mu := |M| = \aleph_{\alpha+\omega}$. In contrast to the general case $\mathrm{cf}(\mu) = \omega$ we do not need further set theoretic assumptions to show that the module has 2^μ submodules. We know already from Theorem 2.4 that for every regular cardinality κ smaller than μ there is a subset of cardinality κ which is linear independent modulo a submodule of smaller cardinality. The idea is now to make this submodule as small as possible.

Definition 4.1 Let M be an R-module, $X \subseteq M$ and $\mu = |M|$. Set

$$\begin{aligned}\Gamma \;:=\; & \{\beta \in \mu;\ \exists I \subseteq X, \exists N \leq M : |I| = |X|, |N| = \beta \\ & \quad \text{and } I \text{ is linear independent modulo } N\}.\end{aligned}$$

Set

$$\tau(X) := \begin{cases} \min(\Gamma), & \text{if } \Gamma \neq \emptyset, \\ |X|^+, & \text{else.} \end{cases}$$

We describe this definition with words: $\tau(X)$ is the smallest cardinality, so that there is a submodule N of this cardinality and a subset I of cardinality $|X|$, which is independent modulo N. We state that $\tau(X)$ depends only on X and not on M.

In the following combinatorial proposition let $[A]^{\leq\rho} := \{B \subseteq A;\ |B| \leq \rho\}$ and $\mathcal{P}(A) := \{B;\ B \subseteq A\}$.

Proposition 4.2 *Let ρ and τ be cardinals, so that $\rho < cf(\tau)$, and μ a regular cardinal with $\mu \leq \tau^\rho$. Let $X \subseteq [\tau]^{\leq \rho}$ with $|X| = \mu$. Then there exists $\alpha \in \tau$, so that $|X \cap \mathcal{P}(\alpha)| = \mu$.*

Proof. Set $X_\alpha := X \cap \mathcal{P}(\alpha)$ for $\alpha \in \tau$. For all $B \in X$, B is not cofinal in τ, thus we have $X = \bigcup_{\alpha \in \tau} X_\alpha$. τ is not cofinal in μ, therefore there is an $\alpha \in \tau$ with $|X_\alpha| = \mu$. □

Proposition 4.3 *Let M be an R-module, $\rho := |R|$ and $\mu := |M| > \rho^+$, μ regular. Then $cf(\tau(M)) \leq \rho$.*

Proof. Because of Remark 2.5 we have $\tau(M) < \mu$. Suppose that $\mathrm{cf}(\tau(M)) > \rho$. Choose I and N, so that I is independent modulo N, $|I| = \mu$ and $|N| = \tau(M)$. For all $x \in I$ define a set $B(x) := \{rx; r \in R \text{ and } rx \in N\} \in [N]^{\leq \rho}$. By Proposition 4.2 we can choose $J \subseteq I$ and $L \subset N$, so that $|J| = |I|, |L| < |N|$ and $B(x) \subseteq L$ for all $x \in J$. But then J is independent modulo $\langle L \rangle$, which is a contradiction to the definition of $\tau(M)$, because $|\langle L \rangle| < \tau(M)$. □

For the construction of large semi-independent sets we need the following proposition.

Proposition 4.4 *Let M be a R-module and $I \subseteq M$ independent modulo N, $I \geq \aleph_0$. Let K be a submodule of M with $|N| \cdot |K| \cdot |R| < |I|$. Then there is $J \subseteq I$ with $|J| = |I|$ and:*

$$\forall n \in \omega \, \forall r_1, \dots, r_n \in R, x_1, \dots, x_n \in J : \\ (\textstyle\sum_{i=1}^n r_i x_i \in N + K \Longrightarrow \forall i = 1, \dots, n : r_i x_i \in N).$$

Proof. As I is independent modulo N, every element of $\langle I \rangle \cap K$ has a presentation as sum of elements of I which is unique modulo N. Let S be the set of elements of I that are needed in such a presentation (that means those $x \in I$ for that there are $r \in R$ and $y \in \langle I - \{x\} \rangle$ with $rx + y \in K$ and $rx \notin N$). Then $|S| \leq |I|$ and we can set $J := I - S$. It is easy to see that J has the desired properties. □

Theorem 4.5 *Let M be a R-module with $|M| = \aleph_{\alpha+\omega} > |R|^+$ (α an ordinal number). Then there is a semi-independent set of cardinality $\aleph_{\alpha+\omega}$ in M.*

Proof. We set $\gamma_0 := \max(\aleph_\alpha, |R|^+)$ and $\gamma_{n+1} := (\gamma_n)^+$ for $n \in \omega$. Now we can choose $I_n \subseteq M$ and $N_n \leq M$ for all $1 < n \in \omega$ so that $|I_n| = \gamma_n, |N_n| \leq \gamma_0$ and I_n is independent modulo N_n. (Choose a submodule of cardinality γ_n. Because of Theorem 2.4 there is a semi-independent set I_n of cardinality γ_n. Because of Proposition 4.3 there is $N_n \leq M$, so that I_n is independent modulo N_n and $|N_n| \leq \gamma_0$.) For $n \in \omega$ set $K_n := \langle (\bigcup_{k\in\omega} N_k) \cup (\bigcup_{k<n} I_k)\rangle$.

As $|K_n| \leq |R| \cdot (\sum_{k\in\omega} |N_k|) + (\sum_{k<n} |I_k|) \leq \gamma_0 \cdot \gamma_{n-1} \leq \gamma_{n-1}$ we can use Proposition 4.4 to find $J_n \subseteq I_n$ with $|J_n| = |I_n|$ and:

$$\forall k \in \omega\, \forall r_1, \ldots, r_k \in R, x_1, \ldots, x_k \in J_n : \quad (\textstyle\sum_{i=1}^k r_i x_i \in N_n + K_n \Longrightarrow \forall i = 1, \ldots, k : r_i x_i \in N_n).$$

Now it is easy to see that $J := \bigcup_{n\in\omega} J_n$ is independent modulo $N := \bigcup_{n\in\omega} N_n$. Therefore J is semi-independent and has cardinality

$$|J| = |\bigcup_{n\in\omega} J_n| = \sup_{n\in\omega} |J_n| = \sup_{n\in\omega} \gamma_n = \aleph_{\alpha+\omega}.$$

□

5 Questions

There are still some cases where we do not know whether a module must have the maximal possible number of submodules. For example let $\rho := |R| = \aleph_0$ and M be a R-module of uncountable cardinality μ with $\mathrm{cf}(\mu) = \omega$. From Theorem 2.4 it is clear that M must have at least $2^{<\mu}$ submodules. On the other hand we know from Theorem 3.3 that if $\kappa^\rho < \mu$ for all $\kappa < \mu$ then M has 2^μ submodules. We have also seen (Theorem 4.5) that if $\rho^+ < \mu = \aleph_{\alpha+\omega}$, then there have to be 2^μ submodules. In [4] Lemma 2.2 STEPRĀNS treats this case without further assumptions. But his proof is unfortunately not complete. In a private communication he changed the final argument of the proof to show that there are at least μ^ω submodules. We can show the latter also, because if there are less then 2^μ submodules, then there is some $\kappa < \mu$ with $\kappa^\omega \geq \mu$ and therefore $\mu^\omega = \kappa^\omega \leq 2^{<\mu}$. The question is still open whether one can show in ZFC that modules with cardinality μ and $\mathrm{cf}(\mu) = \omega$ must have 2^μ submodules.

An example of a module M ($|M| = \aleph_\omega$) over a countable ring can be constructed (under the hypothesis that $2^{\aleph_0} > \aleph_\omega$), so that there is an independence series with $|A_\alpha| < |M|$ for all $\alpha \in \delta$ (where δ is the length of the series).

Another open question due to J. STEPRĀNS is if it is consistent that every uncountable module over a countable commutative ring has an uncountable semi-independent family. He showed in [4] that there is an infinite semi-independent set.

References

[1] R. G. Burns, F. Okoh, H. Smith and J. Wiegold, *On the number of normal subgroups of an uncountable soluble group*, Arch. Math. (Basel) **42** (1984), 289–295.

[2] B. Hartley, *Uncountable artinian modules and uncountable soluble groups satisfying Min-n*, Proc. London Math. Soc. (3) **35** (1977), 55–75.

[3] J. Lawrence, *Independent sets, Jonsson modules and the number of submodules*, J. Indian Math. Soc. (N.S.) **48**(10) (1984), 119–130.

[4] J. Steprāns, *The number of submodules*, Proc. London Math. Soc. (3) **49** (1984), 183–192.

Index

MIX
Papier aus verantwortungsvollen Quellen
Paper from responsible sources
FSC® C105338

If you have any concerns about our products,
you can contact us on
ProductSafety@springernature.com

In case Publisher is established outside the EU,
the EU authorized representative is:
Springer Nature Customer Service Center GmbH
Europaplatz 3, 69115 Heidelberg, Germany

Printed by Libri Plureos GmbH
in Hamburg, Germany